# A Vast Machine

**Infrastructures Series**
Edited by Geoffrey Bowker and Paul N. Edwards

Paul N. Edwards, *A Vast Machine: Computer Models, Climate Data, and the Politics of Global Warming*

# A Vast Machine

Computer Models, Climate Data, and the Politics of Global Warming

Paul N. Edwards

The MIT Press
Cambridge, Massachusetts
London, England

First MIT Press paperback edition, 2013
© 2010 Massachusetts Institute of Technology

All rights reserved. No part of this book may be reproduced in any form by any electronic or mechanical means (including photocopying, recording, or information storage and retrieval) without permission in writing from the publisher.

Set in Stone by Toppan Best-set Premedia Limited.

Library of Congress Cataloging-in-Publication Data

Edwards, Paul N.
A vast machine : computer models, climate data, and the politics of global warming / Paul N. Edwards.
   p. cm.
Includes bibliographical references and index.
ISBN 978-0-262-01392-5 (hardcover : alk. paper)—978-0-262-51863-5 (paperback)
1. Weather forecasting. 2. Climatology—History. 3. Meteorology—History. 4. Climatology—Technological innovation. 5. Global temperature changes. I. Title.
QC995.E296  2010
551.63—dc22
                                            2009030678

in memory of Stephen H. Schneider (1945–2010)


The meteorologist is impotent if alone; his observations are useless; for they are made upon a point, while the speculations to be derived from them must be on space. . . . The Meteorological Society, therefore, has been formed not for a city, nor for a kingdom, but for the world. It wishes to be the central point, the moving power, of a vast machine, and it feels that unless it can be this, it must be powerless; if it cannot do all it can do nothing. It desires to have at its command, at stated periods, perfect systems of methodical and simultaneous observations; it wishes its influence and its power to be omnipresent over the globe so that it may be able to know, at any given instant, the state of the atmosphere on every point on its surface. — John Ruskin (1839)

# Contents

Acknowledgments   ix

Introduction   xiii

1   Thinking Globally   1

2   Global Space, Universal Time: Seeing the Planetary Atmosphere   27

3   Standards and Networks: International Meteorology and the Réseau Mondial   49

4   Climatology and Climate Change before World War II   61

5   Friction   83

6   Numerical Weather Prediction   111

7   The Infinite Forecast   139

8   Making Global Data   187

9   The First WWW   229

10   Making Data Global   251

11   Data Wars   287

12   Reanalysis: The Do-Over   323

13   Parametrics and the Limits of Knowledge   337

14   Simulation Models and Atmospheric Politics, 1960–1992   357

15   Signal and Noise: Consensus, Controversy, and Climate Change   397

Conclusion   431

Notes   441
Index   509

# Acknowledgments

Two people in particular made this book possible. Stephen Schneider brought me into the world of climate science during a professional development fellowship sponsored by the National Science Foundation. He gave me access to his library, involved me in projects, introduced me to colleagues, and taught me how climate models work. He also taught me much about the politics of climate change. I learned more from Steve than from anyone else I've ever worked with. His generosity, both intellectual and personal, is beyond compare.

Gabrielle Hecht read many versions of the manuscript, tearing it apart and helping me put it back together. I have no better editor or colleague. At various points she also took notes for me at interviews and provided other assistance, and she kept her confidence in the project even when I lost my own. Far more than that, though: as the best life partner I can imagine, she taught me the fullest depths of what those words can mean. Without her this book could never have been written.

I owe much to certain colleagues for extended exchanges that helped me develop my ideas. Special thanks go first to Myanna Lahsen, Clark Miller, Simon Shackley, Jim Fleming, and Peter Taylor. Spencer Weart, Sheila Jasanoff, Johan Schot, Timothy Lenoir, and Thomas Parke Hughes provided crucial professional opportunities as well as intellectual frameworks. Others I would like to thank for comments, criticisms, ideas, and other help that shaped this book include Amy Dahan Dalmedico, Manuel Castells, Donna Haraway, Pascal Griset, John Carson, Joel Howell, Alex Stern, Catherine Badgley, Frederick Suppe, John Leslie King, Richard Rood, Erik van der Vleuten, Arne Kaijser, Tineke Egyedi, Wiebe Bijker, Trevor Pinch, Ben Santer, Susan Joy Hassol, Naomi Oreskes, Ted Parson, Geof Bowker, Susan Leigh Star, Steve Jackson, David Ribes, Chris Borgman, Chunglin Kwa, Dharma Akmon, Ted Hanss, Janet Ray Edwards, Dan Hirschman, Jeroen van der Sluijs, Michael Cohen, Rachelle Hollander,

# Introduction

Unless you have been in a coma since 1988, you have certainly heard or read a story that goes something like this: Global warming is a myth. It's all model predictions, nothing but simulations. Before you believe it, wait for real data. "The climate-studies people always tend to overestimate their models," the physicist Freeman Dyson told an interviewer in April 2009. "They forget they are only models."[1] In the countless political controversies over climate change, the debate often shakes out into a contest: models versus data.

This supposed contest is at best an illusion, at worst a deliberate deception—because *without models, there are no data*. I'm not talking about the difference between "raw" and "cooked" data. I mean this literally. Today, no collection of signals or observations—even from satellites, which can "see" the whole planet—becomes global in time and space without first passing through a series of data models.

Since both observing systems and data models evolve, global data also change. We have not one data image of the global climate, but many. The past, or rather what we can know about the past, changes. And it will keep right on changing. I call this reverberation of data images "shimmering." Global data images have proliferated, yet they have also converged. They shimmer around a central line, a trend that tells us that Earth has already warmed by about 0.75°C (1.35°F) since 1900.

Nor is there any such thing as a pure climate simulation. Yes, we get a lot of knowledge from simulation models. But this book will show you that the models we use to project the future of climate are *not* pure theories, ungrounded in observation. Instead, they are filled with data—data that bind the models to measurable realities. Does that guarantee that the models are correct? Of course not. There is still a lot wrong with climate models, and many of the problems may never be solved. But the idea that you can avoid those problems by waiting for (model-independent) data and the idea that climate models are fantasies untethered from atmospheric reality are utterly, completely wrong. *Everything we know about the world's climate—past, present, and future—we know through models.*

This book is a history of how scientists learned to understand the atmosphere, measure it, trace its past, and model its future. It isn't a work of advocacy. I am not going to try to convince you, as a scientist or an activist might, that climate change is real. Still, you will want me to lay my cards on the table, and it would be irresponsible of me not to do that. Yes, I think climate change is real, and I think it's the biggest threat the world faces now and will face for generations to come. Yet what I think about it is completely beside the point. Climate change is not a matter of opinion, belief, or ideology. This book is about how we came to know what we know about climate—how we make climate knowledge.

In the rest of this introduction, I will offer three avenues into this book. First, I will give a short summary of my argument, a quick preview of a very long movie. Next, I will give an idea about how to frame the argument in larger terms. (Think of 'frame' as a verb, not a noun; think of aiming a movie camera and choosing your focus, where you center the scene, and what you leave out of it.) Finally—after some caveats about what might surprise or disappoint you here—I will suggest several different ways to read this book, depending on who you are, what you already know, and what you might want to find out.

## Climate Science as a Global Knowledge Infrastructure

Here is my argument, as briefly as I can make it: Climate is the history of weather—the average state of the atmosphere over periods of years, decades, centuries, and more. You can learn about climate in two ways. First, you can collect records from the past. We have sporadic instrument readings from the surface starting in the seventeenth century, systematic ones from the 1850s on, and good records from the air above the surface, obtained mainly from weather balloons and satellites, starting in the 1950s.[2] This book traces the history of efforts to gather weather and climate records for the whole planet. I call this *making global data,* and I call the effort it involves *data friction.* Second, you can try to understand climate as a physical system. If you succeed, not only can you explain how it works; you can also determine why it changes. And you can predict how it may change in the future. Understanding and predicting the climate is very difficult. In fact, it's one of the hardest challenges science has ever tackled, because it involves many interlocking systems, including the atmosphere, the oceans, the cryosphere (ice and snow), land surfaces (soil, reflectance), and the biosphere (ecosystems, agriculture, etc.). You can't study global systems experimentally; they are too huge and complex. Instead, as I will show you, everything we know about the global climate depends on three types of computer models.

*Simulation models* are based on physical theory. Even after atmospheric physics became adequate to the task early in the twentieth century, *computational friction* prevented serious attempts to simulate weather or climate mathematically. By the late 1940s, with electronic digital computers, this began to change. Weather forecasters built numerical models to calculate the atmosphere's large-scale motions and predict the weather. Climate scientists then used similar techniques to simulate the global climate for long periods (years to decades). By changing the simulated forces and conditions, they also used models to predict how climate will change as human activity alters the composition of the atmosphere and other climate-related systems.

*Reanalysis models* come from weather forecasting. These models also simulate the weather, but unlike pure simulations they constrain their results with actual weather observations. Essentially, they produce a movie-like series of global weather forecasts, blending observations with simulation outputs to produce fully global, uniform data. Climate statistics derived from reanalysis cover the whole planet at all altitudes, unlike data from instruments alone. Since 1990, reanalysis of weather records has created a new source of global climate data.

What I call *data analysis models* (or *data models*, for short) are really a vast family of mathematical techniques, algorithms, and empirically derived adjustments to instrument readings.[3] Philosophers of science use the phrase "models of data"; practicing scientists might say "data analysis." Data analysis models are used to process historical weather and climate records. Observing systems have changed so much and so often that you can only combine long-term records by modeling the effects of different instrument behaviors, data collection practices, weather station site changes, and hundreds of other factors. You also need models to adjust for the tremendous unevenness of observations in space and time. In this process, which I call *making data global*, coherent global data images are created from highly heterogeneous, time-varying observations.

The last part of my brief preview concerns the idea of a *climate knowledge infrastructure*. Systems for observing weather and climate originated in the nineteenth century, for the most part as national weather services. These developed as separate systems, but soon they linked their data reporting through loosely coordinated international networks. The manual forecasting methods of that era focused mainly on regions rather than the whole world. Because processing them by hand would have taken far too long, such methods were not able to use vast quantities of data. When computerized weather forecasting arrived, in the mid 1950s, it required much more data—and soon it would require data from the whole planet. By the early 1960s, those needs, combined with the spur of Cold War politics and the lure of satellite technology, led to the World Weather Watch, an internetwork or web that combined numerous systems and networks into a functional system of global observing, telecommunication, data processing, and forecasting. This pattern of development—from systems to networks to webs—is visible in the histories of most large-scale infrastructures.

Weather forecasting and climatology diverged in the nineteenth century, developing different traditions and even different data sources. In the 1960s, climate modeling began to reunite the two fields. Here too, modeling shifted scientists' focus to the global scale. As in forecasting, existing data systems proved inadequate. One deficiency was the lack of data from the atmosphere above the ground, well sampled by weather balloons only after the late 1950s. Another problem, more complex, was the spotty, inconsistent, and poorly standardized record from surface stations. Investigators now had to make these data global. A long and painful process of *infrastructural inversion* began in the 1970s. Scientists turned the climate record upside down, reexamining every element of the observing system's history, often down to the level of individual measurements. Then, trying to reconstruct a history of the atmosphere, they digitized, interpolated, and processed those elements in many other ways.

Modeling outpaced empirically based knowledge of the global climate. By the 1970s, laboratories specializing in climate modeling had sprung up around the world, and climate models had gained a foothold in energy and environmental policy. As concerns about global warming mounted during the 1980s, scientists and policy makers institutionalized a knowledge-assessment process in the Intergovernmental Panel on Climate Change (IPCC). This process represents the most visible layer of the climate knowledge infrastructure.

Like most true infrastructures, the climate knowledge infrastructure is made up of many interlocking technical systems representing many links and layers of systems and structure, most of which long predate the

IPCC. The assessment process—now in the midst of its fifth cycle since 1990—has created imperatives, structures, and processes that link a vast array of knowledge producers and bring their disparate methods and products to bear on a common project. The assessments compare, combine, and interpret data and models to produce stable, reliable, widely shared knowledge about the global climate. This doesn't mean that controversies are suppressed; in fact, quite the opposite is true. The IPCC *brings controversy within consensus*, capturing the full range of expert opinion.

Climate knowledge is knowledge about the past. It's a form of history—the history of weather—and the infrastructure that creates climate knowledge works in the same way that historians work. What keeps historians in business? Why do they keep on writing new accounts of, say, the French Revolution or the Second World War? Don't we already know everything about those events? In fact we don't. There is always more to learn about the past. Historians continually discover previously unknown documents, letters, drawings, photographs, artifacts, and other kinds of evidence that reveal new aspects even of history's best-known episodes. On top of that, our perspective on the past keeps changing, for many reasons. We argue about how to interpret the evidence, finding flaws in earlier interpretations. And *we* keep changing. What we want to know about the past, what we hope to discover there, depends on who we are now.

Climate knowledge is like this too. People long ago observed climate and weather for their own reasons, within the knowledge frameworks of their times. You would like to use what they observed—not as they used it, but in new ways, with more precise, more powerful tools. How accurate were their measurements? Did they contain systematic errors? The numbers don't speak for themselves. So you dig into the history of data. You fight *metadata friction*, the difficulty of recovering contextual knowledge about old records. If you succeed, you find (perhaps) changes in station siting, faked logbooks, changes in instrumentation, misapplied standards, or a thousand other things that alter your understanding of the numbers in the records. Perhaps you come across a slightly different version of existing records, or a cache of previously unknown ones. You find flaws in previous interpretations of old data sets; perhaps you find new ways to correct them. You build new tools, gain new perspectives, and discover what you still don't know. And afterward, other investigators may follow in your path, discovering yet more about the past and altering your interpretation. In the words of T. S. Eliot,

We shall not cease from exploration
And the end of all our exploring
Will be to arrive where we started
And know the place for the first time.

The climate knowledge infrastructure is constantly opening itself up, reexamining every datum and data set, reanalyzing its data, adding to its metadata. Over time, countless iterations of that process have brought us shimmering data, an ever-expanding collection of global data images that will keep on growing, but never resolve into a single definitive record.

Yet these countless versions of the atmosphere's history have also converged. Could it be that one day some grossly different data image will emerge, in which the planet did not really warm across the period of historical records, or human activity played no significant role in climate change? Sure, it's possible; in science, never say never. But the chances of such a thing happening today are vanishingly small. We have a lot left to learn, but to the extent that anything so complex can ever be known, we know this. The infrastructure that supports climate knowledge is too large, too old, and too well developed.

Where are the politics in all this? Everywhere. A final thread that runs throughout this book is the idea of *infrastructural globalism*. In the context of meteorology, this refers to how the building of technical systems for gathering global data helped to create global institutions and ways of thinking globally. Building global observing systems required creating global intergovernmental organizations, such as the World Meteorological Organization and the Intergovernmental Panel on Climate Change. The Cold War, decolonization, and other aspects of international and world politics shaped the methods and practices of data collection, especially satellite systems.

As global warming rose to the top of the world's political agenda, the climate knowledge infrastructure itself became an object of intense political debates. Sides in these debates often saw the issue as one of models versus data. In the mid 1990s, environmental conservatives and climate-change skeptics promoted the idea that "sound science" must mean "incontrovertible proof by observational data," whereas models were inherently untrustworthy. But in global climate science, at least, this is a false dichotomy. The simplistic "models vs. data" debate lingers on, but in recent years it has been largely replaced by more sophisticated approaches. Amateur scientists and others dig deeply into models, data, and data modeling, sometimes joining the project of climate knowledge and sometimes

# Introduction

seeking to exploit its flaws for partisan purposes. Weblogs and "citizen science" websites now feature volunteer surveys of surface station biases, attempts to rewrite model code, and "audits" of climate data and models—infrastructural inversion, all over again. Outside the standard channels of peer-reviewed science, these sites nonetheless join the "controversy within consensus" model as mainstream science takes up and verifies their most significant results.

No longer under debate, however, are the fundamental frameworks of knowledge about the global climate: *how* we know what we know. Conceiving weather and climate as global phenomena helped promote an understanding of the world as a single physical system. Building the weather and climate knowledge infrastructures spread a specific way of making global knowledge—one whose techniques, values, and implications now extend not only throughout the sciences but far beyond.

Virtually any global thing you try to study will bring you up against the issues of *making global data*, *making data global*, and *data friction*. Studying anything that is planetary in scale—including human systems as well as natural systems—will put you in the business of *infrastructural globalism*. To understand the history of any such object, especially if you go back decades or longer, you will have to do some *infrastructural inversion*, and you will encounter *metadata friction*. Whatever you do, you will be using models of all sorts. These concepts frame this book's larger meaning.

## Monitoring, Modeling, and Memory

Today you can put instruments practically anywhere. Vast numbers of sensors *monitor* an equally vast range of phenomena, on every scale, from elementary particles to individual birds to Antarctic ozone levels to the solar wind. These sensors pour colossal volumes of digitized data into disk drives. Meanwhile, in many fields computer *models* complement or even replace laboratory experiments; analysis and simulation models have become principal means of data collection, prediction, and decision making. Third, vast data resources (scientific *memory*) are now increasingly available, though they are often distributed across thousands of research sites and institutions and in numerous incompatible formats.[4]

Computer models hold the key to transforming these information resources into knowledge. If you use a lot of sensors, you are going to need data models to make their signals into meaningful information. If you want to mine data created by somebody else and blend it with your own, you will need data models. If you want to do experiments on scales you can't access or involving materials you can't handle, you will use a simulation model. If you want to look at long time scales, blending data collected at many places and times by many investigators into a common data set, you will need models to reconcile the differences.

Global knowledge based on global infrastructures for monitoring, modeling, and memory: this path, laid out by weather and climate science from the nineteenth century to the present, has since been followed by many other sciences. Increasingly, these sciences link with one another, sharing digital data and traversing each other's theoretical frameworks by means of computer models. Consider, as just one example, the Group on Earth Observations System of Systems (GEOSS), an initiative that emerged from the 2002 World Summit on Sustainable Development. The GEOSS internetwork links numerous global monitoring systems with modeling and memory of . . . well, practically everything on Earth. GEOSS's ultimate goal is "to transform earth system observations into socio-economic information of value."[5] Anyone interested in any form of globalization, whether political, economic, historical, or cultural, will do well to attend to these new ways of thinking globally.

### A Few Words about Words (and Numbers)

When I talk about *meteorology* and *meteorologists*, I'm not talking about the Weather Channel. I am talking about the spectrum of sciences and scientists that study the atmosphere, including forecasting, climate science, experimental studies, and other disciplines. When I say *anthropogenic*, I mean "of human origin," or "human-caused"; think *anthropos* (Greek for "human") and *genesis* (beginning). When I say *general circulation*, I'm talking about how the atmosphere moves ("circulates") around the planet, its typical patterns of motion on the global scale. *General circulation models* simulate this motion.

*Climate sensitivity* is a widely used benchmark for simulation experiments. Climate sensitivity is short for "how much the global average temperature will change when carbon dioxide concentrations double from their pre-industrial levels." Usually this is expressed as a range; 2–4.5°C is the current IPCC estimate of climate sensitivity. It now appears virtually inevitable that carbon dioxide concentrations will not only double but may triple or even quadruple before they decline. Hence, the climate sensitivity is just a signpost, not a marker for the likely peak concentration.

# Introduction

I use the metric system, the scientific standard. One degree Celsius (°C) is equal to 1.8 degree Fahrenheit (°F); water freezes at 0°C and boils at 100°C. A meter is a little more than a yard. A kilometer is about six tenths of a mile, so 100 kilometers is a little more than 60 miles.

This book has no glossary, but many are readily available. Two good ones are the online Glossary of Meteorology provided by the American Meteorological Society (amsglossary.allenpress.com/glossary/) and the one in the appendix to the IPCC's Working Group I Fourth Assessment Report (available at www.ipcc.ch).

## How to Read This Book

I wanted to write a book that almost anyone could read and understand. At the same time, I wanted the book to appeal to scientists, and my fundamental argument requires going into some depth about weather and climate models, data, and their interactions. As a result, some people will find parts of the book too technical or too detailed, while others will find the same parts not technical or detailed enough. Here I briefly outline the book's structure, then describe three different ways you could read this book, depending on what kind of reader you are.

The book's sequence is roughly but not entirely chronological. Despite its length, it should be thought of as series of vignettes taken from a history so long and so complex that no linear narrative (and no single book) could hope to capture it. After chapter 7, most chapters carry one part of the story from some point in the 1950s to the present. Here is a map.

Chapter 1 outlines the book's conceptual framework and describes the arc of the argument.

Chapters 2–5 treat weather forecasting and climatology before 1945. These chapters provide background and introduce a series of concepts, especially the notions of data friction and computational friction.

Chapters 6 and 7 cover weather prediction and climate modeling from 1945 to 1970. This is the place to find explanations of how weather models and climate models work.

Chapters 8–10 recount how the weather information infrastructure developed between 1950 and 1980: how global weather data were collected ("making global data") and how they were analyzed and modeled to render global forecasts ("making data global"). Chapter 9 is an account of the World Weather Watch and the Global Atmospheric Research Program, the World Meteorological Organization's two major infrastructural achievements through 1980. Chapter 10 develops the concept of *model-data symbiosis*.

Chapters 11 and 12 distinguish weather data from climate data and describe the reconstruction of historical temperature records. Chapter 11, the main subject of which is surface-station records, sweeps from the 1930s to the present. Chapter 12 describes the reunification of weather forecasting and climate science in the reanalysis projects begun in the late 1980s. (Reanalysis creates climate data from historical weather records.)

Chapter 13 returns to climate modeling, focusing on parameterization, tuning, and model validation. Taking a conceptual rather than a chronological approach, it focuses mainly on modeling issues of the period from 1980 to the present.

Chapters 14 and 15 discuss the interaction of models, data, and global atmospheric politics. Chapter 14 covers ozone depletion, nuclear winter, global warming, and other issues of the period 1960–1992, ending with the Framework Convention on Climate Change. Chapter 15 focuses on structural features of the politics of global warming since 1990.

The conclusion revisits the idea of a climate knowledge infrastructure and reflects on its larger meaning.

Most chapters begin and end with a short section written in an informal style. These sections try to draw out each chapter's central lessons briefly and readably, but do not necessarily rehearse the full argument. Reading them may help you decide how deeply to explore the chapter.

Say you are a "general reader"—not a scientist, but somebody who likes to read newspaper science sections or *Scientific American*, or who listens to *Science Friday* on National Public Radio. If you like to delve into historical background, chapters 2–5 and 14 should be relatively accessible. If you are more interested in how we know about climate change, begin with chapter 7 (perhaps skipping its more technical sections, those on pioneering climate models), then read chapters 11–13, chapter 15, and the conclusion.

If you are a scientist, and especially if you work in the atmospheric sciences, you probably already know a lot about current debates on climate change. You may care more about the history of models and data. For you, chapters 6–13 will be the core of the book—especially chapters 6 and 7, which describe the rise of computer modeling in weather forecasting and climate science. Chapter 15 and the conclusion may offer you some new ways to think about current debates.

If you are an academic or a student from a non-scientific discipline (such as history and philosophy of science, science and technology studies, or political science), you will want to read the whole book. Historians will naturally be more interested in the parts relevant to their own periods. People with an epistemological bent will find chapters 10–13 most useful. If you are most interested in climate policy or controversy studies, chapters 11, 14, and 15 may be most rewarding for you.

**What This Book Does Not Do**

This book deliberately violates the received history of meteorology in a way that may upset the expectations of readers already versed in that history. Meteorology includes three main component disciplines: weather forecasting, climatology, and theoretical meteorology. Originally united, these disciplines split apart in the nineteenth century and developed in relative isolation until the advent of computer models after 1950. Computer modeling returned theoretical meteorology to a central role in forecasting, and it transformed climatology into what we now call climate science. Yet operational weather prediction's very different priorities still separated it institutionally and conceptually from climate science. Most histories of meteorology—especially the informal history that meteorologists recount to one another—accept this division at face value. As a consequence, the stories of weather forecasting, climatology, and theoretical meteorology are usually told separately.

In this book I bring the three narratives together in ways that may at first seem puzzling to some scientists. I do this because the arcs of those stories rejoined some time ago. Since 1960, computer models have been the fundamental tool of both weather forecasting and climate science, differing in details and in usage but not in underlying structure. Since 1990, reanalysis projects have reunited the previously separate streams of weather data and climate data, at least to a degree. More recently, operational climate prediction and the Earth System Modeling Framework (which allows model components to be readily exchanged among research labs and operational agencies) have signaled the beginning of a new stage in this reunification. Thus, to understand the infrastructure of climate knowledge you have to understand weather forecasting. Though still quite different in many ways, they are inseparable, and they are increasingly linked.

My deepest regret about the book is that it is not, as I once hoped it would be, a fully international history. Swedish, British, German, Japanese, and Soviet and Russian contributions, in particular, receive much less attention here than they deserve, and it would be possible to come away from this book with an inflated view of the role of the United States in climate science and politics. My defense is that no scholar and no book can do everything. My hope is that I, along with many other writers, can continue to expand this story much as climate scientists have reconstructed climate history: with many iterations, as more evidence appears and as my colleagues find and correct the deficiencies. That's how good history works.

This is not a history of individuals, nor is it an ethnography. It contains few thoroughly drawn characters and few discussions of personal or group interactions. Instead, it is a history of systems, networks, and webs; of data, models, and knowledge flows. My goal is not to document details, but to provide an analytical perspective, a conceptual framework that makes some sense of things. For this reason, I have not attempted to systematically attribute credit for scientific advances, for organizational transformations, or for larger ideas and trends. In the time-honored scientific and scholarly tradition, I cite the most important publications and other works, but inevitably I will have passed over some of these. Many meteorologists maintain a remarkable awareness of their field's history, so these omissions will certainly disappoint some people. If you are one of them, please accept my apology; I intend no slight. I fully understand that everything I have described came about through the tireless work of individual human beings, many of whom dedicated their entire lives to building some part of this gigantic whole. I wish I had found a way to include them all.

My office and my computer's disk drive are crammed with excellent books and articles about meteorology and climate change written by other historians and social scientists. I have made such use of these works as I can, and have tried to gesture in their direction where I can't. In reaching out to a broad audience, I have left to one side what some of my colleagues may regard as important scholarly debates; these I will engage elsewhere.

My final caveat, which I will repeat frequently to avoid misunderstanding, is that this book treats only some pieces of the climate change puzzle. It focuses primarily on atmospheric models and the historical temperature record. These are the two most important ways we know about climate change, but many other lines of evidence, and many other kinds of models, play crucial parts in the knowledge infrastructure. Ultimately it is the convergence of all these lines of evidence, from numerous partly or completely independent disciplines and data sources, that underwrites the scientific consensus on global warming.

# Introduction

## The Book's History

This book has been a long time coming. I first got interested in climate change in the mid 1980s, when I was in graduate school, in the context of global security issues. In 1994, as a junior faculty member at Stanford University, I began studying it in earnest under a National Science Foundation professional development fellowship. I am neither a meteorologist nor a computer scientist by training, although I worked as a computer operator and programmer in the mid 1970s, and although my first book, *The Closed World: Computers and the Politics of Discourse in Cold War America* (1996), took me deep into the history of computing. So I had to learn a lot. I took courses on climatology and studied the scientific literature intensively. I worked closely with my Stanford colleague Stephen Schneider, who provided what amounted to an intensive multi-year tutorial on climate science and politics. Over the years I attended countless scientific meetings.

Originally I planned to write a history of climate modeling, so I visited numerous climate laboratories and other facilities throughout the United States and in the United Kingdom, France, Switzerland, and Australia. During these visits I collected a large archive of documents. More than 800 documents are cited directly in this book, but my research bibliography runs to well over 5000 items. In addition to primary scientific articles, these include letters and other archival documents, email exchanges among scientists, "gray literature," and a variety of photographs, PowerPoint presentations, and materials in other media.

Gray literature—conference proceedings, internal reports from climate laboratories, International Meteorological Organization and World Meteorological Organization publications, and similar items—has considerable importance in meteorology and is often cited in journal publications. Yet laying hands on any of the gray literature published before about 1995 is remarkably difficult. There are only a few well-stocked meteorological libraries in North America, and probably no more than two dozen in the whole world. Very few have the full set of WMO publications—especially operational manuals, whose earlier versions are routinely discarded when updated manuals arrive. Even the WMO's library in Geneva no longer holds copies of some of that organization's own publications. This fact matters in my story. Recently recovered older versions of WMO Publication 47 are now aiding in the reconstruction of weather records from ships (see chapter 11). Therefore, I spent many weeks in meteorological libraries at the US National Center for Atmospheric Research, the Geophysical Fluid Dynamics Laboratory, Environment Canada, the Goddard Institute for Space Studies, UCLA's Department of Meteorology, the European Center for Medium Range Weather Forecasts, and the Australian Bureau of Meteorology Research Centre. I also made use of archives at the National Center for Atmospheric Research, the Massachusetts Institute of Technology, and the American Institute of Physics.

Between 1994 and 2001, I interviewed numerous scientists. In particular, I took lengthy oral histories from the first-generation climate modelers Cecil "Chuck" Leith, Syukuro Manabe, Warren Washington, Akira Kasahara, and Akio Arakawa. (I was not able to take oral histories from two other members of this first generation; Yale Mintz died in 1993, and by 1998, when I interviewed Joseph Smagorinsky, his Parkinson's Disease had progressed to the point that an oral history was no longer practicable. Fortunately, Smagorinsky had already written several excellent historical accounts of the events in which he was involved.) Transcripts of these oral histories are on deposit at the American Institute of Physics' Center for the History of Physics and are available for other researchers to use. The names of all my interviewees are listed in my acknowledgments. In the end, as I learned more and more about the field and as my focus shifted from the history of modeling to the larger topics this book covers, I decided to use most of those interviews as background rather than primary source material.

I made a deliberate decision not to pursue archival sources in great depth. Such sources do exist, particularly for events prior to about 1970, and I explored some of them rather thoroughly. Yet it quickly became clear to me that I needed to choose between ferreting out archival evidence and going where I wanted to go, namely toward a conceptual and long-historical view. Further, I decided that the most credible and relevant sources for this project were the same ones scientists use in their own work: peer-reviewed scientific journals; conference proceedings; documents published by such organizations as the WMO, the IPCC, and national science academies; and other primary professional literature.

Over the years, I have published a series of articles and another book: *Changing the Atmosphere: Expert Knowledge and Environmental Governance* (MIT Press, 2001), co-edited with Clark Miller. Bits and pieces of this previous work, all updated and revised, appear throughout *A Vast Machine*. As time went on, other research work and life events intervened, causing this project about global climate science to travel around the world with me.

I composed substantial parts of it during extended stays in France, the Netherlands, Australia, South Africa, and Namibia. The science continued to grow and change, as did the political context; the huge scale and scope of climate research made merely keeping up with the literature virtually a full-time job. My own perspective continued to shift as I developed the book's framework and further explored the intricate relationships between models and data. No one's understanding of this subject can ever be complete or finished, but here is mine, as done as I can do it and—like all knowledge—merely provisional.

# 1 Thinking Globally

In 1968, three American astronauts became the first human beings ever to see Earth's full disk from space. President Lyndon B. Johnson mailed framed copies of the Apollo mission's photographs to the leaders of every nation as an allegory of the inevitable unity that encompasses all human division and diversity and binds us to the natural world.

By then, of course, representations of Earth as a globe were already centuries old. Nevertheless, many saw a transfiguring power in the awesome beauty of those famous photographs. That small blue ball, spinning alone in darkness: it hit you like a thunderclap, a sudden overwhelming flash of insight. You saw, all at once, the planet's fragility, its limits, and its wholeness, and it took your breath away. The law professor Lawrence Tribe once called it a "fourth discontinuity," as massive a perspectival shift as those brought on by Copernicus, Darwin, and Freud.[1] By 1969, according to rumor, David Brower, founder of Friends of the Earth, had distilled Tribe's "fourth discontinuity" into four words: "Think globally, act locally."[2]

Whatever you think of it as a political principle, "Think globally, act locally" remains arresting in its boldness. It captures an entire philosophy, complete with ontology, epistemology, and ethics, in a bumper-sticker slogan. It asserts an intimate relationship between two vastly different scales: macro, world-scale environmental and economic systems, on the one hand, and the micro sphere of individual choice and action, on the other. It extends an arrow of agency, comprehending macro effects as the results of vast aggregations of micro causes. Thus it locates the meaning of individual action in its relationship to the gigantic whole. Finally, it affirms that global change matters so deeply that it should occupy the intimate corners of everyday awareness and guide each person's every choice.

"Thinking globally" meant seeing the world as a knowable entity—a single, interconnected whole—but in a sense that lacked the secure stasis

**Figure 1.1**
Photograph of Earth taken from Apollo 8, December 1968.
Image courtesy NASA.

of maps, parlor globes, or pre-Darwinian cosmologies. Instead, it meant grasping the planet as a dynamic system: intricately interconnected, articulated, evolving, but ultimately fragile and vulnerable. Network, rather than hierarchy; complex, interlocking feedbacks, rather than central control; ecology, rather than resource: these are the watchwords of the new habit of mind that took Earth's image for its emblem.

Those photographs and that slogan conveyed all this, and more, not just because of what they said but also because of when they said it.[3] They fell directly into an overdetermined semiotic web prepared by (among other things) the post-World War II "One World" movement; the United Nations; the 1957–58 International Geophysical Year, with its scientific internationalism and powerful popular appeal; the Earth-orbiting satellites Sputnik, Telstar, and TIROS; the many variants of systems thinking descending from operations research, cybernetics, and early computer science; scientific ecology; and what I have called the "closed world discourse" of Cold War politics.[4] Long before the astronauts stared down in awe from outer space, notions of a "global Earth" had begun to emerge in language, ideology, technology, and practice.[5]

How did "the world" become a *system*? What made it possible to see local forces as elements of a planetary order, and the planetary order as directly relevant to the tiny scale of ordinary, individual human lives? How did the complex concepts and tools of global thinking become the common sense of an entire Western generation? How has systems thinking shaped, and been shaped by, the world-scale infrastructures that have emerged to support knowledge, communication, and commerce? How did global thinking become a bumper-sticker slogan? No book could ever resolve such huge questions completely. But by exploring one of today's most prominent objects of global knowledge and politics—global warming—in relation to the infrastructure that supports it, I hope to sketch at least the outlines of some answers.

### Global Climate as an Object of Knowledge

If you really want to understand something, I tell my students, you have to ask an elemental question: *How do you know*? At first you may think you have answered that question when you have reviewed the evidence behind the claim. But if you keep asking the question long enough, you will begin to wonder where that evidence came from. If you are talking about a scientific problem, you will begin to care about things like instrument error, sampling techniques, statistical analysis. (*How* do you know?) And if you have the soul of a scientist—or a defense attorney—you will go further still. Who collected that evidence? Why did they see it as evidence, and where did they get the authority to say so? (How do *you* know?) Finally, you will begin to ask how evidence comes to count as evidence in the first place. How do communities interweave data, theories, and models, within their tapestries of culture and commitments, to make what we call knowledge? (How do you *know*?) When you have gone deep enough, you may surrender your Cartesian dreams of total certainty in favor of trust founded in history, reputation, and fully articulated reasoning. Or you may not. Whatever happens, you are going to have to look under the hood.

So how do we know that the world is getting warmer?

First, notice that to say that the global climate has *changed* implies that we know what it *used to be*. At a minimum, we are comparing the present with some period in the past. We would like to know the details, the trend over time. Since we are talking about climate, not weather, we need a long period, ideally 100 years or more. And since we are talking about *global* climate, we need some kind of picture of the whole planet—from the equator to the poles, across the continents, and over the oceans.

How do we get that? Experience isn't enough. No one lives in a "global" climate. Without scientific guidance, not even the most cosmopolitan traveler could perceive a global average temperature change of about +0.75°C, the amount we have seen so far. Extreme weather events—heat waves, hurricanes, droughts, floods—dominate human experience and memory, and they often create false impressions of average conditions. In the winter of 1981–82, the first year I lived in California, it rained in torrents all day, every day, for weeks. Huge mudslides ripped out mountain roads near my house. The San Lorenzo River overflowed, washing away whole neighborhoods. The next winter, I expected the same. It took me most of the decade to really understand that this wasn't normal.

Year to year, weather averages vary naturally. No extreme event or extreme season necessarily reflects a long-term climate change. Rising global average temperatures will not put an end to unusually cold winters, late-spring ice storms, or other episodes that seem to run against the trend. Further, the temperature change that worries us today is an average rise of 2.5–5°C (4.5–9°F) over the next 50–100 years. In terms of human experience, that is far less than the typical difference between daytime highs and nighttime lows in many parts of the world. Every year, the planet's temperate zones endure temperature changes of ten times this magnitude, as frigid winters in the –10°C range bloom into steamy +30°C summers. Thus we can't rely on experience alone.

Let us look for evidence, then. Data should be easy to get. Since the middle of the nineteenth century, meteorologists have been building a global information system of enormous scope and complexity. Each day, weather stations around the planet, on land and sea, generate hundreds of thousands of instrument readings. In addition, satellites, aircraft, radiosondes, and many other instrument platforms measure variables in the vertical dimension, from the surface to the atmosphere's outer edge. Figure 1.2 illustrates the present-day meteorological information system's three

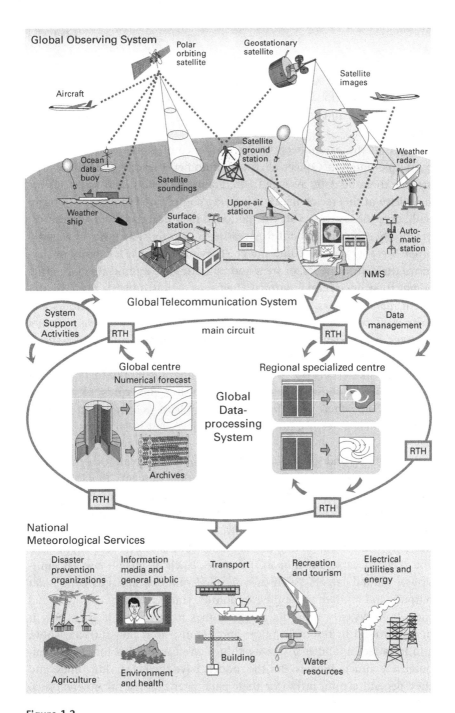

**Figure 1.2**
The global meteorological data, telecommunication, and forecast network as it looks today. RTH stands for regional telecommunications hub and NMS for national meteorological service.
Courtesy of World Meteorological Organization.

components: the Global Observing System, the Global Telecommunication System, and the Global Data Processing and Forecast System. Together they collect, process, and archive hundreds of terabytes[6] of weather and climate data each year.

With all those numbers, finding evidence of global warming should be simple enough. Collect all the thermometer readings for the world, arrange them by date, average them, graph the results, and *voilà*. But the simplicity of this recipe masks some very complicated problems. To begin with, over the last 160 years (the period of historical thermometer records) practically everything about the weather observing system has changed—often. Weather stations come and go. They move to new locations, or they move their instruments, or trees and buildings rise around them, or cities engulf their once rural environs. They get new instruments made by different manufacturers. Weather services change their observing hours and their ways of calculating monthly averages. These and dozens of other changes make today's data different not only from data collected 150 years ago, but also from data collected 20 years ago, or even (sometimes) last week. It's like trying to make a movie out of still photographs shot by millions of different photographers using thousands of different cameras. Can we reconcile the differences, at least well enough to create a coherent image? Yes, we can, scientists believe. But it isn't easy, and it is never finished.

If we go beyond ground-level air temperature to look for evidence of global climate change—including all the other elements that constitute the climate (sea surface temperatures, ocean currents, rainfall, snow, sea ice, etc.)—the data predicament created by constantly changing observing systems goes from very bad to even worse. And if you want to understand not only what is happening to the climate but also why it is happening, you have to parse out human influences (such as greenhouse-gas emissions), natural factors (such as volcanic eruptions and changing solar output), and random variability. To do this, you need a way to understand the whole system: where its energy comes from; where it goes; how it moves around in the atmosphere and oceans; how land surfaces, snow, and ice affect it; and many other things. We can do this too, with computerized climate models. But the models have their own difficulties, and they depend significantly (for their parameterizations) on observational data.

Thus assembling stable, reliable, long-term evidence of climate change is difficult indeed. Nonetheless, a consistent scientific consensus on climate change had developed by the early 1990s. Though some of the details have shifted since then, in general this consensus holds that some global warming has already occurred, and that human activities are responsible for a substantial part of it. A 2007 assessment by the Intergovernmental Panel on Climate Change (IPCC), the world's most authoritative climate knowledge institution, concluded that the planet warmed by about 0.75°C over the period 1906–2005. Models predict that we are in for much more warming (2–6°C) by 2100, depending on greenhouse-gas emissions, deforestation, and many other factors. Some of this future warming is already "committed," in this sense: even if all emissions of greenhouse gases were to cease tomorrow, warming would continue for several decades as the oceans come into equilibrium with the atmosphere. In 2007, the IPCC and former US Vice President Al Gore shared the Nobel Peace Prize for their work in creating knowledge and spreading awareness of the threat of global warming.

"Consensus" does not mean that all scientists agree on every detail, and noisy protests of even the most basic facts continue. Yet most scientists agree on the essential elements just mentioned. Their consensus has held for nearly 20 years now, and the message has sunk in. Today most people in developed countries believe that global warming is happening, and happening to them; they think it will directly affect their lives. Since the turn of the millennium, opinion surveys consistently show that Americans and Europeans believe that global warming is real, that human activity is its principal cause, and that press reports on the issue correctly reflect, or even underestimate, its dangers. Large majorities (four fifths) support the Kyoto Protocol.[7] Not only governments and environmental organizations, but also major insurance, energy, and automobile corporations have publicly accepted the reality of global warming and announced efforts to address it. Toward the end of the twentieth century, then, global warming became an established fact. This book is about how we came to know this fact and what it means to say that we know it.

This should not be the only book you ever read about global warming, because there are many things I do not cover. In particular, I focus only on atmospheric temperature and circulation. There are many other important lines of evidence for global warming, including paleoclimate studies, the rapid melting of glaciers and continental ice sheets, and ocean temperature and circulation. Spencer Weart's book and website *The Discovery of Global Warming* reviews much of this larger range of evidence from a historical perspective. For authoritative scientific treatments, turn first to the IPCC's assessment reports. If you are looking for responsible discussions more accessible to non-scientists, try Stephen Schneider's *Laboratory*

*Earth*, Sir John Houghton's *Global Warming: The Complete Briefing*, Andrew Dessler and Edward Parson's *The Science and Politics of Global Climate Change*, or Joseph DiMento and Pamela Doughman's *Climate Change: What It Means For Us, Our Children, and Our Grandchildren*.

*A Vast Machine* is also not really about scientific uncertainty, the uptake of science into politics, or public understanding of science. Studies of these topics now exist in large numbers, and there are a lot of very good ones, with more appearing almost daily.[8] Nor does it address the impacts of climate change, how to mitigate its effects, or how we can slow its progress, though all of these are, of course, matters of great importance.

Instead, what you are about to read is a *historical account* of *climate science as a global knowledge infrastructure*. Climate science systematically produces knowledge of climate. As Ruskin put it in 1839 (see my epigraph), it is "a vast machine": a sociotechnical system that collects data, models physical processes, tests theories, and ultimately generates a widely shared understanding of climate and climate change. This knowledge production begins with observations, but those are only raw materials. Transforming them into widely accepted knowledge requires complex activity involving scientific expertise, technological systems, political influence, economic interests, mass media, and cultural reception. Even the question of what counts as a valid observation in the first place requires considerable negotiation. This knowledge-production system delivers not only specifics about the past and likely future of Earth's climate, but also the very idea of a planetary climate as something that can be observed, understood, affected by human wastes, debated in political processes, cared about by the general public, and conceivably managed by deliberate interventions such as reforestation or gigantic Earth-orbiting sunshades. Ultimately, this knowledge infrastructure is the reason we can "think globally" about climatic change.

**Dynamics of Infrastructure Development**

To be modern is to live within and by means of infrastructures: basic systems and services that are reliable, standardized, and widely accessible, at least within a community. For us, infrastructures reside in a naturalized background, as ordinary and unremarkable as trees, daylight, and dirt. Our civilizations fundamentally depend on them, yet we notice them mainly when they fail. They are the connective tissues and the circulatory systems of modernity. By linking macro, meso, and micro scales of time, space, and social organization, they form the stable foundation of modern social worlds.⁹

Infrastructure thus exhibits the following features, neatly summarized by Susan Leigh Star and Karen Ruhleder:

• *Embeddedness*. Infrastructure is sunk into, inside of, other structures, social arrangements, and technologies.
• *Transparency*. Infrastructure does not have to be reinvented each time or assembled for each task, but invisibly supports those tasks.
• *Reach or scope* beyond a single event or a local practice.
• *Learned as part of membership*. The taken-for-grantedness of artifacts and organizational arrangements is a sine qua non of membership in a community of practice. Strangers and outsiders encounter infrastructure as a target object to be learned about. New participants acquire a naturalized familiarity with its objects as they become members.
• *Links with conventions of practice*. Infrastructure both shapes and is shaped by the conventions of a community of practice.
• *Embodiment of standards*. Infrastructure takes on transparency by plugging into other infrastructures and tools in a standardized fashion.
• *Built on an installed base*. Infrastructure wrestles with the inertia of the installed base and inherits strengths and limitations from that base.
• *Becomes visible upon breakdown*. The normally invisible quality of working infrastructure becomes visible when it breaks: the server is down, the bridge washes out, there is a power blackout.
• *Is fixed in modular increments, not all at once or globally*. Because infrastructure is big, layered, and complex, and because it means different things locally, it is never changed from above. Changes require time, negotiation, and adjustment with other aspects of the systems involved.¹⁰

Most entities typically classified as "infrastructure," such as railroads, electric power grids, highways, and telephone systems, are network technologies. They channel flows of goods, energy, information, communication, money, and so on. Many infrastructures are transnational, and a few have effectively gone global: for example, by 2008 there were over 4 billion mobile phone accounts and 1.3 billion fixed telephone lines, most of which could (in principle) call any of the others.¹¹

In the 1980s and the 1990s, historians and sociologists of technology began studying the infrastructure phenomenon intensively. These researchers developed a "large technical systems" (LTS) approach to telephone, railroads, air traffic control, electric power, and many other major infrastructures.¹² Around the same time, some scholars began to identify infrastructure as a key analytic category.[13] The LTS school of thought generated new insights into questions of organizational, social, and historical change. Recently, investigators have applied this and related infrastructure-oriented approaches to urban development, European history, globalization, scientific "cyberinfrastructure," and Internet studies.[14]

Where do infrastructures come from? The LTS approach identified a series of common stages in infrastructure development:

- invention
- development and innovation
- technology transfer, growth, and competition
- consolidation
- splintering or fragmentation
- decline.

In the invention, development, and innovation phases, "system builders" create and promote linked sets of devices that fill a functional need. As elaborated by Thomas Parke Hughes, the paradigmatic LTS example of a system builder is Thomas Edison. Neither the light bulb nor electric power alone accounted for Edison's remarkable commercial success. Instead, Hughes argued, Edison conceived and delivered a lighting *system*, comprising DC generators, cables, and light bulbs. Establishing a new LTS such as Edison's demands more than technical ingenuity; it also requires organizational, economic, political, and legal innovation and effort in order to resolve the host of heterogeneous problems that inevitably arise. Finance capital, legal representation, and political and regulatory relationships become indispensable elements of the total system. Over time, the LTS becomes sociotechnical, rather than merely technological.[15]

Technology transfer to other locations (cities or nations) follows the initial system elaboration phase. Typically, developers respond to new local conditions by introducing variations in the system's original design.[16] Hughes, referring to the distinctive look and feel of the "same" LTS in differing local and national contexts, called this "technological style." In the growth phase, the system spreads quickly and opportunities for both profit and innovation peak. New players may create competing systems with dissimilar, incompatible properties (for example, DC vs. AC electric power, or the Windows operating system vs. Macintosh and Linux).

During consolidation, the quasi-final stage of LTS development, competition among technological systems and standards may be resolved by the victory of one over the others. More often, however, "gateway" technologies emerge that can join previously incompatible systems, allowing them to interoperate.¹⁷ AC-DC power converters for consumer electronics and telephone adapters for international travel are examples, as are (in the world of information technology) platform-independent standards such as HTML and PDF. Gateways may be dedicated or improvised (that is, fitted specifically to a particular system), or they may be generic (standardized sockets opening one system to interconnection with others) or meta-generic or "modeled" (protocols for creating new generic standards, without restricting design in detail).¹⁸

Gateway technologies and standards spark the formation of networks. Using gateways, homogeneous and often geographically local systems can be linked to form heterogeneous networks in which top-down control is replaced by distributed coordination processes. The shift from homogeneous systems to heterogeneous networks greatly increases flexibility and creates numerous opportunities for innovation. In a later phase, new gateways may connect heterogeneous networks to one another (as in the Internet, a network of networks whose principal gateway technologies are packet switching and the TCP/IP protocol suite). Container shipping (which joins road, rail, and shipping networks) and the linkage of cellular with land-line telephony are examples of internetworks in other domains. Gateways need not be, and often are not, technological. For example, far more important than hardware in linking global financial markets into a single infrastructure were institutional, legal, and political gateways that permitted trans-border stock trading, currency exchange, and so on.

No system or network can ever fulfill all the requirements users may have. Systems work well because of their limited scope, their relative coherence, and their centralized control. System builders try to expand by simply increasing their systems' scale to reach more potential users, thereby excluding competitors. On the other hand, though users appreciate greater scale, they also want greater scope as well as custom functionality. Therefore, they continually cast about for ways to link incompatible systems and networks. Gateway developers (who may be users themselves) try to find ways to automate these links. When they succeed, gateway innovations and shared standards create *networks* or, at a higher level, *webs* (networks of networks, or internetworks). From the user's viewpoint, a network or a web links stand-alone systems (or networks), providing greater functionality. This was the case with the World Wide Web, which began as a protocol for exchange of hypertext documents but rapidly subsumed numerous pre-existing Internet file sharing mechanisms, including ftp, gopher, and nntp. From the operator's viewpoint, networks or webs shift the focus from control to coordination with the systems or networks on the other side of the gateway. The formation of a network or a web usually benefits users, but it can have unpredictable effects on the owners and operators of underlying systems.[19] The standardization process is a rocky road—even in information technology, where it is often easier than in other domains.[20]

To sum up: System builders seek to find or create well-defined niches that can be served by centrally designed and controlled systems, but users' goals typically include functions that may be best served (for them) by linking separate systems. The fundamental dynamic of infrastructure development can thus be described as a perpetual oscillation between the desire for smooth, system-like behavior and the need to combine capabilities no single system can yet provide. For these reasons, in general *infrastructures are not systems* but networks or webs.[21] This means that, although infrastructures can be coordinated or regulated to some degree, it is difficult or impossible to design or manage them, in the sense of imposing (from above) a single vision, practice, or plan.

Infrastructure formation is never tension-free. Emerging infrastructures invariably create winners and losers. If they are really infrastructures, they eventually make older ways of life extremely difficult to maintain: think of family farms against industrial agriculture, or newspapers against the Internet. Every choice involves tradeoffs and consequences. Infrastructures have victims and "orphans" (people and groups who are unable to use them or to reap their benefits because of their circumstances)—for example, people with rare diseases ignored by pharmaceutical research, blind people unable to navigate graphics-based websites, and the 5 billion people still without access to the Internet.

Even in meteorology, a field in which it is hard to discern many victims, one can find tensions that have real human consequences. When I visited the National Severe Storms Center in Norman, Oklahoma, a few years back, the director could barely contain his bitterness. His research budget had barely held steady even as budgets for climate-change research skyrocketed. From his point of view, advancing tornado or hurricane warnings by even a few hours could save thousands of lives and prevent millions of dollars' worth of property destruction. But the money he needed to improve his prediction models was drained by long-term climate research. Every stage of infrastructure development is marked by struggle.[22]

## Weather and Climate Information Infrastructures

The world weather and climate information infrastructure, described briefly above (and much more extensively throughout this book), exhibits all the classic features of this well-established development pattern.

National weather services inaugurated a system building phase, based on then-new telegraphy, in the latter half of the nineteenth century. With rapid technology transfer and growth, each national weather service created its own technological style, including various systems and standards for data collection and forecasting. Attempts at consolidation began as early as the 1870s, when some meteorologists sought to create a formal international network. They established the International Meteorological Organization (IMO) to negotiate technical standards and promote network development. By 1900, a Réseau Mondial (worldwide network) for real-time weather data exchange via telegraph had been proposed. For decades, however, consolidation remained elusive.

As both system builders and network users, the national weather services experienced conflicting pressures. Answerable to their governments, their highest priority lay in improving national systems and services. Yet as forecasting techniques improved, most nations needed data from beyond their own borders. So coordinating with other nations was in their interest. On the other hand, getting dozens of weather services to agree on and conform to common standards and techniques often cost more in time, money, and annoyance than it seemed to be worth. The tension between sovereign national systems and voluntary international standards severely limited the IMO's potential, and two world wars did nothing to improve the situation. Meanwhile, in the first half of the twentieth century the telegraph-based weather data network rapidly morphed into an tremendously complicated web, integrating both new instruments (such as radiosondes) and new communications media (such as telex and shortwave radio) through a proliferation of improvised gateways. During that period, most data network development was driven by the internal system-building dynamics of national weather services. International data networks remained a secondary priority. IMO standards acted as guidelines, routinely violated but nevertheless producing considerable convergence. As predicted by the LTS model, this phase of technology transfer and growth resulted in numerous different systems, some linked and others not, all governed by a loose patchwork of conflicting national, regional, and international standards. By the 1920s, the klugey pre-World War II network made worldwide data *available* to forecasters almost in real time. But forecasters' ability to use those data remained limited, in part because of the extreme difficulty of sorting out the numerous formats and standards used by various national weather services.

A consolidation phase began around 1955 and lasted for several decades. On the technical side, consolidation was driven by the arrival of computer models for weather forecasting, first used operationally in 1954. Instantly perceived as a superior technique despite early weaknesses, computer modeling brought with it a voracious appetite for data. Weather forecasters adopted computer modeling immediately. Starting out with regional models, they switched to hemispheric models by the early 1960s and global models by that decade's end. As scales grew, these models needed increasingly heroic quantities of data, demanding huge new efforts in standardization, communication systems, and automation. These developments required not only technological but also institutional innovation. The World Meteorological Organization (WMO), founded in 1950 on a base laid by the International Meteorological Organization, gained new authority for standards as an intergovernmental agency of the United Nations. In the 1960s the WMO directed its principal energies toward systems, standards, and institutional mechanisms for the World Weather Watch (WWW), which became operational in the late 1960s.

World Weather Watch planners wisely adopted a network perspective, promoting more unified development within the existing framework of linked national weather services—a web of institutions. The weather information infrastructure is also a web of instrument networks. Weather satellites—capable of observing the entire planet with a single instrument—began to realize the ideal of a fully global observing system. Satellite data sharing and the WWW concept grew directly out of Cold War politics, promoted as a counterweight to military and ideological tensions. Satellites and radiosondes, especially, generate data that differ dramatically in form from data generated by traditional surface stations. In the long run, numerous gateways—primarily in the form of software—made it possible to reconcile disparate forms of data from the many different platforms. Today weather forecast models and data assimilation models serve as the ultimate data gateways, relating each kind of data to every other one through modeled physics, rather than simply correlating unconstrained measurements in space and time. (See chapter 10 for a fuller explanation.)

The networking of national *climate* observing systems into a global climate information infrastructure took much longer. Weather data systems are built for real-time forecasting. Their priority is speed, not precision, and they absorb new instrumentation, standards, and models quite quickly. In contrast, climatology requires high precision and long-term stability—almost the opposite of the rapidly changing weather observing system—as well as certain kinds of data that weather forecasters do not collect. From the nineteenth century on, climate data networks overlapped with weather networks, but also included their own, separate observing systems, such as the US Historical Climatology Network. Central collectors established rudimentary *global* climate data networks in the late nineteenth century. These gained durability in the early twentieth century in the Réseau Mondial and the Smithsonian Institution's *World Weather Records*, which formed the primary basis of most knowledge of global climate until the 1980s.

Meanwhile, climate modeling developed along lines parallel to weather forecasting, but over a longer period. Because they must simulate decades rather than days while remaining realistic, three-dimensional global climate models remained essentially research tools until the late 1970s. They first gained a foothold as predictive tools around 1970 during the controversy over the supersonic transport, but full acceptance did not come until they achieved a rough consensus on greenhouse warming predictions in the late 1970s.

Under the spur of global warming concerns, national climate observing systems finally begin to consolidate into a global internetwork. The WMO initiated a World Climate Programme in 1980 (after the first World Climate Conference, held in 1979), but full consolidation at the technical level did not really begin until 1992, when the Global Climate Observing System (GCOS) was established in support of the Framework Convention on Climate Change. Today GCOS remains, in the eyes of climatologists, at best an incomplete skeleton for what may one day become a fully adequate climate observing network.[23] Very recently, increasing demands from many quarters for reliable climate forecasts have led US agencies to begin discussions on changing the orientation of the climate observing system from a research orientation to an operational—in my terms, infrastructural—one.

Yet while we can consolidate the climate observing system further as time goes on, making sense of the data we already have presents a different, very special issue of consolidation. We are stuck with whatever data we have already collected, in whatever form, from whatever sources with whatever limitations they might have. We have to consolidate a global climate observing network not only prospectively but *retrospectively*, reassembling the motley collection of weather and climate records and reprocessing them *as if* they all existed only to help us in this moment. Since the 1980s, through a meticulous process of infrastructural inversion (see below), scientists have slowly and painfully consolidated these data, unearthing previously uncollected records and metadata (contextual information) and using them to create more comprehensive global datasets, to reduce inhomogeneities, and to render highly heterogeneous data sources into a common form.

The most ambitious version of this consolidation, known as "reanalysis," also represents a consolidation of the weather and climate information infrastructures. In reanalysis, past weather records (not climate data) are run through complex data assimilation models—originally designed for weather forecasting—to produce a single, uniform global data set for 50 years or more. Traditional climate data consist mostly of averages for single variables (temperature, precipitation, etc.) over periods of a month or more. Reanalysis produces a much different kind of data: all-variable, physically consistent data sets containing information for millions of gridpoints every six hours. Although biases in the models prevent them from displacing traditional climate data, climate statistics calculated from reanalysis data can reveal "fingerprints" of climate change not detectable in traditional data.

A final aspect of consolidation in the climate data infrastructure is the Intergovernmental Panel on Climate Change, founded in 1988 to provide periodic assessments of climate knowledge for parties to the UN Framework Convention on Climate Change, who now include almost every country in the world. The IPCC conducts no research of its own. Yet it represents the most important institutional innovation in the history of climate science. The periodic assessments demand regular comparisons of climate models and of all the various climate datasets. Not only does this process reveal weaknesses in both models and data; it also creates a mechanism for surfacing, reviewing, and merging (wherever possible) every element of climate knowledge. Their very title, Synthesis Reports, represents their consolidating role. The ongoing cycles of IPCC assessment also promote increased coupling across domains as diverse as oceanography, ecology, agriculture, and demography. Today the practical outcomes of this coupling are typically suites of linked computer models, such as Earth system models and integrated assessment models, that combine the knowledge and techniques of many disciplines in a single, many-faceted simulation.

An LTS-based analysis, then, helps us to periodize the history of global meteorological networks as technical systems. It also directs us to attend closely to the political, legal, economic, and institutional dimensions of network formation. To understand global warming as an object of knowledge, however, we need more than this. We want to know not just how weather and climate data get moved around—like conversations in the telephone system or electric power through transmission lines—but also how they get created in the first place, how they are transformed into intelligible and reliable information, and, most important, how that information becomes knowledge.

## Knowledge Infrastructures

If we see objects of knowledge as communal, historical products, we can readily extend the concept of infrastructure we have just discussed. The LTS approach to infrastructure always began with a technology base. Yet it also invariably found that to explain a technical system's development one had to understand its social elements—hence its cornerstone phrase, "*socio*technical systems."

If we take this notion seriously, it applies directly to knowledge. Instead of thinking about knowledge as pure facts, theories, and ideas—mental things carried around in people's heads, or written down in textbooks—an infrastructure perspective views knowledge as an enduring, widely shared sociotechnical system. Here is a definition: *Knowledge infrastructures comprise robust networks of people, artifacts, and institutions that generate, share, and maintain specific knowledge about the human and natural worlds.*

Consider how we produce the specific type of knowledge we call science.[24] If you want to be a scientist, you probably are going to need some technological items, such as instruments and computers. If you are going to share what you learn, you will want the Internet, or at least the telegraph (postal mail, in a pinch). By themselves, however, such tools and media will only get you started. Once you have a result, you still need to convince people that it is true, useful, and consistent with other things they already know. To do that, you need authority and trust. Those can come only from being connected—in both a present-tense sense and a historical sense—with a community that understands what you have found and what you think it means. Thus, if you want to create and maintain scientific *knowledge*, you are also going to need at least the following:

- enduring communities with shared standards, norms, and values
- enduring organizations and institutions, such as libraries, academic departments, national science foundations, and publishers
- mathematics
- specialized vocabularies
- conventions and laws regarding intellectual property
- theories, frameworks, and models
- physical facilities such as classrooms, laboratories, and offices
- "support" staff: computer operators, technicians, secretaries

Let me elaborate this a bit further. Science emanates from a set of respected institutions, including university departments, research laboratories, national academies, and professional organizations. These institutions evolve and sometimes fail, but many of them have endured over long periods of time (decades or even centuries). In addition to instruments, computers, libraries, and so on, these institutions rely on suites of well-accepted models and theories. They generate specialized vocabularies and mathematical techniques, which all practitioners must learn. Professional training in science is long and demanding. It teaches would-be scientists to think according to prevailing standards of logic, to interpret and judge evidence according to disciplinary norms, to use and trust particular instruments and research methods (and reject others), to design experiments around well-established (often community-specific) principles, and to communicate with others according to certain kinds of protocols and conventions. Those who cannot master these practices cannot receive crucial credentials (typically a PhD degree).

Scientific knowledge is transmitted through a variety of material and human forms—journals, conferences, websites, students, postdoctoral fellows, and professional organizations, among others. Only some of this communication is ever condensed into formal publications. Libraries and online depositories store and provide access not only to published results but also (increasingly) to raw data, preprints, and other intermediate products of scientific investigation; this storage and maintenance activity represents a major commitment of human and financial resources. To keep the whole thing going, large institutions such as national science foundations transfer money from taxpayers to researchers. Along the way, they impose numerous practices and policies regarding peer review, ethical behavior, data sharing, credentialing, and so on. Vast legal structures govern and enforce intellectual property rights, informed consent for human subjects, and other forms of scientific integrity.

The infrastructural quality of this edifice appears vividly in the daily routines of scientific work. Consider writing grant applications, posting a new result to an Internet preprint site, keeping track of recent journal articles via Internet connection to a library, attending professional meetings, manipulating experimental data with computerized statistics packages or modeling software, getting one's laptop repaired by a local technician, and teaching a class how to use a simple model. Each of these activities both relies upon and helps reproduce the knowledge infrastructure. That infrastructure is a production, communication, storage, and maintenance web with both social and technical dimensions. Instruments, disk drives, and Internet links blend seamlessly with thinking, talking, and writing. Journals and websites mirror community life. The complex forms required for grant proposals reflect the routines of funding organizations and act as gatekeepers to reduce the number of proposals. Computer software embodies theories and facts. All the features of infrastructure discussed above appear here: embedded in everyday life, transparent to users, wide reach and scope, learned as part of membership, linked with conventions of practice, built on an installed base, and so on.

I intend the notion of knowledge infrastructure to signal parallels with other infrastructures, such as those of communication, transport, and energy distribution. Yet this is no mere analogy or metaphor. It is a precise, literal description of the sociotechnical supports that invariably undergird facts and well-accepted theories.

Get rid of the infrastructure and you are left with claims you can't back up, facts you can't verify, comprehension you can't share, and data you can't trust. Without the infrastructure, knowledge can decay or even disappear. Build up a knowledge infrastructure, maintain it well, and you get stable, reliable, widely shared understanding. The concept of knowledge infrastructure resembles the venerable notion of scientific paradigms, but it reaches well beyond that, capturing the continuity of modern science, which keeps on functioning as a production system even while particular theories, instruments, and models rise and fall within it.[25] This is not an entirely new idea in science and technology studies, where scholars sometimes use the word 'technoscience' to capture the technological dimension of science as a knowledge practice. Ethnographic studies of laboratories and "epistemic cultures" have looked at science as a production system characterized by "inscription devices," document flows, and other material-technical features.[26] I prefer the language of infrastructure, because it brings home fundamental qualities of endurance, reliability, and the taken-for-grantedness of a technical and institutional base supporting everyday work and action.

Further, and perhaps most important, the idea of infrastructure captures the notion of extensibility. Climate knowledge once came from a few relatively uniform and similar scientific disciplines, but that has not been true for decades. Since the 1960s the climate knowledge infrastructure has been extending itself by building gateways linking different fields. Computer models are its most important technical gateways; since the 1960s modelers have progressively linked models of the atmosphere to models of the oceans, the cryosphere, the biosphere, and human activities. Since the late 1980s, the primary institutional gateway joining disparate elements of the climate knowledge infrastructure has been the Intergovernmental Panel on Climate Change, whose regular cycles of comparison, assessment, and integration link numerous scientific fields in a common knowledge project.

### Infrastructural Inversion

To understand an infrastructure, you have to invert it. You turn it upside down and look at the "bottom"—the parts you don't normally think about precisely because they have become standard, routine, transparent, invisible. These disappearing elements are only figuratively "below" the surface, of course; in fact they *are* the surface. But as with anything that is always present, we stop seeing them after a while.[27]

This book is going to invert the climate knowledge infrastructure for you, but scientists themselves do it often. Infrastructural inversion is, in fact, fundamental to how scientists handle data. Climate scientists put it this way:

> For long-term climate analyses—particularly climate change analyses—to be accurate, the climate data used must be *homogeneous*. A homogeneous climate time series is defined as one where variations *are caused only by variations in weather and climate*. Unfortunately, most long-term climatological time series have been affected by a number of non-climatic factors that make these data unrepresentative of the actual climate variation occurring over time. These factors include changes in: instruments, observing practices, station locations, formulae used to calculate means, and station environment.[28]

In other words, data aren't data until you have turned the infrastructure upside down to find out how it works. Other "non-climatic factors" in historical data stem from garbled communication, coding errors, and other noise. To decide whether you are seeing homogeneous data or "non-climatic factors," you need to examine the history of the infrastructure station by station, year by year, and data point by data point, all in the context of changing standards, institutions, and communication techniques.

That history, as I intimated earlier, has been deeply problematic for climatology. By the early twentieth century, weather forecasting and climatology had diverged. Most national weather services, focused on providing short-term forecasts, paid scant attention to the observational needs of climatology. New observing stations often did not measure important climatological variables, such as precipitation. Meanwhile, existing stations changed location, replaced old instruments with new ones of a different type, disappeared, or saw their originally rural settings slowly transformed into (warmer) urban ones. These changes and many more affected the continuity, stability, and quality of their data records. As a result, only about one fourth of stations in the US Cooperative Observer Network meet the US Historical Climatology Network's standard that a station have provided "at least 80 years of high-quality data in a stable environment."[29]

Since the 1950s, standardization and automation have helped to reduce the effect of "non-climatic factors" on data collection, and modeling techniques have allowed climatologists to generate relatively homogeneous data sets from heterogeneous sources.[30] But it is impossible to eliminate confounding factors completely. Indeed, since the late 1990s the temporal and spatial consistency of surface weather data has been undermined by technological changes and by a reduction in the number of surface stations and ocean platforms.[31] Such an infrastructural change produces not only quantitative but also qualitative effects. For example, today's climate information system collects much more information than was collected in the past. Surface data for (say) 1890–1900 were produced by a much smaller, much less well-distributed station network than data for (say) 1990–2000. In addition, however, today's data network collects new *kinds* of data, including measurements from radiosondes (weather balloons) and satellite radiometers, which monitor the atmosphere's vertical dimension. Among other things, this means that no 1890–1900 time series will have any data at all from high above the ground, whereas a 1990–2000 series might have much more data from the upper air than from the surface. As we will see, climate scientists have found numerous ingenious ways to confirm, correct, combine, and reject data. Yet these methods, too, have evolved. With each iteration in the cycle of reexamination, correction, and analysis, *the climate data record changes*. As a result, we have not one data image of the planetary climate, but many—very many.

How can this be? Aren't data supposed to be the stable cornerstone of the entire edifice of knowledge? In the strange and wonderful world of computational meteorology, the answer can, in fact, be "not quite." In modern weather forecasting, for example, only about ten percent of the data used by global weather prediction models originate in actual instrument readings. The remaining ninety percent are synthesized by another computer model: the analysis or "4-dimensional data assimilation" model, which creates values for all the points on a high-resolution, three-dimensional global grid. This isn't as crazy as it sounds. Put very simply, the analysis model starts with the previous weather forecast, then corrects that forecast with current observations, producing values for every gridpoint. At the same time, the analysis model checks the observations for errors and inconsistencies, rejecting some observations and modifying others. Thus the raw readings from the observing system constrain, *but never fully determine,* the data that serve as forecast inputs. In empirical tests, these synthetic data sets produce far better weather forecasts than could be achieved using observations alone.[32]

This strange situation raised hopes among climatologists for a technique known as "reanalysis." Analyzed weather data aren't of much use to climatologists because forecasters frequently revise their analysis models (as often as every six months in some cases). Each change in the analysis model renders the data it produces incommensurable with those produced by the previous model. Reanalysis eliminates this problem by using a single "frozen" model to analyze historical observational data over some long period (40–50 years or even more). Because analysis models are built to combine readings from all available observing systems, reanalysis also overcomes the otherwise thorny problem of comparing instruments such as radiosondes and satellite radiometers. The result is a physically self-consistent global data set for the entire reanalysis period. Potentially, this synthetic data set would be more accurate than any individual observing system.[33]

Reanalysis would deal in one fell swoop with many of the data inconsistencies caused by infrastructural change. Yet climatologists currently regard reanalysis data sets as problematic for climate trend studies. Biases in the analysis models—too small to matter in forecasting—accumulate to produce significant errors when applied over the long periods needed to track climatic change. Nonetheless, some scientists hope that reanalysis will eventually generate definitive data sets, useable for climate trend analysis, that will be better than raw observational records. For the moment, however, they are stuck with infrastructural inversion—that is, with probing every detail of every record, linking changes in the data record to social and technical changes in the infrastructure that created it, and revising past data to bring them into line with present standards and systems.

Inverting the weather and climate knowledge infrastructures and tracing their history reveal profound relationships, interdependencies, and conflicts among their scientific, technological, social, and political elements. Over time, as knowledge production becomes infrastructural, these relationships become increasingly invisible, even as they continue to evolve. The difference between controversial claims and settled knowledge often lies in the degree to which the production process is submerged. Thus, *an established fact is one supported by an infrastructure.* In the rest of this book,

I explore the meaning and implications of this claim for knowledge about weather, climate, and global warming.

## Meteorology as Infrastructural Globalism

Clearly what I am talking about belongs with the larger phenomenon of globalization, a subject that has consumed scholarship, historiography, and political discourse in recent years. Is globalization old or new, a long slow trend or a sharp discontinuity? What are its causes and its consequences? Was the Age of Empire more "global" than the present? Does globalization really link the whole world, or does it disconnect and disenfranchise the poorest people and their nations? No one who has followed these debates can fail to notice the prominence of information and communication technologies in virtually all accounts. Marshall McLuhan long ago described the "global village," the shrinkage of space and time through printing, literacy, and mass media.[34] More recently, Manuel Castells defined the global economy as one "whose core components have the institutional, organizational, and technological capacity to work *as a unit in real time, or chosen time, on a planetary scale*" through information and communication infrastructures.[35] Every chapter in a recent survey of the literatures on political, economic, and cultural globalization systematically addressed the role of communication infrastructures.[36] Similar examples could be multiplied ad infinitum.

In an important variation on this theme, Martin Hewson proposed a notion of "informational globalism." The concept refers simultaneously to systems and institutions for transmitting information around the world, and to systems and institutions for creating information about the world as a whole.[37] Hewson sees informational globalism as developing in three phases. First, during the nineteenth century, national information infrastructures such as telegraph systems, postal services, and journalism were linked into interregional and intercontinental (if not fully global) networks. Between 1914 and 1960 (Hewson's second phase), the pace of infrastructural linking diminished, and some delinking occurred. Yet simultaneously, world organizations such as the League of Nations and the International Monetary Fund "established the legitimacy of producing globalist information"—that is, information about the whole world—in such areas as health, armaments, and public finance (although they did not yet achieve that goal). Hewson's third phase brought generalized attainment of the two previous eras' aspirations, beginning with worldwide civil communication networks (from the 1967 inauguration of the Intelsat system) and global environmental monitoring (from the UN Conference on the Human Environment, 1972). Hewson sees global governance institutions such as the United Nations and the International Telecommunications Union, rather than an autonomous technological juggernaut, as chiefly responsible for the rise of informational globalism.

The story this book tells confirms the pattern Hewson discerned, but it also has special characteristics. The weather data network, along with its cousins in the other geophysical sciences, especially seismology and oceanography, is arguably the oldest of all systems for producing globalist information in Hewson's sense. When Ruskin wrote, in 1839, that meteorology "desires to have at its command, at stated periods, perfect systems of methodical and simultaneous observations . . . to know, at any given instant, the state of the atmosphere on every point on its surface," he was only giving voice to his contemporaries' grandest vision. By 1853 the Brussels Convention on naval meteorology had created a international standard meteorological logbook for ships at sea; these logs now constitute the oldest continuous quasi-global meteorological record. The International Meteorological Organization, despite its endemic weakness, represents an early international governance institution, while the Réseau Mondial and its successors reflected the ambition to build a global weather information infrastructure.

By 1950 the informational-globalist imperative was already far stronger in meteorology than in many other putatively "global" systems that emerged around the same time. Though rudimentary, a planetary monitoring network had been functioning for decades, and had gained speed and scope in the 1920s with the arrival of shortwave radio, which untethered the data network from telegraph cables. Computerized weather forecast models, operational in 1955, covered continental and (soon) hemispheric scales, displaying an insatiable thirst for data from every corner of the world. By the early 1960s, satellites brought the once unthinkable realization of the God's-eye view: the ability to observe the entire planet with a single instrument. Unifying the existing global observing system, improving the communication network, and preparing meteorology for satellite data became the World Meteorological Organization's fundamental goals in its World Weather Watch program. Simultaneously, as a global governance institution operating within the UN system, the WMO actualized a new commitment by governments throughout the world to link their weather services through shared investment in a worldwide weather network. Intergovernmental status meant national government involvement, bringing the political dimensions of weather science into the open.

I contend that the history of meteorology from the 1850s to the present illustrates a profoundly important, albeit messy and incomplete transition: from voluntarist internationalism, based on an often temporary confluence of shared *interests*, to quasi-obligatory globalism based on more permanent shared *infrastructure*. Therefore, I will speak not only of informational globalism but also of *infrastructural globalism*: projects for permanent, unified, world-scale institutional-technological complexes that generate globalist information not merely by accident, as a byproduct of other goals, but by design.[38] Enduring, reliable global infrastructures build scientific, social, and political legitimacy for the globalist information they produce. Meteorology as infrastructural globalism sought to establish permanent sociotechnical systems for monitoring the weather, modeling its processes, and preserving planetary data as scientific memory.[39]

Infrastructural globalism is about creating sociotechnical systems that produce knowledge about the whole world. It may be driven by beliefs about what such knowledge can offer to science or to society, but it is not principally an ideology. Instead it is a *project*: a structured, goal-directed, long-term practice to build a world-spanning network, always including a worldwide epistemic community as well as a technical base.[40] If such a project succeeds, it creates an infrastructure that endures far beyond individual careers, social movements, or political trends. This endurance itself legitimizes the knowledge it produces, and becomes self-perpetuating.

Such projects were never, of course, unique to meteorology. The other geophysical sciences (seismology, oceanography, etc.), the epidemiology of infectious diseases, financial markets, and the American and Soviet intelligence agencies of the Cold War era exemplify other disciplines and organizations that built infrastructures for generating globalist information. They too built monitoring and communication networks, created models of large-scale system behavior, and kept long-term records for the purpose of identifying global trends. Few, however, had either meteorology's need to engage the entire planet, or its great age.

# 2 Global Space, Universal Time: Seeing the Planetary Atmosphere

Today we see world maps almost everywhere we go. Backdrops to the nightly news, they appear transparent, obvious, unmediated. We seem to grasp their God's-eye view intuitively, without thought. GPS receivers in our phones and our cars pinpoint us precisely on the global grid. In all their incarnations, from Mercator projections to parlor globes to interactive GPS, maps are information technologies of the first order. They are "objects to think with," in Sherry Turkle's felicitous phrase.[1]

Behind the seeming immediacy of global maps and images lie vast bodies of complex and expensive collective and collaborative work and social learning accomplished over many centuries. This labor and this learning included not only invention, exploration, and surveying but also the slow spread, through practical use and formal education, of the graphical conventions, iconography, and social meaning of global maps. Projections of Earth's spherical surface onto a rectangular page, systems of latitude and longitude, the North Pole as the world's "top," ways of depicting geographical features—these conventions and many more evolved and spread along with Western empires. While learning to "see" the whole world with maps, people also imagined traveling its farthest reaches, flying high above its surface, or peering down on it from space long before they could actually do so.[2] World maps undergird our ability to conceive global space. They are an infrastructural technology, a principal material support for "thinking globally."

Like other cartographic concepts, the idea of mapping weather data took centuries to develop. Graphical conventions for showing weather relationships in space emerged many decades after the first international weather data networks. Drawings illustrating the global circulation—the prevailing structure of atmospheric motion—appeared rather suddenly in the middle of the nineteenth century. From then on, graphical representations of weather, climate, and global circulation became core technologies of the emerging climate knowledge infrastructure.

Universal time is basic to the texture of modern life. Seconds before your morning alarm goes off, you awaken to an inner clock nearly as accurate as the one beside your bed: you have internalized the infrastructure. Your world runs on time, in both senses. Like the latitude-longitude grid, today's universal time is a widely shared convention that exists in and through technology (watches, clocks), political decisions (adopting a national standard time, the Greenwich prime meridian), commercial interests (railroads, airlines), and social practices (the different meanings of "on time" in Switzerland, Brazil, and South Africa). We constantly re-create and reaffirm the infrastructure of universal time simply by using it, whenever somebody, somewhere, checks her watch to keep an appointment or catch a train.

Yet for most of human history, the only time that really mattered was the one marked by the sun. In medieval Europe, one daytime "hour" equaled one-twelfth of the time between sunup and sunset: accordion-like, hours shrank in the winter and expanded in the summer.[3] Mechanical clocks fixed the length of the hour, but "noon" still meant the moment when the sun crossed the zenith wherever you were. The present system of universal time, where "noon" is kept by the clock within a broad zone running (mostly) along lines of longitude from pole to pole, was not even conceived until the late nineteenth century. It did not become a worldwide infrastructure until the second half of the twentieth. The momentous change from a plethora of local times to a single universal time began when one emerging infrastructure, the telegraph, made it possible to synchronize clocks across large distances, while another, the railroad, made that synchronization necessary.

Similarly, modern meteorology arose when new infrastructures, including a potent combination of widely shared mapping conventions, telegraphic data transmission, and new time standards, made it possible to create data images of the atmosphere in motion—permitting wide-area forecasting for the first time. Because this capability resonated with national military and commercial interests, it led quickly to the formation of national weather services and telegraph-based international data networks. This chapter examines the origins of these elements of the weather and climate knowledge infrastructure.

## Global Space

Eratosthenes (third century BCE), Ptolemy (second century CE), and other ancient astronomer-geographers deduced Earth's spherical shape and accurately estimated its circumference. They also devised latitude-and-longitude systems and mapped the lands they were aware of according to those coordinates. Ptolemy's famous map of his known world covered about 80 degrees of latitude, from the equator to the Arctic, and 180 degrees of longitude, from China to the Canary Islands.[4]

The ancients tied their ideas of climate directly to their conceptions of global space. The English word 'climate' derives from the Greek word *klima*, which is also the root of 'inclination'. *Klima* means "sloping surface of the earth," linking climate to latitude, which governs the inclination of the sun's rays. Ptolemy based his system of fifteen climatic zones on the lengths of their longest day—a quantity that also served him to express latitude, taking the place of degrees.[5] Then, and in succeeding centuries, many natural histories began with descriptions of local and regional climates. Even today, common language terms such as 'tropical', 'desert', and 'temperate' refer interchangeably to geographic regions and their typical weather patterns. Often these are directly associated with latitude (e.g., "the tropics").

In 1686, the British astronomer Edmond Halley published one of the first theories to go beyond the Ptolemaic view of climate as a simple function of latitude. Halley sought to understand the physics of the trade winds, which blow from the northeast in the northern hemisphere and from the southeast in the southern. He proposed a planetary-scale, three-dimensional explanation. "Having had the opportunity of conversing with Navigators acquainted with all parts of India, and having lived a considerable time between the Tropicks, and there made [his] own remarks," Halley theorized that solar heating caused air to rise near the equator.[6] This "rareified" air caused denser air from higher latitudes to "rush in," creating the trade winds.

Halley had identified a fundamental mechanism of weather: the movement of heat from the equator toward the poles. Scientists today still use Halley's term 'circulation' to describe global patterns of air movement, and still use his notion that the atmosphere must "preserve the Æquilibrium." Standard texts cite Halley as the originator of these ideas, but Halley's own discussion makes clear that even earlier work by the geographer Bernhard Varen and by "several" unnamed others sparked his thinking.[7] Halley

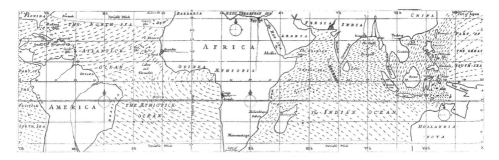

**Figure 2.1**
Halley's 1686 map of the trade winds in the Atlantic and Indian Oceans.
*Source*: E. Halley, "An Historical Account of the Trade Winds, and Monsoons, Observable in the Seas Between and Near the Tropicks, With an Attempt to Assign the Phisical Cause of the Said Winds," *Philosophical Transactions of the Royal Society of London* 16, no. 183 (1686), opposite 151.

included a map of the trade winds—reputedly the first meteorological chart ever published—to bolster his explanation (figure 2.1).

Half a century later, when George Hadley modified Halley's explanation to take account of the Coriolis effect, he employed no maps or diagrams of any kind.[8] The absence of graphical elements typified seventeenth-century and eighteenth-century meteorology. Even simple graphs—among today's most ubiquitous tools of data analysis—saw little use before about 1890. Instead, early weather scientists typically published their observations in long tables, which often combined quantitative with qualitative information.[9] Poring over such tables, early meteorologists hoped to discover regularities in weather phenomena. They were sorely disappointed. As Frederik Nebeker put it, "the tabulation of data was forceful in *diminishing* belief in virtually all simple correlations involving meteorological phenomena" (emphasis added).

The nineteenth century witnessed a virtual explosion of scientific cartography, including the first systematic use of mapping as a tool of data analysis (what we would now call "scientific visualization"). In the early 1800s the German scientist Alexander von Humboldt traveled much of the world, over land and sea, measuring, recording, and classifying nearly everything he saw. Humboldt's famous 1817 chart of the northern hemisphere deployed a new graphical technique: "isotherms," smooth lines demarcating zones of similar average temperature (figure 2.2). This chart showed average temperatures curving away from latitude lines, thus defying the ancient theory of *klimata* and posing a climatological problem.[10]

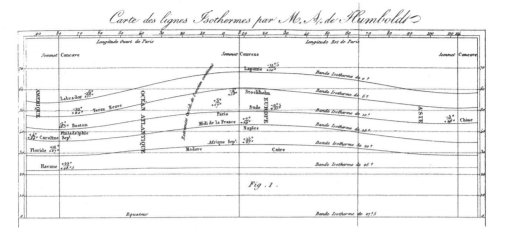

**Figure 2.2**
Humboldt's "chart of isothermic lines," covering the northern hemisphere from approximately 100°W (the Mississippi river) to 100°E (Thailand).
Source: A. von Humboldt, "Sur Les Lignes Isothermes," *Annales de Chimie et de Physique* 5 (1817): 102–12.

Almost simultaneously (between 1816 and 1819), independent of Humboldt, another German physicist, Heinrich Wilhelm Brandes, produced what were probably the first weather maps. Brandes's original maps are lost, but evidence suggests that they employed "isobars," lines indicating regions of similar barometric pressure.[11]

The meteorologist Heinrich Wilhelm Dove, a close colleague of Humboldt's, soon adopted his method and used it, in part, to discover the relationship of atmospheric pressure to wind direction during the passage of storms.[12] In 1852 Dove published isothermal charts for the entire Earth (figure 2.3).

The isoline—the general term for this technique of mapping relationships among data points—was a crucial innovation in weather and climate data analysis. Its importance can hardly be exaggerated. Isolines were the first practical technique for visualizing weather patterns *from data* over large areas: pictures worth a thousand numbers. Little changed since the days of Humboldt, Brandes, and Dove, isolines remain a basic convention of weather and climate maps today.

The data tables created by seventeenth- and eighteenth-century scientists recorded information from points—that is, individual places. By contrast, isotherms, isobars, and similar cartographical tools displayed

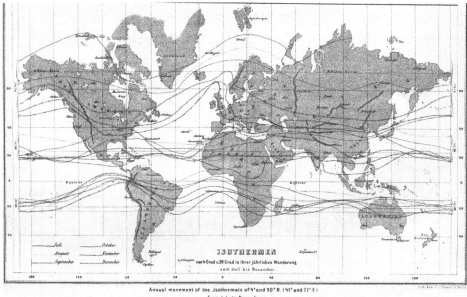

**Figure 2.3**
Dove's map showing isotherms for July through December at three latitudes. Lines above 40°N are isotherms for 4°C (one line for each monthly average). Lines around 20°N and 20°S are isotherms for 20°C.
*Source*: H. W. Dove, *The Distribution of Heat Over the Surface of the Globe* (Taylor and Francis, 1853), opposite 27.

spatial continuities or regions. At the same time, however, isolines brought awareness of the stark limitations of available data. One drew a smooth line connecting measurements—but what was actually going on between the points, often hundreds of kilometers apart? To bring the tool closer to the reality it represented, meteorologists knew, they were going to need more data.

In the seventeenth and eighteenth centuries, meteorological observing networks arose sporadically across Europe, America, and Russia, including northern Asia. (See box 2.1.) Until the middle of the nineteenth century, however, none of these networks endured for more than about 20 years. Before the telegraph, these networks communicated via postal mail, taking weeks or months to assemble a data set. Such data had no economic or military value, since they could not be used in forecasting. With every reason to exchange information and none to keep it secret, early meteorologists shared weather data freely. Data sharing became a deeply entrenched norm, which Nebeker named the "communality of data."[13]

**Box 2.1**
Pre-Nineteenth-Century Meteorological Data Networks

### The Accademia del Cimento

From 1654 to 1667, under the patronage of Ferdinand II, Grand Duke of Tuscany, the Accademia del Cimento organized a pan-European weather network using comparable (and in some cases even redundant) instrumentation. The ten stations in this network extended across Italy from its base in Florence, as well as northward to Paris, Warsaw, Innsbruck, and Osnabrück in what is now northern Germany.

### James Jurin's Network

Jurin invited European observers to submit weather records for publication, based on his recommended observing scheme.[a] He collected data submitted for the period 1724–1735, publishing them in the British Royal Society's *Philosophical Transactions* between 1732 and 1742. These data included some from the American colonies.[b]

### The Great Northern Expedition

Scientists participating in the 1733 Great Northern Expedition, which explored northern Russia seeking sea trade routes, established a network of meteorological stations as far east as Yakutsk at 130°E.[c]

### The Palatine Meteorological Society

From 1780 to 1795 the Societas Meteorologica Palatina, based in Mannheim, organized a network of 37 weather stations scattered across Europe and the United States. Thirty-one of these stations carried out synchronous observations.

a. J. Jurin, "Invitatio Ad Observationes Meteorologicas Communi Consilio Instituendas," *Philosophical Transactions of the Royal Society of London* 32, no. 379, 1723: 422–27.
b. A. K. Khrgian, *Meteorology: A Historical Survey* (Israel Program for Scientific Translations, 1970), 71–73.
c. D. C. Cassidy, "Meteorology in Mannheim: The Palatine Meteorological Society, 1780–1795," *Sudhoffs Archiv* 69, 1985: 8–25.

This communality of data helped make meteorology among the most open and cosmopolitan of sciences.

The communality of data set meteorology and climatology apart from the laboratory or "bench" sciences. Laboratory experiments produce data. Usually, numerous failures and false starts precede a "successful" experiment (one whose data confirm a hypothesis). Data from "failed" experiments are mostly discarded, and no one outside the laboratory ever sees them. During this process, the laboratory functions as a private space; no one need know how many mistakes you made along the way. Only data that can be explained by theory get published, and even these usually appear only in highly processed form.[14] This special power not only to isolate and concentrate natural forces but also to multiply mistakes and conceal its own internal processes has made the laboratory one of modernity's most potent inventions.[15] In meteorology and in other field sciences, by contrast, data can't be generated in some closed, private room. Instead, meteorology has to spread itself through large-scale geographical space, distributing its network of people, instruments, and knowledge widely. Few sciences have had such fundamental reasons to make themselves "omnipresent over the globe," as Ruskin put it in 1839.

## "Data Guys": The Network Structure of Meteorology

A network such as that of meteorology is known to historians of technology as an "accumulative" infrastructure or infrasystem.[16] Its goal is to accumulate many observations at some central point (or points), where they can be analyzed, charted, and then distributed. In meteorology, in contrast with the laboratory sciences, one can gain professional recognition simply for accumulating a substantial data set. Even in present-day weather science, in which theory and modeling have taken pride of place and gigantic data sets circulate effortlessly across the Internet, the men and women responsible for collecting, "cleaning," and archiving large data sets are held in high esteem, as I learned while interviewing scientists for this book. Climatologists call them "data guys."

Among the first "data guys" to try to build a global data set was Matthew Maury, a US Navy officer. Maury made it his mission to collect and map ships' logs of winds and currents at sea. Beginning in 1848, he cranked out a prodigious series of publications—some 200 volumes.[17] They remained standard works well into the twentieth century.

In the beginning, Maury's project suffered mightily from the lack of standard metrics and measuring practices in existing ships' logs. The metric system, then gaining popularity among scientists, competed with other units of measure—especially the British imperial system adopted by (or forced upon) many countries around the world. Fahrenheit and Celsius temperature scales both remained in common use. Later, when he became director of the US Naval Observatory, Maury used the office to promote "an universal system of meteorological observations by sea and land," to which every government in the world would (he hoped) contribute.[18] He organized the first intergovernmental conference on standardized observing systems, held at Brussels in 1853 and attended by representatives from nine European nations and the United States.

Participants in the Brussels conference settled on a standard logbook format and a standard set of instructions for taking observations. By 1858, nineteen countries had joined the Brussels convention. As a global standard, the agreement saw only partial success. It required, for example, that vessels using the Fahrenheit scale also record temperatures in Celsius; many captains simply ignored this and other inconvenient obligations.[19] Nonetheless, the US Naval Observatory, the British Meteorological Office, and the Netherlands Meteorological Institute each processed the collected naval data to produce important series of marine maps.[20] Guided by such maps, ships shaved 33 days off the average ocean transit from New York to San Francisco.[21] Though Maury became justly famous for his superhuman collecting effort, even more important in the long run was the data network he established. Naval logs remain the longest continuous quasi-global data record.

Two decades later, *HMS Challenger's* four-year scientific voyage renewed Maury's vision of data-based global charts of prevailing winds, currents, and temperatures. Between 1872 and 1876 the *Challenger* sailed more than 127,000 km, traversing most of the world's oceans, with the explicit mission of comprehending "the terraqueous globe taken as one whole." Observers on the *Challenger* recorded weather data every two hours throughout the entire mission. The scientific reports and charts they produced occupied some 50 volumes. This was the nineteenth-century equivalent of a weather satellite: an attempt to observe the entire planet from a single platform using well-calibrated instruments and consistent techniques.

Another "data guy," a meteorologist named Alexander Buchan, took up the task of analyzing the *Challenger* data. Any serious discussion of the oceanic circulation, Buchan wrote, would require "maps showing for the various months of the year the mean temperature, mean pressure, and prevailing winds of the globe, with carefully prepared and extensive tables of the observational data required for the graphic representation of the results." *Challenger* data could contribute to that project, but the majority of data would have to come from weather stations. Yet, as Buchan lamented, "the only works [previously] available were Dove's isothermals, 1852; Buchan's isobars and prevailing winds, 1869; and Coffin and Wojekof's winds of the globe, 1875—*all of which were based necessarily, when written, on defective data*."[22] Buchan was already inverting the data infrastructure, reviewing and revising existing knowledge.

In a story endlessly repeated in subsequent decades, Buchan's effort proved a monumental project. It took him and his assistants seven years to complete their *Report on Atmospheric Circulation*. He tabulated temperature data from 1620 surface stations covering the period 1870–1884.[23] The 52 beautiful color maps illustrated global and hemispheric average temperatures, pressures, and winds (figure 2.4).

In subsequent decades the project of collecting, filtering, and mapping global climate data would be repeated over and over. Each new collector

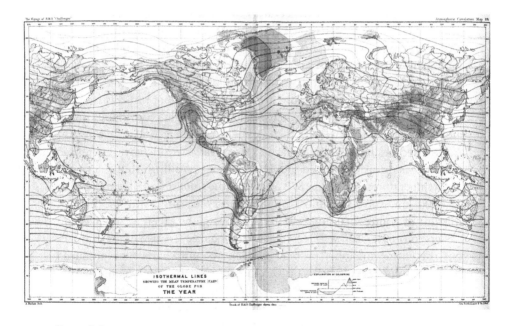

**Figure 2.4**
Global isothermal lines of mean annual temperatures (°F), constructed from HMS *Challenger* data plus annual averages at 1620 surface stations in 1870–1884 (many series incomplete).
*Source*: A. Buchan, *Report on Atmospheric Circulation Based on the Observations Made on Board HMS* Challenger *During the Years 1873–1876, and Other Meteorological Observations* (HMSO, 1889).

would invert the infrastructure anew, adding some data and rejecting others. Often collectors would frame some new way to refine the data, correct for systematic errors, or create a set more evenly distributed in global space.

These early maps and charts were strictly climatological. They were attempts to chart average conditions over years or decades—whatever the available data would support. Meteorologists also began to visualize—for the first time—the vertical motions of the planetary atmosphere as well as its horizontal ones. For example, Maury combined the theoretical work of Halley, Hadley, and others with his own knowledge of surface winds to create the diagram in figure 2.5.[24] Both the tropical Hadley cells and the

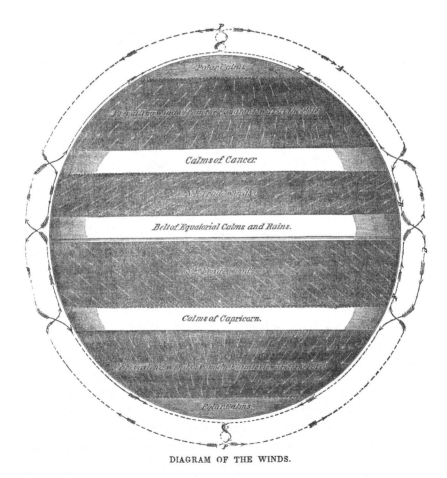

DIAGRAM OF THE WINDS.

**Figure 2.5**
Maury's two-cell diagram of the global circulation.
*Source*: M. F. Maury, *The Physical Geography of the Sea* (Harper, 1855), 70.

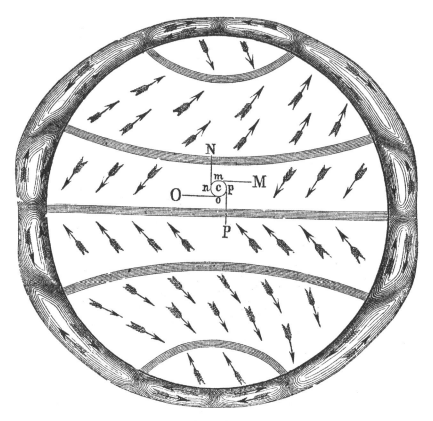

**Figure 2.6**
Ferrel's three-cell global circulation diagram.
*Source*: W. Ferrel, "An Essay on the Winds and Currents of the Ocean," *Nashville Journal of Medicine and Surgery* 11, no. 4–5 (1856), 290.

high-latitude circulatory cells proposed by Dove are readily visible.[25] The diagram relates the vertical circulation to the prevailing winds, and also to the calmer areas between the circulatory cells.

William Ferrel soon modified Maury's two-cell diagram, proposing a third cell to account for prevailing winds near the poles.[26] Meteorology's picture of the atmospheric circulation has changed little since then, as one can see by comparing Ferrel's diagram (figure 2.6) with the recent representation shown here in figure 2.7. These images captured the prevailing understanding of large-scale atmospheric motion, explaining the prevailing winds, the horse latitudes (or "calms of Cancer and Capricorn"), and some other fundamental phenomena. Unlike the climatological maps dis-

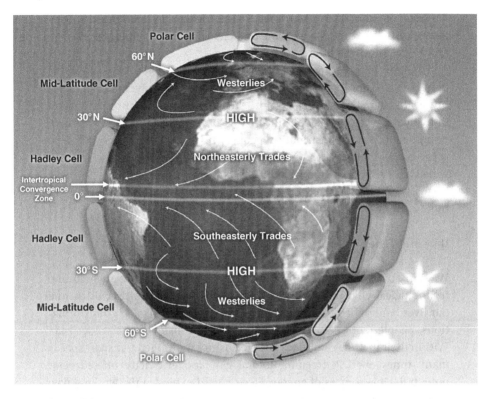

**Figure 2.7**
An idealized picture of the global circulation. Note the strong similarity to Ferrel's 1856 diagram.
Image courtesy of NASA.

cussed above, however, these were pictures of *theory*. Meteorologists had no way to test these theories, since all their measurements were taken at the surface. Although the physics were compelling, the existence of these vertical structures could only be inferred. Confirming their presence and charting their details would have to wait for the radiosonde, the airplane, and the satellite.

Thus, by the middle of the nineteenth century meteorology had created two crucially important ways to visualize the global atmosphere. The two-dimensional *data image*, as I will call it, sketched continuities in space by mapping averages of instrument readings from weather stations on land and from ships at sea. Meanwhile, the much cruder *theory image* pictured a three-dimensional, constantly moving circulation driven by solar heat; it was literally imagined rather than measured. The seemingly simple conventions of the climate data image, in particular, are technical achievements of the first order—fundamental elements of the emerging climate knowledge infrastructure.

Through these images, meteorology participated in the larger scientific project of envisioning "the world" as a whole—a single, dynamic, coherent physical system, knowable as a unit even though far beyond the scale of individual perception. In pursuit of this project, meteorology sought to occupy global space, distributing people, instruments, and knowledge to every corner of the Earth and its seas. Many other sciences also traveled widely in the Age of Empire, but few had such strong reasons to make themselves "omnipresent over the globe," as Ruskin put it in 1839—or, in more current terminology, "distributed" and "networked."

**Universal Time**

We think of the present as an era of extremely rapid, even overwhelming, technological change. A perpetual, inexorable acceleration seems to rule our lives.[27] Yet when it comes to overwhelming technological transformations, people living between 1840 and 1890 probably had a wilder ride. Within their lifetimes, railroads and steamships multiplied transport speeds many times over. Petroleum, natural gas, and electricity offered new, potent, flexible sources of energy. Electric light extended the working day. By comparison, almost everything that has happened since looks merely incremental.[28]

The most mind-bending new technology of all was the telegraph. Instantaneous communication over very long distances broke the ancient link between the speed of information and the speed of bodies moving through space. Invented simultaneously during the 1830s in Europe and the United States, the electric telegraph quickly came into worldwide use not long after Samuel Morse demonstrated a practical long-distance technique in 1844. The annihilation of distance and the rise of global virtual community, trumpeted today as consequences of television and the Internet, were felt even more vividly by those who experienced, for the first time ever, the arrival of news from across the oceans on the same day events occurred.[29]

At the dawn of electric telegraphy, scientific meteorology remained chiefly a pastime for gentlemen, academics, and amateur scientists. Forecasting was one goal, but not yet the main focus. Available forecasting techniques offered very limited accuracy and scope. Barometers gave generic clues to imminent local weather shifts, but beyond a day or so their skill was little better than chance. Rather than predict the future, then, most meteorologists studied the past, simply recording temperature, pressure, rainfall, wind direction, wind speed, and so on.

The telegraph, meteorologists immediately realized, would change all that. Now they could begin to realize Ruskin's dream of "perfect systems of methodical and simultaneous observations." Exchanging data across far-flung observing networks almost in real time, they could map the weather over very large areas. The resulting "synoptic" maps of simultaneous observations functioned like snapshots.[30] The maps charted pressure, temperature, and other weather conditions at each observing station. Wind direction and speed told which way the weather was moving, and how fast. All this provided a basis for at least a rational guess at what would happen next, and where. "Weather telegraphy," as it was known, proved considerably more effective than the barometer alone.

At the dawn of the telegraph era, meteorology was not yet professionalized. Except for a few university professors, mostly in Europe, there were no full-time meteorologists. Meteorology had little money, organization, or political power. So when the telegraph appeared on the scene, meteorologists never contemplated a dedicated weather telegraphy system. Instead, they planned to piggyback on the networks built by railroads, armies, and public or private telegraph companies.

Only governments and military forces, with their geographic reach, financial resources, and interests in practical weather prediction, could provide the necessary stability, scale, and funding for serious weather telegraphy. These networks exemplified an emerging social contract between science and the state. The United States' network was established in 1849 by Joseph Henry as one of the new Smithsonian Institution's first projects. Henry secured the agreement of commercial telegraph companies to transmit weather data free of charge. Ten years later, the Smithsonian Institution's weather telegraph network comprised about 500 observing stations.[31]

Europeans developed weather telegraphy around the same time, in similar ways. In 1854, during the Crimean War, a disastrous storm destroyed a French fleet near Balaklava on the Black Sea. Since observers had seen the same storm moving across the Mediterranean the previous day, it was clear that advance warning (by telegraph) might have prevented the debacle. In response, the French astronomer-meteorologist Urbain Jean-Joseph le Verrier proposed to Napoleon III that France sponsor an international weather telegraphy network. A few months later, in mid 1855, the network began operating in France. By 1857, Paris was receiving daily telegraph reports from Russia, Austria, Italy, Belgium, Switzerland, Spain, and Portugal. Not to be outdone, Japan developed a weather telegraphy network in the 1870s.[32]

In 1870, the US Congress established a Division of Telegrams and Reports for the Benefit of Commerce (also known as the Signal Service) within the War Department. It subsumed the Smithsonian's weather telegraphy network, marking the increasing value of weather forecasts for both military and commercial interests. The military network soon spanned the continent, comprising around 200 stations by the 1880s. As early as the 1880s, the Signal Service employed facsimile transmission techniques, such as "autographic telegraphy" and a crude facsimile cipher system, to transmit weather maps directly over telegraph lines (figure 2.8). In 1891, Congress placed the network on an entirely civilian basis, under the US

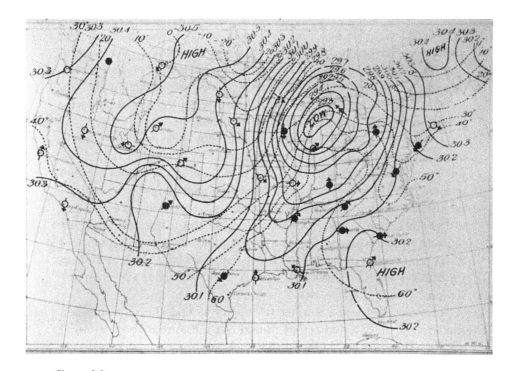

**Figure 2.8**
A US Signal Service weather map depicting a huge storm over the Great Lakes region of the United States in 1889. The map charts isobars (solid lines), isotherms (dashed lines), and wind direction.
*Source*: Signal Service, U.S. War Department, "Weather Map of United States." Image courtesy of National Weather Service Historical Photo Collection.

Weather Bureau within the Department of Agriculture. The Signal Service soon distributed daily synoptic charts of the continental United States as well as individual station data. Early in the twentieth century, some city newspapers began publishing these charts on a daily basis. However, high costs and technical problems with reproduction gradually caused newspaper weather maps to die out in the United States. Publication ceased almost entirely during World War I.[33]

By 1900, many countries possessing substantial telegraph networks also sponsored national weather services that conducted empirically based synoptic weather forecasting. Telegraph organizations throughout the world, both public and private, contributed to this project by transmitting weather data at no cost.[34] Forecasts based on weather telegraphy were not very accurate by today's standards, but they were better than local barometers alone, especially for storm warnings. They served important public interests, especially in shipping and agriculture. With the stability of state financial backing and the increased accuracy provided by weather telegraphy, meteorology began to professionalize, moving away from primarily amateur volunteer networks and toward a state-sponsored, technology-based, institutionalized infrastructure.

Telegraphy also catalyzed another event of major importance for the future of meteorology: the general standardization of time. Before about 1800, the norm was solar time, with noon meaning literally the middle of the day—i.e., the precise moment at which the sun crosses an observer's zenith. Time thus varied with longitude. Furthermore, though highly accurate pendulum clocks and chronometers did exist, most ordinary clocks and watches remained quite imprecise. For most purposes, "what time it is" was settled locally, marked by church bells and public clocks. The small differences in solar time between nearby locations caused few difficulties. Moving at relatively low speeds, long-distance travelers on land had little need of universal time.[35] With the advent of railroads, however, transportation speeds reached a point where even small differences mattered not simply for passengers' convenience, but for network coordination and safety. In early single-track rail systems, a southbound train would wait on a siding for the northbound train to pass; when trains failed to meet their schedules, terrible accidents often occurred. For these reasons, most of Great Britain's railways adopted Greenwich Mean Time for their operating schedules in 1847. Expanding telegraph networks made this possible by allowing near-instantaneous transmission of time signals by which conductors and stationmasters throughout the network could set their timepieces. Indeed, railroad companies built many telegraph lines themselves, their railbeds serving as ready-made long-distance rights-of-way. Railroad and telegraph rapidly became tightly coupled, taking on an interdependent web structure. Integrating transport and communication systems helped railroads to become the first widely distributed, large-scale business organizations.[36]

A different set of interests stemmed from the developing needs of science. Traditionally, astronomical observatories had performed a number of important practical functions in addition to scientific activities. These included recording the weather and signaling the precise moment of solar noon to their local communities by means of "time balls" (figure 2.9) or guns. These signals worked well if you were somewhere nearby, but if your town had no observatory your watch was probably going to be wrong. In

**Figure 2.9**
A time ball atop the Charing Cross telegraph station in London, 1852. The time ball would be raised to the top of the mast, then dropped to mark the exact hour. Note the public clock in the square outside the station.

the 1850s, soon after telegraph networks began to spread, observatories teamed with telegraph companies throughout the United States to sell time signals much further afield. Clients for this service included railroads, jewelers (who sold watches), and municipalities. Observatories calibrated their highly accurate clocks to solar time at their locations, then used them to deliver regular signals marking the hour. Before long, some 100 entities in the United States sold time-signal services. A cacophony of different time standards emerged, including local solar time, local civil time, railroad standard time, and Washington mean time. For a while, local knowledge sufficed to manage the various parallel systems; people simply added or subtracted minutes to accomplish the necessary conversions.

Meteorologists had long recognized the desirability of uniform observing times. Yet despite repeated calls for a consistent international system, each national weather service still established its own observing hours. Further, in these systems weather observers generally used local (solar) time. Before weather telegraphy, this didn't matter much. But once it became practical to gather weather data from a large area in near real time, differences in observing times could affect forecast quality. Among the first to recognize this was Cleveland Abbe. Formerly director of the Cincinnati Observatory (which sold time signals), Abbe became first head of the new Signal Service's weather department in 1870. Abbe often emphasized the need for *simultaneous* observation (same universal time), as opposed to the then common practice of *synchronous* observation.[37] When Abbe took over the Signal Service, standing instructions fixed observing hours by local solar time. Abbe immediately instituted a new system. The Naval Observatory in Washington would now telegraph time signals to Signal Service observers across the United States, thus establishing a uniform continental time standard in meteorology.

Abbe initiated a general campaign for time reform beyond the Signal Service. He chaired the American Metrological Society's committee on the subject.[38] Eventually he also chose, strategically, to advance his efforts through the powerful railroad companies (most of which had already implemented their own time standard), telegraph companies, and the American Association for the Advancement of Science. Abbe promoted standard, simultaneous observing times not only in the United States but internationally. One result was the Signal Service's monthly *Bulletin of International Observations Taken Simultaneously*, published for the entire northern hemisphere from 1875 to 1884.[39] Abbe later called this work "undoubtedly the finest piece of international cooperation that the world has ever seen."[40] In 1883 most of the United States adopted the present system of standard time zones, with Greenwich as the prime meridian. In subsequent decades the rest of the world gradually embraced this time standard.[41]

In any infrastructure, replacing an old standard with a new one requires overcoming the "inertia of the installed base."[42] And indeed, despite Abbe's efforts, international meteorology proved very slow to accept standard time. Just a year before Abbe's time-reform campaign began, the first 1873 International Meteorological Congress had agreed to fix observing hours according to the mean solar day at each station.[43] So synchronous observation became the norm, and it remained the international standard in meteorology well into the twentieth century. The International Meteorological Organization's 1910 Conference of Directors noted the potential value of global simultaneous observations, but merely recommended using Greenwich Mean Time (GMT) on international balloon days, when simultaneous observations aloft were attempted. On all other days, according to this recommendation, observers should simply note which time system they used.[44]

Not all scientists shared Abbe's enthusiasm for simultaneous observation. In 1918, Sir Napier Shaw, director of the British Meteorological Office, explained the scientific rationale for solar time in meteorology: ". . . the diurnal variations of weather are controlled by the sun, and for climatological purposes the fundamental principle of meteorological work is to note the conditions day by day at the same interval before or after true [solar] noon. . . ."[45] Daylight saving time, introduced during World War I as "summer time," injected what many meteorologists viewed as another dangerous confusion. Charles Marvin, chief of the US Weather Bureau, wrote of the "grave doubts surrounding the chronology and history of events resulting from the arbitrary advancement and retardation of clocks involved in [this] scheme. . . ."[46] Shaw agreed, noting that "in spite of very careful instructions a great deal of confusion arose with the observers . . . , and the continuity of many series of observations has been interrupted . . . There is now no possibility of placing beyond dispute the exact time of any event, except those dealt with by the telegraph, which occurred between May 21 and September 30, 1917."[47] Not until 1946 did the International Meteorological Organization officially designate standard observing hours using Greenwich Mean Time.

Global standard time bound continuous space to simultaneous time. It positioned people and communities relative to huge regions defined purely by their longitude. Its adoption required the infrastructures of instantaneous communication and accurate clocks. Standard time also marked a deep conceptual shift in the fundamental meaning and experience of time. Where solar time was tied to one's exact location, universal standard time created large, abstract zones within which time would be the same at all latitudes and all times of year. Time itself became a fundamentally globalist form of information, well suited to a meteorology already far along the road to globalism.

# 3 Standards and Networks: International Meteorology and the Réseau Mondial

Universal time was only one among many standards sought and achieved by scientists, engineers, commercial enterprises, and governments during the latter half of the nineteenth century. Indeed, standardization itself should be seen as a major characteristic of this historical period. For one thing, the period witnessed the first widespread manufacture of interchangeable parts, the most important innovation of the industrial age. Yet a great deal of standardization occurred in the realms of organization, technique, and practice, rather than in the realm of technology *per se*. Industrial mass production depended not only on machine tools capable of the precision necessary to manufacture interchangeable parts but also on forms of work organization that divided production into simple steps, collected them under a single roof, and systematized assembly of finished products.[1] Standardized techniques for processing, presenting, and storing information (such as filing systems, paper forms, the Dewey decimal system), as well as genre conventions (such as the memorandum and the annual report), were also crucial to what the historical sociologist James Beniger termed the "control revolution."[2]

As they consolidated, electric power grids, railroad networks, and telephone systems created standards for line voltage, track gauge, connectors, duplexing, and hundreds of other technical issues. Forms, schedules, operator languages, legal agreements, and other non-technical standards were equally important. Standards enabled local networks to connect more readily with neighboring ones, leading quickly to national and international networks and internetworks.[3]

By the 1880s, many countries sought to formalize and control this process, creating national bureaus of standards. Yet burgeoning global trade and communication had already short-circuited strictly national efforts, and many de facto international standards entrenched themselves long before 1900. Science, with its internationalist traditions, helped promote this transformation. For example, scientists from many disciplines began to adopt metric units, sometimes going against their own national traditions. This groundswell of de facto standardization helped create momentum for the 1875 Meter Convention. The Convention established the International Bureau of Weights and Measures, whose principal purpose was to compare and verify national standards for meters, kilograms, and other measures, including thermometers.[4]

International bodies soon emerged to formalize standard-setting processes. In an era that regarded national sovereignty as absolute, these were among the first intergovernmental and international non-governmental organizations of any kind. In the colonial era, however, "international" rarely meant "universal." The competing standards of former colonial powers can still be seen today in the shape of infrastructure in their former colonies: electrical sockets, telephone plugs, right-hand-drive- vs. left-hand-drive vehicles, and so on. At least until World War II, national and colonial standards dominated the politically weaker efforts at international standard setting, leading to an uneven patchwork whose legacy persists today. Economic, material, and social investment in particular infrastructural forms—including technical standards—created what Thomas P. Hughes termed "technological momentum," a "velocity and direction" that inhibited the convergence of national technological systems.[5]

Electronic communication was among the first infrastructures to see widespread standardization across borders. Telegraphy quickly led to the general adoption of International Morse Code, starting in 1851. By 1865, twenty countries had formed the International Telegraph Union to promote and develop technical standards for international telegraphy. This organization—among the first intergovernmental bodies to operate on a world scale—evolved into the International Telecommunication Union.

The last quarter of the nineteenth century saw a flowering of international organizations aimed at improving trans-border communication, standard setting, and (in the case of science) data sharing. Examples include the International Geodetic Association (founded in 1864 and concerned with establishing the size and shape of the Earth), the Universal Postal Union (founded in 1874), and the International Bureau of Weights and Measures (founded in 1875). The emergence of international meteorological observing systems must be seen against the background of this vast and powerful but incomplete and uneven convergence toward common languages, metrics, technological systems, and scientific understanding.

As meteorologists began to share information across national borders in real time, the long-established norm of communality of data took on larger significance. Like the telegraph whose wires they traveled, meteorological data gained in value as data networks grew. And wherever national borders stopped or delayed the flow of data (whether for technical or political reasons), value was diminished. In the United States, with its great size and lack of neighbors to the east and west, this did not matter so much; the national weather network could function reasonably well on its own. But the much smaller nations of Europe (and elsewhere) stood to gain a great deal from cooperative data exchange using common standards.

It is always one thing to set a standard and quite another to implement it. Inertia, resistance, ignorance, competing standards, lack of resources, legal barriers, and dozens of other problems must be overcome. Even if everyone wants the same standard and no one is standing in the way, it is usually difficult to avoid local variation—especially when standards require human beings to act in a uniform way across expanses of space and time. The "two steps forward, one step back" history of international standards in meteorology exemplifies this slow and painful process.

### The International Meteorological Organization

With so many standardization efforts already underway, standards were very much in the air in 1873, when the First International Meteorological Congress convened in Vienna. Indeed, parties to the congress saw the potential for better standards as the main reason to participate: the invitation to a preparatory meeting noted that "if there is any branch of science in which it is especially advantageous to work according to a uniform system, then that branch is the study of the laws of the weather."[6] Thirty-two representatives of twenty countries, most of them European, met for two weeks. Most delegates were scientists, but many were also the directors of national weather services and thus in some sense represented their governments. Whether the formal relationships they established should be scientific or governmental in nature would remain a controversial and difficult subject for the next 75 years.

Several delegates to the First International Meteorological Congress pleaded passionately for a genuinely global observing network. Christophorus Buys Ballot, director of the Royal Netherlands Meteorological Institute, proposed to fund weather stations placed on islands and other "distant points" lacking weather observations.[7] But in the 1870s the ambitions of meteorologists far exceeded their governments' willingness to pay for such stations, or for the expensive telegraph connections that conceptual revolution in weather forecasting. The gigantic air forces of World War II brought with them further expansion of data networks, new tools (especially radar and computers), and the training of thousands of new meteorologists.

The rise of international air travel figured in one major outcome of the 1919 Paris Peace Conference that concluded World War I: the Convention relating to the Regulation of Aerial Navigation, which laid the legal basis for international air traffic and effectively codified the vertical extent of the nation-state. Under this convention, each nation retained sovereign rights over its own airspace. The issue of national airspace would later become a crucial object of Cold War maneuver and diplomacy regarding overflight by artificial satellites.[11] Among other things, the convention specified guidelines for international meteorological data exchange, to be carried out several times daily by radiotelegraph. The Convention also established an intergovernmental International Commission for Air Navigation (ICAN), charged in part with implementing these meteorological standards.[12]

Thus, in the early twentieth century air travel's effect on meteorology resembled that of shipping in the middle of the nineteenth century, when a vast increase in sea traffic related to a major technological change (steam power) boosted the need for weather knowledge. But in the 1850s, when Maury pleaded for standard ships' logs, meteorological data could be used only long after they were collected, when ships returned from their voyages of weeks or months. By the 1920s, new communication technologies permitted the incorporation of aircraft data into daily synoptic forecasting. Radio obviated fixed cables, permitting cheaper, faster data exchange both within and among nations, while the new international codes facilitated the use of data from beyond national borders. New airports became observing stations, increasing the density of the network. Airlines employed numerous meteorologists. Research and military aircraft, and eventually commercial aircraft too, recorded observations at altitude. In the interwar era, most national weather services, both civil and military, benefited from these changes.[13] Under these circumstances, and in the relatively optimistic diplomatic atmosphere that prevailed in the years immediately after World War I, formal intergovernmental cooperation in meteorology seemed within reach.

Yet until after World War II, such cooperation remained limited specifically to *aeronautical* meteorology, where it was driven by the nascent commercial airlines. In 1919, the IMO established a Technical Commission for the Application of Meteorology to Aerial Navigation, a body that competed, in a sense, with ICAN. But governments officially recognized only ICAN, not the IMO's technical commission. By 1935, this led the IMO to transform its technical commission into an International Commission for Aeronautical Meteorology (known by its French acronym, CIMAé) with members appointed by governments. CIMAé was the first, and until after World War II the only, IMO entity to acquire official intergovernmental status. In the event, most CIMAé members also sat on ICAN, so it functioned more as a liaison than as an independent organization.[14]

The IMO's failure to dominate the field of aeronautical meteorology reflected both its endemic institutional weakness and the relative infancy of professional meteorology. Despite the vast scientific, technological, and political changes sweeping around it, the IMO administrative structure constructed in 1889 remained largely unchanged until after World War II. As before, the Conference of Directors of national weather services did most of the detail work, now supplemented by IMO Technical Commissions covering specific areas. A larger, broadly inclusive International Meteorological Committee met infrequently to discuss general policy and directions. Both of these bodies met as scientists and forecasters, rather than as government representatives. The IMO had no policy-making powers, serving only as an advisory and consensus body. Between meetings, the organization itself did little.

In this era research meteorologists themselves remained divided over the desirability of government involvement. In this respect they participated in what Paul Forman identified as a tension inherent in the very old ideology of "scientific internationalism," defined as "propositions and rhetoric asserting the reality and necessity of supranational agreement on scientific doctrine, of transnational social intercourse among scientists, and of international collaboration in scientific work." These essentially cooperative tenets conflict with the simultaneous, often intense competitiveness of scientists. Such competition is organized partly through "scientific nationalism," by which individual scientists' prowess confers glory on their home country. Scientific nationalism need not involve any direct relationship between scientists and governments. It exists, as Forman noted, because the most meaningful praise comes from "parties with a negative bias," so that scientists regard honors bestowed by other countries as having great prestige.[15] The interwar period saw renewed eruption of this underlying tension, connected with the extreme nationalism that had reemerged, especially in Europe.

Meteorology may have suffered less from this than many sciences because of its inherent need for shared data.[16] Yet the tension was nevertheless reflected in the conflicting loyalties of the directors of national weather services. As in other organizations that followed the internationalist model, the IMO's constituents—national weather services—cooperated when it served their mutual interests but ignored IMO directives when their goals diverged. The weather agencies served national governments, on the one hand, but on the other they saw their primary identity as scientific, and they regarded IMO meetings as apolitical spaces for scientific discussion. Intergovernmental status, they feared, might change this, turning them into representatives of their governments, reducing their independence and prerogatives, and perhaps subverting IMO proceedings by introducing political agendas. For this group, in other words, scientific internationalism served as a means to *bypass* the nation-state and to separate science from politics. But another faction saw governmental commitment as the only road to permanent, fully integrated international data exchange. As long as the IMO lacked official status, its decisions could not bind government weather services. As a result, many standardization problems remained unresolved, or progressed only slowly toward solutions. For this group, the road to better science lay *through* political commitment.

By 1929, the desire for official recognition prevailed. The IMC issued a letter to governments seeking intergovernmental status. Arriving on the eve of the Great Depression, the proposal was generally ignored by governments managing domestic crises. The IMC revisited this issue with renewed vigor at its 1935 meeting in Warsaw. This time, in an attempt to achieve government endorsement by stealth, the IMC decided to submit future invitations to meetings of the Conference of Directors directly to governments, asking them to designate their weather service directors as official government representatives. Beyond this, the IMO, led by France and Norway, began drafting a World Meteorological Convention that would secure intergovernmental status. A preliminary draft of this convention was presented to the 1939 meeting of the IMC, held in Berlin. World War II intervened, preventing further consideration of intergovernmental status until 1945.

### The Réseau Mondial

Data standards such as recording forms, temperature and pressure scales, observing hours, telegraph codes, and other minutiae had occupied the IMO from the beginning. Indeed, they were the primary motive for creating the organization in the first place. But although IMO congresses debated many standards, implementation was anything but automatic, even when members agreed. Many IMO standards took decades to gain general acceptance, especially at locations remote from the European and American centers of data collection and calculation. The plaintive tone of typical IMO resolutions reflected the organization's powerlessness to enforce change. For example:

> The Conference is of opinion that the subject is of the highest importance, but that it presents great difficulties, and that a general approach to the adoption of the most usual [observing] hours (viz., 7h., 2h., 9h.) is in the highest degree desirable.
>
> The Congress declares it to be most desirable, if it be not possible to introduce uniform measures at present, to use henceforth only metric and English units (with Centigrade and Fahrenheit temperature scales). All action is to be supported which tends to the introduction of the uniform metric system.[17]

This lack of standing sometimes made even minor disputes over best practices quite difficult to resolve. In 1919, the IMO's Conference of Directors of Meteorological Services formally adopted the standards outlined in the International Meteorological Codex, but that was only a beginning. Many national weather services maintained their own, somewhat divergent standards into the 1970s and even beyond.

The wide gap between the early IMO's global ambitions and the *Realpolitik* of international cooperation before World War II may be seen in the fate of the rather grandly named Réseau Mondial (worldwide network) for global climatology. In 1905, the French meteorologist Léon Teisserenc de Bort began to advocate collecting daily data via telegraph from a set of stations representing the entire globe. Two years later, he succeeded in getting the IMO to appoint a Commission for the Réseau Mondial. The commission immediately scaled back Teisserenc de Bort's grand plan, reducing its goal to publishing monthly and annual averages for pressure, temperature, and precipitation from a well-distributed sample of meteorological stations on land. The standard for distribution was two stations within each 10° latitude-longitude square (an area about twice the size of France). Ultimately, the network included about 500 land stations, from 80°N to 61°S; the oceans were not covered at all.

Operating even this seemingly modest version of the Réseau Mondial proved extremely difficult. Today it may be hard to comprehend how simply gathering data from a small sample of the world's meteorological stations and calculating a few simple averages could be challenging. The explanation lies in the lack of settled standards and the limits of communications infrastructure before World War II. Most data exchange for the Réseau Mondial took place by mail, which often took months to arrive from distant locations. The problem of non-standard observing and recording techniques also remained considerable. Napier Shaw's prefaces to the volumes for 1910 and 1914 (not published until 1920 and 1921, respectively) hint at the swarm of frustrations he experienced in presiding over this work for many years:

> ... any meteorologist who has attempted to put together information for the whole globe will realize that the differences of practices of the various Governments in respect of the material and method of arrangement, the units employed for its expression, and the time, if any, at which publication of the several contributions may be expected, place such an enterprise outside the limits of possibility for any but a few individuals, who must have at their disposal the facilities of such a library as that of the Meteorological Office. There are probably not a dozen such libraries in the world. The alternative that the few workers who deal with the meteorology of the globe should each one of them separately and severally have to go through an identical process of laborious compilation, reduction, and tabulation in order to attain a result which is of itself an indispensable stepping-stone to a comprehension of the meteorology of the globe, is sufficient to justify any establishment in making public a compilation for the benefit of the world at large.
>
> [The demand for this compilation was] generally regarded as urgent, [but it proved] so difficult to satisfy on account of the number of obstacles, each in itself trivial but practically deterrent by their number.... Those who have experience in supervising work carried on at a great distance from the base will recognize that the headings of columns on a form for the entry of observations are sometimes misunderstood or disregarded by the observer until his attention is called thereto; and unless the examination is done promptly and regularly it is too late.[18]

As a result of such difficulties, the mere collection and compilation of data for each edition of the Réseau Mondial—whose activity consisted in an annual publication rather than in what we might think of today as a network—took several years. The first annual data, for 1911, were not published until 1917, World War I having interrupted work for 3 years. Hiatuses and delays of up to 13 years marked the publication of subsequent volumes. In 1953, the first meeting of the WMO Commission on Climatology thanked the UK Meteorological Office for publishing the Réseau Mondial through 1932 and essentially recognized the network's demise.[19]

In the meantime, the Smithsonian Institution had embarked upon a similar series of publications, the *World Weather Records*, for a smaller set of stations (about 380). Initially somewhat less detailed than the Réseau Mondial, this global data set also covered land stations only. The Smithsonian and later the US Weather Bureau (at the WMO's request) continued to produce the publication approximately every 10 years.[20] By the 1950s it had superseded the Réseau Mondial as the accepted standard source for world climatological data.

Though the IMO remains important in the history of international scientific cooperation, its significance has sometimes been exaggerated, not least by its own official historians.[21] In the absence of government support, the grand ambitions of the 1870s soon faded. For most of its history the IMO remained a skeleton organization, supported primarily by small, voluntary contributions from France, Great Britain, and the United States. Most of its members worked for national weather services, but their participation was strictly unofficial. Most IMO publications were printed by one of the participating weather services.

Data sharing proceeded anyway, of course, through publications, regional telegraphy, and informal scientific networks. But it remained quite basic. The Réseau Mondial—which Napier Shaw called "the guiding principle of international co-operation between the meteorological establishments of the world since the [1873] Vienna Congress"—remained limited to whatever individual nations might see as in their interest and within their budgets to support.[22] Movements to improve the data network gained considerable ground at the 1929 Copenhagen Conference of Directors, which issued new standards for radio transmission of weather data using revised meteorological codes. These were widely adopted in Europe and the United States. Not until the 1960s did the IMO's successor, the World Meteorological Organization, begin to implement the World Weather Watch, the near-real-time observing and data exchange system of which Teisserenc de Bort, 60 years earlier, could only dream.

# 4 Climatology and Climate Change before World War II

As we saw in chapter 2, the basic structure of the global circulation was well established by the middle of the nineteenth century. So were the fundamental forces driving that motion. Yet the causal relationship between the circulation and the climate remained poorly understood. Even as late as World War II, meteorologists could still say little about this relationship with any certainty.

There were two main reasons for this. First, until the 1930s virtually all weather and climate data had been collected at the surface. Few direct measurements existed that might be used to chart the details of circulatory structures hundreds or thousands of meters above the ground. Second, climatologists understood only a few general principles of large-scale atmospheric movement, such as the Hadley and Ferrel cells. These principles proved inadequate to the task of explaining the wide variation of local and regional climates across the world's land surface. As a result, climatology before World War I was heavily dominated by qualitative approaches and regional studies. Mathematical approaches based on statistical analysis gained ground in the interwar period, owing in part to mechanical aids to computation. Nonetheless, *global* climatology emerged only slowly. This chapter sketches, in broad, non-technical terms, how climatologists thought about climate and climate change before World War II.

## Climatology as Geography and Statistical Law

As Frederik Nebeker showed in *Calculating the Weather*, after the rise of weather telegraphy in the nineteenth century meteorology began to divide into three relatively separate subfields. Each subfield had its own methods. With these differing methods came differing relationships to numerical data.

The *forecasters* (discussed in detail in the following chapter) developed one set of methods, aimed chiefly at improving the timeliness and accuracy of weather prediction. Explanation concerned them only insofar as it might improve their predictions. Tightly constrained by the need to prepare daily forecasts quickly, their methods relied heavily on synoptic charts and isolines as visualization techniques, and on the trained judgment of experienced forecasters. Few numerical calculations were involved.

By contrast, the applied physicists, or *theoretical meteorologists*, sought to understand weather deductively, through physical theory. They wanted to ground their principles in fluid dynamics, gas physics, and other basic sciences. This branch of meteorology sought to apply the growing body of mathematically expressed laws, building on a solid framework of established physics. It proceeded in the tradition of reductionism, hoping to identify the many processes involved in weather, construct an explanatory theory of each process, and finally combine them to achieve full understanding. Here too calculation *per se* was relatively unimportant, albeit for a different reason: the mathematics of atmospheric motion centered on calculus, but numerical methods for integrating differential equations did not yet exist. Hence equations that could not be solved analytically usually could not be solved at all.

The third branch, the *empiricists,* believed that the atmosphere's tremendous complexity might prevent the application of physical theories derived from simpler phenomena. Therefore, they sought to obtain laws of weather behavior inductively from data. The empiricists did concern themselves with explanation, but they sought such explanations in the close analysis of observations rather than in fundamental mechanics of motion, heat, and gas behavior.[1]

In the final decades of the nineteenth century, the empiricist strand gradually evolved into the subdiscipline of climatology. Adopting meteorology's traditional concern with analyzing large collections of recorded data, the new subfield began to develop its own, primarily statistical methods. As mathematics, statistical methods occupied a middle ground between the arithmetical minimalism of the forecasters and the complex calculus of the theoretical meteorologists. Statistical calculations were laborious and time consuming, but unlike differential equations many of them could be solved by hand or with simple calculating aids, such as logarithmic tables and adding machines. Therefore, the empiricists were the first to deploy calculation as a primary technique.

Scientific discussions of climate first appeared mainly in the context of natural history and geography. Descriptions of climate, topography, and the other physical features of regional environments accompanied narratives and catalogs of flora and fauna. Though climatic description sometimes included data analysis, more often it took the form of experience-based, qualitative narrative, perhaps with a few measurements thrown in for support. By 1900, however, techniques of statistical analysis provided the wherewithal to make more direct use of the rapidly accumulating data.

Today the statistical language of averages, percentages, and probabilities, and the visual grammar of graphs and statistical charts, seem timeless and transparent. But in the nineteenth century these methods represented a radical departure. According to the Oxford English Dictionary, the term 'statistics' originally referred to "that branch of political science dealing with the collection, classification, and discussion of facts (especially of a numerical kind) bearing on the condition of a state or community." In the post-revolutionary era of the late eighteenth and early nineteenth centuries, new national interests in such matters as voting, agriculture, market performance, taxation, and the size, distribution, health, race, and wealth of national populations led to new, more direct relations between states and the scientists and institutions that could provide such information.[2] Nebeker notes that early government statistical offices often collected meteorological data in order to explore the latter's relationship to public health and national economies, the principal subjects of early statistical investigation.[3] With tuberculosis ravaging the world, and with tropical diseases such as malaria and yellow fever strongly affecting colonial empire building, theories of a relationship between climate and health proliferated.

Armed with statistical methods, empirical climatologists became the "number crunchers" of nineteenth-century meteorology. They also began to seek ways to automate the processing of weather records. Ultimately, climatologists sought to discover general laws of climate inductively, through data analysis. For that reason, many of them also came to promote improvements in the slowly emerging global data network. (See box 4.1.)

As a date for the founding of professional climatology, we might as well choose 1883. In that year, the Austrian meteorologist Julius von Hann published the first edition of his *Handbook of Climatology*, which would remain the standard textbook in theoretical climatology for 50 years.[4] Hann defined climate as the "sum total of the meteorological conditions insofar as they affect animal or vegetable life." Responding to the then dominant tradition of natural history, where most previous discussions of climate had occurred, Hann felt compelled to emphasize that climate could also be understood as a purely physical phenomenon, independent of

**Box 4.1**
Quetelet's Climatology and the Analytical Engine

A striking premonition of meteorology's future as a computational science appears in an 1835 letter from Charles Babbage to the Belgian polymath Adolphe Quetelet.

Quetelet (one of the foremost early statisticians) promoted the unification of the human and natural sciences through mathematics. To social science, he contributed the controversial concept of *l'homme moyen*—the "average man"—whose features could be discerned in the bell curves of normal probability distributions. Applying similar techniques to climatology, he offered treatises on how Belgium's climate affected human health. As founding director of the Royal Brussels Observatory, Quetelet initiated standardized meteorological observations in 1833. He also authored one of the first texts on geophysics.[a]

Babbage, a British scientist and mathematician of equally prodigious talents, shared Quetelet's interest in statistics. Obsessed with accuracy, Babbage wanted to automate the complex, labor-intensive, error-prone process of large-scale calculation. In pursuit of this goal, Babbage designed an "Analytical Engine": a steam-powered, gear-based, fully programmable calculating machine. Although Babbage never completed the machine, historians of technology regard his design for it as the first fully formed conception of a digital computer.[b]

Babbage's earliest recorded mention of the Analytical Engine appeared in an 1835 letter to Quetelet: "I am myself astonished at the power which I have been enabled to give [the Engine], and which I would not have believed possible a year ago. This machine is intended to understand one hundred [25-digit] variables. . . ." In the same letter, Babbage informed Quetelet that Sir John Herschel had written him about "some curious results concerning a general motion of the atmosphere" and asked Quetelet to distribute Herschel's call for a worldwide network of weather observers—a goal Quetelet shared.[c] In the 1860s, Quetelet advocated strongly for what would later become the International Meteorological Organization.

a. A. Quetelet, *Sur l'homme et le développement de se facultés, ou essai de physique sociale* (Bachelier, 1835); *De l'influence des saisons sur la mortalité aux différens ages dans la Belgique* (Hayez, 1838); *Météorologie de la Belgique comparée à celle du globe* (Muquardt, 1867); *Sur la physique du globe* (Hayez, 1861).
b. M. R. Williams, *A History of Computing Technology* (Prentice-Hall, 1985); D. D. Swade, "Redeeming Charles Babbage's Mechanical Computer," *Scientific American* 267, no. 8, 1993: 86–91.
c. A. W. van Sinderen, "Babbage's Letter to Quetelet, May 1835," *Annals of the History of Computing* 5, no. 3, 1983: 263–67.

ecosystems. Still, unlike theoretical meteorology, climatology remained chiefly descriptive, "a branch of knowledge which is in part subordinate to other sciences and to practical ends."[5]

In much of his book, Hann focused on physical, planetary-scale features, such as the zonal (latitudinal) distribution of temperature. He noted that as early as 1852 climatologists had determined (from data) that the global average temperature does not remain stable over the entire year, but instead rises from January to June.[6] Since the amount of insolation (incoming solar energy) is constant over the globe, this result meant that northern-hemisphere temperatures dominated the global average. Hann reviewed a series of theoretical studies predicting how temperature should vary by latitude, then compared these with data from Dove, Buchan, and others. These comparisons yielded "quite close" agreement between theory and observations. In another chapter, Hann discussed "solar or mathematical climate"—i.e., the climate of an abstract globe, attributable only to differential insolation. These calculations constituted early versions of what are known today as "energy budget" models of climate. Yet for Hann these forays into physical theory did not place climatology in the same arena as theoretical meteorology. Hann saw the field mainly as "a science auxiliary to geography,"[7] and the bulk of his *Handbook of Climatology* treated climate as a matter of statistical description.

In other chapters, Hann discussed three kinds of climatic change: geological, periodic (cyclical), and "secular" (within historical rather than geological time scales). For secular climate change, Hann included a chart of temperature fluctuations for the whole Earth, calculated from 280 stations (figure 4.1). The table gave five-year "departures" from the average for ten selected periods, the first of which was 1736–1740 and the last of which was 1881–1885. (This technique, which compares station temperatures with their own average value for some given period, is now known as a temperature anomaly calculation.) The range of anomalies, both positive and negative, was about 1°C.

Hann expressed considerable skepticism about the accuracy of calculated global averages on the secular time scale. Even the figures in his own table, he wrote, derived from only 280 stations "and naturally do not really represent the temperature conditions of the whole world."[8] Nevertheless, it is clear that Hann—in the tradition of Maury, Dove, Ferrel, and the *Challenger* expedition before him—saw the climate as a global system. In 1896 he published a three-volume work entitled *The Earth as a Whole: Its Atmosphere and Hydrosphere*.[9]

TEMPERATURE DEPARTURES DURING PERIODS OF HIGH AND LOW TEMPERATURES.

| | | | | | | | |
|---|---|---|---|---|---|---|---|
| 1736-1740, | - | - | −0·43 | 1821-1825, | - | - | +0·56 |
| 1746-1750, | - | - | +0·45 | 1836-1840, | - | - | −0·39 |
| 1766-1770, | - | - | −0·42 | 1851-1855, | - | - | +0·11 |
| 1791-1795, | - | - | +0·46 | 1866-1870, | - | - | +0·11 |
| 1811-1815, | - | - | −0·46 | 1881-1885, | - | - | −0·08 |

**Figure 4.1**
Global temperature anomalies expressed as departures from the mean for an unspecified period.
*Source*: J. von Hann, *Handbook of Climatology* (Macmillan, 1903), 411.

As regards climate *change*, Hann argued (in effect) that one had to invert the infrastructure before accepting any conclusion:

> Whenever it has been supposed that . . . records did show an increase in temperature or in rainfall, it has always turned out that this increase may have been due to the method of exposing the thermometer, or the rain-gauge. . . . Dufour, who made a thorough study [in 1870] of the available evidence concerning a change in climate, concludes that the uncertainties connected with this evidence make it impossible to regard a change of climate as proved. The question, nevertheless, remains an open one, and the common assertion that the climate is not changing is, under the circumstances, a no more legitimate consequence of known facts than is the opposite view.[10]

Some of Hann's contemporaries, however, were less skeptical, considering the case for climate change settled and even occasionally advocating a political response. (See box 4.2.)

Many "laws" put forward by statistical climatology fell somewhere between the poles of description and explanation. For example, reviewing climatological knowledge in 1924, Stephen Visher noted that Earth's surface temperature decreases regularly with latitude.[11] This law has a general aspect: on any planet, the angle of insolation (incoming solar radiation) changes from the equator, where solar rays are perpendicular to the surface, to the poles, where they are parallel to the surface. This creates a systematic variation in the amount of solar heat per unit of surface area. On Earth this gradient is approximately 1°F per degree of latitude, according to Visher. But this zonal temperature gradient also depends on the volume and chemical composition of the atmosphere, the presence of oceans, the shape of continents, the inclination of Earth's axis, and many other features particular to our planet and especially to its land surfaces.

Box 4.2
Early Climate Politics

Studying the meteorological record of the nineteenth century, the German geographer Eduard Brückner noted an approximately 35-year periodic cycle of climatic variation. Brückner viewed this cycle as a "universal occurrence" of "global importance," arguing that it caused crop failures, economic crises, and disease epidemics.[a]

Brückner also claimed that human-induced warming, desertification, and drought were already occurring in regions of Europe and North America as a result of deforestation.[b] He noted that the governments of Prussia, Italy, Austria, Russia, and France had all debated proposals involving reforestation to counteract climatic change. Although nothing came of these proposals, Brückner's work as an "issue entrepreneur," promoting political action on the basis of scientific evidence, stands as a remarkable precursor to modern climate politics.[c] Julius von Hann, while skeptical of most claims regarding near-term climatic change, guardedly endorsed Brückner's finding of a 35-year cycle.

Brückner was exceptional in making climate change a political issue, yet his views on anthropogenic climate change were far from unique. Indeed, historians have cataloged numerous episodes in which people perceived climatic change, sometimes ascribing it to human causes. Aristotle's student Theophrastus attributed local climate changes to swamp drainage and agriculture, and in the Middle Ages the Church sometimes explained climate anomalies as a divine response to human sin.[d] From the seventeenth century until the mid 1800s, many people—including such luminaries as David Hume and Thomas Jefferson—believed that clearing land for agriculture could favorably alter local or regional climates.[e] Some nineteenth-century colonial forest policies were predicated on a "dessicationist" theory that deforestation caused local, regional, and even continental drought, and some modern forest-conservation agendas descend directly from these colonial policies.[f] Until relatively recent times, however, most concerns about anthropogenic climate change focused on scales far smaller than the planet as a whole.

a. Brückner, "How Constant Is Today's Climate?" in *Eduard Brückner*, ed. N. Stehr and H. von Storch (Kluwer, 1889), 74.
b. H. von Storch and N. Stehr, "Climate Change in Perspective: Our Concerns About Global Warming Have an Age-Old Resonance," *Nature* 405 (2000): 615.
c. N. Stehr and H. von Storch, "Eduard Brückner's Ideas: Relevant in His Time and Today," in *Eduard Brückner*, ed. N. Stehr and H. von Storch (Kluwer, 2000).
d. N. Stehr et al., "The 19th Century Discussion of Climate Variability and Climate Change: Analogies for Present Debate?" *World Resources Review* 7 (1995): 589–604.
e. J. R. Fleming, *Historical Perspectives on Climate Change* (Oxford University Press, 1998).
f. R. Grove, *Ecology, Climate and Empire* (White Horse, 1997).

For example, the climate of Paris (at 48°N) is colder than that of Madrid (at 40°N); this fits Visher's law of latitude. Yet Madrid's climate is substantially warmer than that of New York, which also lies at 40°N. This is in part because the prevailing winds blow westward across the warm Gulf Stream current, keeping temperatures in both Madrid and Paris well above the global averages for their latitudes. Specific relationships such as these can only be discovered empirically.

Nevertheless, climatologists found that by treating the nineteenth century's rapidly accumulating data with newly developed statistical methods they *could* discover empirical "laws"—and that they could do this more easily than their counterparts in theoretical meteorology could discover the physical principles governing atmospheric motion. Thus climatologists came to see their goals and methods as substantially different from those of theoretical meteorology. As Hann put it, climatology "must treat the different atmospheric processes separately only insofar as this is unavoidable. . . . Climatology must give us a mosaic-like picture of the different climates of the world; but it must also present these facts in a systematic way, by grouping together climates which are naturally related. Thus order and uniformity are secured, the mutual interactions of the different climates are made clear, and climatology becomes a scientific branch of learning."[12] Here we hear again the echoes of the natural history tradition in the value placed on classification, ordering, and holistic description. Theoretical meteorology, by contrast, would reduce weather and climate to individual elements that could be treated separately. Though parts of Hann's *Handbook* are theoretical in nature, most climatology would remain within this primarily descriptive tradition for another 50 years. Climatologists made increasingly extensive use of statistical mathematics, but little use of physical theory. The educational backgrounds of most professional climatologists were in geography rather than in physics; the heavily mathematical discussions that increasingly characterized dynamical (theoretical) meteorology from 1900 on remained largely absent from climatology.

A look at typical handbooks from the interwar period illustrates the point. W. G. Kendrew's *The Climates of the Continents* appeared in five editions between 1922 and 1961.[13] This text was widely accepted as "the most lucid description of the normal distribution of the elements over the land masses of the world."[14] Yet the book's entire theoretical content—four pages—consisted of a narrative, non-mathematical sketch of global pressure and wind systems. In the remainder of the text, Kendrew delivered qualitative, descriptive accounts of typical weather patterns on each continent (except Antarctica). In his opinion the notion of a global climate made little sense, because "between the Tropics and the Poles the weather is so variable . . . that it is difficult to form a conception of the climate unless it be the idea of something very changeable."[15] (Kendrew did refer the reader to other sources, notably Hann's *Handbook*, for theories of global climate.)

The absence of theory in climatology was due less to ignorance than to an understanding of the problem's fearsome complexity. In 1914, Cleveland Abbe celebrated the regular publication of US Weather Service daily charts of the northern hemisphere, mapped on a polar projection. Abbe's elation stemmed from the new prominence these maps accorded to the atmospheric general circulation. To him, these maps demonstrated a long-held truth that "the atmosphere must be studied as a unit." Like many of his colleagues, Abbe eagerly awaited some practical method for analyzing the general circulation using physical theory. As he saw it, William Ferrel had already isolated the major factors: Earth's rotation, gravity, surface friction, moisture, radiation and its absorption, and the thermodynamics of rising and falling air masses. Yet Abbe despaired of near-term success:

> So far as we know no one has as yet dared to begin the discussion of the motions of the atmosphere under the combined influences of all these seven factors [identified by Ferrel], and yet these must be gathered into one set of systematic equations or graphic charts. . . . May we not adopt the enthusiastic words of the immortal Kepler, in his *Harmonies of the World*: 'The die is cast! The book is written! It can well afford to wait a century for a reader, since God has waited 6000 years for the astronomer'.[16]

Clearly, Abbe thought that meteorology would be waiting for a long time. In the absence of a general circulation theory, synoptic charts would have to serve as an empirical substitute. And indeed, tables and charts such as those shown in figures 4.2 and 4.3 remained the principal technique of climatological data analysis into the latter half of the twentieth century. These figures appeared in the *Monthly Weather Review*, a publication founded in 1871 as the principal venue for American weather and climate data. Each issue included narrative summaries of weather as well as circulation "highlights." The journal continued to publish these descriptive summaries well into the 1980s, though by then such discussions were increasingly accompanied by theoretical explanation.[17]

The gradual maturation of statistical climatology in the early twentieth century is evident from a later handbook: Victor Conrad's *Methods in Climatology*, published in several editions.[18] Reflecting the discipline's

TABLE 1.—*Condensed climatological summary of temperature and precipitation by sections, January 1938*

| Section | Section average (°F) | Departure from the normal (°F) | Station | Highest (°F) | Date | Station | Lowest (°F) | Date | Section average (In.) | Departure from the normal (In.) | Station | Amount (In.) | Station | Amount (In.) |
|---|---|---|---|---|---|---|---|---|---|---|---|---|---|---|
| Alabama | 48.1 | +1.7 | 2 stations | 81 | 23 | Tuscumbia | 11 | 28 | 3.39 | −1.45 | Bishop | 7.14 | Maxwell Field | 1.75 |
| Arizona | 44.5 | +2.7 | do | 83 | ¹3 | Springerville | −4 | 26 | .66 | −.67 | Bright Angel | 3.40 | Yuma Citrus | .00 |
| Arkansas | 42.5 | +1.1 | El Dorado | 81 | 21 | Devil Knob | 4 | ¹16 | 6.48 | +2.04 | Grannis | 11.58 | Calico Rock | 2.87 |
| California | 46.2 | +1.6 | Yorba Linda | 90 | 9 | Twin Lakes | −10 | 20 | 3.72 | −1.06 | Scales | 13.64 | Brawley | .00 |
| Colorado | 26.9 | +3.0 | Holly | 73 | 18 | Fraser | −30 | 7 | .82 | +.06 | Pagosa Springs (near) | 6.77 | 2 stations | T |
| Florida | 58.6 | −.6 | Lake Placid | 88 | 7 | Garniers (near) | 15 | 27 | 2.04 | −.73 | Jacksonville | 5.21 | Lake Placid | .10 |
| Georgia | 47.4 | +.3 | Stillmore | 82 | 23 | 2 stations | 8 | 28 | 2.01 | −2.26 | Lafayette | 4.03 | Glenville | .61 |
| Idaho | 27.7 | +3.8 | 2 stations | 62 | ¹12 | Obsidian | −19 | 25 | 1.80 | −.39 | Roland | 7.47 | Arco | .10 |
| Illinois | 28.5 | +.8 | Harrisburg | 62 | 24 | Freeport | −7 | 31 | 2.57 | +.18 | Fairview | 5.02 | Hillsboro | 1.05 |
| Indiana | 29.6 | +.5 | Shoals | 66 | 24 | Goshen | −11 | 28 | 1.72 | −1.42 | Laporte | 5.17 | Shelbyville | .43 |
| Iowa | 21.0 | +2.5 | Thurman | 59 | 22 | Marshalltown | −25 | 10 | 1.14 | +.04 | Burlington | 3.79 | 2 stations | .15 |
| Kansas | 34.6 | +4.9 | Liberal | 75 | 15 | Burr Oak | −12 | 31 | .45 | −.22 | Oswego | 2.73 | 11 stations | .00 |
| Kentucky | 35.9 | −.1 | Pikeville | 72 | 24 | Greenville | −4 | 27 | 3.56 | −.99 | Cumberland | 6.25 | Grant | 1.57 |
| Louisiana | 52.7 | +.9 | Donaldsonville | 84 | 22 | 4 stations | 20 | ¹11 | 4.98 | +.05 | Jonesville | 7.58 | Shreveport | 2.25 |
| Maryland-Delaware | 33.6 | −.1 | Cumberland, Md | 67 | 30 | Oakland, Md | −2 | 29 | 2.32 | −.98 | Snow Hill, Md | 3.90 | Keedysville, Md | .91 |
| Michigan | 19.5 | −1.5 | Monroe | 55 | 24 | Iron Mountain | −32 | 28 | 2.54 | +.67 | Munising | 6.73 | Port Huron | .43 |
| Minnesota | 8.7 | −.5 | 2 stations | 47 | 23 | Warroad | −41 | 31 | .60 | −.16 | Duluth | 1.55 | Alexandria | .07 |
| Mississippi | 45.2 | −.7 | Port Gibson | 85 | 22 | Forest | 15 | 27 | 4.85 | −.31 | Holly Springs | 10.97 | Forest | 2.39 |
| Missouri | 33.1 | +2.3 | Garber | 75 | 16 | Tarkio | −9 | 31 | 2.93 | +.53 | Mountain Grove | 5.55 | Tarkio | .39 |
| Montana | 24.6 | +5.2 | 2 stations | 61 | 27 | Summit | −47 | 30 | .60 | −.27 | Haugan | 3.87 | Bridger | T |
| Nebraska | 28.1 | +5.1 | Broken Bow | 72 | 15 | Nenzel (near) | −22 | 31 | .35 | −.17 | Hartington | 1.10 | 4 stations | T |
| Nevada | 35.2 | +5.8 | Las Vegas | 80 | 22 | Elko | −5 | 24 | .74 | −.44 | Marlette Lake | 3.30 | Mina | .00 |
| New England | 21.4 | −1.3 | 4 stations | 59 | ¹7 | First Conn. Lake, N.H. | −34 | ¹18 | 4.16 | +.63 | Danbury, Conn | 7.68 | Presque Isle, Maine | 1.38 |
| New Jersey | 31.0 | +.1 | Canoe Brook | 68 | 25 | Layton | −19 | 19 | 3.64 | +.07 | Paterson | 5.73 | Camden | 2.58 |
| New Mexico | 34.1 | +.5 | Tularosa | 89 | 6 | Eagle Nest | −28 | 7 | .57 | +.01 | Chama | 2.90 | 4 stations | .00 |
| New York | 22.4 | −.9 | Albany | 61 | 25 | Indian Lake | −31 | 19 | 2.91 | −.06 | Bedford Hills | 5.57 | Avon | .96 |
| North Carolina | 41.1 | −.6 | 2 stations | 76 | ¹23 | Mount Mitchell | −10 | ¹28 | 2.89 | −.90 | Siler City | 5.15 | Red Springs | 1.46 |
| North Dakota | 9.5 | +3.4 | Carrington | 50 | 15 | Hannah | −40 | 31 | .48 | −.01 | Carson | 1.47 | Mayville | .01 |
| Ohio | 30.0 | +1.5 | 3 stations | 72 | 24 | Xenia | −5 | 28 | 1.62 | −1.50 | Chilo | 2.99 | Put-in-Bay | .48 |
| Oklahoma | 42.2 | +3.9 | Seminole | 82 | 17 | 2 stations | 0 | 31 | 1.44 | +.01 | Idabel | 11.49 | 2 stations | .00 |
| Oregon | 34.7 | +3.0 | Brookings | 71 | 9 | 2 stations | −8 | ¹7 | 3.39 | −.47 | Valsetz | 16.96 | Mitchell | .41 |
| Pennsylvania | 28.9 | +.4 | Claysville | 66 | ¹24 | Gouldsboro | −20 | 19 | 2.51 | −.75 | Mt. Pocono | 5.95 | Vandergrift | .87 |
| South Carolina | 45.5 | −.4 | 5 stations | 82 | 22 | 2 stations | 5 | 28 | 1.50 | −2.28 | Landrum | 3.53 | Wedgefield | .55 |
| South Dakota | 19.5 | +2.0 | Ardmore | 65 | 4 | Pollock | −30 | 31 | .52 | −.03 | Dummont | 3.12 | Ornan | .06 |
| Tennessee | 39.6 | +.4 | Newport | 74 | 24 | Rugby | −4 | 28 | 5.38 | +.46 | Ashwood | 8.73 | Dresden | 2.74 |
| Texas | 50.2 | +2.0 | Laredo | 92 | 18 | Spearman | −8 | 31 | 2.81 | +1.12 | Naples | 10.98 | 3 stations | .00 |
| Utah | 30.8 | +5.7 | St. George | 66 | 28 | Loa | −11 | 21 | .91 | −.29 | Silver Lake | 4.05 | Myton | .00 |
| Virginia | 36.3 | −.3 | 2 stations | 69 | ¹1 | Mountain Lake | −3 | 27 | 2.73 | −.60 | Pennington Gap | 4.65 | Riverton | .87 |
| Washington | 34.0 | +.4 | Signal Peak | 63 | 10 | Stockdill Ranch | −6 | 29 | 3.53 | −1.62 | Wynoochee Oxbow | 14.64 | Rock Island (near) | .29 |
| West Virginia | 33.5 | +.7 | London | 77 | 24 | Flat Top | −2 | 27 | 2.22 | −1.48 | Pickens | 4.41 | Wardensville | .66 |
| Wisconsin | 15.0 | .0 | 3 stations | 48 | 23 | Solon Springs | −40 | 31 | 2.03 | +.79 | Milwaukee | 4.60 | Holcombe | .70 |
| Wyoming | 22.2 | +2.3 | Fort Laramie | 63 | 15 | Buffalo Ranch | −37 | 30 | .79 | .00 | Bechler River | 5.29 | Evanston | .04 |
| Alaska (December) | 6.5 | +1.9 | Wrangell | 56 | ¹5 | 2 stations | −54 | ¹29 | 2.11 | −.34 | Little Port Walter | 21.02 | Anchorage | T |
| Hawaii | 70.3 | +1.5 | Kohala | 89 | 13 | Kula Sanitarium | 45 | 9 | 10.09 | +1.59 | Piihonua | 40.01 | Napoopoo | .72 |
| Puerto Rico | 72.8 | −.1 | Juncos | 92 | 13 | Lares | 51 | 26 | 2.00 | −1.68 | La Mina (El Yunque) | 12.39 | 5 stations | .00 |

¹ Other dates also.

**Figure 4.2**
US annual climatological summary for 1937, based on data from about 200 stations.
*Source*: J. P. Kohler, "Condensed Climatological Summary," *Monthly Weather Review* 66, no. 1 (1938): 27–28.

increasingly sophisticated mathematical footing, this text grounded climatology in data analysis. Conrad saw climatology's chief purpose as the creation of a larger picture from specifics: ". . . observations are made at isolated points. Only by comparing these data can the climate of the whole region be interpolated from place to place. *It is therefore the principal and fundamental aim of climatological methods to make the climatological series comparable.*"[19]

By Conrad's time, mathematics had replaced narrative as the dominant framework of climate studies. Hence *Methods in Climatology* focused on such topics as statistics, computational aids (graphs, nomograms, punch-

# Before World War II 71

**Figure 4.3**
Departures of mean temperature from normal, January 1938. Shaded and unshaded portions respectively represent areas where temperatures were higher and lower than normal.
*Source*: J. P. Kohler, "Condensed Climatological Summary," *Monthly Weather Review* 66, no. 1 (1938): c1.

card machines), curve fitting, and harmonic analysis of cyclical phenomena. Further, unlike Kendrew, who stressed their differences, Conrad emphasized the *connectedness* of the world's climates. His comments on this point give an important clue to climatology's future direction. The purpose of comparing relationships among climatic variables in different places, Conrad wrote, was "to determine the circulation of the atmosphere (physics of the *actual* state of the atmosphere), and to get full information about average and extreme values of climatic elements . . . (physics of the *average* state of the atmosphere). For these large-scale purposes, there must be a network, internationally administered, that covers the entire globe."[20]

After World War I, as climatologists gradually discovered more about the general circulation from upper-air observations, they came to realize that even a century's worth of records from ground level would contribute little to their understanding of the circulation higher up, especially in the absence of theory. In 1933 the Scandinavian meteorologists who had developed polar-front and air-mass theory—Vilhelm Bjerknes, Jakob Bjerknes, Tor Bergeron, and others (see chapter 5)—produced an early dynamical treatment of the general circulation.[21] In 1939, Carl-Gustav Rossby, working with others, extended these ideas in a breakthrough analysis of very long stationary atmospheric waves (wavelengths between 3000 and 7000 km), created by topographic influences such as airflow over major mountain ranges.[22] Yet with these few exceptions, until the middle 1950s most climatologists would probably have agreed with British meteorologist C. S. Durst, who in 1951 wrote: "Climatology, as at present practiced, is primarily a statistical study without the basis of physical understanding which is essential to progress."[23]

### Physical Theories of Global Climate Change

Nowhere was the split between theorists and empiricists more significant than in the study of global climate change. As we have seen, speculations on anthropogenic local, regional, and even continental climate change were not uncommon in the eighteenth and nineteenth centuries, and even before. But concerns about anthropogenic *global* climate change were new.

The evolution of physical theories of global climate change has been recounted so often that the story has become a "potted history" (including a number of frequently repeated errors of attribution and date).[24] For this reason, I will review that story only briefly here. A complete account may be found in Spencer Weart's *The Discovery of Global Warming*.[25]

Scientific theories of natural global climatic change date to the middle of the nineteenth century. They originated in physics, chemistry, and geology, and especially the new subfield of historical geology. Scientists had just begun to suspect that Earth was very old. James Hutton first proposed a "uniformitarian" geology in 1785, but his ideas did not achieve broad acceptance until the 1830s. Only after Thomas Lyell's work did scientists begin to interpret fossil evidence as indicating dramatic climatic oscillations in Earth's distant past. Over eons, they now saw, recurrent ice ages had sometimes covered much of the planet's continental surface with mile-thick glaciers. In other periods, climates far warmer than today's had prevailed. In the Cretaceous, for example, global average surface temperatures had been some 6°C warmer. All this and more was first discovered in the middle of the nineteenth century, not long before the publication of Charles Darwin's *On the Origin of Species* (1859).

Also in the first half of the nineteenth century, physicists began to explore the mechanisms by which Earth's atmosphere absorbs solar heat. By 1817, and perhaps as early as 1807, Joseph Fourier had worked out a theory of how air in enclosed spaces traps radiant heat. This was part of Fourier's general theory of heat, which he applied to many subjects, including the heating of the atmosphere by the sun. Fourier hypothesized that by retaining heat the atmosphere keeps Earth's surface temperature far higher than it would otherwise be. For this reason, standard histories of global warming invariably reference his use of the word *serre* (French for "greenhouse") as the source of the present-day term "greenhouse effect."[26] Fourier also described the principle of radiative equilibrium, which (put simply) states that Earth maintains a balance between the energy it receives from the sun and the energy it re-radiates to space. Today radiative equilibrium is seen as the ultimate driver of Earth's climate system, which transports heat from the equator, where more heat is received than re-radiated, toward the poles, where the opposite is true.

Other physicists and chemists investigated the capacities of different substances to absorb and retain it. In 1859, in Great Britain, John Tyndall began experiments to determine the radiative potential of various gases, including water vapor, carbon dioxide, ozone, and hydrocarbons. He concluded that water vapor's enormous capacity for heat retention made it the most important of what we know today as "greenhouse gases" —i.e., heat-trapping atmospheric constituents. By 1861, Tyndall had decided that these gases could be responsible for *"all the mutations of climate which the researches of geologists reveal. . . . They constitute true causes, the extent alone of the operation remaining doubtful."*[27] He speculated that a decrease in atmospheric carbon dioxide ($CO_2$) might have caused glacial periods.

Tyndall's results (and others) paved the way for the work of Svante Arrhenius, a Swedish scientist whose research in electrochemistry eventually garnered him a Nobel Prize. In 1895, Arrhenius announced to the Stockholm Physical Society his first calculation of how much the heat retained by "carbonic acid" ($CO_2$) and water vapor contributes to Earth's surface temperature, using an energy budget model much like Hann's "solar or mathematical climate." Arrhenius's now famous paper on the subject appeared in 1896.[28] In that paper he calculated that doubling the amount of $CO_2$ in the atmosphere would raise the global average temperature by 5–6°C.

Arrhenius may have been the first scientist to imagine that human activities might cause global climatic change. In public presentations, he even conjectured that combustion of fossil fuels might eventually raise $CO_2$ concentrations enough to change the Earth's temperature substantially.[29] Yet in his day world consumption of fossil fuels remained so low that this seemed merely a speculation, not a threatening near-term possibility. Indeed, Arrhenius estimated that such a change would take about 3000 years.[30] Perhaps because he lived in a very cold place, Arrhenius thought that global warming might be a good thing.

In recent years, Arrhenius's 1896 publication has acquired iconic status; it is routinely cited as the principal origin of modern climate-change concerns. Less well known is the fact that like most scientists of his day, Arrhenius was less interested in global warming than in the global *cooling* that caused ice ages. Hence he first computed the probable effects of *decreasing* $CO_2$. He calculated the effects of increased $CO_2$ mainly to explain the high global temperatures of the Tertiary period, not to predict the future. Until the 1950s, in fact, scientific discussions of global climate change focused more on paleoclimate (climates of the geological past) than on historical time.

Working around the same time as Arrhenius, the American geologist Thomas Chamberlin began to combine gas-physics theories of Earth's temperature with geological ones. He produced a sweeping explanation of global climatic changes on geological time scales, with $CO_2$ the fundamental driver.[31] Chamberlin argued that the vast quantities of $CO_2$ released during periods of high volcanic activity raised Earth's temperature, causing greater evaporation from the oceans, thus adding water vapor to the atmosphere. The additional water vapor, he theorized, further increased the atmosphere's heat-retention capacity, raising temperatures even higher.

Water vapor makes up a far larger percentage of the atmosphere than does carbon dioxide. Water vapor also absorbs much more heat; it is, in fact, Earth's principal radiatively active gas, responsible for the greatest part of the greenhouse effect. However, concentrations of water vapor fluctuate dramatically on very short time scales (days), while carbon dioxide's lifetime in the atmosphere is on the order of centuries. From these facts, Chamberlin concluded that although the water-vapor feedback vastly amplifies carbon dioxide's radiative warming effects, it is actually the latter, not the former, that causes climatic change. On geological time scales, he surmised, carbon dioxide in the atmosphere combines with calcium in igneous rock, forming calcium carbonate; this weathering process gradually absorbs the $CO_2$ released by volcanoes. Chamberlin's carbon cycle also had an organic component, with living things, especially plants, sequestering carbon from the air in their bodies, but for him the geological cycle was far more important. In periods of low volcanic activity, Earth's surface would absorb more $CO_2$ than volcanoes released, causing global temperatures to cool.

Chamberlin's theory that the carbon cycle was the principal driver of global climate oscillations was discussed widely—becoming, in fact, much better known than Arrhenius's work. Yet it fell from favor in just a few years, after other research seemed to show that $CO_2$ could not play the role Chamberlin and Arrhenius had ascribed to it. Between 1900 and 1905, research by Knut Ångström and others on radiatively active gases led most scientists to conclude that water vapor's effect on Earth's temperature overwhelmed that of carbon dioxide. Thus, they believed, additional $CO_2$ would have virtually no effect on global temperature.[32]

By 1903, when the second edition of Hann's *Handbook of Climatology* was translated into English, it echoed the general rejection of the carbon dioxide theory. The *Handbook's* chapter on "Geological and Secular Changes of Climate" included three paragraphs on the theories of Arrhenius and Chamberlin, but immediately dismissed them on the basis of Ångström's research. Even Chamberlin himself became convinced, by 1913, that his $CO_2$ theory was incorrect, and eventually he came to see his enthusiastic response to Arrhenius's 1896 paper as an overreaction.[33] After that, the carbon dioxide theory essentially disappeared from mainstream climatology until the late 1930s, when G. S. Callendar revived it. With a few exceptions, the idea of anthropogenic climate change went into hibernation along with it.[34]

In the 1920s, astronomical explanations of geological climate change rose to prominence, such as those of the Serbian geophysicist Milutin Milanković.[35] Several scientists had previously explored these ideas, but Milanković was first to complete a full calculation.[36] He showed that the confluence of three major astronomical cycles could explain the periodicity of recurring large-scale climatic changes such as ice ages. The three Milanković cycles are the eccentricity of Earth's orbit (a 100,000-year period), the planet's axial tilt (a 41,000-year period), and the precession of its axis (a 26,000-year period). These cycles interact, producing large variations—up to 30 percent—in the amount of solar energy Earth's surface receives. In combination, the timing of the Milanković cycles accounts well for major changes in Earth's climate (as dated from geological and fossil records). Because of this, they were for a time widely believed to close the book on terrestrial factors such as carbon dioxide. Other periodic and random astronomical factors, including sunspot cycles and variations in solar output, also affect Earth's temperature.

All the theories mentioned above addressed global climate chiefly on geological time scales. Furthermore, though they could explain large changes in global temperature, such as the ±5–10°C global average changes that produced the ice ages or the Cretaceous heat, these theories offered little insight into the causes of smaller changes. Nor could they explain fluctuations of climate on the scale of decades to centuries, the scale significant in human historical time. For climatologists, they represented useful steps toward identifying the ultimate drivers of the climate system, but they did nothing to explain the specific patterns of climate over the globe. Such understanding would have to await a genuine theory of the global circulation.

### The "Callendar Effect"

After lying dormant for three decades, the carbon dioxide theory resurfaced in 1938, when Guy Stewart Callendar, a British steam engineer and self-educated meteorologist, decided to reexamine it in the light of new research on the radiative behavior of carbon dioxide and water vapor. Callendar's role in resurrecting this theory was so important, James Fleming has argued, that the "greenhouse effect" should be renamed the "Callendar effect."[37]

As we have seen, Knut Ångström had demonstrated that carbon dioxide and water vapor absorb infrared radiation in the same parts of the spectrum. His result implied that $CO_2$ contributed almost nothing to the total radiation absorbed by the atmosphere, since water vapor—whose concentration in the atmosphere dwarfs that of $CO_2$—would absorb all radiation in those spectral regions. Callendar was the first person to notice the implications of subsequent studies using new, more precise instruments, which revealed spectral details that Ångström's instruments could not resolve. These studies showed that $CO_2$ absorbed radiation in spectral regions where water vapor did not. Because of this, Callendar realized, $CO_2$ might after all prove very important as an influence on global climate. In 1938, Callendar calculated that $CO_2$ might be responsible for 5–15 percent of the total "sky radiation" (i.e., heat radiated downward from the air), depending on latitude.[38] This was at most half of Arrhenius's 1903 estimate of 30 percent, but it still made $CO_2$'s contribution to Earth's heat balance much larger than its minuscule concentration might suggest.

Meanwhile, many things had changed dramatically since Arrhenius's day. With exploding populations of people as well as of automobiles, power plants, and other oil-burning and coal-burning technologies, Callendar realized, the world's fossil fuel consumption had already multiplied many times over. He calculated that fossil fuel combustion injected some 4.3 billion tons of carbon dioxide into the atmosphere each year. These emissions, he estimated, should have raised the atmosphere's $CO_2$ levels by about 6 percent between 1900 and 1936. To test this hypothesis empirically, he chose as a baseline a "very accurate" set of observations, made in 1900, that put the $CO_2$ concentration at about 274 parts per million (ppm). Meanwhile, observations taken in Paris between 1930 and 1936 gave an average value of 310 ppm. This increase—about 12 percent over 1900—agreed with Callendar's general hypothesis, though not with his more conservative estimate. Callendar took it as confirmation that human activities were, in fact, rapidly raising the concentration of carbon dioxide in the atmosphere. If his calculations were correct, Callendar then reasoned, this increase should already have produced a detectable rise in Earth's temperature, so he set out to determine whether global average temperatures had increased.

In the 1930s most climatologists still shared Hann's skepticism about global climate measurements. Few saw more than natural variability in short-term trends (years to decades). Some statistical studies purported to show warming trends since the nineteenth century in the eastern United States and in Europe, and in the United States the eccentric geographer Ellsworth Huntington and a few other popular writers had raised fears of regional warming in the 1930s. But in most scientists' view these remained irresponsible speculations.[39] Even those who did accept the reality of these trends could not explain them, since the field lacked any well-accepted causal theory for trends on the decadal scale.

Consulting the Smithsonian *World Weather Records*, Callendar conducted an exercise similar to Hann's 1897 climate-change calculation (figure 4.1), but with the generally more precise and reliable data recorded since 1880. He found only eighteen stations possessing records longer than 100 years. Of these, only the ones at Oxford and Copenhagen "could be classed as continuous throughout."[40] Therefore, Callendar restricted his analysis to the period 1880–1934. He chose 147 station records with good continuity and precision. From these he calculated temperature trends for large latitude zones and for the Earth as a whole. To ensure a reliable result, he checked every station's reliability by comparing it against others nearby, performing the calculation more than once on several subsets of the data.

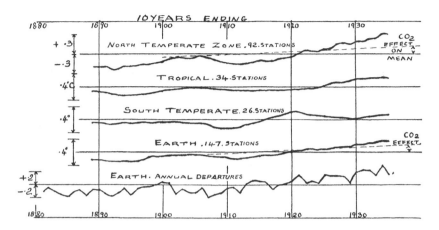

**Figure 4.4**
Global temperature anomalies, 1890–1935, expressed as a ten-year moving average departure from the 1901–1930 mean (global averages computed from records for 147 stations, grouped and weighted according to the surface area represented). Dashed lines represent Callendar's estimate of the contribution of $CO_2$ increases to the temperature rise.
*Source*: G. S. Callendar, "The Artificial Production of Carbon Dioxide and Its Influence on Temperature," *Quarterly Journal of the Royal Meteorological Society* 64, no. 275 (1938), 233.

His results indicated a global average temperature increase of 0.3°–0.4°C since 1890 (figure 4.4). He concluded that the observed rise in temperature was consistent with, and probably caused by, that to be expected from the observed rise in $CO_2$ concentrations. Ending on an optimistic note, Callendar pointed out that rising temperatures might "prove beneficial in several ways," such as bettering agricultural conditions and staving off the prospect of another "deadly" ice age.[41] Yet his final sentence noted that "the reserves of [fossil] fuel ... would be sufficient to give at least ten times as much carbon dioxide as there is in the air at present."

Callendar presented his results before the Royal Meteorological Society at a meeting attended by some of the foremost meteorologists of the 1930s. The published paper includes a summary of the ensuing discussion, which makes fascinating reading in light of later events. Although Callendar's efforts were applauded, and although the society agreed to publish his paper, every single meteorologist in Callendar's audience expressed grave doubts about his results. Sir George Simpson said "the rise in $CO_2$ content and temperature during the last fifty years must be taken as rather a coincidence," and that "the rise in temperature was probably only one phase of one of the peculiar variations which all meteorological elements experienced." Another interlocutor doubted that past measurements of either $CO_2$ or temperature were precise enough to support Callendar's analysis. C. E. P. Brooks agreed with Callendar's finding of rising global temperatures, but objected that this could be "explained, qualitatively if not quantitatively, by changes in the atmospheric circulation, and in those regions where a change in circulation would be expected to cause a fall of temperature, there had actually been a fall." David Brunt agreed with Simpson "that the effect of an increase in the absorbing power of the atmosphere would not be a simple change of temperature, but would modify the general circulation, and so yield a very complicated series of changes in conditions."[42]

If Callendar knew Brunt's textbook *Physical and Dynamical Meteorology* (1934, second edition 1939), he must have found this opinion extremely discouraging. In that book, Brunt expressed the general view of most theoretical meteorologists:

... *it is impossible to derive a theory of the general circulation.* . . . Not only are the laws which determine the transfer of energy by radiation too complicated to permit this, but the transport of heat by advection through the medium of the general circulation, and the interrelationships of cloud amount, radiation transfer and the general circulation, whose precise nature are unknown, make it impossible to derive any simple theory. . . . Increasing recognition of the tremendous complexity of the problem has led to an increasing disinclination to attempt a general theory.[43]

Brooks, too, remained unconvinced by Callendar's analysis. In 1951, reviewing knowledge of climatic change in the authoritative *Compendium of Meteorology*, he dismissed the $CO_2$ theory: "Callendar sees in [rising $CO_2$ concentrations] an explanation of the recent rise of world temperature. But during the past 7000 years there have been greater fluctuations of temperature without the intervention of man, and there seems no reason to regard the recent rise as more than a coincidence. This theory is not considered further."[44]

From today's perspective it may seem easy to dismiss the Royal Society's skepticism about the carbon dioxide theory as misguided. That would be a mistake. In fact Callendar's cool reception accurately reflected the state of the art in global climatology on the eve of World War II. To experienced climatologists, Callendar's crude methods revealed little. Even if a generalized global warming really could be deduced from only 200 station records, how did *surface* warming relate to the temperature structure of the entire three-dimensional atmosphere? Did a 30-year or even a 50-year warming trend represent a permanent climatic change, or just a temporary fluctuation around some normal zone? In view of the many other factors shaping climate trends, including solar output, sunspot cycles, orbital eccentricity, and occasional volcanic eruptions, how much could a trace gas such as carbon dioxide really matter? In 1938, none of these questions could be answered with authority. Many climatologists still thought of climate as a relatively fixed feature of local or regional geography, rather than a global system. Those who did take the global view, including David Brunt, despaired at the limits of understanding imposed by their lack of knowledge of the general circulation. They regarded Callendar's effort as an interesting speculation, and his evidence—no matter how carefully selected and filtered—as insufficient. Yet the "Callendar effect" gained ground steadily among professional climatologists in succeeding years (thanks largely to further work by Callendar himself), and its eventual acceptance led ultimately to modern scientific concern with anthropogenic global warming.[45] As a result, Callendar's stature among climatologists has steadily grown.[46]

I conclude this chapter with two points about how Callendar's 1938 paper reflects the relationship among data, computing, and climate knowledge on the eve of the computer age.

First, even though Callendar's calculation represents one of the most careful efforts ever made to estimate global temperature trends from data, in principle he (or someone else) could have done far more. For one thing, data sets far more substantial than the *World Weather Records* were already available. By 1938 some weather agencies had transcribed vast quantities of climatological and daily weather station records onto punch cards. The punch-card data processing equipment of Callendar's day could easily perform calculations as simple as temperature averages, even in very large volumes. So it would have been possible, even then, to recalculate monthly and annual averages from original daily records, or to use many more station records in the global calculation. Yet neither Callendar nor anyone else appears to have attempted such detailed climate calculations until several decades later. Why? One reason is what I call "data friction": the great difficulty, cost, and slow speed of gathering large numbers of records in one place in a form suitable for massive calculation.

Second, and related, one effect of Callendar's strategy was to reduce the computational load to a level easily handled by one individual working with simple calculating aids. This may not have been Callendar's goal (no one could accuse the man of shirking work), but the fact remains that using the pre-calculated averages in the *World Weather Records*, and evaluating these for only a couple of hundred stations, made it conceivable for him to carry out, by hand, the numerous calculations reflected in his seminal article. "Computational friction," as I will call it, prevented him from doing more.

# 5 Friction

You toss your laptop into a backpack, tuck your Blackberry into a shirt pocket, email your colleague a spreadsheet from the back seat of a taxicab. Once-inconceivable computer power is now ubiquitous. That power drives a colossal, networked information infrastructure through which pass terabytes of data and communication (not to mention vast amounts of "spam") each day. When one is caught up in the cascade of words, images, and numbers, in the frenetic traffic from screen to screen, it is easy to lose sight of the infrastructure—to forget that, underneath that glistening surface of free-flowing information, computing remains a material process.

Before computers were tiny flakes of neatly etched silicon, they were machines with gears, disks, levers, and dials. Before computers were machines, they were people with pencils (sometimes aided by simple machines). Computing involved, and still involves, a lot of moving parts. Some parts may be people, some may be electrons, and some may be pieces of metal or plastic. Operating on numbers (or on any other form of information) always involves a series of transformations, only some of which occur inside the chip, the machine, or the person. The terms 'input' and 'output' express the moments at which numbers pass from inside the computer to outside it, but many things happen to those numbers before they become input and after they become output. Every calculation requires time, energy, and human attention.[1] These expenditures of energy and limited resources in the processing of numbers can be called *computational friction*.

'Friction' is a metaphor, of course, but it is an apt and a deep one. In physical systems, friction means resistance. It occurs at the interfaces between objects or surfaces. It consumes kinetic energy and produces heat. Friction between moving parts consumes substantial amounts of the energy required to operate any mechanical device. Machines transform energy into work; friction reduces the amount of work they can do with a given input. Information systems transform data (among other things) into information and knowledge. Computational friction opposes this transformation; it expresses the resistance that must always be overcome, the sociotechnical struggle with numbers that always precedes reward. Computational friction reduces the amount of information and knowledge that can be extracted from a given input; engineers, mathematicians, computer scientists, and others spend a great deal of time seeking ways to reduce it.

Computational friction includes not only the physical and economic limits on processor speed and memory capacity, but also the human work involved in programming, operating, debugging, and repairing computers. Friction arises from round-off errors in repetitive calculations and from program bugs. It also appears in crucial but often ignored social work—for example, the effort involved in convincing others to accept as valid the results of calculations so extensive they could never be directly reviewed by human beings (e.g. global circulation model outputs), and the secondary calculations (e.g. statistical tests) used to validate such results.

Whereas computational friction expresses the struggle involved in transforming data into information and knowledge, the complementary concept of *data friction* expresses a more primitive form of resistance. Like computation, data always have a material aspect. Data are *things*.[2] They are not just numbers but also numerals, with dimensionality, weight, and texture. 'Data friction' refers to the costs in time, energy, and attention required simply to collect, check, store, move, receive, and access data. Whenever data travel—whether from one place on Earth to another, from one machine (or computer) to another, or from one medium (e.g. punch cards) to another (e.g. magnetic tape)—data friction impedes their movement. Since computation is one kind of operation on data, computational friction and data friction often interact.

Climatology requires long-term data from many locations, consistent across both space and time. This requirement implies a lengthy chain of operations, including observation, recording, collection, transmission, quality control, reconciliation, storage, cataloguing, and access. Every link in this chain represents an information interface subject to data friction. Every point at which data are moved or transformed represents an opportunity for data loss or corruption; one example of frictional cost is the effort involved in attempts to recover or correct lost or corrupted data. Interfaces between human beings and machines are points of special vulnerability, as are interfaces between organizations. Questions of trust, data quality, and access concern not only the numbers but also the people and institutions who recorded and transmitted them and the policies and practices of those who hold data and those who control access to data. To turn the metaphor back on the atmosphere: friction can also create turbulence. In social systems, friction means conflict or disagreement, which (metaphorically) consume energy and produce turbulence and heat. Both computational friction and data friction have both physical and social aspects, consuming both physical and human energy.

This chapter looks at computational friction and data friction in meteorology in the days before electronic computers, when the materiality of both computation and data was more salient. It focuses chiefly on the first half of the twentieth century, when the mathematical techniques of numerical weather prediction were first identified and when scientists began to commit vast quantities of data to punch cards for machine processing. Like the rest of this book, the chapter's approach is semi-chronological, with occasional fast forwards and rewinds. The goal is an evocative sketch rather than an exhaustive account.

## Computing the Weather

Soon after 1900, atmospheric physics made major advances that would eventually lead to the resolution of the three-way split between forecasting, theoretical meteorology, and empirical climatology. A Norwegian scientist, Vilhelm Bjerknes, showed that large-scale weather dynamics could be described by what are today known as the *primitive equations* of motion and state. For any given individual parcel of air, these equations describe how mass, momentum, energy, and moisture are conserved during the parcel's interactions with neighboring parcels. Established by various branches of physics during the eighteenth and nineteenth centuries, the primitive equations include Newton's laws of motion, the hydrodynamic state equation, mass conservation, and the thermodynamic energy equation.[3] This breakthrough paved the way for "weather by the numbers."[4]

Bjerknes was hardly the first to see basic physics as the road to a powerful theory of meteorology. Indeed, a theoretical tradition known as *dynamical meteorology* had been slowly developing in Europe and the United States ever since William Ferrel's work on general circulation in the 1850s. Bjerknes's achievement was to create a remarkably parsimonious theory of atmospheric behavior from the known laws of fluid dynamics.

Bjerknes's system of equations required only seven variables to describe the primary meteorological features: pressure, temperature, density, water content, and three variables for motion (the three dimensions). Differential forms of these equations could be solved to project an initial state—a set of simultaneous weather observations for a given area—forward in time. For the first time, a weather forecast could be calculated objectively (at least in principle) from basic physical laws, rather than inferred from maps on the basis of subjective experience. With this achievement, Bjerknes believed, both observational and theoretical meteorology had attained the status of mature sciences. They lacked only two things: sufficiently dense observational coverage and computational methods capable of solving the equations. Bjerknes recognized that both of these lacks would be exceedingly difficult to fill.[5]

On the computational side, the mathematical methods available to Bjerknes made solving his equations prohibitively difficult. In Bjerknes's day, physicists typically approached complex systems of partial differential equations either by seeking "closed-form" solutions or by using analog methods. Closed-form solutions exist when equations can be reduced to a system that can be solved with a bounded number of well-known operations. Such solutions are, however, rarely possible for complex physical equations.

Analog methods are essentially ways of substituting measurement for analysis or calculation by modeling, or simulating, the system in question. One analog technique involves building a physical scale model. Examples include the wind tunnels used to observe airflow around scale model aircraft or automobile bodies and the huge concrete models of the San Francisco Bay and the Zuider Zee built to study water flows in tidal basins.[6] If you can make such a model, you can bypass most calculation. Instead, you simply measure the model's behavior. If the model is a good analog—that is, if it behaves much like the system you want to know about—you can simply scale up your measurements of the model. But many of the forces that govern planetary atmospheres, such as gravitation, simply cannot be modeled on a small scale.

Another analog technique involves physical models that simulate some but not all of the target system's characteristics. In the 1940s and the 1950s, atmospheric scientists did experiment with physical models, such as "dishpan" analogs that simulated atmospheric flows by means of a rotating bowl or tank (like the Earth) filled with a viscous fluid (like the atmosphere) and exposed to a heat source (like the sun).[7] Here too, the large differences between the physical model and the real atmosphere made the technique too unrealistic for accurate simulation.

A third, less direct analog method involves creating physical or electrical components whose behavior resembles that of a desired mathematical function. For example, an amplifier that triples the strength of an incoming signal can be used to simulate multiplication by 3. By combining a series of components, complex equations can often be simulated without the need for calculation with numbers. One simply measures the final result (e.g. by means of a needle that indicates signal strength).

Numerical approximation was another alternative, but it posed formidable challenges. This technique requires working out finite-difference equations that approximate the differential equations, then solving them numerically. In Bjerknes's day, numerical techniques required vast amounts of tedious hand calculation. Pre-calculated, carefully checked tables listing numerical solutions to common functions could relieve some but by no means all of this work. Mechanical calculating aids, such as adding machines, could also help. Yet in Bjerknes's day there was usually no way to avoid a very large number of steps requiring human computers.[8] So numerical methods were labor intensive, time consuming, and prone to error. On top of that, they often resulted only in crude approximations, since mathematicians were just beginning to work out accurate methods of numerical analysis. This meant that no scientist used computational techniques for complex equations unless he or she lacked any other alternative. As we will see, the problem of numerical methods soon became a fundamental concern of meteorology.

Bjerknes and his colleagues attempted to simplify the primitive equations and, at the same time, elaborated an analog "graphical calculus" that would eliminate much of the calculation involved. This technique employed maps specially prepared from observational data. Forecasters applied graphical tools, similar to many other such tools then in use, to compute future changes from the maps. One such tool was the *nomograph*, consisting of two or more straight or curved graduated lines, each representing a variable. To determine an unknown value, the user would connect known values on two of these lines with a straightedge, then read the solution given by the intersection of the straightedge with another line on the nomograph. The slide rule, the most basic scientific computing device of the first half of the twentieth century, used a similar process. Such methods were quicker and easier, in many cases, than pre-calculated function tables, though less accurate.

Using the graphical method, forecasters could apply relatively complex mathematics with little calculation. The degree of precision such procedures could offer was low, typically two or three significant figures at best. However, in most cases they were little less accurate than the data to which they were being applied, and there was no real alternative. For these reasons and others, analog computing—graphical aids, nomographs, slide rules, mechanical and electromechanical differential analyzers, and other technologies—remained the preferred method of scientific computation until the advent of digital computers in the 1950s, and even beyond.[9]

Unfortunately, Bjerknes's graphical calculus proved difficult to use. Despite extensive efforts to develop a usable version, it was not applied successfully to forecasting until the 1950s, when it was soon superseded by computer-based numerical methods. Largely for this reason, Bjerknes's theoretical meteorology—though widely hailed by other scientists—failed to beget the institutional success he sought for his scientific program and his own career. In his intellectual biography of Bjerknes, Robert Marc Friedman has argued that this disappointment led the scientist to develop a new, pragmatically oriented forecasting system during World War I.[10] This system, which did not depend on the primitive equations, resulted in part from improvements in observing networks, the other remaining "detail" in a mature dynamical meteorology.

## World War I and Meteorology

During much of World War I, Bjerknes lived in Leipzig, where the Germans dragooned his small meteorological institute into military service. From this vantage point, Bjerknes observed—and to some extent assisted—the transformation of German meteorology from a prewar network with only twenty upper-air stations to "an intricate, dense network of stations behind all the fronts," taking both surface and upper-air observations at least three times a day and exchanging these by wireless telegraphy and field telephone. According to Friedman, the war "transformed meteorology" not only in Germany but in all combatant nations, and after 1914, Friedman argued, "weather began to be perceived and used differently."[11]

Detailed weather information was helpful to ground troops and to naval units, of course, but its greatest utility lay in artillery rangefinding and in flight assistance for zeppelins and airplanes. Since these operations required information about wind speed and direction at altitude, meteorologists engaged in systematic, fine-grained observation of the upper air for the first time. Instruments carried on board military aircraft returned an unprecedented number and frequency of upper-air measurements. In 1900, "aerology" (the study of the atmosphere above the surface) was an emerging subfield. But its chief observing platforms—instrumented balloons and kites—were difficult to use, and only a few dozen aerological stations operated routinely before World War I. Wartime meteorologists expanded and standardized these upper-air networks.

Most international exchange of weather data ceased immediately upon the outbreak of war in 1914. By forcing nations to rely solely on their own meteorological resources, this interruption helped to increase the density of observing networks. Military forces on both sides—not only on the muddy battlefields of France but also on the North Sea and the Atlantic Ocean, and eventually throughout the colonial world—needed accurate weather forecasts more than ever. Therefore, all the major combatants created or vastly augmented their military weather services. The US Army Signal Corps, for example, delivered three-dimensional weather reports to its parent service every two hours, a previously unprecedented frequency and level of detail.[12] The urgency of military needs led to heavy national investments in training and data collection. Owing to the concentration on practical techniques at the expense of theoretical work, that same urgency further promoted the split between forecasters and theorists.

The experience of the United States illustrates the war's effects. In 1904 the US Weather Bureau employed hundreds of observers. It had more than 200 paid weather stations, about 3700 unpaid observing stations, and an overall budget of almost $1.4 million.[13] Yet forecasting itself remained the highly specialized occupation of a small number of people. A handful of central forecasting centers collected observations and issued synoptic predictions for the entire country, distributing them by telegraph to local Weather Bureau offices.[14]

With the outbreak of World War I, the United States, like other combatant nations, scrambled to train new forecasters. The US Army Signal Service and the Blue Hill Observatory (near Boston) trained several hundred meteorologists during the war, in crash courses of a few months' duration. Many of these trainees continued in the field after the war's end.[15] The courses covered some physics, and many recruits already had some scientific education. But the essential skills taught were pragmatic: how to plot a weather map from data, and how to make a forecast from a map.

The most popular forecasting method in use during this period was an analog technique. Forecasters would chart the current set of observations, then look through a library of past maps to find the one that most resembled the new chart. Once you had found a reasonably similar map, you looked at how the past situation had evolved and based your forecast on that. Instructors directed forecasters in training to study the map library intensively. Choosing from among thousands of maps the one that most nearly resembled the current situation was a pattern-matching skill requiring years of practice and a prodigious memory. The emphasis on practical training and experience, rather than physics, reflected the ongoing split between forecasters and theorists in meteorology.

Before World War I, academic research institutions in the United States generally considered meteorology a peripheral, if important, element of either geology or geography. As late as 1910, US universities had awarded a grand total of only three doctorates in meteorology or climatology. By 1919, some 70 of 433 US institutions of higher education offered courses in meteorology or climatology, but only eight of these colleges and universities listed more than two courses in these fields. These figures reflect the general view of meteorology as a practical profession or a component of a larger field rather than a research science in its own right. Even into the 1930s, only three US Weather Bureau staff members held PhDs. Scientific meteorology was somewhat better established in European universities, and research goals were more significant to European weather services. But everywhere national weather services, rather than research institutions, remained by far the largest employers of meteorologists. Within these services, weather prediction was regarded as a form of practical work, rather than a research science.

Bjerknes's own professional career reflected this trend. Failing to garner substantial support for his theoretical research, he had turned his attention to forecasting. In 1917 he returned to Norway at the invitation of the government, which offered him the chance to build a major meteorological institute. There he applied the lessons he had learned in wartime Germany. Bjerknes and his collaborators—most notably his son Jacob, Tor Bergeron, and Halvor Solberg—redefined the basic concepts of weather prediction.

The Victorian meteorologists had emphasized the location and movement of high-pressure and low-pressure zones at the surface, where virtually all the data then available were collected. To these concepts, the Bergen School (as it came to be known) added the vertical dimension, now under systematic instrumental observation for the first time. Bjerknes's group came to visualize weather as the collision and conflict of discontinuous "masses" of air.[16] Each mass was marked by different characteristics, especially temperature, pressure, and humidity. Storms and other significant phenomena occurred where two masses collided, along a "front" (a term whose military resonance reflects its context of origin).[17] Much of the cartographic symbolism and much of the vocabulary of air-mass analysis (cold, warm, and stationary fronts; polar fronts; occlusion) remains in use today.

Developed in 1919 and 1920, the Bergen School's notion of the "polar front" proved especially significant to the globalization of meteorology. The polar front, as the Bergen School conceived it, was a single major surface of discontinuity where cold polar air collided with the warmer air of lower latitudes (figure 5.1). Typically, the polar front encircled the North Pole at around the latitude of northern Europe. Smaller-scale weather phenomena developed around this larger one, which functioned as a prime mover in the Bergen system. Since it implied that weather formed a unified system across the globe's high latitudes, the concept of the polar front added urgency to calls for international exchange of weather data, especially between North America and Europe. The concept also began to focus the attention of forecasters (as opposed to theorists) on the global circulation, although it would be several decades before the latter became a dominant theme in forecasting.

Most of the upper-air networks developed during World War I collapsed when the war ended, leaving Bergen School meteorologists without access to data they needed for their physics-based approach. Eventually, they had to accept that "without upper-air observations, nothing useful could be

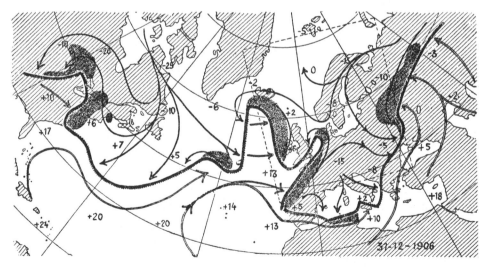

**Figure 5.1**
A representation of a polar front, from an early publication on the idea.
*Source*: V. Bjerknes, "The Meteorology of the Temperate Zone and the General Atmospheric Circulation," *Monthly Weather Review* 49, no. 1 (1921), 2.

done by starting out from the laws of physics, for all parts of the atmosphere interact, but the weather charts at that time showed only the conditions at sea level." So they developed a set of simple, practical mathematical techniques for forecasting in the absence of such data. Sverre Petterssen—later famous for his role in forecasting weather for the 1944 invasion of Normandy—recalled:

> What I did was develop a series of simple mathematical expressions for the velocity, acceleration, and rate of development of weather fronts and storm centers, without asking *why* and *wherefore*. I had to be satisfied with just, *it is so*. . . . My Bergen colleagues were skeptical: any 'method' that was not rooted in the laws of physics was hardly a method at all. True! But what little I had done was useful, and it was the only thing that could be done at the time.[18]

In the interwar years, the success of Bergen School predictive methods actually deepened the division between forecasting and theoretical meteorology. Bergen School techniques dominated forecast meteorology from the 1920s until the 1950s and beyond (although a few national weather services—most notably that of the United States—did not adopt them until the 1930s). Its three-dimensional approach made the Bergen method especially valuable in the 1920s as the age of commercial air travel dawned. It could forecast things pilots needed to know about, such as visibility, cloud ceilings, and cloud forms, which previous prediction systems had ignored.

Though Bjerknes's earlier work had helped to place meteorology on a physical-theoretical footing, the techniques and concepts promoted by the Bergen School in the 1920s were descriptive, not theoretical. The primary forecasting techniques of frontal analysis were in fact cartographical rather than mathematical. They vindicated the empiricist approach to meteorology. Bjerknes, as Frederik Nebeker puts it, "the man who became known as the advocate of calculating the weather, and as the advocate of meteorology based on the laws of physics," was also (ironically) "the man who initiated the development of a set of effective techniques that were neither algorithmic nor based on the laws of physics."[19] In retrospect, much later, Tor Bergeron himself argued that the Bergen method's success, by focusing attention on prediction at the expense of theoretical understanding, probably delayed meteorology's unification by as much as several decades.[20]

## The Forecast-Factory

The triumph of the Bergen practical program did not mean that Bjerknes's theoretical researches went entirely ignored. Indeed, they became the subject of a full-scale test not long after the Great War's end. In 1922, the English mathematician Lewis Fry Richardson attempted the first actual forecast based on a numerical (rather than analytical) solution of Bjerknes's primitive equations for the seven basic variables: pressure, density, temperature, water vapor, and velocity in three dimensions.[21] To do this, Richardson invented new mathematical methods involving finite-difference techniques.

Differential equations involve derivatives, which express a relationship among variables as one of them—in physics, usually time—approaches zero (e.g. $dx/dt$, the derivative of distance $x$ with respect to time $t$). Analytical solutions, when they are possible, resolve derivatives "to the limit," that is, across an infinitely small interval or "difference"—hence the phrase "differential equation." By contrast, finite-difference methods replace these derivatives with ratios between *finite* numbers. For example, $dx/dt$ might be replaced with $\Delta x/\Delta t$: a finite change in distance $x$ divided by a finite change in time $t$. A common quantity expressing such a relationship is "miles per hour."

In effect, finite-difference methods reduce calculus to arithmetic. Because they transform operations on variables into operations on numbers, methods like these are called *numerical*. Finite-difference methods can generate only approximate solutions, however, because the time interval is finite rather than infinitesimal. Crucially, they assume that the value of $\Delta x/\Delta t$ remains constant during the interval $\Delta t$, although in real physical systems it often does not. Continuing the miles-per-hour example, consider a one-hour trip through a city. Over a $\Delta t$ of one hour, if $\Delta x$ is 20 miles we could say that the driver's speed was 20 mph. Yet during any given $\Delta t$ of one minute during that hour, her speed could be anywhere between zero (at a stoplight) and 50 mph or more. Indeed, if our driver happened to join an impromptu drag race, her speed could rise from zero to 120 mph during that minute. Such an absence of constant or smoothly varying rates of change in physical systems is called *nonlinearity* and is a chief source of the great mathematical complexity of most physics. In general, for nonlinear behaviors, the smaller the time interval used in finite-difference methods, the better the mathematical approximation.

Richardson's forecast system employed a *finite-difference grid*. On a map of Europe, he divided the forecast area into a regular grid along latitude/longitude lines. This resulted in 18 rectangles with sides around 200 km long (2° latitude by 3° longitude), representing an area somewhat larger than Germany and centered, roughly, on Göttingen. The grid also extended vertically, with one layer at the surface and four more above it up to about

12 km. In this way Richardson obtained 90 three-dimensional grid cells. Making the assumption (for simplicity) that within each cell the values of the seven variables were constant, he set himself the task of calculating the change in these variables for a single time interval of six hours—that is, of computing a retrospective weather forecast.[22]

Richardson began with data from an "international balloon day" in 1910. A number of European stations had gathered simultaneous observations from weather balloons, providing an unusually complete set of data that included the vertical variables Richardson needed. But the irregular spacing of actual stations did not match the abstraction of Richardson's regular grid. Nor were observations available from every grid cell, even for this unusually well-covered time interval. This condition forced Richardson first to interpolate values at the center of each grid square from the available observations. As we will see, the task of interpolating real-world observations to the abstract grids of models remains a crucial element of meteorological practice today.

Richardson developed many ingenious simplifications to make his calculations easier. A practical man, he constructed 23 "computing forms" that laid out, step by step, the arithmetical operations to be performed on the data. A hypothetical forecaster would simply enter the required data and carry out operations as instructed. The forms, in effect, constituted a program, i.e., an *algorithm* by which forecasting could be reduced to a mechanical series of operations on numerical data. After about six weeks of tedious calculation, Richardson arrived at the calculated six-hour forecast for May 20, 1910.

The burden of calculation led Richardson to propose a fanciful solution which he called the "forecast-factory." The "factory"—really more like a numerical orchestra for global weather prediction—would have filled a vast theater with 64,000 human computers:

Imagine a large hall like a theatre, except that the circles and galleries go right round through the space usually occupied by the stage. The walls of this chamber are painted to form a map of the globe. The ceiling represents the north polar regions, England is in the gallery, the tropics in the upper circle, Australia on the dress circle and the Antarctic in the pit. A myriad computers are at work upon the weather of the part of the map where each sits, but each computer attends only to one equation or part of an equation. The work of each region is coordinated by an official of higher rank. Numerous little 'night signs' display the instantaneous values so that neighboring computers can read them. Each number is thus displayed in three adjacent zones so as to maintain communication to the North and South on the map. From the floor of the pit a tall pillar rises to half the height of the hall. It carries a large pulpit on its top. In this sits the man in charge of the whole theatre; he is surrounded by several assistants and messengers. One of his duties is to maintain a uniform speed of progress in all parts of the globe. In this respect he is like the conductor of an orchestra in which the instruments are slide-rules and calculating machines. But instead of waving a baton he turns a beam of rosy light upon any region that is running ahead of the rest, and a beam of blue light upon those who are behindhand.

Four senior clerks in the central pulpit are collecting the future weather as fast as it is being computed, and despatching it by pneumatic carrier to a quiet room. There it will be coded and telephoned to the radio transmitting station. Messengers carry piles of used computing forms down to a storehouse in the cellar.

In a neighbouring building there is a research department, where they invent improvements. But there is much experimenting on a small scale before any change is made in the complex routine of the computing theatre. In a basement an enthusiast is observing eddies in the liquid lining of a huge spinning bowl, but so far the arithmetic proves the better way. In another building are all the usual financial, correspondence and administrative offices. Outside are playing fields, houses, mountains, and lakes, for it was thought that those who compute the weather should breathe of it freely.[23]

Noteworthy in Richardson's beautiful fantasy—depicted here in a recent artist's conception (figure 5.2)—are the forecast-factory's global coverage and his description of a multiple-medium telecommunications network for disseminating forecasts. This was truly the world in a machine.

Nonetheless, Richardson thought that even this impossible apparatus would permit calculating the weather only about as fast as it actually happens, rendering it useless as a forecast technique. This sobering assessment of the computational requirements of numerical modeling discouraged meteorologists from further attempts to apply Richardson's technique until the advent of electronic digital computers in the 1940s. Perhaps even more damaging than the impracticality of Richardson's method was the complete failure of his test forecast. An error in his equations led to a surface pressure prediction 150 times larger than the actual observed change. Paradoxically, then, the most time-consuming, precisely calculated forecast in history was also among the least accurate ever prepared by any method. Like Napier Shaw, who also remarked on the significance of this forecast failure, Frederik Nebeker sees this as a principal reason why meteorologists abandoned the numerical approach for the next 25 years:

Most people today who know of [Richardson's] work see it as an important piece of science out of its place in time and see Richardson as an ignored genius. Neither

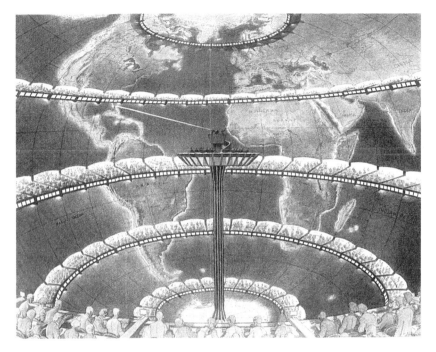

**Figure 5.2**
A contemporary illustration, by François Schuiten, of Richardson's "forecast-factory."

view is correct. Far from being out of its place in time, Richardson's work was a full trial of the leading research program of his time. Far from being ignored, Richardson's work was widely noticed and highly regarded, and as a result it had a highly important effect—it directed meteorologists elsewhere. In short, Bjerknes pointed out a new road, Richardson traveled a little way down it, and his example dissuaded anyone else from going in that direction until they had electronic computers to accompany them.[24]

Richardson's striking metaphors of calculation as factory, theater, church, and orchestra reach to the heart of computing as a coordinated human activity that harmonizes machines, equations, people, data, and communication systems in a frenetic ballet of numerical transformation. At the same time, they stand in stark contrast to today's dominant metaphors of computation, which are mostly individual: the brain, memory, neurons, intelligence. Richardson's forecast-factory remains a better description of the practical reality of computing. The limits of computer power, even today, stem from these human and material dimensions.

## "The Complete Statistical Machine"

Getting one instrument reading is easy enough. It is when you are trying to collect a lot of readings, from a lot of places, that the moving parts begin to rub together and you start losing energy to the process. Even ink on paper weighs a great deal when you're talking about millions or billions of numbers, and moving paper around the planet is slow and expensive. So instead of publishing tables, you try to move the data some other way: by telegraph or teletype, for example, if you lived in the early part of the twentieth century. Then you begin to get more interfaces: between the original measurement and a meteorological code (for economical telegraph transmission), between the telegraph form and the telegraph key operator. There are more: between the sending operator and the receiving operator, the receiving operator and another paper form, the Morse code transmission and the decoded message, and so on. At every interface, dissimilar data surfaces make contact. Some of these surfaces are human; they make mistakes, argue, and negotiate. Every interface slows you down and eats up energy. All that friction generates errors and noise, like sparks flying off a concrete saw.

By the early twentieth century, data friction helped to create, and then to widen, a split between the data used by forecasters and those required by climatologists. In the pre-computer era, the two subfields labored under quite different constraints. Forecasters did not want or need large volumes of data. The forecast techniques at their disposal did not benefit much from higher resolution. Even more important, the forecasters themselves could not process larger data volumes in the few hours available. For them, a well-distributed, reliable network that could bring in data quickly was far more important. Synoptic stations required direct access to telegraph, teletype, and other electronic communication systems in order to transmit their data immediately to central forecast centers. Because of these constraints, until World War II American forecasters used only data from the roughly 300 stations in the synoptic network staffed by professional Weather Bureau observers.[25]

The opposite was true of climatology. Not constrained by real-time operations, climatologists wanted every scrap of data they could get, and from a much denser network of stations. Climatological stations kept their own records and filed periodic reports by mail (generally monthly); by 1939 the US cooperative observer network included some 5000 stations manned by volunteers. As with the Réseau Mondial, climatological reports arrived at centers of calculation more slowly and required far longer (months) to process and publish.

In climatology, time itself is an interface between two data surfaces: the present and the past. Thus, while both forecasters and climatologists wanted observations standardized across geographical space, climatologists also wanted data to be standardized over time. Changes in observing systems created serious discontinuities in climatological data series. For example, the replacement of one kind of barometer with another, the construction of large buildings near a station, or the alteration of standards such as observing hours nearly always produces changes in a station's instrument readings. No matter how slight, such changes could affect climatological averages, sometimes rather dramatically. (See chapter 11.) Forecasters might have to adjust their systems to accommodate such a change, but having done so they could more or less forget about it. By contrast, climatologists were faced with reconciling data across time. To combine ten years' readings taken by one kind of instrument (say, an unshielded thermometer) with another ten years' readings taken by another kind of instrument (say, a shielded thermometer) required adjustments (for example, adding half a degree to the unshielded thermometer readings to account for the cooling effect of wind blowing across the thermometer bulb). As more and more network elements and standards were repeatedly altered in the name of improving the forecasting system, these adjustments became increasingly complex.

These differing needs and approaches were reflected in an institutional split between forecasting and climatology. In the United States, the advent of professional Signal Service forecasting in the 1880s orphaned the Smithsonian Institution's cooperative observer network, leaving once-enthusiastic observers with nowhere to report their measurements. As a result, "purely climatological work practically ceased" for most of that decade.[26] By 1890 the Weather Bureau had revived the cooperative observer network. By 1908 the climatological network comprised more than 2000 stations; by 1941 there were almost 5000 stations, staffed almost entirely by volunteers. Thus, one consequence of the split between forecasting and climatology was the rise of parallel overlapping networks for collecting and handling weather data.

Statistical climatologists made the most use of the ever-growing stores of climate data. As data accumulated and international communication infrastructures improved during the interwar years, statistical techniques grew more refined and complex. Manual handling and analysis of data became increasingly unwieldy. A major innovation in data processing arrived with punch-card tabulating equipment. Invented for the US Census Bureau around 1890 by Herman Hollerith, punch-card machines were initially capable only of recording, counting, and sorting. By the 1920s, however, they had grown much more sophisticated, adding multiplication and other mathematical capabilities. Meanwhile, Hollerith's successful company evolved into the business machine giant IBM, with divisions throughout Europe.

European climatologists first began using punch-card equipment for statistical calculation in the 1920s, primarily for analysis of ships' weather logs. In 1927, the Czech meteorologist L. W. Pollak developed an inexpensive hand punch. He distributed one to every Czech weather station, enabling it to record and submit data directly on punch cards. In 1928 the Czechoslovak Republic's State Statistical Office pressed the International Meteorological Organization to introduce punch-card recording on a worldwide basis; it even offered to supply uniform cards to IMO-nominated weather stations and to process the cards for the whole world. Reiterating dreams expressed in the earlier project of the Réseau Mondial, Pollak proposed to "make the necessary arrangements for an expedient, unified observational and publishing activity," a proposition approved by the International Meteorological Committee in 1929. However, his vision of a "Central Meteorological or at least World Climatological Office" never materialized.[27] (Had the Great Depression and World War II not intervened, perhaps Czechoslovakia might have become the world's leading center for climatology.) Though punch-card data processing never was quite the universal medium Pollak hoped it would become, by the 1930s systems had been deployed in Britain, Holland, Norway, France, and Germany as well as Czechoslovakia.[28] Those countries' climatological services occasionally shared their huge card decks with their counterparts in other countries.

By the 1930s the United States had also adopted the punch-card technology. In one of the first Depression-era government make-work projects, Civil Works Administration workers punched some 2 million ship log observations for the period 1880–1933; these were used to prepare an atlas of ocean climates. A subsequent Work Projects Administration effort punched about 20 million upper-air observations taken by balloon and radiosonde.[29] By 1939, many national climatological services employed punch-card systems routinely, and the trend gained momentum after World War II.

Punch-card devices could be potent calculating aids. Electromechanical card duplicators, sorters, tabulators, and adders were commonplace office equipment by the 1920s. They could deliver output to printers, or punch it directly onto new card decks for input into another set of calculations.

By the 1940s, the IBM 601 multiplying punch could perform more complex calculations. Used in sequence, a series of punch-card machines could solve quite difficult problems. For example, Manhattan Project physicists used them to solve equations for designing the first atomic bombs.[30] Durable, inexpensive, easily duplicated, and readable by humans as well as machines, punch-card data processing became the technology of choice for input and output in virtually all early computers. (See figure 5.3.) For example, each step of the ENIAC forecasts was punched to a card deck, which then became input for subsequent operations.[31] Teletype machines and paper tape devices were also used in early computers, but to a lesser degree and primarily as a way to store programs rather than data.

By the late 1940s, some punch-card machines could be programmed, using plugboards or punch cards, to carry out long sequences of mathematical operations (including division, which was important for scientific

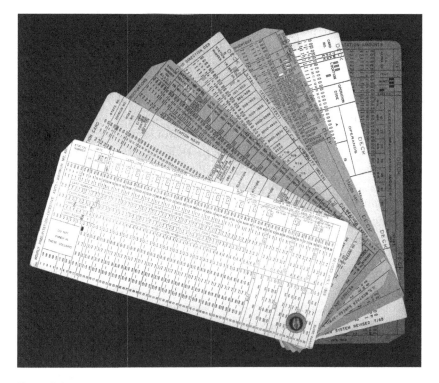

**Figure 5.3**
Meteorological punch cards. Image courtesy of National Oceanic and Atmospheric Administration.

needs). The IBM 604 Electronic Calculating Punch, introduced in 1948, could execute a program of up to 60 steps. Unable to perform conditional branching, these machines lacked the flexibility of true computers. Nonetheless, in the 1950s these highly reliable, powerful devices became the workhorses of most scientific computing. However, their slow speed—only a few times faster than hand desk calculators, albeit more convenient and more accurate for large volumes of data—restricted their use to non-real-time work, such as climatology and other kinds of statistical analysis. Electronic computers largely took over their calculating functions during the 1960s, but card readers and punches remained the principal method of data entry well into the 1970s.

According to a 1948 US military manual, World War II created a genuine crisis for climatological data processing. Commanders planning operations around the world sought information about local climates. Indeed, the military value of long-term climatological information was at least as great as that of short-term weather forecasts—and much more likely to be accurate. For many operations planned weeks or months in advance, knowing the likely weather conditions troops would encounter was critical. Military climatologists purchased a wide variety of special punch-card equipment. Yet they recognized that "even with the flexibility of these machines, the wide range of requirements presents many problems, whose solutions are impractical on present day equipment," and that these devices were "still far short of the complete statistical machine."[32]

Between 1941 and 1945, the Air Force climatology program recorded 26,000 station-months of records on about 20 million cards. The program acquired a major addition to its library at the end of World War II, when Allied forces captured two large card decks from German weather services. These decks, representing areas that had been inaccessible to Allied weather services during the war, had enormous climatological value. Before being shipped to the United States, the 21-ton Kopenhagener Schlüssel card deck, containing some 7 million cards, sat outdoors in crates on a Dutch dock for several weeks. It arrived at its destination waterlogged. In a heroic effort, the deck was resurrected by carefully drying the spongy cards and feeding them through card duplicators by hand. Keypunch operators re-punched thousands of cards too damaged to be read automatically. Similar punch-card databases of varying sizes existed throughout Western Europe and in China, Turkey, Manchuria, Russia, Poland, Finland, Estonia, and elsewhere.[33]

By 1960, the data library at the US National Weather Records Center contained over 400 million cards and was growing at the rate of 40 million cards per year. These cards represented all US states and possessions plus US-operated weather stations in Antarctica, Korea, France, Germany, Greenland, and Japan and on numerous Pacific islands.[34] By 1966 the cards occupied so much space that the Center began to fill its main entrance hall with card storage cabinets (figure 5.4). Officials became seriously concerned that the building might collapse under their weight.[35]

The 1949 fourth edition of *Machine Methods of Weather Statistics* projected a global climatological program much like Pollak's 1920s vision of a "World Climatological Office." The program would have three phases:

Phase One . . . , the unification of reporting forms and placement of all current observations on cards, is an accomplished fact.

**Figure 5.4**
Punch-card storage cabinets in the main entrance hall of the US National Weather Records Center in Asheville, North Carolina, early 1960s.
Image courtesy of National Oceanic and Atmospheric Administration.

Phase Two, or placing historical data on cards, is progressing quite rapidly, considering the magnitude of the task. This phase also includes indexing all climatological data, in all publications, and securing exchange of punched cards or data with other countries.

Phase Three, which is just beginning, envisions the establishment of a central weather records repository for the receipt, filing and servicing of all weather records, whatever the originating service or source. Such a repository . . . would be capable of servicing the needs of all users of weather material. This would eliminate the present confusing and time-consuming situation of having to go to several repositories. . . .[36]

Earlier, Victor Conrad's climatology textbook had expressed the similar hope that "all nations will eventually agree to an international exchange of punched cards, just as they now exchange coded messages and publications."[37]

Through the 1950s, armies of keypunch operators created vast card libraries manually by reading off data from printed or handwritten records. In 1952 one meteorologist scratched his head over the paradox that "the relatively simple and routine operation of transcription between various forms of record is still performed manually, while relatively complicated operations such as communication and involved calculations can be performed automatically by machines."[38] This remained generally true well into the 1960s. As the US Weather Bureau put it in *Climatology at Work*,

> . . . the simple card punching machine has not been matched economically. Certain types of data which are produced by analogue means, or which can be recorded and read graphically, can be converted to high-speed media [for electronic computers, i.e. magnetic tape, disk, or drum] by specialized equipment. However, *the bulk of weather data used in climatology is still produced manually, and the cold facts of budgets require that the first translation be to the punched card.*[39]

Thus, although computers had begun reducing computational friction, data friction due to manual processing remained high.

Why wasn't the slow, error-prone process of data transcription automated before routine numerical weather prediction? For example, why weren't teletype tapes—already encoded for machine reading—punched directly onto cards? This kind of automation was certainly *technically* possible by the mid 1930s, yet for several reasons no one attempted it on a large scale. First and foremost, forecasters cared much more about speed than about data volume. They took incoming data directly off the wire, selecting only the data their particular methods required. Although many weather stations reported readings every two or three hours, forecasters typically used only data from the two main synoptic hours, 00 and 12 GMT. For synopticians, punch-card machines remained far too slow and cumbersome. Climatologists *did* want all the data, but they could wait for manual punch-card transcription. Second, the wide variety of meteorological codes in use meant that most incoming data still required interpretation by a human being. Neither punch cards nor teletype tapes lent themselves easily to this kind of inspection. Also, not all data arrived by teletype; short-wave radio broadcasts in Morse code, which had to be transcribed by hand, remained the norm in many parts of the world. Third, large amounts of "noise" in the incoming data required constant quality control. In the pre-1950s routines, obvious errors (obvious, that is, to those with some experience and training) could be rejected without ever being transcribed, which saved precious time. Conversely, incomplete or garbled messages—routinely rejected by automated systems—could sometimes be salvaged by experienced transcribers, who could recognize the source of a partial message and supply the missing information. Finally, on a more general level, communication and computing mostly remained functionally and technologically separate until the late 1950s, despite considerable interplay between their underlying components.[40] With these systems still being invented, little if any commercial equipment existed that might have helped meteorologists to automate data entry within the severe time constraints of forecasting.

Simple, reliable, and ubiquitous, punch cards remained the input method of choice for the new electronic computers for several decades. Thus, although punch-card data processing itself never became the "complete statistical machine" envisioned by the Air Force, the significance of the vast punch-card data libraries must not be underestimated. They lay ready to hand when electronic forms of data storage, such as magnetic tape and disks, came on the scene in the following decade. Laboriously hand corrected and "migrated" to magnetic tape, these data became fundamental resources for global climatology.[41]

**From Analog to Digital . . . and Back Again**

Meteorologists and climatologists of the early 1950s faced not only an explosion of data but also a plethora of alternatives for coping with it. They were well aware that their choices about how to collect and transmit data would also affect how data could be stored, retrieved, and processed. Not all of these techniques were digital.

Until recently, if you wanted to know the temperature, you would look at a thin glass tube containing a few drops of mercury (or some other fluid). You'd sight across the column to the numerical scale beside it, and record the closest number as the temperature. Rising and falling as it expands and contracts, the mercury column is an analog of the heat contained in the atmosphere. It's a physical thing, so it changes continuously, and it knows nothing about numbers. Your act of reading the thermometer transforms a continuous, infinitely variable analog quantity into a discrete number, such as 75°F: a set of digits, which vary discretely or discontinuously. In fact the fluid is almost never precisely on the line marked on the scale, but instead slightly above or below it. By setting aside that fact, *you* make the observation digital.

Well into the 1960s, and even beyond, few sensors of any kind produced numerical data directly.[42] Instead, most instruments were analog devices. Radiosondes (an instrument package attached to a balloon) provide an interesting example. Early radiosondes broadcast readings in the form of a single modulated radio frequency. At a ground station, reception equipment demodulated the signal and charted it on specially prepared graph paper. As the balloon rose, a sensor detected a series of predetermined pressure levels. At each level, the radiosonde would switch first its thermometer and then its humidity sensors into the circuit, broadcasting them on the same frequency. Thus the record produced at the ground station showed only periodic changes in the modulated frequency. Observers on the ground could tell when the sonde reached a new pressure level by the sudden change in the signal. Because they knew the order in which the sensors were switched into the circuit, they knew that the first change represented temperature and the second one humidity. At the ground station, observers created graphical representations of temperature and moisture variation across pressure levels by charting the data points and drawing lines between them. These and many other analog-digital and digital-analog conversion processes were entirely ordinary in science before computers. As we saw earlier, analog computers and computing aids, some of them extremely complex and sophisticated, helped scientists model physical systems and solve mathematical problems.[43] Even today, despite often ideological commitments to the notion that digital is inherently better than analog, the most common visual representation of data is as a curve on a graph, a continuous rather than discrete (digital) form. Likewise, the ultimate output of weather forecasting—the weather map, with its continuous curves—is an analog representation.

Still, converting data from analog to digital and back again always creates friction. By the early 1950s, the many conversion points in the atmospheric data network provoked reconsideration of the system as a whole. Conversions took time, as did even simple transcription from one medium to another. New possibilities for digital data processing inspired numerous publications addressing issues in the collection, transmission, and processing of data. Consider figure 5.5, which appeared in a 1952 analysis of geophysical data processing. The diagram outlines the wide variety of paths from measurement to final analysis and data record.

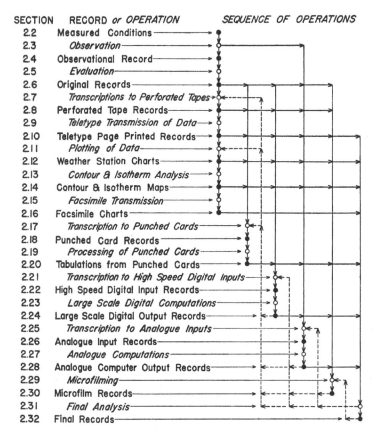

Figure 5.5
Operations involved in processing geophysical data. Arrows indicate possible data paths, some actual and some speculative.
Source: J. C. Bellamy, "Automatic Processing of Geophysical Data," *Advances in Geophysics* 1 (1952), 4.

Analog isotherm maps, for example, might be transmitted by (analog) facsimile machine before being gridded and transcribed to punch cards. These ideas amounted to early thoughts about automating the weather observing system. Perhaps the moving parts in this machine could be lubricated. Maybe some of them could even be eliminated altogether.

### The Third Dimension

Bjerknes, the Bergen School, and Richardson's numerical experiment clearly established that computed weather forecasts would require two things that meteorologists did not have: better, faster computing aids, and data from the vertical dimension. Although meteorologists of the interwar period regarded aircraft, kite, and balloon sounding networks as extremely important, in fact they remained quite limited. As late as 1937 the total number of daily airplane soundings available to the US Weather Bureau was only about thirty.[44] Radiosondes were first developed in the late 1920s and began to replace aircraft sounding for routine weather forecasting around 1938. During World War II, when international broadcasts of weather data generally ceased, upper-air reports from military aircraft and radiosondes became crucial to the combatants. As a result, the number of upper-air networks grew rapidly. The US radiosonde network went from just six stations in 1938 to 51 stations in 1941 and 335 stations in 1945. In the 1950s, the number of stations settled at about 200.[45] Many American military airbases built during the war continued to operate afterward, generating both weather reports and demands for forecasts at ever-higher altitudes as the capabilities of military aircraft increased.

Unlike the network of ground stations, whose location is fixed, upper-air networks collect data along the variable flight paths of carrier platforms. Balloons could rise quickly, exposing the vertical structure of the atmosphere, such as its temperature profile. Tracked by radar, their flight paths revealed wind speed and direction at different altitudes. Airplanes could cover hundreds of kilometers in a few hours. During World War II, high-flying military aircraft first encountered the previously unknown jet streams: powerful, relatively narrow high-speed air currents located between 10 km and 50 km above the ground. The dangers and opportunities jet streams posed for air travel made them the object of intensive study. By the mid 1940s, upper-air data had confirmed the theories of Rossby, Bjerknes, and others that ground-level weather is largely controlled by airflows higher up. For these reasons, upper-air data were actually more valuable than data from ground stations.

In addition to military needs, the postwar boom in commercial air travel amplified the demand for weather reports and forecasts at flying altitudes. These developments caused a major shift of emphasis in weather forecasting. Rather suddenly, understanding the large-scale circulation at altitude acquired real urgency. "The general circulation of the Earth's atmosphere," Hurd Willett noted in 1951, "is the primary problem of meteorology on which scarcely a beginning has been made. *It is on the solution of this that all basic improvement of weather forecasting is now waiting.*"[46] Upper-air data from the northern hemisphere began to provide, for the first time, a detailed window onto the three-dimensional global circulation.

## Managing Friction in Forecasting and Climatology

By the late 1930s, a real-time weather data network could be said to exist in the northern hemisphere between the Arctic Circle and the tropics. Throughout Europe and the United States, weather services used teletype (an automated form of telegraphy widely adopted in the 1920s) to exchange data. Data from more remote locations began to arrive via shortwave radio: from ships at sea, from remote island and land stations, and from other locations beyond the reach of the telegraph network. Shortwave signals, discovered by radio amateurs around 1920, could "bounce" off the ionosphere, traveling thousands of kilometers—even across oceans, if weather conditions were right. Although noise in these broadcasts often caused errors or incomplete transmissions, in many places shortwave radio rapidly displaced telegraph for long-distance communication. The technological underpinnings of a global data network were rapidly coming together.

The institutional and organizational aspects of data exchange, however, lagged far behind. More data might arrive, but in a bewildering variety of forms. In the 1945 *Handbook of Meteorology*, a review article declined even to attempt a worldwide survey of meteorological data communication:

> . . . the currently used codes are far too numerous. There are many reasons for the complexity in weather codes and reports: the diversity of elements to be observed, the various techniques of observation, variation in the information desired by analysts in different parts of the word, and lack of uniformity in the codes adopted by separate political units are some of the reasons. . . . Many political units use International [IMO] codes, [but] others use portions of these codes or have devised forms of their own.[47]

Transmission techniques exhibited similar diversity. Teletype over the telegraph cable network was the highly reliable standard, but ordinary telegraph, radio teletype, shortwave radio, point-to-point microwave radio, and other systems were also in use. The many interfaces in this multimodal network generated tremendous data friction.

By the 1930s, the rapidly increasing quantities of data flowing through the global network widened the division between forecasting and climatology. The synoptic (weather) data network continued to function as a high-speed, centralized system based primarily on professional observers working for national weather services and military forces. The forecasting system managed computational friction in three ways: by limiting data input, by centralizing calculation at forecast centers, and by limiting the amount of calculation to what could be done in a couple of hours.

Friction also affected climatologists, but in a much different way. They wanted densely spaced networks, and they did not need instantaneous reporting, so their data network became a low-speed, partially decentralized system. Observations could be sent in by postal mail and published months or years later. With thousands of volunteer observers as well as professionals contributing data, the climatological data network eventually grew much denser (at least in some countries) than the weather observing system.

Climatology decentralized some of its essential calculations. Rather than submit the entire weather record, climatological observers typically computed monthly average temperature and rainfall, for example. Central collectors, including national climatological services, the World Weather Records, and a few individuals, gathered and published these figures, rarely if ever re-computing them from the original observations. The advent of punch-card systems seemed, in the 1930s, to promise a new age, automating much of the calculation and permitting far more extensive and detailed statistical processing. But at the same time, the sheer weight and volume of punch-card libraries soon threatened to overwhelm the capacity of weather records centers.

In the age of the World Wide Web, it is easy to forget that data are never an abstraction, never just "out there." We speak of "collecting" data, as if they were apples or clams, but in fact we literally *make* data: marks on paper, microscopic pits on an optical disk, electrical charges in a silicon chip. With instrument design and automation, we put the production of data beyond subjective influences. But data remain a human creation, and they are always material; they always exist in a medium. Every interface between one data process and another—collecting, recording, transmitting, receiving, correcting, storing—has a cost in time, effort, and potential error: data friction. Computing, too, is always material; there are always moving parts in the overall system. Every calculation costs time, energy, and human labor. Computing and automation can reduce those costs, yet no matter how smooth and inexpensive calculation may become, there will always be some scale of activity at which computational friction can slow things to a crawl. These forms of friction helped maintain the separation of forecasting from climatology, and the separation of "weather data" from "climate data," well into the computer age.

# 6 Numerical Weather Prediction

Between 1945 and 1965, digital computers revolutionized weather forecasting, transforming an intuitive art into the first computational science. Unlike many scientific revolutions, this one was planned. Numerical weather prediction became the civilian showcase for a machine invented in wartime to support specifically military needs. Scientists conceived and carried out the first experiments with numerical forecasting in the earliest days of electronic computing, years before commercial computers became widely available, as a joint project of American military research agencies and the US Weather Bureau. A principal architect of that project was John von Neumann, who saw parallels between the science of nuclear weapons and the nonlinear physics of weather.

This chapter reviews that story not only as revolution, but also as a step toward infrastructural globalism. Computer models for weather forecasting rapidly came to require hemispheric data, and later global data. Acquiring these data with sufficient speed demanded automatic techniques for data input, quality control, interpolation, and "bogusing" of missing data points in sparsely covered regions. The computer itself helped solve these problems, but their full resolution required substantial changes to the global data network. Like all infrastructure projects, these changes involved not only scientific and technological innovation, but also institutional transformation.

By the mid 1950s, weather services had begun to deploy computerized numerical weather prediction models in operational forecasting. Having established the potential of numerical models for short-term prediction, researchers moved on to create general circulation models, which could simulate global atmospheric motions not only over periods of days and weeks, but over months, years, decades, or more. In the long run these models made it possible for forecasting and climatology to base their methods in physical theory, to use each other's data, and eventually to converge upon an intimate scientific connection they had not enjoyed since the nineteenth century. Because of this close relationship between numerical weather prediction and climate modeling, this chapter covers the early years of numerical weather prediction in considerable detail.

**Weather as a Weapon**

After World War II, with Europe partitioned and the Cold War looming, many scientists and engineers who had been involved in urgent, interdisciplinary, and fascinating wartime projects continued to consider military problems a natural part of their research agenda.[1] The fact that military agencies provided most of the available research funding in the postwar United States amplified this effect. In a process I have called "mutual orientation," scientists and engineers oriented their military sponsors toward new techniques and technologies, while the agencies oriented their grantees toward military applications. This relationship mostly produced general directions rather than precise goals; rarely did military funders require scientists to specify exactly how their research might be used by the armed forces. Nonetheless, military funders did expect that at least some of the work they paid for would ultimately lead to weaponry or to other forms of strategic advantage, including useful practical knowledge.[2]

Mutual orientation played a substantial role in the origins of numerical weather prediction. Since weather affects virtually every aspect of battlefield operations, accurate weather forecasts have great military value. The vast increase in upper-air data collection during World War II occurred because military aviators needed to know about conditions in the upper atmosphere; with missiles and high-flying jet aircraft joining the arsenal, these needs would only grow. Radar, another product of World War II, also increased the observational abilities of meteorologists.[3]

If weather could be predicted, some thought, it might also be controlled. This too could have profound military implications. General George C. Kennedy of the Strategic Air Command claimed in 1953 that "the nation which first learns to plot the paths of air masses accurately and learns to control the time and place of precipitation will dominate the globe."[4] Rainmaking by seeding clouds with dry ice or silver iodide, discovered and developed in 1946–47, seemed to some—including the physicist and chemist Irving Langmuir, a major proponent—to offer the near-term prospect of controlling precipitation. Respected scientists, both American and Soviet, believed in the mid 1950s that the Cold War arms race would include a new struggle for "meteorological mastery." Proponents of weather control often compared the tremendous amounts of energy released by atomic weapons with the even greater energy contained in weather systems, and they sought the ability to alter climate as a possible weapon of war. By January 1958, in the aftermath of Sputnik, *Newsweek* warned readers of a "new race with the Reds" in weather prediction and control.[5]

The key to numerical weather prediction was the electronic digital computer, itself a product of World War II military needs. The principal wartime American computer project, the University of Pennsylvania's ENIAC, had been designed under contract to the Army's Aberdeen Proving Ground, where the production of ballistics tables had fallen far behind schedule. But the machine was not completed until the final months of 1945. John von Neumann, among the twentieth century's most famous mathematicians and a major figure in World War II–era science, joined the ENIAC team when he recognized its relevance to the Manhattan Project, to which he served as a consultant. He contributed to the design of ENIAC's successor, the EDVAC, and also suggested the ENIAC's first application: mathematical simulation of a hydrogen bomb explosion.[6]

Von Neumann also conceived the computer's application to weather prediction. Both the hydrogen-bomb problem and the issue of weather prediction were essentially problems of fluid dynamics, an area of particular scientific interest to von Neumann.[7] But von Neumann's interest in weather prediction was connected to the bomb in other ways too. As a Hungarian émigré, he felt a deep personal commitment to American military prowess in the emerging Cold War. During hearings on his nomination to the postwar Atomic Energy Commission, von Neumann announced: "I am violently anti-Communist, and I was probably a good deal more militaristic than most. . . . My opinions have been violently opposed to Marxism ever since I remember, and quite in particular since I had about a three-month taste of it in Hungary in 1919."[8] If weather could become a weapon, von Neumann wanted to see that weapon in American hands.

## Weather Forecasting as a Demonstration of Digital Computer Power

At the end of World War II, electronic digital computing was in its earliest infancy. Only a handful of prototype machines existed in the United States, Great Britain, and Germany.[9] No computer meeting the full modern definition—electronic and digital, with both programs and data stored in main memory—actually operated until 1948. Commercial computer manufacturers did not make a sale until about 1950, and military customers snapped up most of the first machines. A robust civilian market did not develop until the latter half of the 1950s.

The celebrated Electronic Numerical Integrator and Calculator (ENIAC) was really a proto-computer.[10] Programmers had to set up its operations using plugboards and switches in a tedious process that required a new, painstakingly worked out set of connections for every problem (figure 6.1). ENIAC contained some 18,000 fragile vacuum tubes, consumed 140 kilowatts of electric power, required a forced-air cooling system to keep from catching fire, and suffered downtime of about 50 percent due to tube failures and other technical problems. Though ENIAC gained tremendous fame for its unprecedented speed, it and its immediate successors were far from ready to handle routine computation, especially for problems (such as weather prediction) for which time pressure was severe. Nonetheless, news of the ENIAC galvanized the international scientific community.

Von Neumann's most important contribution to computing lay not in the ENIAC itself, but rather in the design of a successor machine known as the Electronic Discrete Variable Automatic Computer (EDVAC). The EDVAC would store both programs and data in main memory, the crucial

**Figure 6.1**
ENIAC programmers setting racks of switches and plugboards to enter a program. Image courtesy of US Army.

innovation separating true computers from their predecessors. The EDVAC design, perhaps the most famous document in the history of computing, circulated widely under von Neumann's name.[11,12] Von Neumann secured commitments for an "Electronic Computer Project" from the Institute for Advanced Study (located in Princeton) and from the Radio Corporation of America. The project culminated in a computer, usually known simply as "the IAS machine," that was completed in 1952.

Von Neumann's intellectual interests lay mainly in advanced applications of mathematics, such as those found in physics. In late 1945, with the war over and the Electronic Computer Project well underway, he began casting about for a future application of electronic digital computing that would meet three criteria. First, it had to be of strong mathematical and scientific interest, and ideally it would solve some previously insoluble problem by replacing analytical mathematics with numerical techniques. Second, it had to possess significant military value, not only to attract funding but also to contribute to the emerging Cold War deterrent against communism. Finally, the chosen application should demonstrate the power and importance of computing to the widest possible audience.

As we saw in chapter 5, meteorologists had understood the mathematics of a calculated forecast, at least in its bare outlines, since the work of Bjerknes and Richardson early in the century. One numerical forecast (Richardson's) had actually been completed, albeit with disastrous results. The approach had been abandoned because two vital ingredients remained unavailable: numerical methods for solving nonlinear equations, and massive computing power. This was, then, an area ripe for new mathematics of precisely the sort von Neumann was already working on, and in a field of physics (fluid dynamics) with which he already had some acquaintance from his atomic weapons work.[13] Furthermore, von Neumann saw, computers might permit substituting numerical simulations for traditional experiments. In an early postwar funding request, von Neumann mentioned "high speed calculation *to replace certain experimental procedures* in some selected parts of mathematical physics."[14] Like nuclear explosions, weather defies both close observation and laboratory experiment; thus, both are ideal candidates for simulation studies. In a well-designed numerical simulation, one could manipulate variables at will and observe their effects on outcomes in a way that no other technique could hope to match.[15]

In addition, military commanders now regarded accurate weather prediction as a critical capability. The experience of World War II had shown that forecast quality could have decisive effects on a military campaign.

For example, Operation Overlord—the 1944 Allied invasion of Normandy—succeeded in part because General Dwight D. Eisenhower trusted his meteorologists' prediction of a one-day interval of good weather in the midst of a long series of intense storms over the English Channel. On the basis of this forecast, Eisenhower delayed the start of the invasion, originally planned for the small hours of June 5, until late that night. The operation began under cover of wind, rain and high seas, but the weather then cleared as predicted, and the Allied forces landed before dawn on June 6th. German commanders were caught by surprise, their forecasters having failed to foresee the interlude of calm.[16]

The widely recounted D-Day story, along with many others, entrenched meteorology as a military science. For example, commanders used "applied climatology" in siting new bases, choosing transport routes, and deciding when to launch operations.[17] Statistical climatology had also assisted in scheduling the Normandy invasion: analysis had revealed that May and July would probably be worse than June for operations in the English Channel. Ironically, had they considered the full story of the D-Day episode, commanders might have been far less sanguine about weather forecasting's potential. Three teams of Allied forecasters had applied entirely different techniques, including the analog map method and upper-air analysis. In the months preceding the invasion, these teams had disagreed far more often than they agreed. Forecasters involved in the June 6 forecast regarded their consensus on that day as a minor miracle.[18]

John von Neumann settled on weather prediction as the showcase problem for electronic computers after a November 1945 encounter with Vladimir Zworykin, the RCA electrical engineer who had developed the scanning television camera. By then the idea of computing the weather was already in the air. John Mauchly had approached the Weather Bureau in April 1945 about using the ENIAC and the planned EDVAC for weather prediction, but the response had been lukewarm. (Soon afterward, however, Mauchly met with Army Weather Service meteorologist Harry Wexler, who showed much greater enthusiasm for the idea; Wexler became head of the Weather Bureau's Scientific Services Division in 1946.) Zworykin, too, had proposed weather prediction by computer to Weather Bureau chief Francis W. Reichelderfer. Reichelderfer invited Zworykin for a formal meeting. He added von Neumann's name to the invitation list at the suggestion of Edward Condon, director of the National Bureau of Standards, who wanted to collaborate with the Weather Bureau to assess the new machines' potential.[19] The meeting, held early in 1946, included staff members of the

# Numerical Weather Prediction 117

Weather Bureau and the National Bureau of Standards as well as representatives of the military services. The meeting was supposedly confidential, but the story leaked to the *New York Times*, whose somewhat breathless account focused on the possibility of weather control.[20]

Zworykin, von Neumann, and others were at least willing to speculate that weather simulation could lead to weather control, which might become a weapon of war.[21] For example, the Soviet Union's crops might be ruined by inducing a drought.[22] Indeed, Kristine Harper speculates that the 1946 *New York Times* story about meteorological computing was leaked deliberately to promote the project among top Navy brass who controlled the bottomless military purse: "The leak indicated that a comprehensive meteorological theory existed (when it most certainly did not) and emphasized the weather control aspects." In order to sell a project that could forecast, or control, the weather, the meteorologists needed to have a plausible theory to back it up."[23]

Several interests thus converged on the idea of computerized weather forecasting: theoretical meteorologists' concern to develop atmospheric physics, military goals of accurate battlefield forecasting and potential weather control, the mathematical and technological objectives of computer pioneers such as Mauchly and von Neumann, and the Weather Bureau's and the American public's interest in improving civilian forecasts.

## The IAS Meteorology Project

Meteorology met all of von Neumann's criteria for a major new push in the application of computers. After meeting with Zworykin and Reichelderfer, von Neumann began to familiarize himself further with meteorological theory. He met with the Swedish émigré Carl-Gustav Rossby, perhaps the world's leading theoretical meteorologist, who during World War II had organized the training of US military meteorologists from his base at the University of Chicago. Rossby provided von Neumann with theoretical background for weather simulations, laying out the known mathematics and defining the unsolved problems. Within a few months the Weather Bureau, together with the Navy and Air Force weather services, granted funds for a Meteorology Project at the Institute for Advanced Studies to develop computer forecast methods. The grant proposal anticipated that numerical weather prediction might lead to weather control: "finally the first step towards influencing the weather by rational, human intervention will have been made—since the effects of any hypothetical intervention will have become calculable."[24] Rossby himself soon became a proponent of cloud seeding for military weather control.[25]

The scale of the Meteorology Project's ambitions and the amount of confidence it expressed in the future of computer technology would be difficult to overstate. The project began more than two years before the first stored-program electronic computer (at Manchester University in England) would actually operate, and six years before the IAS computer would be completed. In August 1946, to build momentum for the project, von Neumann's group invited some twenty meteorologists to Princeton to discuss the possibility of numerical weather prediction.

Rossby and von Neumann assembled a group of theoretical meteorologists at the IAS, but things did not go well. Von Neumann, in high demand from many quarters during the postwar years, was unable to devote much of his own time to the Meteorology Project. Few meteorologists had the requisite training in advanced physics and mathematics. Though Rossby persuaded a number of excellent young meteorologists to join the project, none of them seemed to have the leadership abilities needed to move the vision forward. The work devolved into small individual projects without much coherence.

The turning point came in mid 1948, when Jule Charney took the project's helm. Charney had earned a PhD in 1946 at the University of California at Los Angeles, where Jacob Bjerknes—Vilhelm Bjerknes's son and the inheritor of his theoretical tradition—chaired the meteorology department. Between 1946 and 1948, Charney had studied with Rossby at Chicago and with Arnt Eliassen, another leading theoretician, at the University of Oslo. He had attended the 1946 initial meeting of the Meteorology Project, which had stimulated him to work on ways of simplifying Bjerknes's primitive equations to create a model capable of solution by numerical methods.

Charney's unique combination of skills and traits made him an ideal leader for the Meteorology Project. He possessed not only mathematical sophistication, state-of-the-art knowledge of physical theory, and a concern to match theory with observational evidence, but also youth, tremendous energy, and organizational skill. Bringing Eliassen with him from Norway, Charney immediately began work on "a step by step investigation of a series of models approximating more and more the real state of the atmosphere."[26] As he put it later, "[our] philosophy . . . has been to construct a hierarchy of atmospheric models of increasing complexity, the features of each successive model being determined by an analysis of the shortcomings of the previous model."[27] Climbing the "hierarchy of models," Charney and his team hoped, would lead directly to the goal: operational forecasting by numerical process.

**The ENIAC Experimental Forecasts**

By late 1949 the Meteorology Project was ready to carry out the first computerized weather forecast (actually a "hindcast" of weather that had already occurred). Von Neumann secured use of the ENIAC, now modified to store programs internally and installed at the Army's Aberdeen Proving Ground in Maryland. He also helped devise methods for programming Charney's model on the machine.

The ENIAC's very limited memory and low speed put an extreme premium on simplicity in the calculations. For this reason, a simplifying technique Charney had developed previously, the "quasi-geostrophic" model, proved crucial to this effort's success. "Geostrophic" winds move parallel to the isobars (lines of equal pressure) surrounding a low-pressure zone. The word names a balance between two forces: the pressure-gradient force, which produces horizontal winds moving directly toward the low-pressure center (perpendicular to the isobars), and the Coriolis force generated by the planet's rotation, which deflects wind motion to the right in the northern hemisphere and to the left in the southern. Charney's model assumed that these two forces approximately cancel each other, making large-scale wind motion "quasi-geostrophic." This assumption permitted him to reduce the three equations of motion to a single one with pressure as the only dependent variable. This had the salutary effect of eliminating from the calculations high-frequency atmospheric motions, such as sound waves, which had no meteorological importance but vastly magnified the computational task. The ENIAC possessed only 10 words of read/write memory and another 600 words of read-only memory, and it could perform only about 400 multiplications per second. This was so slow that the machine sometimes produced chugging sounds as it churned through repetitive processes. Only a much-simplified weather model could be run on such a limited machine.[28]

In March and April of 1950, Jule Charney, Ragnar Fjørtoft, George Platzman, Joseph Smagorinsky, and John Freeman spent five weeks at Aberdeen. Von Neumann's wife, Klara, taught the team to code for the ENIAC and checked the final program. Von Neumann himself rarely appeared, but called in frequently by telephone. Working around the clock for 33 days, the team at Aberdeen carried out two 12-hour and four 24-hour retrospective forecasts. A second ENIAC expedition took place a year later, but the group never published its results.[29]

The Meteorology Project's model used a two-dimensional 15×18 grid (figure 6.2)—270 gridpoints, spaced 736 km apart—covering North America and much of the surrounding oceans. The coarse grid permitted them to use a three-hour time step. The chosen forecast level was 500 millibars, roughly corresponding to an altitude about 6 km above the surface.[30] This choice reflected the relatively recent shift from surface to upper-air analysis as the principal basis of forecasting. By the 1920s the Bergen School had introduced the notion of a "steering line" guiding air masses. As airplane

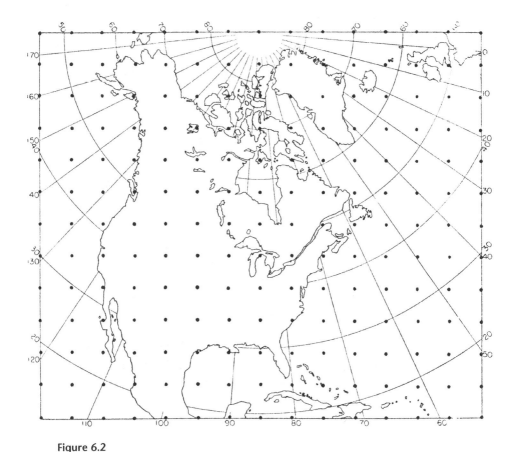

**Figure 6.2**
The finite-difference grid used in the 1950 ENIAC calculations.
*Source*: J. G. Charney et al., "Numerical Integration of the Barotropic Vorticity Equation," *Tellus* 2, no. 4 (1950), 245.

flight and radiosondes brought better knowledge of the upper air—especially the phenomenon of the jet stream and its persistence—Jacob Bjerknes, Carl-Gustav Rossby, and others developed the idea that winds high above the surface steered weather below. Under this theory, if you knew the velocity and direction of winds at the "steering level," you could predict where weather would move. Strong differences of opinion remained about where the steering level was located and even about whether it existed at all, but the idea had shown good results in practice.[31]

Beginning in 1939, Rossby, working with MIT synopticians engaged in the first detailed observational studies of the atmosphere's vertical structure, developed the mathematics of very long atmospheric waves—wavelengths of several thousand kilometers, now known as Rossby waves—and demonstrated their importance for the large-scale features of the planetary circulation.[32] After World War II, Rossby and others developed a concept of "group velocity" that explained the physical relationship between these slow or stationary long waves and the much faster propagation of energy at shorter wavelengths. They estimated the speed of this propagation at about 25°–30° of longitude (roughly 1500–3000 km, depending on latitude) per day. Under Charney's quasi-geostrophic assumption, winds could be determined from the pressure field. Thus, in principle a pressure forecast for the "steering level" could be turned into a weather forecast for the surface. Charney's group settled on the bartotropic vorticity equation[33] for the ENIAC forecasts, solving for the height of the 500-millibar pressure surface. They chose 500 mb in part because Air Force weather reconnaissance aircraft already flew at that altitude. As a military-supported group, the Meteorology Project had ready access to these observations.[34]

Figure 6.3 shows the relationship between one forecast and the observed situation after 24 hours. Though inaccurate, the computed results clearly bore a substantial resemblance to the observed outcomes. This came as something of a surprise to the group. In tribute, Charney mailed copies of the forecasts to Lewis Richardson, who was still living in England at the time.[35] This early triumph encouraged the team to proceed with further experiments, and it became a crucial argument for further funding.

## Computational Friction, Data Friction, and the ENIAC Forecasts

The ENIAC forecasters were among the first people to confront the radical cost-benefit tradeoffs involved in computerization. Achieving the benefits—in this case, the ability to calculate a forecast within a reasonable amount of time—required reducing a few elegant equations to a vast

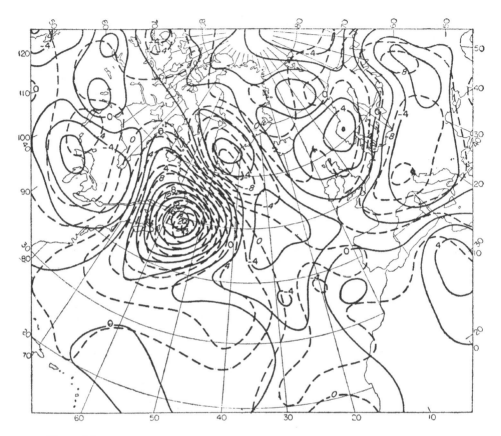

**Figure 6.3**
An ENIAC forecast of the height of the 500-mb pressure surface for January 31, 1949. Solid lines represent observed height at 24 hours. Broken lines represent the computer forecast.
*Source*: J. G. Charney et al., "Numerical Integration of the Barotropic Vorticity Equation," *Tellus* 2, no. 4 (1950), 248.

number of step-by-step instructions to the computer. Computational friction occurred when the computer or its programs required attention or caused problems. Debugging programs, dealing with hardware failures, and working through the many manual processes still required between program steps all took much longer than expected. Even the computer's calculating speed of roughly 400 multiplications per second created significant drag in a process whose goal was to forecast weather before it occurred. Data friction appeared in such areas as data quality control and interpolation between the locations of actual observations and the forecast

# Numerical Weather Prediction

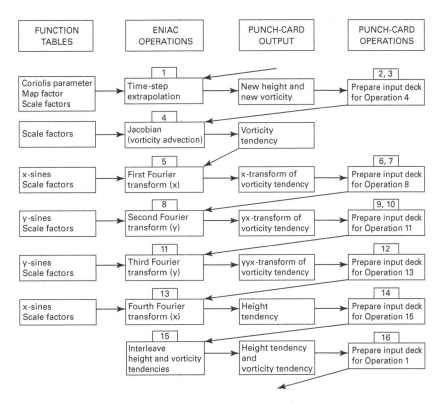

**Figure 6.4**
Flow chart of the sixteen input-output operations required for each time step of the ENIAC forecasts. Many operations involved manual handling of punch cards.
*Source*: G. W. Platzman, "The ENIAC Computations of 1950—Gateway to Numerical Weather Prediction," *Bulletin of the American Meteorological Society* 60 (1979), 309.

models' regular grids. As we will see in chapter 10, in subsequent decades meteorologists would devote at least as much energy to new data-handling techniques and new mathematical methods of analyzing observations as to the numerical forecast models themselves.

A 24-hour forecast took the ENIAC at least 24 hours to perform. Manual, card-punch, and card-reader processes involving some 100,000 IBM cards—necessitated by the ENIAC's limited internal memory—absorbed a considerable portion of this time. (See figure 6.4.) Charney's group guessed that, with practice in these techniques, the same 24-hour forecasts might be accomplished in 12 hours. They also estimated that once the much faster IAS computer was ready, 24-hour forecast computations might be accomplished in only one hour, even with a grid twice as dense as their experimental grid. "One has reason to hope," they wrote, "that Richardson's dream (1922) of advancing the computation faster than the weather may soon be realized, at least for a two-dimensional model."[36]

In 1978, one member of the Aberdeen team, George Platzman, delivered a lecture memorializing the ENIAC forecasts. Using original grant proposals, his own logbooks from the weeks at Aberdeen, and other contemporary documents, Platzman recalled the stupefyingly tedious hand coding and data entry, the computer's frequent breakdowns, and the uneasy mix of military and academic norms. At the same time, he evoked those days' tremendous excitement, calling it "an initiation into the harsh realities of an undeveloped technology that exacted much toil and torment but ultimately yielded rich rewards."[37]

The computational challenges of the ENIAC forecasts have been widely remarked, but the data friction it involved has received little attention. The ENIAC experimenters initialized their model with the Weather Bureau's 500-millibar data analyses for January 5, 30, and 31 and February 13, 1949. On these dates, huge pressure systems moved across North America, giving clearly marked phenomena for the "forecasts" to reproduce. The scale of these large weather systems made them easier to represent on the model's coarse grid, which could not resolve smaller phenomena. Finally, "insofar as possible, weather situations were chosen in which the changes of interest occurred over North America or Europe, the areas with the best data coverage." Despite this choice of dates with "best coverage," forecasts for the western coasts of the United States and Europe were "reduced somewhat in accuracy by lack of data in the Pacific and Atlantic oceans."[38]

The Weather Bureau's analysis was "accepted without modification," though the Meteorology Project team naturally knew that they contained substantial errors introduced by the analysis routines of the day. The term 'analysis' has a special meaning in meteorology, referring to the process of creating a coherent representation of weather systems from data. Analysis maps connected data points using the familiar smoothly curved isolines for pressure, temperature, and other variables. In this period, a different map was created for each pressure level; from the full set of maps, forecasters could develop a mental model of three-dimensional weather systems. Though this connect-the-dots exercise was far from random, its basis in physical theory was limited. The direction, shape, and spacing of isolines away from the data points relied heavily on heuristic principles and the forecaster's experience.

The Meteorology Project team essentially overlaid its finite-difference grid on the Weather Bureau's 500 mb analysis map and interpolated the values of variables to each gridpoint. In some cases, actual points of observation coincided closely with gridpoints. But at most locations, especially over the oceans, the data entered on the finite-difference grid came from the Weather Bureau analysis isolines or from the team's interpolation rather than from weather stations. At one point the team remarked that "too great consideration should not be given to the [ENIAC] Atlantic forecast, since data for this area were virtually non-existent."[39]

These kinds of data friction were hardly new. Many sciences had long interpolated intermediate values from known ones. In meteorology the practice dated at least to the early nineteenth century, when it arose as an almost unconscious effect of the isoline technique in weather and climate mapping. By the 1940s, numerous interpolation methods were available, including nomographs and numerical techniques, although "subjective" interpolation based on forecasters' experience remained common in synoptic meteorology.

The regularly spaced, abstract grids of computer models made data voids and interpolation errors far more salient and significant than they had been in the era of hand-drawn synoptic maps. They shifted forecasting from a fundamentally qualitative, analog principle (isolines) to a fundamentally quantitative, digital one (precise numbers at gridpoints). Experienced meteorologists knew about data voids and areas of likely error, and could compensate through judgment. But in a numerical model, substantial errors *even in individual data points* could propagate throughout a forecast. The Meteorology Project team noted that "the conventional analyst pays more attention to wind direction than to wind speed and more attention to directional smoothness of the height contours than to their spacing. . . . [We] thought that the more or less random errors introduced in this way would be smoothed out in the [ENIAC] integration. Unhappily this was not always so. . . ."[40] These problems marked new kinds of data friction specific to numerical weather prediction, and they would soon lead to new techniques of data analysis. (See chapter 10.)

At the same time, the model grids implicitly represented an ideal observing network—"ideal," that is, with respect to the model, though not necessarily to the physical reality it was supposed to reflect. Von Neumann had anticipated this consequence from the beginning: ". . . a new, rational basis will have been secured for the planning of physical measurements and of field observations in meteorology, since complete mathematical theories and the methods to test them by comparing experience with the rigorously calculated consequences of these theories will have been obtained."[41] In subsequent decades, the needs of numerical modeling would increasingly drive agendas in data collection, processing, and communication.

News of the ENIAC forecasts, published in *Tellus* in 1950, electrified the meteorological community. Within five years, weather agencies around the world had initiated numerous parallel projects. Rossby's group in Sweden quickly took the lead, becoming first in the world to introduce operational numerical weather prediction. In the United States, forecast modeling efforts began at the Air Force Cambridge Research Center in Massachusetts, the Joint Numerical Weather Prediction Unit (discussed below), and the Departments of Meteorology at the University of Chicago and the University of California at Los Angeles, as well as other institutions. By 1953, the Deutscher Wetterdienst (German Weather Service) and the Japanese Meteorological Agency had also built and tested two-level quasi-geostrophic models using Charney's approach.[42] In the United Kingdom, the Napier Shaw Laboratory of the British Meteorological Office experimented with numerical forecasting early on; however, delays in acquiring a suitable computer halted implementation of operational NWP in the UK until 1965.[43]

Soviet meteorologists developed geostrophic forecast models more or less independently of Western science.[44] By 1954 they had implemented these models on the BESM computer, also a product of quasi-independent Soviet design.[45] Although operational numerical weather prediction did not emerge in the Soviet Union until the 1960s, the 1950s did witness a remarkable series of parallel, semi-independent Soviet experiments with computer models.[46] Meteorological agencies in Belgium, Israel, Canada, and Australia also initiated or planned new NWP projects.[47]

**Climbing the Hierarchy of Models**

The period 1950–1960 saw the rapid exploitation of Charney's strategy of a "hierarchy of models." Two-dimensional models gave way to three-dimensional ones, and model grids expanded to include the entire northern hemisphere. Elaboration of meteorological theory and new mathematical methods contributed to this evolution, but technological change was almost equally important. Increasing mathematical complexity and higher model resolutions devoured computer memory and processing power, adding up very fast indeed. For example, doubling the resolution of a three-dimensional grid calls for eight times ($2^3$) the number of gridpoints, and can also require a smaller time step. Thus numerical forecasters immediately confronted the heavy drag of computational friction on their

# Numerical Weather Prediction 127

efforts, and the lockstep growth of model sophistication, model resolution, and computer power rapidly became a trope of the meteorological literature. Virtually every early article on modeling reported technical details of the experimenters' computers along with the equations represented by the models.

Most early efforts at numerical weather prediction deployed barotropic models similar to the one used in the ENIAC experiments. The next move in modeling therefore incorporated baroclinic equations, in which atmospheric density depends on temperature as well as on pressure. In baroclinic models, winds, temperature, and other variables change with altitude, as they do in the real atmosphere. Unlike the two-dimensional barotropic models, baroclinic models simulate vertical motions, which enables them to capture the cyclonic patterns—large air masses rotating around a low-pressure center—responsible for many kinds of storms, from hurricanes to mid-latitude rains.

Representing a baroclinic model on a finite-difference grid required multiple vertical levels. In 1952, with von Neumann's IAS computer operational at last, Charney's Princeton group tested a baroclinic model. Adding the third dimension exponentially increased computational requirements. On the IAS machine, a 24-hour forecast now required about 2½ hours of computer time "at full speed; however, the machine usually operated at half speed."[48] The team successfully "forecast" a major storm that had occurred on Thanksgiving Day in 1950. Since conventional techniques had failed to predict this event, the group considered the baroclinic model's success a major demonstration of the potential superiority of NWP over traditional methods.

As we have seen, the earliest NWP models covered a limited area, about the size of the North American continent. For forecasts beyond 24 hours, this proved highly problematic for two reasons, one physical and the other computational. The physical reason was that since the energy flows responsible for weather move quickly around the globe, limited-area grids could not detect the influence of weather systems moving in from beyond the grid's edges during longer forecast periods. The computational reason was equally important. Since early models did not represent the full global circulation, they suffered from computational difficulties near their edges, where they resorted to ad hoc techniques to dissipate energy. Larger model grids therefore increased the usable forecast area. As figure 6.5 illustrates, even the hemispheric grid required a boundary area in which "meteorologically meaningless" but computationally necessary model calculations could be performed.[49] Hemispheric grids remained the state of the art until

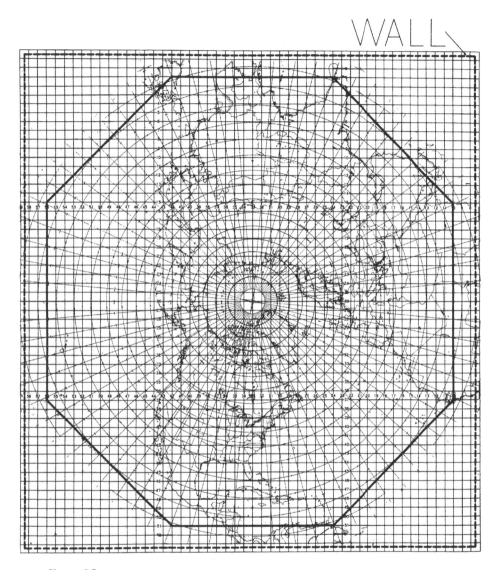

**Figure 6.5**
Quasi-hemispheric 53×57 model grid used by the JNWP Unit starting in October 1957. The octagonal region represents the forecast area, extending from 10°N to near the equator. Gridpoints were roughly 400 km apart. Gridpoints outside the octagon served as buffers to handle computational problems resulting from the artificial "wall" around the outer edge.
*Source*: F. G. Shuman and J. B. Hovermale, "An Operational Six-Layer Primitive Equation Model," *Journal of Applied Meteorology* 7, no. 4 (1968), 528.

the mid 1970s, when the first global NWP models arrived, eliminating the problem of artificial boundaries.

## Operational NWP

The first group in the world to begin routine real-time numerical weather forecasting was a Swedish-Norwegian collaboration involving the Institute of Meteorology at the University of Stockholm, the Royal Swedish Air Force Weather Service, and the University of Oslo. This precedence was no accident. Jule Charney's mentor, Carl-Gustav Rossby, had helped to initiate the IAS Meteorology Project. While the latter was starting up, however, Rossby had accepted an invitation to return to his native land to head the Swedish Institute of Meteorology. There he remained in close touch with his Princeton colleagues, following their progress on NWP with intense interest. He managed this in part through a "Scandinavian tag team" of colleagues and students, including Arnt Eliassen, Ragnar Fjørtoft, and others, who traveled to Princeton for varying periods to maintain contact between the two groups.[50]

In Sweden, Rossby—then perhaps the world's foremost theoretical meteorologist—commanded a suite of resources that included a strong team of theorists and computing facilities roughly comparable with those available in the United States. Sweden had moved quickly to develop its own digital computers, completing the BARK, a plugboard-controlled relay machine used primarily by the military, in 1950. By 1953, the BESK, a fully electronic computer modeled on the IAS machine and the British EDSAC, had replaced the BARK. While a number of outside groups were permitted use of the BESK, the NWP project received priority second only to military aircraft design.

According to Anders Carlsson, NWP became so central to Swedish computer research and development that the meteorologists and the BESK's sponsor, the Board of Computing Machinery, "merged into one practice, a working model similar to the one at IAS."[51] Beginning in December 1954, this group made forecasts for the North Atlantic region three times a week, using a barotropic model.[52] As in the United States, modelers were pleasantly surprised by the relative accuracy of their barotropic forecast. Though deemed successful by participants, these forecasts were discontinued briefly before resuming on a permanent, daily basis. The more famous American effort did not match the Swedish model's predictive skill (a measure of forecast accuracy, discussed below) until 1958.

In the United States, von Neumann, Charney, and others convinced the Weather Bureau, the Air Force, and the Navy to establish a Joint Numerical Weather Prediction (JNWP) Unit. The JNWP Unit opened in 1954 in Suitland, Maryland, under George Cressman, and began routine real-time weather forecasting in May 1955. Establishing this project required significant institutional and financial commitments. Annual budgets for the project were projected at over $415,000 in 1953 (equivalent to about $3 million today). Rental fees for the project's IBM 701 computer devoured nearly half of this amount. Political and financial backing from the military services was therefore essential. The Air Force and the Navy contributed most of the money. The Weather Bureau, under budgetary pressure from its parent Commerce Department, closed some twenty field stations in order to come up with its share of this commitment.[53]

The JNWP Unit began routine forecasting with a three-level baroclinic model first developed by Jule Charney's group in 1954. Early experience with this model proved disappointing. According to George Cressman, the model's results "were not nearly as good as the earlier tests had indicated. It had not occurred to us that we could have success with [the] very difficult baroclinic events [studied during testing] and then have severe problems with many routine, day-by-day weather situations."[54]

The Unit retreated to a two-layer model, which also failed to produce acceptable results. Finally it reverted to a simple barotropic model based on the one used in the original ENIAC forecasts. This model still did not provide a useful level of forecast skill, but its simplicity enabled researchers to diagnose the reasons for its failures while working to improve it. Work continued, in parallel, on a better three-level baroclinic model. Reintroduced in 1962, this model initially offered only slight improvements over the barotropic forecast.

**Predictive Skill in Early NWP**

How good were early NWP models? The answer to this question is more complex than one might imagine. To understand that answer, we must first note just how little forecasts had improved since the early days of "weather telegraphy." At first, simply extending the observing network provided the chief means of improvement. However, empirically based synoptic methods soon reached their limits. Hurd Willett of the Massachusetts Institute of Technology, writing around 1950, bemoaned the absence of noticeable improvement: ". . . probably there is no other field of applied science in which so much money has been spent to effect so little real progress as in weather forecasting. . . . In spite of . . . [the] great expansion of forecasting activity, there has been little or no real progress made during the past forty years in the verification skill" of basic surface forecasts.[55] Richard Reed later observed that the introduction of frontal analysis in the 1920s and of upper-air analysis in the late 1930s "should have resulted in some increase in predictive ability, and likely it did. However, the increase must have been small, since it is hard to find factual support for this belief, and there are even those who question whether any improvement took place at all. . . ."[56]

In this context, any skill enhancement at all would certainly have been cause for celebration. Three years after operations began, the barotropic model first began to deliver measurable improvement over "subjective" methods.

Figure 6.6 illustrates the steady improvement of forecast skill at the US National Meteorological Center, beginning with the introduction of the

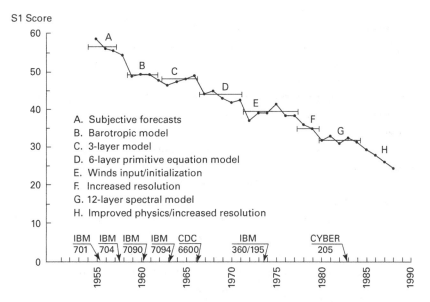

**Figure 6.6**
$S_1$ skill scores for US National Meteorological Center 36-hour 500-mb geopotential height forecast for North America, 1955–1988. On the $S_1$ scale, a standard measure of predictive skill, a score of 20 represents a perfect forecast; a score of 70 represents a useless one.
*Source*: F. G. Shuman, "History of Numerical Weather Prediction at the National Meteorological Center," *Weather and Forecasting* 4 (1989), 288.

barotropic model in 1958. The 500-millibar height field shown in the figure—the same variable predicted by the experimental ENIAC forecasts—is not in itself a weather forecast. As we have seen, winds at this altitude guide, but do not completely determine, weather on the ground. In practice, National Meteorological Center forecasters used NWP model products as just one of several inputs to their surface forecasts. *Surface-level* NWP did not begin to outperform "subjective" forecasts until about 1962.

Even in 1989, when 95 percent of forecast products in the United States were automatically generated NWP outputs, "the remaining 5 percent consist[ed] primarily of predictions of weather itself: rain, snow, heavy precipitation, etc. These remain[ed] intractable to automatic methods alone, although the analysts producing them rely heavily on numerical weather predictions of circulation, temperature, humidity, and other parameters. . . ."[57] In 1995, a study concluded that experienced human forecasters still retained a small but significant skill advantage over numerical models alone, although this advantage "continues to diminish and now largely reflects the human ability to recognize relatively infrequent (about 10 percent of the cases studied) departures" from typical norms.[58] Phaedra Daipha's dissertation on US Weather Service regional forecast centers showed that as recently as 2007 forecasters often altered raw model predictions when composing their final forecasts. Indeed, Daipha points out, "providing a good forecast in the eyes of the NWS bureaucracy literally presumes 'beating the models,'" so forecasters must work against computer models even as they work with them.[59]

The history of NWP is often presented as one of continuous success and steady progress. Indeed, many scientists with whom I spoke while researching this book expressed surprise or outright skepticism when I noted that it took several decades for computer models to approach human forecast skill. Yet, as we have seen, climbing the hierarchy of models did *not* lead instantaneously to better forecasts; in fact, initially the opposite was true. Simple barotropic models remained extremely popular with working forecasters well into the 1970s, long after baroclinic and primitive-equation model forecasts had become routinely available.

Why, then, did the national weather services of most developed countries commit themselves so rapidly and completely to this new technological paradigm? First, older techniques had reached their limits; no one put forward any competing vision for major progress in forecast quality and scope. Second, it marked a generational change. As meteorology's scientific sophistication increased, and as the field became professionalized during and after World War II, consensus had developed around the desirability of grounding forecasting in physical theory. Roger Turner has argued that a small group of theoretically savvy meteorologists, led most notably by Carl-Gustav Rossby and Francis Reichelderfer, "actively constructed" this consensus around what Turner calls "universal meteorology": ". . . between the 1920s and the 1940s, Rossby, Reichelderfer and their allies designed the institutions, established the curriculum, and cultivated the values that guided the weather cadets trained during World War II."[60] This new cadre of theorists stood ready, on both sides of the Atlantic, to tackle the risky physics and thorny mathematics of numerical modeling. Finally, Charney, Rossby, and von Neumann together articulated a clear research and development program for NWP. As computer power increased, researchers would climb the hierarchy of models, increasing grid resolution and adding ever more realistic physics. Linking meteorological R&D to the simultaneous improvement of electronic computing allowed them to project the success of both projects forward in time. With the ENIAC as proof of concept, von Neumann staked his considerable reputation on the belief that electronic digital computing would develop rapidly and inexorably. In this regard, military support for both Swedish and American efforts proved crucial, funding not only the large research effort but also the expensive computer equipment, then still years away from commercial viability.

From the beginning, Charney, Rossby, and von Neumann framed numerical weather prediction as a plan, not a gamble. In the end it worked, of course. But 20-20 hindsight makes it easy to miss the many ways in which it might have failed. For example, as I and others have shown, between 1945 and 1955 the rapid progress of electronic digital computing was by no means assured. Funders hedged their bets, supporting analog as well as digital machines. Meanwhile, the extreme cost and high failure rates of internal random-access memory based on vacuum tubes made early machines both expensive and unreliable; without magnetic core storage, invented in 1949–1951, the cost of memory might have made progress unaffordable.[61] Further, had early attempts at numerical forecasting ended in disaster owing to errors in either the mathematics or the programming, the long-term effort might never have gotten off the ground. Even after the triumph of the ENIAC experiments, nothing could guarantee that more complex models would translate into better forecasts than the crude approximations of barotropic models, and in fact they did not (at first) do so. Nonetheless, as a research program, the hierarchy of models and the assumption of steadily increasing computer power permitted early achievements to be easily converted into credible promises of future success. This combination of factors brought the dawn of operational NWP long before the computer alone could produce results better than any experienced human forecaster.

Meteorologists of the 1950s were hardly the only people who expected better, faster results from computing than it could in fact deliver. Indeed, historians of computing observe this pattern in almost every field where visionaries promised "computer revolutions" in the modeling or automation of complex systems, including automatic language translation, speech recognition, and artificial intelligence. In virtually every case, early success with simplified systems led pioneers to trumpet extravagant claims that dramatic advances on more complicated problems lay just ahead. Yet further work invariably revealed previously unknown complexity in the phenomena. (Entire disciplines, among them cognitive psychology, owe their existence to the issues uncovered by attempts at computer simulation.) At least meteorologists, stymied for centuries in predicting the weather, cannot be accused of expecting too much simplicity in the phenomena.

## Organizational Effects of Computer Modeling in Meteorology

Measurable improvement in surface-forecast skill came slowly, yet optimism about NWP's potential remained enormous and widely shared. Manual methods seemed to have reached their limits. By the mid 1960s, routine NWP had changed meteorology more than anything since the advent of the telegraph more than 100 years earlier. As Nebeker has argued, it reunified a field long divided between practical forecasting and theoretical study. This reunification affected meteorological institutions in three important ways.

First, and most obviously, reduced computational friction smoothed the path to calculated forecasts based in physical theory. As a direct complement to this change, however, data friction increased, immediately and severely. The needs of hemispheric models for data from fully half of the planet made data voids over the oceans and in the vertical dimension much more critical concerns than they had been before. By the early 1960s, as we will see in chapter 9, some national weather services and the World Meteorological Organization responded to the emerging data gap with major efforts not only to increase data collection but also to streamline, standardize, centralize, and automate it on a global scale.

Second, computer modeling required a considerably different range of expertise than previous forecasting techniques. The ideal modeler had a strong background in physics and mathematics as well as in meteorology. Anticipating this requirement in 1952, John von Neumann pointed out that an "educational problem" blocked the path to operational NWP:

> There is an educational problem because *there are practically no people available at the present time capable of supervising and operating such a program.* Synoptic meteorologists who are capable of understanding the physical reasoning behind the numerical forecast are needed to evaluate the forecasts. . . . Mathematicians are needed to formulate the numerical aspects of the computations. During the first several years of the program the meteorological and mathematical aspects probably cannot be separated and personnel familiar with both aspects are needed. An intense educational program could conceivably produce enough people in about three years.[62]

Von Neumann did not mention—and may not have anticipated—that veritable armies of computer technicians, programmers, and other support personnel would also be required.

Academic institutions were not entirely unprepared for the changed technological paradigm. By the end of the 1930s, meteorology had begun to build a stronger base in physical theory, especially in Scandinavia, England, and Germany. American universities followed suit in the early 1940s, largely under the influence of Carl-Gustav Rossby, Jacob Bjerknes, and other Scandinavian émigrés.[63] During the formation of UCLA's Meteorology Department, one faculty member remarked that "to attempt to prepare young men for research careers in meteorology in an atmosphere not reeking with physics and mathematics would . . . be a mistake of the first magnitude."[64] Francis Reichelderfer, a former Navy officer who had studied in Bergen in the early 1930s, had brought the Bergen methods and sympathy for a physics-based approach to forecasting to the Weather Bureau when he became its director in 1938.

Nonetheless, despite the boom during World War II, when 7000–10,000 American students received meteorological training, by the end of the war few practitioners could claim the breadth and depth of knowledge required to advance NWP work.[65] For most of the 1950s, American institutions graduated only a handful of new meteorology PhDs each year, and more than half of them were produced by just three universities (MIT, the University of Chicago, and UCLA). Of these, only MIT's doctoral program predated 1940. Furthermore, those institutions ready to educate research meteorologists in the new paradigm found neither strong support nor interested graduate students. In the United States between 1950 and 1955, the total number of undergraduate and graduate degrees awarded annually in meteorology declined from a high of 206 in 1950 to a low of 107 in

1955, though it began to climb again thereafter. About half of the students studying meteorology were military personnel.[66]

In 1958 and again in 1960, committees of the US National Academy of Sciences reviewed the state of the field and concluded that it had been severely neglected. The Academy asserted that "the shortages of scientific manpower . . . are more acute in meteorology than elsewhere. The number of meteorologists graduated from universities in recent years has been inadequate to offset losses, much less to build up a strong complement of manpower."[67] It recommended funding increases of 50–100 percent for university education and research, and it backed the establishment of a National Center for Atmospheric Research. These institutional adjustments refocused the discipline on the new combination of theory and computer simulation.

Finally, computer modeling altered meteorological institutions by establishing computer power as a critical—and scarce—resource. Even more than before, national weather services now became what Bruno Latour has called "centers of calculation": points through which data must pass to become widely accepted knowledge.[68] Throughout the 1950s, computers remained expensive, unreliable, unfamiliar, and extremely difficult to program. (The first high-level programming language, FORTRAN, did not appear until 1957.) As late as 1964, advice to weather centers noted that "the typical computer used in [operational] meteorological work costs about $1.5 million"[69] plus another $60,000 a year for machine maintenance and technical personnel. On top of that, the machines would have to be replaced every three years.[70] Research facilities required even more expensive computers. In this environment, only a few "centers of calculation" could hope to keep pace with the state of the art. These price tags alone effectively prohibited most countries from participating directly in meteorology's computer revolution. Even those that could afford them found the costs too high to support widespread distribution of the new tool. And price was only one factor. Scientific programmers, skilled technical support personnel, and expertise in numerical methods all remained in short supply. The most successful institutions paired theorists with technical wizards, many of them meteorologists who had discovered an aptitude for programming. Without the latter, the crucial, highly complex translation of mathematics into computer code could not proceed. For technical support, many institutions came to rely on exceptionally close relationships with computer manufacturers.

This situation spawned an organizational pattern that persists into the present. Since the most complex meteorological computer models—then as now—always demanded the maximum available computer power, possessing the newest, fastest, best-designed "supercomputers" became not only a mark of prestige but a *sine qua non* even to participate in the fast-changing world of numerical modeling. From the beginning, and increasingly over time, modelers working outside the major centers had to beg, borrow, or share computer time. Indeed, the possibility that many researchers might share a single supercomputer motivated both the US National Center for Atmospheric Research (founded in 1960) and the European Centre for Medium-Range Weather Forecasts (planned from the late 1960s and established in 1975). For this reason, meteorology has historically been the foremost civilian consumer of supercomputer power. Only the designers of nuclear weapons have laid a greater claim to the world's most advanced computers.[71]

# 7 The Infinite Forecast

The concentration of computing resources at a few institutions probably affected no field more than it affected climatology. In the 1960s and the 1970s, this data-driven, regionally oriented, descriptive field would be transformed into a theory-driven, globally oriented discipline increasingly focused on forecasting the future. The change would be wrought not by traditional climatologists, but by scientists based in theoretical meteorology and computer programming, working at a handful of institutions endowed with what were then enormous computing resources. Unlike traditional climatologists, who searched for regularities in data from the past, this new generation of scientists sought to simulate the climate, building models from the same techniques used in numerical weather prediction. From there, they moved gradually toward simulating the entire Earth system, replicating the world in a machine.

If you can simulate the climate, you can do experiments. God-like, you can move the continents, make the sun flare up or dim, add or subtract greenhouse gases, or fill the stratosphere with dust. You can cook the Earth, or freeze it, and nobody will even complain. Then you can watch and see what happens.

For a scientist, experiments are all-important. You use them to find out what really matters in a complicated system. In a laboratory experiment, you create a simplified situation, blocking out most of the real world's complexity while retaining a few variables you can manipulate. Then you compare the outcome with an unmodified "control" that leaves those variables at their ordinary values.

In the geophysical sciences, though, the controlled-experiment strategy generally doesn't work. The system you are dealing with is just too large and too complex. You can isolate some things, such as the radiative properties of gases, in a laboratory, but to understand how those things affect the climate you need to know how they interact with everything else.

There is no "control Earth" that you can hold constant while twisting the dials on a different, experimental Earth, changing carbon dioxide or aerosols or solar input to find out how they interact, or which one affects the climate most, or how much difference a change in one variable might make.

Sometimes you get a "natural experiment," such as a major volcanic eruption that measurably cools the whole planet for a year or two. Similarly, the oceanographer Roger Revelle famously called human carbon dioxide emissions a "large scale geophysical experiment."[1,2] Yet these aren't really experiments, precisely because there is no control Earth. You don't know with certainty what *would have happened* without the eruption, or the $CO_2$ increase. You can assume that without the change things would have remained much the same, which seems like a pretty good assumption until you realize that climate varies naturally over a rather wide range. So unless the effect is very large and sudden—for example, a massive volcanic eruption or a gigantic meteor impact—you can't know for sure that you are seeing a signal and not just more noise.

Simulation modeling opened up a way out of this quandary. Only through simulation can you systematically and repeatedly test variations in the "forcings" (the variables that control the climate system). Even more important, only through modeling can you create a control—a simulated Earth with pre-industrial levels of greenhouse gases, or without the chlorofluorocarbons that erode the ozone layer, or without aerosols from fossil fuel and agricultural waste combustion—against which to analyze what is happening on the real Earth. As we saw in chapter 6, John von Neumann and others understood the significance of this power "to replace certain experimental procedures" almost immediately, although the full extent of it would take some time to dawn on anyone.[3]

A convergence of technical capabilities and theoretical understanding made climate simulation possible. By the late 1930s, theoretical meteorology had set its sights firmly on the planetary scale as the most fundamental level of explanation. Rossby and others had established a theory of large-scale circulation based on very long waves, confirming and detailing the three-cell circulatory structure described by Ferrel almost a century earlier. (See figure 2.6.) Rossby's 1941 summary of "the scientific basis of modern meteorology" already focused on the planetary circulation as the ultimate cause of weather patterns.[4] World War II military aviators, flying frequently at high altitudes, accumulated substantial experience of the polar jet stream at 50–60° latitude, near the northern boundary of the mid-latitude

Ferrel cell.[5] Rossby's long waves explained the jet stream's meanders, which in turn helped explain the movements of weather systems closer to the ground.

Therefore, models of *global* atmospheric motion occupied the pinnacle of Charney's hierarchy of models. Today such models are known as GCMs, an acronym standing interchangeably for "general circulation model" and "global circulation model."[6] GCMs represented the last step in von Neumann's meteorological research program, which would culminate in what he called the "infinite forecast."[7] By this phrase, von Neumann did not intend deterministic prediction of weather over long or "infinite" periods. Instead, he had in mind the statistically "ordinary circulation pattern" that would emerge when "atmospheric conditions . . . have become, due to the lapse of very long time intervals, causally and statistically independent of whatever initial conditions may have existed." This phrase sounds remarkably like the mathematical concept of chaos, by which minute variations in initial conditions rapidly generate extreme divergences in outcomes: a butterfly flaps its wings over Brazil and causes a tornado in Texas. In fact, the theoretical meteorologist Edward Lorenz first discovered what we now call chaos theory while working with atmospheric models in the early 1960s.[8] But in the mid 1950s, these results, and the idea of chaos itself, remained unknown. In 1955, von Neumann's "infinite forecast" expressed the widespread belief that global atmospheric flows might display predictable symmetry, stability, and/or periodicity. Research aimed at finding such predictable features remained active throughout the 1950s.[9]

But how could modelers verify an "infinite forecast"? In other words, what did meteorologists actually know *from data* about the general circulation in the mid 1950s? An image of the state of the art may be found in the work of MIT's General Circulation Project, which began in 1948 and continued for some 20 years under the leadership of Victor Starr, supported principally by the Air Force and the Atomic Energy Commission. (UCLA conducted similar, independent work under the same auspices.) The General Circulation Project collected every available data source that might reveal more details of the circulation's structure. Articles describing these data almost invariably opened with extensive caveats regarding the small quantity and poor quality of upper-air measurements then available. Nonetheless, a crude data-based picture began to emerge. One study, by Starr and Robert White (of the Travelers Insurance Corporation), included all available upper-air observations for 1950 from 75 stations along five latitude bands, each with 10–19 stations. This "involved altogether a total of 176,000 individual wind readings, 57,000 humidity readings, and 77,000 temperature readings."[10] In the mid 1950s, such a quantity of observations seemed gigantic. Yet these data remained insufficient for all but the crudest possible picture of global flows. After ten years of intensive effort, Starr would write: ". . . whether we like it or not, meteorologists have been struggling with problems of the most primitive kind concerning the motions of the atmosphere. . . . The questions at issue have not been such as relate to some fine points concerning a general scheme that is accepted as sound. Rather [they concern] whether some crucial portion of the system operates forward or in reverse."

Starr was talking about the basic mechanism of the large-scale circulatory cells. (See figure 2.7.) Did those cells produce a positive energy release, converting potential energy into kinetic energy (as in Ferrel's scheme), or a negative one, converting kinetic energy into potential energy (as General Circulation Project studies of the mid 1950s seemed to show)? Starr wrote that even this fundamental question could not yet be answered because general circulation theory remained inadequate.

> The limiting factor always has been, and will continue to be, the completeness and accuracy of the de facto pictures of the general circulation. . . . What counts are not mere tabulations of data; it is their intelligent organization according to physical laws so as to lead to physical depiction of relevant processes and schemes of motion. . . . Due no doubt to the complexity of the system considered, correct processes have thus far found their incorporation into theoretical models, almost without exception, only *after* their empirical discovery.[11]

Further details of this issue need not concern us here. What is of interest to us is Starr's conviction that the absence of sufficient data, especially in the vertical dimension, still rendered theories of the general circulation fundamentally speculative, even on very basic issues. Starr's *cri de coeur* for "intelligent organization" of data leading to "physical depiction of relevant processes and schemes of motion" would be answered through general circulation modeling.

The first experiments with GCMs came only months after the operational launch of numerical weather prediction. In these models, the limits of predictability, numerical methods, and computer power would be tested together. As Nebeker has shown, GCMs would reunify meteorology's three main strands—forecasting, dynamical-theoretical meteorology, and empirical-statistical climatology—within a single research program.[12] They would transform climatology from a statistical science oriented toward the particularity of regional climates into a theoretical science more focused on the global scale. Ultimately, they would also guide vast changes in the global data infrastructure, and they would transform the sources and the very meaning of the word 'data'.

In this chapter, I discuss the basic principles of climate simulation, focusing on general circulation models. I then tell the story of the first generation of GCMs and the modeling groups that created them, ending with a genealogy that links them to subsequent groups and models. Finally, I discuss early studies of carbon dioxide doubling (a paradigmatic experiment that later became a benchmark for climate-change studies) and the role of data friction in delaying the empirical evaluation of climate simulations.

**How Climate Models Work**

Charney's model hierarchy sketched a path from simple to complex: from two-dimensional, regional forecast models to three-dimensional hemispheric and global circulation models. As climate modeling matured, another model hierarchy emerged, this one aimed not at forecasting but at characterizing how the atmosphere and the oceans process solar energy. This section departs from the historical approach for purposes of exposition, briefly describing how climate models work before returning to the history of GCMs.

Earth is bathed in a constant flood of solar energy. The laws of physics dictate that over time the planet must remain, on average, in thermal equilibrium. In other words, it must ultimately re-radiate all the energy it receives from the sun back into space. The atmosphere's blanket of gases—primarily nitrogen, oxygen, and water vapor—absorbs much of the incoming radiation as heat. The oceans absorb solar energy directly; they also transfer heat to and from the atmosphere. With their great mass, the oceans retain much more energy than the air, functioning as a gigantic heat sink and moderating changes of temperature in the atmosphere. If Earth had no atmosphere or oceans, its average surface temperature would be about –19°C. Instead, the heat retained in the atmosphere and oceans maintains it at the current global average of about 15°C.

At the equator, Earth receives more heat than it can re-radiate to space; at the poles, it re-radiates more heat than it receives. Thus the climate system, as a thermodynamic engine, serves to transport heat from the equator toward the poles. The most fundamental climatological questions regard exactly how much heat is retained (the global average temperature), where the energy resides (in the oceans, atmosphere, land surfaces, etc.), how it is distributed, and how it circulates around the globe in the course of moving poleward.

Fundamentally, then, Earth's temperature is a matter of what climatologists call "energy balance." Modeling the climate therefore begins with the "energy budget": the relationship between incoming solar radiation and outgoing Earth radiation. Energy balance models (EBMs) use measured or calculated values for such factors as solar radiation, albedo (reflectance), and atmospheric absorption and radiation to compute a single global radiative temperature. The simplest, "zero-dimensional" EBMs treat Earth as if it were a point mass. These models involve just a few equations and can be solved by hand, with no need for computers. Energy balance can also be calculated one-dimensionally, by latitude bands or "zones," to study how the relationship changes between the equator and the poles, or two-dimensionally, including both zonal and longitudinal or "meridional" energy flows. Svante Arrhenius's calculations of carbon dioxide's effect on Earth's temperature constituted one of the earliest one-dimensional (zonal) EBMs.[13]

A second type of climate model, the radiative-convective (RC) model, focuses on vertical transfers of energy in the atmosphere through radiative and convective processes. Such models typically simulate the atmosphere's temperature profile in either one dimension (vertical) or two (vertical and zonal). When Callendar revived the carbon dioxide theory of climate change in 1938, he used a one-dimensional radiative model that divided the atmosphere into twelve vertical layers.[14]

A third type is the two-dimensional statistical-dynamical model, employed primarily to study the circulatory cells. In these models the dimensions are vertical and meridional. Schneider and Dickinson, who first described the hierarchy of climate models, identified sub-classes of EBMs and RC models, as well as several kinds of dynamical models less computationally demanding than GCMs.[15]

In contrast with the way NWP models developed, work on computerized climate models did not follow a straightforward path up the hierarchy from simple to complex. Instead, climate modeling *began* with GCMs, the most complex models of all. In the course of this work, new radiative-convective models were developed—sometimes by the same people, such as Syukuro Manabe, who also pioneered general circulation modeling at the Geophysical Fluid Dynamics Laboratory (see below).[16] Major work on new EBMs—the simplest models—followed later, in the late 1960s.[17] Work on RC models and EBMs helped GCM builders conjure up code for vertical transfers of heat, and also provided "reality checks" on GCM results.[18] Climate scientists continue to use these simpler models for many purposes, but this book will not discuss them further.[19]

General circulation models (GCMs) are three-dimensional simulations, integrated over time. The earliest such models used simplified equations, but virtually all modern GCMs are based on the atmospheric primitive equations developed by Vilhelm Bjerknes, Lewis Fry Richardson, and others, which were in turn based on the Navier-Stokes fluid dynamics equations. GCMs simulate not only the planetary heat exchange described by EBMs and RC models, but also the planetary circulation or movement of the atmosphere: patterns of flow, such as the jet streams and Hadley cells, resulting from the interaction of planetary rotation, insolation, gravitation, heat exchange, humidity, orography (surface features such as mountain ranges), sea surface temperature and friction, and many other factors. The oceans also circulate, with dynamics of their own. Transfers of energy between the oceans and the atmosphere play an enormous role in both weather and climate. For this reason, today's atmospheric GCMs (AGCMs) are typically linked to ocean GCMs in "coupled atmosphere-ocean models" (AOGCMs).

All GCMs consist of two major elements. The "dynamical core" simulates large-scale movement using the primitive equations of fluid motion. The underlying mathematics of dynamical cores are relatively well understood. However, like any equations, they may be expressed in different forms for different purposes. Further, solving the equations on a computer requires representing them numerically, and there are numerous ways to do this, each with characteristic computational benefits and drawbacks. For example, the rectangular latitude-longitude grid is only one way to represent the Earth's surface, and it is not the best one for modeling a sphere, because the convergence of longitude lines near the poles inconveniently shortens the model time step. To avoid this, modelers have experimented with many kinds of grids, including triangular, hexagonal, icosahedral, and Gaussian ones.

The second element, the "model physics," includes all the other major processes that occur in the atmosphere. Many of these processes involve transfers of heat. Others involve transfers of moisture—for example, from lakes, rivers, and oceans to the atmosphere and vice versa. In the real atmosphere these processes generally occur at scales much smaller than the model grids—all the way down to the molecular level, at which heat is absorbed and re-radiated by gases; modelers call these "sub-grid scale" processes. Model physics also includes friction between land or ocean surfaces and the air, transfers of heat between the ocean and the atmosphere, cloud formation, and many other processes. Modelers represent sub-grid-scale physics indirectly by means of parameters (mathematical functions and constants that capture the large-scale effects of smaller-scale processes without modeling them directly). Parameterizing physical processes accurately is the most difficult aspect of climate modeling and is a source of considerable scientific and political controversy.

As climate modeling evolved and spread, modelers worked to improve their simulations along three major fronts at once. First, they developed better numerical schemes for integrating model equations, reducing the amount of error incurred by approximating the solutions. Second, to resolve smaller-scale phenomena, they decreased the distance between model gridpoints. Finally, they added more physical processes to the models. Major additions to climate model physics included radiative transfer, cloud formation, ocean circulation, albedo, sulfate emissions, and particulate aerosols. In the 1980s, modelers sought to expand climate studies by adding other elements of the overall climate system, such as sea ice, vegetation, snow, and agriculture, to create "Earth system models" (ESMs). By the mid 1990s, some models included the entire carbon cycle, including uptake by plants, absorption by the oceans, and release through fossil fuel combustion and the decay of plant matter. Figure 7.1 illustrates additions to climate model physics from the mid 1970s to about 2005, based on the climate models used in the four assessment reports of the Intergovernmental Panel on Climate Change. (These include virtually all of the world's GCMs.) In chapter 13, we will explore model physics and the controversies surrounding parameterization in much more detail.

At this writing, in 2009, typical climate AGCM grid resolutions at the surface are 1–5° latitude by 4–8° longitude, translating roughly as rectangles with sides between 100 and 500 km in length. Layers of varying depth represent the vertical dimension to a height of around 20 km, with more layers at lower altitudes, where the atmosphere is denser; modern GCMs typically have 30–50 layers. Early GCMs used a Cartesian grid structure (figure 7.2), computing vertical and horizontal mass and energy transfers between grid boxes at each time step (typically 10–15 minutes). Later GCMs used spectral transform techniques to carry out some computations in "wave space" (see below). Whereas converting early models into computer programs required only a few thousand lines of FORTRAN, today's most sophisticated models contain more than a million lines of program code.

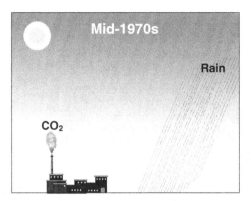
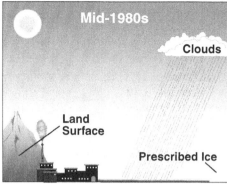

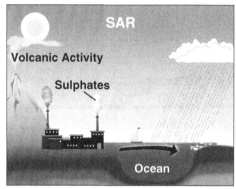
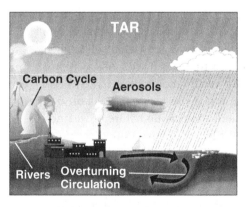
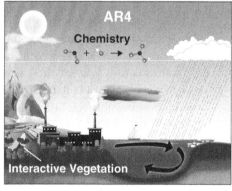

**Figure 7.1**
Processes incorporated in generations of GCMs beginning in the mid 1970s. Acronyms refer to the four assessment reports (AR) of the Intergovernmental Panel on Climate Change, released in 1990 (FAR), 1995 (SAR), 2001 (TAR), and 2007 (AR4).
*Source: Climate Change 2007: The Physical Science Basis* (Cambridge University Press, 2007). Image courtesy of IPCC.

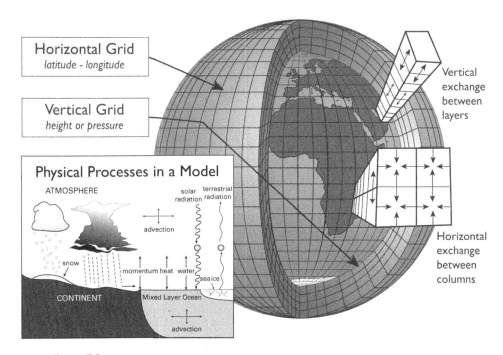

**Figure 7.2**
A schematic representation of the Cartesian grids used in finite-difference GCMs. Graphic by Courtney Ritz and Trevor Burnham.

GCMs can be used both to forecast the weather and to simulate the climate. With a forecast GCM, you want to know exactly what will happen tomorrow, or next week, so you initialize the model with observations. (This means that you set the values for each variable in every grid box using the best available data.) Then you run the model for a week or so at the highest possible resolution. Because of the chaotic nature of weather, however, forecast GCMs lose almost all their predictive skill after two to three weeks. Climate GCMs are used quite differently. You want to see the model's climate—its average behavior over long periods, not the specific weather it generates on any particular day. So you need to run the model for years, decades, or even longer. That can add up to a lot of very expensive supercomputer time. To reduce the amount of computer time each run required, early climate modelers often initialized GCMs with climatological averages—observational data. Today, however, models typically are *not* initialized with data. Instead, they are simply allowed to "spin up" from a resting state, with the various forces driving the climate—solar radiation, gravity, evaporation, the Coriolis effect, and so on—generating the model's circulation. The spin-up phase, which takes years to decades of simulated time, ends when the model comes into a relatively steady state known as "equilibrium." At equilibrium the model has developed its own "climate," with Hadley, Ferrel, and polar cells, circulatory patterns such as jet streams, seasonal changes, a vertical temperature profile, and other characteristic features.

Only when the model comes to equilibrium does the actual climate simulation begin. After that, you run the model for many simulated years (20, 100, perhaps even 1000) and measure its averages, just as you would average real weather data to characterize the real climate. For example, at each gridpoint you might take the maximum and minimum temperature for each day in a simulated January, add them and divide by 2, then add the averages for 50 simulated Januarys and divide by 50 to get the model's average temperature in January (and so on). To evaluate the simulation, you would compare its averages with real climatological data. If your model is a good one, it should more or less reproduce, for example, the seasons and the global distribution of temperature and rainfall.

No climate model, past or present, can precisely reproduce every detail of the real climate. Furthermore, since models represent so many processes and variables, it is commonplace for a GCM to reproduce one variable (e.g. temperature) quite well even as it does poorly on another variable (e.g. winds or rainfall). Because the models are complex, diagnosing the reasons for these performance differences can be extremely difficult. The goal is to get as close as is possible to a realistic climate, using a minimum of ad hoc adjustments, in order to understand why the system behaves as it does. Ultimately you want to know how changes in "forcings"—the variables that alter system behavior, such as solar output, greenhouse gases, and particulate aerosols—affect the global climate. We will return to these matters below, but for the moment we will once again follow the historical thread, beginning with the earliest effort to carry out an "infinite forecast."

The next five sections of this chapter describe pioneering efforts in climate modeling. Readers versed in meteorology and/or computer modeling should have no trouble following the discussion, but others' eyes may glaze over. If you are among the latter, I invite you to skip ahead to the section titled "The General Circulation of Circulation Models."

## The Prototype: Norman Phillips's Model

The first person to attempt a computerized general circulation model was Norman Phillips. During World War II, as an Army Air Force officer, Phillips had studied meteorology. He went on to graduate training at Chicago under George Platzman (a member of the ENIAC experiment team) and Carl-Gustav Rossby. Phillips joined the IAS Meteorology Project in 1951 and worked with Rossby's group in Stockholm during 1953 and 1954. While in Stockholm, he participated in the world's first operational numerical forecasts. Thus Phillips found himself at the very center of early NWP research. By 1954 he had already worked with his generation's foremost dynamic meteorologists, and they had encouraged him in his ambition. By mid 1955, Phillips had completed a two-layer, hemispheric, quasi-geostrophic computer model.[20]

As we saw in chapter 6, results from the initial ENIAC tests of numerical weather prediction had elated the experimenters, who hadn't expected their highly simplified model to work so well. Having worked closely with the group, Phillips shared their euphoria. The historian John Lewis notes that Phillips also paid close attention to another line of research on circulation: "dishpan" model experiments.[21] These analog models consisted of a rotating tank filled with viscous fluid. When heated at the rim and cooled in the center, the tank produced flow patterns resembling those observed in the atmosphere. Watching these flows, Phillips felt "almost forced to the conclusion that at least the gross features of the general circulation of the atmosphere can be predicted without having to specify the heating and cooling in great detail."[22]

Phillips's published study began with a discussion of the chief features of the general circulation then known from observation. These included the following:

- the vertical temperature profile of the atmosphere and the existence of the stratosphere
- the decrease of temperature between the equator and the poles, with the largest effect occurring in middle latitudes
- average wind direction and speed at different latitudes and altitudes, including such features as the trade winds and jet streams
- cyclones (low pressure systems), anticyclones (high pressure systems), and the associated fronts, major features of the circulation above the tropical latitudes
- the absence of an "organized" meridional (equator-to-pole) circulation outside the trade wind regions.[23]

Much of this knowledge had been acquired in the previous 15 years, as upper-air data networks emerged.

For theorists, the above-mentioned features and their interactions demanded explanation. For example, cyclones and anticyclones are large-scale eddies, much like those that form near rocks in a river as water flows over and around them. Globally, the circulation transports solar heat from the equator to the poles. If there is no organized meridional (longitudinal) circulation, such eddies must be involved in this heat transport process—but how? And how did they interact with the average zonal (latitudinal) winds?[24]

Phillips applied nascent NWP techniques, testing whether the geostrophic approximation could reproduce these features of the general circulation. He described his approach as a "numerical experiment" whose "ideal form would . . . be to start with an atmosphere at rest . . . and simply make a numerical forecast for a long time." He hoped that "ultimately some sort of quasi-steady state would be reached, in which the input of energy would be balanced by frictional losses, and the typical flow patterns would have some resemblance to those in the actual atmosphere."

Phillips carried out his first "numerical experiment" on the IAS computer at Princeton, a machine with just 1 kilobyte of main memory and 2 kilobytes of magnetic drum storage. To navigate between these memory constraints and his mathematical model, he chose a 17×16 finite-difference grid, with gridpoints spaced 625 km apart in the $y$ coordinate (latitude) and 375 km apart along the $x$ coordinate (longitude). This rendered a model surface that was 10,000 km by 6000 km. The $y$ axis represented roughly the actual distance from equator to pole, while the $x$ axis approximated the size of a single large eddy and was thus much smaller than Earth's circumference at the equator (about 40,000 km). To simulate the circulatory flow, eddies moving off the eastern edge of the model re-entered it on the west, making the model's topology effectively cylindrical. Two pressure levels represented the vertical dimension. The model used a one-day time step during a "spin-up" of 130 simulated days. After that, the time step was reduced to around 2 hours and the model was run for 31 simulated days.[25]

Phillips used numerous simplifying techniques to make his model tractable. For example, the model contained no moisture or clouds. It failed to replicate many characteristics of the observed circulation, but it did reproduce some fundamental features, including the easterly-westerly-easterly zonal distribution of surface winds, a jet stream, and a net transport of heat from equator to pole. It is now generally regarded as the first working numerical general circulation model.

Almost in passing, Phillips made note of what would become a crucial issue for all future GCMs. "The experiment," he wrote, "*contains empirical elements* in that the representation of certain physical effects is based on meteorological experience with the actual atmosphere, *rather than being predicted from fundamental laws of physics.*" For example, rather than attempt to simulate evaporation and condensation directly, Phillips simply fixed his model's stability factor at 80 percent of its measured value in the atmosphere. Phillips hoped that "a more complete theory of the atmospheric motions [would] eventually explain these quantities also." This tension between "empirical elements" and prediction from physical theory has persisted throughout the subsequent history of atmospheric modeling. Even as modelers eliminated some empirical elements in favor of theory-based calculation, they introduced others as models became more complex.

Phillips's model provoked enormous excitement. In the short term, it sparked action by John von Neumann, who immediately called a conference at Princeton on "Application of Numerical Integration Techniques to the Problem of the General Circulation" and composed a research proposal on "Dynamics of the General Circulation." As John Lewis points out, by framing the project as a contribution to forecasting von Neumann positioned it strategically to gain further funding from the Weather Bureau, the Air Force, and the Navy.[26] In the long term, Phillips's model inspired an entire generation of dynamical meteorologists and climate scientists, not only in the United States but also in the United Kingdom and Japan. In 1956, his paper received the first Napier Shaw Prize of the UK's Royal Meteorological Society. Akio Arakawa, later a leader in general circulation modeling, has noted more than once that his "excitement about the new developments" in studies of the general circulation "reached its climax when Phillips's paper appeared in 1956." In order to "share this excitement with other Japanese meteorologists" Arakawa, then working at the Japan Meteorological Agency, published a monograph in Japanese describing Phillips's work and its relevance to circulation theories.[27]

With Phillips's experiment as proof of concept, a new generation of theoretical meteorologists began to envision models based directly on the primitive equations. Phillips marked out the path: start with simplifying assumptions, such as barotropy and quasi-geostrophy, then eliminate them, one by one, until nothing remained but the primary physics. Putting this program into practice would require not only a lot more computer power, but also further refinement of numerical methods for solving differential equations. Perfecting primitive-equation GCMs became something of a "holy grail" for both forecasting and climatology. In the following sections, I describe the first four efforts to meet this challenge. All four were based in the United States, but Japanese émigré scientists played major roles in three of them.

## The Geophysical Fluid Dynamics Laboratory

Responding positively to von Neumann's proposal, the US Weather Bureau created a General Circulation Research Section under the direction of Joseph Smagorinsky in 1955. Smagorinsky saw his charge as completing the final step of the von Neumann–Charney computer modeling program: a three-dimensional, global, primitive-equation general circulation model of the atmosphere.[28] The General Circulation Research Section initially found laboratory space in Suitland, Maryland, near the Weather Bureau's JNWP unit. Subsequently this lab endured several changes of name and location, moving first to Washington, D.C. as the General Circulation Research Laboratory. Renamed the Geophysical Fluid Dynamics Laboratory in 1963, the lab decamped to Princeton University in 1968, where it remains. (For consistency, I will refer to all these incarnations by the lab's current acronym, GFDL.)

In 1955–56, as lab operations commenced, Smagorinsky collaborated with von Neumann, Charney, and Phillips to develop a two-level baroclinic model based on a subset of the primitive equations. Like Phillips's, this model reduced complexity by using "walls" to mark off a section of the planetary sphere. This time the walls were latitudinal instead of longitudinal, so that the model could simulate zonal circulation across about 65° of latitude, roughly two thirds of one hemisphere.[29]

Like virtually all GCMs, GFDL's early models were developed by collaborative teams consisting chiefly of dynamical meteorologists and computer programmers, many of whom, including Leith Holloway and Richard Wetherald of GFDL, had begun their careers as meteorologists. Often wrongly seen as purely mechanical work, in the late 1950s and early 1960s scientific computer programming was an arcane and difficult art. Scientists expressed GCMs as sets of variables, parameters, and systems of equations, many of them nonlinear. Solving those equations with a computer program required choosing from among multiple possible ways to translate the differential equations into finite-difference equations. These then had to be translated into computer code, which required another series of choices, almost all of which would influence both the efficiency and the accuracy of computer processing.

Writing code for GCMs required great skill and ingenuity not only with programming itself, but also with the complex mathematics of the theoretical models and with numerical and computational methods. Therefore, GCM laboratory teams often consulted, and sometimes included, mathematicians specializing in such techniques as numerical analysis and nonlinear computational stability. (For example, GFDL employed Douglas Lilly.) At first these partners in model development received recognition only as assistants, but within a few years GFDL and some other laboratories acknowledged the scientific importance of their technical contributions by listing the main programmers as co-authors.

In 1959, Smagorinsky invited Syukuro Manabe of the Tokyo NWP Group to join the General Circulation Research Laboratory. Impressed by Manabe's early publications, Smagorinsky assigned Manabe to GCM development. Smagorinsky provided Manabe with a programming staff, allowing him to focus on the mathematics of the models without writing code.[30] Initially, they worked on re-coding Smagorinsky's two-level baroclinic model for the experimental IBM STRETCH supercomputer, then still under development. When the STRETCH finally arrived in 1962, GFDL shared the machine with the Weather Bureau's numerical weather prediction unit. In an interview with me, programmer Richard Wetherald described this arrangement as "disastrous" for GFDL, which had second priority on the breakdown-prone machine. By 1965, Smagorinsky, Manabe, Holloway, Wetherald, and other collaborators had completed a nine-level, hemispheric GCM using the full set of primitive equations.[31] They began by re-coding relevant parts of the old two-level model in FORTRAN to avoid the difficult STRAP assembly language used by the IBM STRETCH.

From the beginning, GFDL took a long-range view of the circulation modeling effort. Lab members expected a slow path to realistic simulations, and they remained acutely aware of the pitfall of getting the right results for the wrong reasons. Their research strategy used GCMs to diagnose what remained poorly understood or poorly modeled, and simpler models to refine both theoretical approaches and modeling techniques, in an iterative process:

> We first constructed with considerable care, and in fact programmed, the most general of a hierarchy of models in order to uncover in some detail the body of physics needed, to determine where the obvious weaknesses were, and to give us some idea of the computational limitations we could expect. The perspective thus gained was invaluable. We then laid out a program of simplified models which can be constructed as a sub-set of the most general one. The main requirements were (1) that each model represent a physically realizable state, (2) that they could be

constructed computationally . . . , and (3) that they collectively would provide a step-by-step study of the behavior of new processes and their influence on the interactive system. Hence, many of the intermediate models in themselves may lack detailed similitude to the atmosphere but provide the insight necessary for careful and systematic scientific inquiry.[32]

This strict attention to correcting physical theory and numerical methods before seeking verisimilitude became a hallmark of the GFDL modeling approach.[33] The full primitive-equation GCM ("the most general" model) served as a conceptual framework, driving work on simpler models which led to refinements of the GCM.

Exchanges of energy between the oceans and the atmosphere are fundamental to general circulation physics. Sea water, far denser than the atmosphere, retains much more heat. Ocean currents such as the Gulf Stream absorb, release, and circulate this heat around the planet, strongly affecting global atmospheric temperatures and flows. Smagorinsky foresaw, very early, the need to couple ocean circulation models to atmospheric GCMs. In 1961 he brought the ocean modeler Kirk Bryan to the GFDL to begin this work.[34] At first, GFDL used a highly simplified one-layer ocean model, widely known as the "Manabe swamp ocean." Ultimately, however, better climate simulations would require coupling atmospheric GCMs to ocean GCMs. In 1969, Manabe and Bryan published the first results from a coupled atmosphere-ocean general circulation model (AOGCM).[35] However, this model used a highly idealized continent-ocean configuration that did not much resemble the real Earth. The first results from an AOGCM with more realistic continent-ocean configurations did not appear until 1975.[36]

By the mid 1960s, Smagorinsky had taken a leading role in planning the gigantic Global Atmospheric Research Program (GARP), a project which would continue into the 1980s. GARP occupied an increasing share of Smagorinsky's time, and Manabe emerged as the de facto leader of GFDL's GCM effort. Manabe's work style was always highly collaborative. Manabe's group was among the first to perform carbon dioxide doubling experiments with GCMs,[37] to couple atmospheric GCMs with ocean models,[38] and to perform very long runs of GCMs under carbon dioxide doubling.[39] Box 7.1 sketches the major GFDL model series through the mid 1990s.

**The UCLA Department of Meteorology**

Jacob Bjerknes founded the UCLA Department of Meteorology in 1940. He soon established a General Circulation Project as a central focus of the department's research. Yale Mintz, a Bjerknes graduate student who

**Box 7.1**
The GFDL GCM Series

> The following list, derived from interviews with various GFDL staff members in the late 1990s, describes the major GFDL models in the informal terms used at the laboratory.
>
> **MARKFORT**
>
> MARKFORT was GFDL's first "production" model. The prototype for the MARKFORT series was the original nine-level Smagorinsky-Manabe hemispheric model described in the text.[a] Used well into the 1960s, a two-level version of the model was initially run on the IBM STRETCH.
>
> **Zodiac**
>
> The Zodiac finite-difference model series was the second major GFDL GCM and its first fully global one. Used throughout the 1970s, its most important innovation was a spherical coordinate system developed by Yoshio Kurihara.[b]
>
> **Sector**
>
> Not a separate GCM but a subset of the GFDL global model series. To conserve computer time (especially for coupled atmosphere-ocean modeling), integrations were performed on a longitudinal "sector" of the globe (e.g. 60° or 120°) with a symmetry assumption for conversion to global results. The early Sector models employed highly idealized land-ocean distributions.[c]
>
> **Skyhigh**
>
> Work on Skyhigh, a GCM with high vertical resolution covering the troposphere, the stratosphere, and the mesosphere, began in 1975.[d]
>
> **GFDL Spectral Model**
>
> In the mid 1970s, GFDL imported a copy of the spectral GCM code developed by William Bourke at the Australian Numerical Meteorological Research Centre.[e] Bourke and Barrie Hunt had originally worked out the spectral modeling techniques while visiting GFDL in the early 1970s.

**Box 7.1**
(continued)

---

**Supersource**

In the late 1970s, Leith Holloway began to re-code the GFDL spectral model to add modularity and user-specifiable options. The resulting model, Supersource, remained in use at GFDL through the 1990s. Supersource physics descend from Manabe et al.'s Zodiac grid model series. Users can specify code components and options. Supersource has often been used as the atmospheric component in coupled atmosphere-ocean GCM studies.[f]

---

a. J. Smagorinsky et al., "Numerical Results from a Nine-Level General Circulation Model of the Atmosphere," *Monthly Weather Review* 93, 1965: 727–68.

b. Yoshio Kurihara, "Numerical Integration of the Primitive Equations on a Spherical Grid," *Monthly Weather Review* 93, no. 7, 1965: 399–415.

c. See e.g. S. Manabe and K. Bryan, "Climate Calculations with a Combined Ocean-Atmosphere Model," *Journal of the Atmospheric Sciences* 26, no. 4, 1969: 786–89.

d. J. D. Mahlman et al., "Simulated Response of the Atmospheric Circulation to a Large Ozone Reduction," in *Proceedings of the WMO Symposium on the Geophysical Aspects and Consequences of Changes in the Composition of the Stratosphere*, Toronto, 1978.

e. W. Bourke, "A Multi-Level Spectral Model. I. Formulation and Hemispheric Integrations," *Monthly Weather Review* 102 (1974): 687–701; T. Gordon and B. Stern, "Spectral Modeling at GFDL," in *Report of the International Symposium on Spectral Methods in Numerical Weather Prediction*, 1974; C. T. Gordon, "Verification of the GFDL Spectral Model," in *Weather Forecasting and Weather Forecasts*, ed. D. L. Williamson et al., Advanced Study Program, National Center for Atmospheric Research, 1976, volume 2.

f. S. Manabe and R. J. Stouffer, "Two Stable Equilibria of a Coupled Ocean-Atmosphere Model," *Journal of Climate* 1, 1988: 841–65; S. Manabe et al., "Response of a Coupled Ocean-Atmosphere Model to Increasing Atmospheric Carbon Dioxide," *Ambio* 23, no. 1, 1994: 44–49.

received his PhD in 1949, stayed at UCLA as Bjerknes's associate project director. Among other things, the General Circulation Project carried out extensive data analysis for a climatological atlas. An obituary later recalled Mintz's "heroic efforts . . . in the earlier phase of this project during which he orchestrated an army of student helpers and amateur programmers to feed a prodigious amount of data through paper tape to SWAC, the earliest computer on campus."[40]

In the late 1950s, Mintz began to design numerical GCMs.[41] As Smagorinsky had done, Mintz recruited a Tokyo University meteorologist, Akio Arakawa, to help him build general circulation models. Arakawa, known for his wizardry with numerical methods, was particularly interested in building robust schemes for parameterizing cumulus convection, a major but poorly understood process of vertical heat transport in the atmosphere. Beginning in 1961, Mintz and Arakawa constructed a series of increasingly sophisticated GCMs. Arakawa persuaded Mintz to pay more attention to designing model dynamics that could sustain long-term integration.[42] The first UCLA GCM, completed in 1963, was a two-level global primitive-equation model with 7° latitude by 9° longitude horizontal resolution. It included realistic land-sea distributions and surface topography. Mintz never learned to program computers; Arakawa did all the model coding. With this project completed, Arakawa returned to Japan, but Mintz persuaded him to return to UCLA permanently in 1965.

Of all the world's general circulation modeling groups, the UCLA laboratory probably had the greatest influence on others, especially in the 1960s and the 1970s. This influence resulted not only from continuing innovation, particularly in cumulus parameterization, but also from the UCLA group's exceptional openness to collaboration and sharing. Whereas GFDL was a pure-research institution, UCLA operated in the mode of an academic graduate program, with a mission that included training and knowledge diffusion. Also more typical of the academic tradition, until the 1980s the UCLA group focused primarily on model development, leaving "production" uses of the models (e.g. experimental studies) to other institutions.

Because the UCLA group emphasized advancing the state of the art rather than perfecting the models in detail, its models developed somewhat more rapidly than those of more experiment-focused GCM groups. In addition, the more open nature of the institution encouraged migration of the model to other laboratories. UCLA Department of Meteorology graduates carried the model with them to numerous other institutions, and visitors from around the world spent time at the group's laboratories. This pattern is vividly apparent in the history of the UCLA model series, described by Arakawa in a festschrift celebrating his work.[43] Box 7.2, based on Arakawa's account, my interviews with him, and model documentation, summarizes the models' characteristics and their migration to other laboratories.

## The Livermore Atmospheric Model (LAM)

In 1960, Cecil E. "Chuck" Leith began work on a GCM at the Lawrence Livermore National Laboratory in Livermore, California. Trained as a physicist, Leith became interested in atmospheric dynamics through his discussions with Joseph Knox, a meteorologist at Livermore. As Leith recalled it in a 1997 interview, Knox was there because of the lab's interest in nuclear

**Box 7.2**
The UCLA GCM Series

(N.B.: The numbering in this box follows Arakawa, "A Personal History.")

**UCLA I** The initial model, completed in 1963.

**UCLA II** When Arakawa returned from Japan in 1965, he and Mintz abandoned the UCLA prototype and began work on the first "production" GCM (UCLA II). It increased model resolution to 4° latitude by 5° longitude (although it still had only two vertical levels), and it introduced a new horizontal grid structure. In the latter half of the 1960s, IBM's Large Scale Scientific Computation Department in San Jose provided important computational assistance and wrote a manual describing the model.[a] Around 1970, Lawrence Gates, a UCLA graduate, carried the model with him to the RAND Corporation, where he deployed it in studies sponsored by the Advanced Research Projects Agency of the US Department of Defense. The RAND version of the model eventually migrated to Oregon State University.[b] A three-level version of UCLA II, developed around 1968, soon traveled to three NASA laboratories. In 1972, the Goddard Institute for Space Studies (GISS) adopted the model. Later in the 1970s, it migrated to the Goddard Laboratory for Atmospheric Sciences and the Goddard Laboratory for Atmospheres.

**UCLA III** Extended vertical resolution to six or twelve levels and incorporated the Arakawa/Lamb "C" finite-difference horizontal grid, used in all subsequent UCLA models. Two versions of this model, with slightly different sets of prognostic variables, were built in the mid 1970s. One version was exported to the US Naval Environment Prediction Research Facility and the Fleet Numerical Oceanographic Center, both in Monterey, California, where it evolved into an operational forecasting system called NOGAPS.[c]

**Box 7.2**
(continued)

> It also traveled to the Meteorological Research Institute in Tsukuba, Japan, where it continues to be used in a wide variety of forecasting and climate studies.
>
> **UCLA IV** Begun in the late 1970s, UCLA IV employed a new vertical coordinate system which used the top of the planetary boundary layer as a coordinate surface and extended vertical resolution to 15 layers. This model was adopted by the Navy research centers mentioned above, as well as by the Goddard Laboratory for Atmospheres. Versions also made their way to Lawrence Livermore National Laboratory and the Central Weather Bureau of the Republic of China. In 1988, David Randall, a former student of Arakawa's, introduced the model at Colorado State University.
>
> **UCLA V** An improved version of UCLA III with up to 29 vertical levels. Begun around 1990, UCLA V included new schemes for radiation, cloud prediction, cumulus convection, and other parameters.

a. W. E. Langlois and H. C. W. Kwok, "Description of the Mintz-Arakawa Numerical General Circulation Model," in *Numerical Simulation of Weather and Climate* (IBM-UCLA technical report, 1969); "Numerical Simulation of Weather and Climate Part II: Computational Aspects," n.d.

b. D. Randall, "Colorado State University General Circulation Model: Introduction," kiwi.atmos.colostate.edu.

c. T. F. Hogan and T. E. Rosmond, "The Description of the Navy Operational Global Atmospheric Prediction System's Spectral Forecast Model," *Monthly Weather Review* 119, no. 8, 1991: 1786–815.

fallout. When Leith expressed interest in numerical simulations of the atmosphere, Knox invited him to go to MIT to visit Charney and Norman Phillips. They made the trip together in the spring of 1960.[44]

With Charney's encouragement and the blessing of the Livermore Laboratory's director, Edward Teller, who had long been interested in weather modification, Leith spent the summer of 1960 at the Swedish Institute of Meteorology, studying the literature on global circulation and learning about numerical simulation methods. By the end of the summer he had coded a five-level GCM for Livermore's newest computer, the Livermore Automatic Research Calculator (LARC), due to be delivered in the fall of 1960. Leith wrote the code in Stockholm, basing it on the manual for the new machine. His initial model, like Smagorinsky's first effort, covered only the northern hemisphere, with a "slippery wall" at

60°N. It had five vertical levels, a 5°×5° horizontal grid, and a five-minute time step. At John von Neumann's suggestion, Leith introduced an artificially high viscosity to damp the effects of small-scale atmospheric waves. (This caused serious problems and helped to stimulate Leith's career-long interest in turbulence.) By the end of 1960, Leith's five-level simulation was running on the LARC. The model came to be known as the LAM (for Leith Atmospheric Model or Livermore Atmospheric Model).

The fate of Leith's work contrasts starkly with that of the GFDL and UCLA modeling efforts, illustrating significant institutional contrasts of Cold War–era American science. Livermore and GFDL were both government laboratories, but whereas GFDL served a pure research function in a civilian environment, most of Livermore's work was on secret military projects related to the design of atomic weapons. Although unclassified work such as the LAM *could* have been published in the open literature, Livermore's culture of secrecy did not reward such publication. Leith's other work (like most Livermore research) appeared only in classified reports circulated internally. As a result, Leith's first nonclassified publication did not appear until 1965, a long delay in a fast-evolving field.[45] Around that time, Leith abandoned work on the model. As a result, the LAM's mathematics and computer code did not have much direct effect on GCM development.

However, by 1963 Leith had already presented his model in numerous talks. These talks made a deep impression on their audiences because Leith often screened an animated film of LAM results. In that era, with the field of computer graphics in its earliest infancy, most output took the form of printouts. Leith collaborated with a Hollywood company called Pacific Title to turn these printouts into a film, using color tints to highlight particular features.

The film showed a hemispheric projection, with the North Pole in the center of the frame. Sixty days' worth of simulation results appeared. One second's worth of film equaled one simulated model day. In the first few minutes, atmospheric features appeared one by one; in the final segment, all the features appeared superimposed. Diurnal tides—the 12-hourly rise and fall of the atmosphere under the influence of solar heating and gravity—were among the features that showed up quite clearly. "I remember I drove places to give talks about what I was doing," Leith recalled, "and people would be watching the film with interest. I remember once somebody came up to me afterwards and said, 'I'm from Israel, and I noticed the remarkably realistic way in which you've got these things tracking across Israel.' And of course I don't know what's going on in Israel, I had never paid any attention to it, but he had spotted the fact that . . . it was doing the right sort of things as far as he was concerned."[46] In interviews with me, several pioneering climate modelers who had seen Leith's talks in the 1960s mentioned the elation they experienced on watching his movie. For the first time they could actually witness the dynamic activity of their models, rather than having to imagine it.

Leith became increasingly interested in statistical modeling of turbulence, one of the many points of commonality between atmospheric science and nuclear weapons design. Although Leith ceased work on his own GCM, he became involved with the nascent GCM group at the National Center for Atmospheric Research. In the summers of 1965 and 1966, as the NCAR team awaited delivery of a new CDC 3600 computer, Leith offered use of Livermore's computer to Warren Washington of NCAR, who described Leith to me as "a kind of father figure" for his own modeling work. Leith began to visit NCAR frequently, contributing especially in the area of mathematical methods.[47] In 1968, Leith left Livermore permanently to join NCAR, where he played instrumental roles in several climate modeling projects, both as an administrator and as a turbulence specialist.

### The National Center for Atmospheric Research

The US National Center for Atmospheric Research, established in 1960, initiated its own GCM effort in 1964 under Akira Kasahara and Warren Washington. Kasahara—like Syukuro Manabe, a veteran of the Tokyo University meteorology department under Shigekata Syono—had arrived in the United States in 1954 to join the Department of Oceanography and Meteorology at Texas A&M. He moved to NCAR in 1963 following a year at the Courant Institute of Mathematical Sciences, where numerical analysis for nonlinear equations, with particular attention to shock waves, was a specialty. Washington, still writing his PhD thesis on objective analysis techniques in numerical weather prediction, arrived soon afterward. NCAR's founding director, Walter Orr Roberts, informed them that he wanted to mount a global circulation modeling effort. Roberts told Washington: "I really want you to work about half the time on helping us get started on modeling. The other half, work on what you think is important."[48]

Kasahara and Washington began by studying the advantages and drawbacks of the three existing GCM efforts (GFDL, UCLA, and LAM), which both already knew well. They also reconsidered Lewis Richardson's 1922 effort, looking at it "more carefully, because we thought that Richardson's model didn't have all the problems [of] the sigma-type models."[49] They made an early decision to "go global," and they adopted a z-coordinate system, in which the vertical coordinate is height rather than a pressure-related quantity. This allowed their model to function more realistically with orography (mountain ranges).

The Kasahara-Washington modeling group focused a great deal of attention on numerical schemes for finite-difference approximations. In addition, much work was done on problems of computational error arising from truncation. Two major GCM series were eventually constructed; these are summarized in boxes 7.3 and 7.4.

**Box 7.3**
The Kasahara-Washington GCM Series

> **NCAR 1** A two-layer global model with a 5° horizontal resolution.[a]
> **NCAR 2** Completed around 1970, this version added considerable flexibility. The basic model had a 5° horizontal resolution and six vertical layers, but it could also be run at resolutions as fine as 0.625° horizontal over a limited domain, with up to 24 vertical layers.[b]
> **NCAR 3** Around 1972, NCAR began work on a third-generation GCM incorporating improved finite-difference schemes. This model also allowed multiple resolutions, including a user-specifiable vertical increment. Although under evaluation as early as 1975, this model did not see "production" use until the end of the decade.[c]

a. A. Kasahara and W. M. Washington, "NCAR Global General Circulation Model of the Atmosphere," *Monthly Weather Review* 95, no. 7, 1967: 389–402.
b. J. E. Oliger et al., *Description of NCAR Global Circulation Model*, National Center for Atmospheric Research, 1970; A. Kasahara and W. M. Washington, "General Circulation Experiments with a Six-Layer NCAR Model, Including Orography, Cloudiness and Surface Temperature Calculations," *Journal of the Atmospheric Sciences* 28, no. 5, 1971: 657–701; A. Kasahara et al., "Simulation Experiments with a 12-Layer Stratospheric Global Circulation Model. I. Dynamical Effect of the Earth's Orography and Thermal Influence of Continentality," *Journal of Atmospheric Sciences* 30, no. 7, 1973: 1229–51.
c. W. M. Washington et al., "Preliminary Atmospheric Simulation with the Third-Generation NCAR General Circulation Model: January and July," in *Report of the JOC Conference on Climate Models: Performance, Intercomparison, and Sensitivity Studies*, ed. W. Lawrence, WMO/ICSU Joint Organizing Committee and Global Atmospheric Research Programme, 1979.

**Box 7.4**
The NCAR Community Climate Model Series

> **CCM-0A** The initial version of the Community Climate Model was based on the spectral model of the Australian Numerical Meteorological Research Centre.[a] A member of the ANMRC team, Kamal Puri, brought the model to NCAR during an extended visit. Later it was extensively revised.
>
> **CCM-0B** A second version of the Community Climate Model was developed in 1981. This version combined medium-range and long-range global forecasting (from three days to two weeks) and climate simulation in a single set of model codes. A modular design permitted flexible choices of resolution and other features. Initial code for CCM-0B came from an early version of the ECMWF model. Physical parameterizations (including the radiation and cloud routines of Ramanathan) and numerical approximations were added from CCM-0A.[b] Energy balance and flux prescriptions similar to GFDL models were used. The vertical and temporal finite-difference schemes were derived from the Australian spectral model that was also the basis for CCM-0A.[c]
>
> **CCM-1, 2, and 3** Evolved from CCM-0B, coming into use in 1987. The primary differences were changed parameterizations, new horizontal and vertical diffusion schemes, and changes to moisture adjustment and condensation schemes. CCM versions 2 and 3 were developed in the early 1990s.
>
> **CCSM** In 1994, NCAR initiated work on a Community Climate System Model (CCSM), coupling atmosphere, ocean, land surface, and sea ice models. Greater efforts were made to involve NCAR's extended community of users in model development.

a. W. Bourke, "A Multi-Level Spectral Model. I. Formulation and Hemispheric Integrations," *Monthly Weather Review* 102 (1974): 687–701; Bourke et al., "Global Modeling of Atmospheric Flow by Spectral Methods," in *General Circulation Models of the Atmosphere*, ed. J. Chang (Academic Press, 1977); B. J. McAvaney et al., "A Global Spectral Model for Simulation of the General Circulation," *Journal of Atmospheric Sciences* 35, no. 9 (1978): 1557–83.
b. V. Ramanathan et al., "The Response of a Spectral General Circulation Model to Refinements in Radiative Processes," *Journal of Atmospheric Sciences* 40, 1983: 605–30.
c. D. L. Williamson, ed., *Report of the Second Workshop on the Community Climate Model* (NCAR, 1988).

Along with most other major GCM groups, after 1975 NCAR gradually abandoned its finite-difference models in favor of spectral models. Spectral models ameliorate one of the most difficult problems in Earth system modeling: representing wave motion on a sphere. All early weather and climate models used rectangular latitude-longitude grids. However, these grids encountered difficulties similar to the well-known distortions of Mercator and other world map projections, especially at higher latitudes, where the distance between longitude lines shrinks—eventually to zero at the poles. Many solutions were devised, including stereographic projections and spherical, hexagonal, and icosahedral grids, but none proved ideal.[50] Spectral transform methods offered a useful alternative to grid-point schemes.

The technical details of spectral mathematics lie beyond the scope of this book, but their essence may be expressed briefly as follows. Atmospheric motion can be conceived as numerous waves of varying frequency and amplitude; the superimposition of these waves upon one another produces highly complex patterns. To visualize this, imagine dropping a stone into a completely still pond, then dropping several other stones of various sizes into the same pond at other points. The first stone produces a single set of simple, concentric waves, but as other stones fall the interacting waves rapidly create a complex surface with much more complicated patterns.

Horizontal atmospheric motion can be described as a set of interacting waves, like ripples crossing each other on the surface of a pond. These can be analyzed in "wave space." Mathematical techniques—Fourier transforms, reverse transforms, and others—convert model variables back and forth between physical space (the familiar Cartesian grid) and wave space, a mathematical construct that is difficult to visualize (figure 7.3). Spectral techniques can also handle the vertical dimension (interaction between horizontal model layers), but usually modelers retain physical grids for this part of the analysis.

The idea of using spectral methods to analyze planetary atmospheric waves was explored as early as 1954.[51] Their mathematical advantages rapidly became obvious. As noted above, all grid-point schemes inevitably face problems related to Earth's spherical shape. Since they do not compute wave interaction in physical space, spectral methods avoid these difficulties. Other mathematical properties of this technique, including the simplification of certain nonlinear partial differential equations and the reduction of nonlinear computational instability, also offer substantial advantages over finite-difference schemes. By the latter half of the 1950s experiments with two-dimensional spectral models had begun, but three-

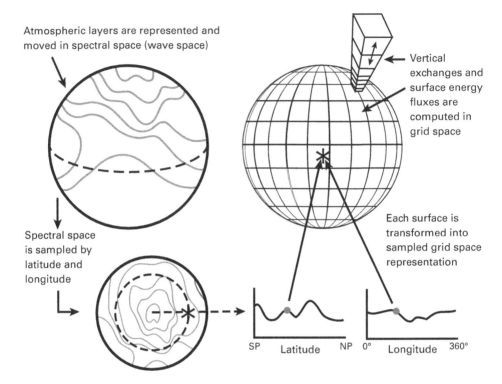

**Figure 7.3**
Spectral models handle horizontal motion in mathematical "wave space" and vertical motions in physical grid space. Gridpoint values are computed by sampling the wave space.
Graphic by Courtney Ritz and Trevor Burnham.

dimensional models posed greater difficulties.[52] Initially the method required far more calculation than finite-difference techniques, so it was not favored for use in GCMs. However, by 1973 spectral methods had become more efficient than finite-difference schemes, since computers had gotten faster and algorithms had been improved.[53] In the early 1970s both NWP and GCM developers began to adopt these techniques.

One spectral model, NCAR's Community Climate Model (CCM) series, has been especially important because a relatively large number of researchers were able to use it. As its name implies, NCAR intended the CCM to serve not only modelers working at NCAR but also its large constituency at universities affiliated with its parent organization, the University Corporation for Atmospheric Research. The CCM's construction was highly collaborative and international. The first two model versions, CCM-0A and CCM-0B, were based on, respectively, a spectral model constructed at the Australian Numerical Meteorological Research Center model and an early version of the European Centre for Medium-Range Weather Forecasts model. Several other groups adopted versions of the CCM in the late 1980s; NCAR's strong focus on documentation and modularity made this relatively easy. In an early manifestation of what might today be called "open source" development, NCAR made user manuals and code documentation available for all elements of the models beginning with CCM-0B.

## The General Circulation of Circulation Models

Modelers, dynamical cores, model physics, numerical methods, and computer code soon began to circulate around the world, like ripples moving outward from the three pioneering climate modeling groups (the Geophysical Fluid Dynamics Laboratory, the US National Center for Atmospheric Research, and UCLA's Department of Meteorology). By the early 1970s, a large number of institutions had established new programs in general circulation modeling. Computer power had grown to the point that major weather forecast centers began moving to hemispheric, and later global, GCMs for operational use. Figure 7.4 illustrates the genealogy of atmospheric GCMs from Phillips's prototype up to the early 1990s.

Figure 7.4 is merely a sketch, of course. It tells only part of the story, it includes only three weather forecast GCMs (NMC, ECMWF, and UKMO), and it does not capture any details of relationships among models. In some cases, one lab imported another's computer code with only minor changes; in others, one lab adopted another's *mathematical* model but programmed it for a different computer. Often labs imported only one part of an existing model (for example, a dynamical core, a grid scheme, or a cloud parameterization) and built the rest themselves. An exhaustive account of modeling groups, model variations, and the relationships among them after the 1960s would require a volume of its own. Here I will simply highlight certain interesting features of the circulation of circulation models (so to speak), as they traveled from lab to lab around the world.

The Australian GCM story, and especially Australia's contribution to spectral modeling, is perhaps the most dramatic case in point. Several labs had briefly considered spectral transform techniques in the 1950s, but computer capacity and numerical methods limitations rendered them impractical until the late 1960s, when André Robert began developing spectral models at the Canadian Meteorological Centre (CMC). In 1970–71,

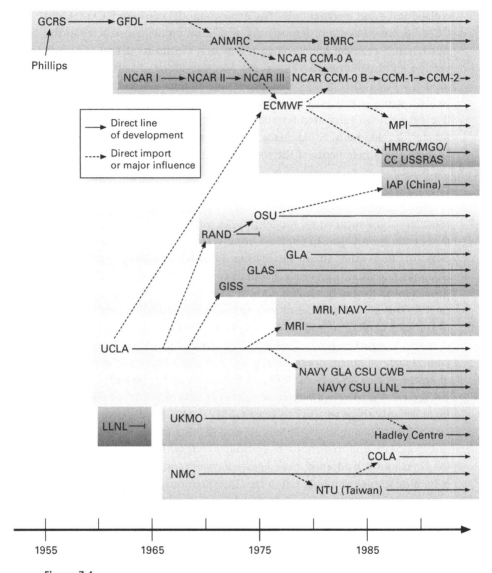

**Figure 7.4**
The AGCM family tree. Revised and expanded from P. N. Edwards, "A Brief History of Atmospheric General Circulation Modeling," in *General Circulation Development, Past Present and Future*, ed. D. A. Randall (Academic Press, 2000).
Graphic by Trevor Burnham.

seeking to deepen its expertise in weather modeling, Australia's new Commonwealth Meteorology Research Centre (CMRC) seconded its modeler William Bourke to McGill University and the CMC. At the CMC, Bourke learned Robert's spectral techniques.[54] Bourke recalled the ninth Stanstead Seminar, held in Montreal in 1971,[55] as a watershed for spectral modeling techniques.[56]

A couple of years earlier, in 1969, another Australian researcher, Doug Gauntlett, had visited GFDL and the US National Meteorological Center. Gauntlett had returned to Australia with code for the GFDL N30 nine-level hemispheric GCM (a finite-difference model). The CMRC adapted this to the southern hemisphere in hopes of using it for NWP, but it took four times as long to make a prediction as the existing baroclinic model, so it was never used operationally.[57] When Bourke returned to Australia in 1971, he joined with Gauntlett and others to build the CMRC's seven-level spectral GCM, which replaced its gridpoint model for operational forecasting in 1976.[58] Bourke, Bryant McAvaney, Kamal Puri, and Robert Thurling also developed the first global spectral GCM.[59] McAvaney recalled that this work was mostly completed in 1974–1975, but that "it took a while to publish."[60] Later, Puri carried code for the Australian spectral model to NCAR, where it became the principal basis of CCM0-A.

McAvaney's six-month visit to the Soviet Union in 1977 also contributed to spectral modeling. Though the Soviet modelers lacked the computer capacity to implement their models operationally, on the level of theory and mathematical methods "they were as good, definitely as good as anybody," McAvaney recalled when I interviewed him in 2001. "Certainly their numerical techniques were up there with anything . . . in the US or anything that we'd done. . . . They were using a spectral approach for diagnostic work, and had done a lot of spectral decomposition of atmospheric fields and particularly interaction between atmospheric fields."[61]

A similar story surrounds GCM development at the European Centre for Medium-Range Weather Forecasts. Conceived in the latter half of the 1960s, the ECMWF opened its doors at Shinfield Park, England, in 1974, with Aksel Wiin-Nielsen as director. Wiin-Nielsen, a Dane, had joined Rossby's International Meteorological Institute in Stockholm in 1955. He moved to the JNWP Unit in the United States in 1959 and, soon afterward, worked at the US National Center for Atmospheric Research during its earliest years. His career thus embodied the international movement of people and ideas typical of meteorology during this period.

From the outset, the ECMWF's goal was to forecast weather for up to 10 days: the "medium range." This would require a global GCM. The US

National Meteorological Center had introduced a hemispheric forecast GCM several years earlier, but after periods beyond a few days this and other hemispheric models developed problems related to the artificial handling of computations at their equatorial "edge." Using a global model would avoid such problems. Rather than build the ECMWF's global forecast GCM from scratch, Wiin-Nielsen contacted Smagorinsky at GFDL and Arakawa and Mintz at UCLA, requesting copies of their models. Both institutions agreed to share their code, with no conditions attached other than appropriate credit—impressive generosity, considering that each model had taken more than a decade to develop.

Acquiring the code called for in-person visits. In 1975, Robert Sadourny, a French modeler who had studied with Arakawa and Mintz in the 1960s, spent four weeks at UCLA. Meanwhile, Tony Hollingsworth made his way to GFDL. Both returned to ECMWF bearing model code and documentation, as well as personal knowledge gained during the visits. ECMWF comparison tested the two models before settling on the GFDL scheme. Soon, however, the Centre replaced the GFDL model physics with a new physics package of its own, retaining only the dynamical core. Later this too was replaced with a spectral core developed internally.[62]

This circuitous exchange of concepts, mathematical techniques, and computer code became entirely typical of computational meteorology and climatology after the mid 1960s. Rather than start from scratch, virtually all the new modeling groups began with some version of another group's model. Veterans and graduate students from the three original GCM groups left to form new groups of their own, taking computer code with them. Others coded new models, most often modified versions of the physical equations and parameterizations used in existing models. The availability of a widely shared, well-standardized scientific computer language (FORTRAN, short for FORmula TRANslation) facilitated these exchanges substantially, as did the scientific-internationalist culture of meteorology.

Yet, as I have mentioned, the number of modeling groups remained small. The count can be expanded or contracted under various criteria, but a reasonable upper limit might be the 33 groups that submitted GCM output to the Atmospheric Model Intercomparison Project (AMIP), an ongoing effort to compare climate models systematically.[63] A smaller figure comes from the Coupled Model Intercomparison Project (CMIP), an AMIP follow-on, which evaluated 18 coupled atmosphere-ocean GCMs.[64]

Notably, all the CMIP simulations came from modeling groups based in Europe, Japan, Australia, and the United States, the historical leaders in climate modeling. The AMIP models (simpler and less computer-intensive than coupled AOGCMs) also included entries from Russia, Canada, Taiwan, China, and Korea. But the elite world of global climate simulation still includes no members from South or Central America, Africa, the Middle East, or southern Asia. The barriers to entry in climate modeling remain high. They include not only the cost and complexity of supercomputers, but also the human infrastructure required to support advanced research of this nature. In the political arena, this fact contributes to a widespread perception that the issue of climate change "belongs" to the developed countries, not only because they are the initial (and still principal) sources of fossil fuel emissions but also because they are the "owners" of knowledge about the problem.

## Climate Modeling and Computational Friction

The mathematical complexity of general circulation modeling was one reason it did not spread further or faster. But a more important reason was access to supercomputers. These expensive, highly advanced machines used new techniques such as parallel processing (multiple instructions handled at the same time), reduced instruction sets, and vector processing (multiple data items handled at the same time). Today's personal computers employ such methods routinely, but in the 1960s and the 1970s those techniques were seen as highly specialized, uniquely suited to scientific computing and not to much else. For programmers, working with these machines required specialized knowledge and training. Their operating systems were minimal; sometimes a new-model supercomputer was delivered to eager customers with no operating system at all. Typically, manufacturers sold at most a few dozen copies of any given model. Little or no commercial software was available for this minuscule customer base. While supercomputer manufacturers such as Control Data Corporation and Cray provided high levels of support, the highly specialized scientific programming of GCMs remained the province of a tiny elite.

As modelers sought to increase model resolution and include more physical processes directly in the models, their models required more and more computer power. Every group building GCMs either owned or had access to the largest, fastest supercomputers available. Greater computer power allowed longer runs, higher resolution, and larger numbers of runs.

Because modelers' appetite for computer power constantly outstripped the available capacity, climate laboratories endured a nearly continuous cycle of re-purchasing, re-learning, and re-coding as successive generations of supercomputers arrived, typically every 3–5 years. The machines required a substantial, highly trained staff of operators, programmers, and technical support personnel. They also needed air-conditioned rooms and consumed prodigious quantities of electric power. (The Cray 1-A processor, for example, sucked down 115 kilowatts, and its cooling and disk storage systems devoured a similar amount.)

With a handful of exceptions, such operations lay beyond the means of academic institutions, which might otherwise have been expected to develop their own modeling programs.[65] Instead, until the 1990s most climate modeling activity was confined to national laboratories, weather services, and a few other large, well-financed institutions. As a result, the number of GCMs remained small. In the mid 1990s—four decades after Phillips's pioneering prototype—just 33 groups worldwide submitted GCMs for the first Atmospheric Model Intercomparison Project.[66]

Tables 7.1 and 7.2 illustrate the rapid growth of computer power at major modeling centers. As table 7.1 shows, the capabilities of GFDL's best computers increased by a factor of 3000 between 1956 and 1974. Table 7.2 catalogs the vast performance improvements of ECMWF's IBM POWER5+ computer (installed in 2006) over its Cray 1-A (installed in 1978), including a 90,000-fold increase in sustained computational performance. Virtually all climate laboratories could produce similar charts.

Table 7.1

Computers in use at GFDL, 1956–1982. Data from Geophysical Fluid Dynamics Laboratory, *Activities—FY 80, Plans—FY 81: With a Review of Twenty-Five Years of Research 1955–1980* (Geophysical Fluid Dynamics Laboratory, 1980).

|  | Time period | Relative performance |
|---|---|---|
| IBM 701 | 1956–1957 | 1 |
| IBM 704 | 1958–1960 | 3 |
| IBM 7090 | 1961–1962 | 20 |
| IBM 7030 "STRETCH" | 1963–1965 | 40 |
| CDC 6600 | 1965–1967 | 200 |
| UNIVAC 1108 | 1967–1973 | 80 |
| IBM 360/91 | 1969–1973 | 400 |
| IBM 360/195 | 1974–1975 | 800 |
| Texas Instruments X4ASC | 1974–1982 | 3000 |

**Table 7.2**
ECMWF's latest supercomputer vs. its first one.

| Specification | Cray-1A | IBM POWER5S+ system | Approximate ratio |
| --- | --- | --- | --- |
| Year installed | 1978 | 2006 | |
| Architecture | Vector processor | Dual cluster of scalar CPUs | |
| Number of CPUs | 1 | ~5000 | 5000:1 |
| Clock speed | 80 megahertz | 1.9 gigahertz | 24:1 |
| Peak performance of each CPU | 160 megaflops | 7.6 gigaflops | 48:1 |
| Peak performance of whole system | 160 megaflops | ~34 teraflops | 200,000:1 |
| Sustained performance | ~50 megaflops | ~4.5 teraflops | 90,000:1 |
| Memory | 8 megabytes | ~9 terabytes | 1,000,000:1 |
| Disk space | 2.5 gigabytes | ~100 terabytes | 40,000:1 |

*Source*: European Centre for Medium-Range Weather Forecasts, "ECMWF Supercomputer History," 2006, www.ecmwf.int. Flops are "floating-point operations per second," a measure of calculating speed.

No fields other than nuclear weapons research and high-energy physics have ever demanded so much calculating capacity. As a result, like the nuclear weapons labs, the world's major climate and weather modeling centers have consistently maintained state-of-the-art supercomputer facilities, and have significantly influenced the development path of the supercomputer industry. For example, in 1977 NCAR purchased the first production supercomputer from Cray Research, a Cray 1-A with serial number 3.[67] (A test model, serial number 1, had been delivered to the nuclear weapons laboratory at Los Alamos the previous year.) Toward the end of the 1970s, NCAR insisted for the first time that its computer suppliers deliver operating systems, compilers, and other basic software for the machines. Previously this had been the responsibility of customers rather than manufacturers. As a result, users—including scientists themselves as well as laboratory computing staff—wrote their own software, sometimes even operating systems. Manufacturers considered themselves responsible only for technical assistance with software development, while customers generally preferred to retain control of their highly specialized, lab-specific software.[68] According to Elzen and MacKenzie, NCAR's break with this traditional social contract helped force Cray Research to supply system software. A former Cray vice president, Margaret Loftus, recalled: "Before, we had difficulty selling software to the sophisticated scientific labs. Now, we have difficulty selling a Cray without software."[69] Packaged weather modeling software for workstations and personal computers has appeared in recent years,[70] but in meteorology the craft tradition persists even today, especially in the relatively small field of climate modeling.

The prestige attached to having built the fastest computer in the world, combined with the small customer base, sometimes injected a heavy dose of nationalism into the competition among supercomputer manufacturers. Between 1994 and 1996, NCAR held a competition for the contract to replace its aging Crays with new supercomputers. For the first time in history, the winning bidder was a non-US firm: the Japanese firm NEC, whose SX-4 registered the highest performance of any computer NCAR had ever evaluated.[71] NCAR planned to lease four SX-4s at a price near $35 million, but US-based Cray Computing challenged NEC's bid before the US Department of Commerce. Cray accused NEC of illegally "dumping" its machines below their cost. The Commerce Department ruled in favor of Cray, imposing a 454 percent tariff on NEC, a move which effectively prevented NCAR from acquiring the SX-4s. Many observers viewed the Commerce Department's decision as motivated by nationalism, under behind-the-scenes pressure from Congress (Edwards interviews). NEC retaliated by suing the Commerce Department in the US Court of International Trade, alleging that the agency had "revealed itself as a partisan ally of NEC's competitor [Cray] for the UCAR contract."[72] The suit went to the Supreme Court, which in 1999 upheld the Commerce Department ruling without comment. NCAR mourned the decision: "We are denied access to the most powerful vector computing systems in the world."[73] Eventually it adopted an IBM machine, which it clearly viewed as a stopgap.

Because computers are so fundamental to modeling work, climate laboratories typically display a certain fetishism regarding their machines. In the late 1990s I spent time at more than a dozen weather and climate modeling centers, including NCAR, GFDL, ECMWF, the UK's Hadley Centre for Climate Prediction and Research, and the Australian Bureau of Meteorology Research Centre (BMRC). Almost invariably, and usually within an hour of my arrival, my hosts would offer a tour of the computer facility. Often such a facility occupies an entire floor, housing at least one supercomputer and numerous large disk drives, robotic tape libraries, and other peripherals. Some labs, such as NCAR and the BMRC, feature prominent viewing windows through which visitors can admire the machines, quietly humming in their spotless rooms. Likewise, the websites and publications of climate laboratories virtually always highlight their computer capacity.[74]

After flaunting the beautiful machines, climate modelers typically proceed to explain the difficult balancing act they must perform. In their accounts, a kind of Heisenberg principle of modeling results from computational friction. Modelers can increase model resolution by sacrificing complexity, or they can increase model complexity by decreasing resolution—but they cannot do both at once. Other tradeoffs can balance these limits differently; for example, modelers can "nest" a higher-resolution limited-area model within a lower-resolution global model. All solutions depend on the availability of central processing unit (CPU) time, data storage (for the very large model outputs), CPU speed, and other technical aspects of the machines and the software. This discourse of insufficient computer power is a basic trope of climatology, widely reported in journalistic treatments of the climate-change issue.

Yet computational friction appears in many other less public but more interesting and subtle forms. As one of many possible examples, consider the problem of round-off error in GCMs. In order to handle a vast range of numerical values, from the extremely small to the very large, scientific mathematics conventionally represents numbers in two parts: a mantissa and an exponent. The mantissa expresses the value, while an exponent of 10 (positive or negative) gives the scale. Thus 256859.6235 becomes $2.568596235 \times 10^5$, and 0.000164757852 is rendered as $1.64757852 \times 10^{-4}$. Computer arithmetic employs a similar strategy. A computer's word length is fixed at the hardware level, as the basic unit of addressable memory.[75] Floating-point notation divides each word into two parts, one for the mantissa and one for the exponent. Software techniques can increase the functional word length to almost any desired (finite) number of digits, but since they must work around the hardware constraint, these techniques impose a prohibitive computational cost on calculation-intensive operations such as GCMs. In modern supercomputers, word length is typically 64 or 128 bits, corresponding roughly to a precision limit of 16 or 32 decimal digits respectively.

Floating-point arithmetic handles irrational numbers such as $\pi$, and other numbers with mantissas longer than the available word length, by rounding them off at the last available digit. This limit on precision produces no noticeable effects for most purposes. But in simulations that require extremely long series of calculations, the tiny errors that occur with each rounding can accumulate to a problematic degree. Furthermore, although standards issued by the Institute of Electrical and Electronics Engineers govern most computers' floating-point arithmetic, nonstandard conventions also exist. And as with most standards, *implementations* of IEEE standards vary slightly among manufacturers, so that different machines produce slightly different results. Scholars have argued that such differences can never be entirely eliminated, because designers base computer arithmetic on human arithmetic, whose conventions remain subject to debate and potentially to change.[76] Intel demonstrated one aspect of this problem to mass consumers in 1995 when it released its Pentium chip with a faulty floating-point unit that resulted in substantial division errors under a few rare circumstances. The floating-point algorithm used in the chip had been subjected to mathematical proof, but a (human) mistake in the proof led to errors in the chip design. The mistake led to a recall that cost Intel about $475 million.[77]

GCMs' enormously long series of interdependent calculations make them sensitive to round-off error. In 1973, NCAR considered adopting new supercomputers that would operate most efficiently at a lower precision than its existing CDC 6600 and 7600 machines (24-bit or 21-bit mantissas vs. the CDC's 48-bit mantissa). David Williamson and Warren Washington tested the effects of this lower precision on a GCM and found that predictability errors in the model itself—essentially, the effects of nonlinear chaotic behavior—exceeded round-off error so greatly that the lower precision of the new machines would not appreciably affect NCAR model results.[78] Errors in observational data also exceeded round-off error. Therefore, Williamson and Washington argued, the round-off error problem could be ignored for the time being. However, as model quality improved and model codes began to spread from one group to another, this and other issues of computational error emerged once again.

Between the 1960s and the 1990s, supercomputer architecture underwent two major changes. Cray computers, first released in 1975, embodied the first major innovation: vector processing, which permits the same operation to be performed simultaneously on many data elements. A second innovation followed in the late 1980s: massively parallel computers, with dozens to hundreds of processing units permitting multiple instruction streams to proceed simultaneously. Both changes presented difficult challenges—especially parallel processing, which required programmers to reconfigure instruction streams to take advantage of many processors. Supercomputing thus remained a highly specialized niche, requiring knowledge and technical skill entirely different from those well known (and well understood) in business and consumer computing. In these difficult conditions, climate modelers—or rather the programmers who worked on climate modeling teams—had to "port" old GCM code to the new machines or else reprogram them from scratch, working from the mathematical model rather than the existing computer code.

Meanwhile, GCMs grew ever more complex. Resolution increased, and the typical length of GCM runs rose from a few months to 20–100 years or more. In the mid 1990s, as they prepared to convert to massively parallel architectures, James Rosinski and David Williamson of NCAR investigated the effects of round-off error, code branches, and other computer-related problems on GCMs, including the porting of old GCMs to new computers. They ran NCAR's CCM2 twice on the same computer, the two runs identical except for a tiny difference in the specified initial temperature. Comparing the results, they saw "small, rounding-sized differences between model solutions grow rapidly, much faster than expected from the predictability error growth associated with nonlinear fluid flow." They isolated this particular problem to effects of computer round-off error on an algorithm associated with cloud parameterization, and speculated that similar errors would affect other parameterizations. Round-off error also produced differences between the same model run on the same computer from identical initial conditions but using two different versions of a FORTRAN compiler and a code library of intrinsic functions (figure 7.5, two Cray lines). Finally, Rosinski and Williamson demonstrated differences between two runs of the same GCM with the same initial data on two different computers (figure 7.5, thick solid line).

Thus hardware and low-level software differences each created idiosyncratic errors that accumulated to a small but significant degree. Since each climate laboratory develops much of its own low-level software (such as the "new" and "old" libraries of intrinsic function codes indicated in figure 7.5), their individual computers have calculating "styles" or "personalities" (my terms) that show up as small differences in outcomes over long model runs. The effect is that the "same" model, when ported to a different computer, will behave somewhat differently for reasons related to how the machine implements computational processes, rather than to the model's mathematics. Rosinski and Williamson suggested that although these computational differences cannot be eliminated, ported models can be statistically validated against original model results to ensure that computer "personalities" do not substantially alter model behavior.

### 2×$CO_2$: A Paradigmatic Modeling Experiment

As GCMs added vertical levels, going from two in Phillips's prototype to nine levels or more by the mid 1960s, modelers had to work out how to

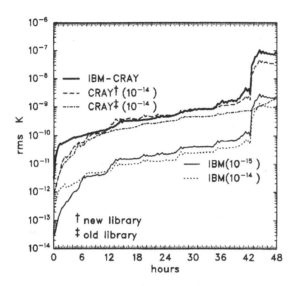

Figure 7.5
An illustration of differences between IBM and Cray runs of NCAR CCM2, differences between control and perturbed versions of CCM2 Cray and IBM runs, and differences between Cray runs using two versions (new and old) of an intrinsic code library. Original legend: "Global RMS [root mean square] temperature differences between IBM and Cray versions of CCM2 and various perturbation pairs on the IBM and Cray. The notation ($10^{-x}$) denotes an initial perturbation taken from a rectangular distribution bounded by $\pm 1.0 \times 10^{-x}$."
Source: J. M. Rosinski and D. L. Williamson, "The Accumulation of Rounding Errors and Port Validation for Global Atmospheric Models," SIAM Journal on Scientific Computing 18, no. 2 (1997), 560.

simulate transfers of energy, mass, and moisture from one level to another. Among other things, without such transfers GCMs would not reproduce large-scale vertical circulatory features, such as the giant Hadley, Ferrel, and polar cells (see figure 2.7). These features, which span tens of degrees of latitude and encircle the entire planet, could not simply be entered into the models. Instead, they would have to emerge from grid-scale processes. Research on vertical structures and processes became critical to the scientific agenda of modelers during this period.

Modelers focused much of their attention on *radiative transfer*.[79] Atmospheric gases absorb and re-radiate solar energy, heating and cooling the atmosphere. These energy transfers drive the entire climate system. The atmosphere consists mostly of nitrogen (78 percent), oxygen (21 percent), and argon (1 percent). However, none of these gases absorbs much radiation in the infrared portion of the spectrum responsible for heating. Instead, the principal "radiatively active" gases are water vapor, ozone, and carbon dioxide. These and other radiatively active trace gases create the greenhouse effect, which maintains Earth's average temperature at around 15°C.

Recall from chapter 4 that, just before World War II, G. S. Callendar had revived the carbon dioxide theory of climate change, first proposed by Tyndall, Arrhenius, and others in the nineteenth century. Climatologists initially received Callendar's results quite skeptically, but Callendar doggedly continued to promote the idea despite its chilly reception.[80] By the 1950s, numerous scientists—many trained in fields other than meteorology—began to reconsider the carbon dioxide theory.

Among others, the physicist Gilbert Plass, inspired by World War II and early Cold War military research on radiation (he had worked on the Manhattan Project), revisited the work of Arrhenius and Callendar. In 1956, Plass published a series of articles reviewing the $CO_2$ theory of climate change.[81] Unlike his predecessors, Plass had access to a computer (the MIDAC at the University of Michigan) and thus was able to calculate $CO_2$'s radiative transfer effects without resorting to the "many approximations . . . used . . . in earlier attempts to solve this complex problem." Plass calculated radiation fluxes at 1-km intervals from the surface to an altitude of 75 km. He performed these calculations for the existing concentration of $CO_2$, then repeated them for half and for double the existing $CO_2$ concentration. His result: "The surface temperature must rise 3.6°C if the $CO_2$ concentration is doubled."[82] Plass warned of significant and enduring global warming by the twentieth century's end. In 1957, Hans Suess and Roger Revelle published their famous paper on the carbon cycle, indicating that not all the carbon dioxide from combustion of fossil fuels could be absorbed by the oceans, so that the gas's concentration in the atmosphere would continue to rise.[83] I discuss this work further in chapter 8.

As they began to construct their nine-level GCM a few years later, Manabe and his colleagues at the GFDL also examined the effects of doubling or halving $CO_2$ concentrations (using the one-dimensional model, not the GCM). Smagorinsky had assigned Manabe to create radiative transfer code for the model. Manabe recalled it this way when I interviewed him:

My original motivation for studying the greenhouse effect had very little to do with concern over environmental problems. But greenhouse gases . . . are the second most important factor for climate next to the sun. Greenhouse gases on this planet change Earth's surface temperature by as much as 30°C. So in order to test the radiative computation algorithm, how effective it is to simulate the thermal structure of the atmosphere, you have to put this greenhouse gas in. . . . Here my job requirement to try to develop a radiation algorithm came in very handy. . . . So I was an opportunist. I took this opportunity to do this problem.

But then a funny thing happened. Fritz Möller, who was working with me, . . . found out that the methods of Callendar and Plass . . . start failing when . . . the temperature warms up in the atmosphere. When the temperature warms up, you get more water vapor in the atmosphere. And that means more downward flux of radiation. So when Möller started putting in what we call water vapor feedback, [previous] methods started failing miserably. . . . Sometimes when you doubled $CO_2$ you got a cooling of 10 degrees, depending upon temperature, but at another temperature you doubled $CO_2$ and you got a 15 degree warming. All kinds of crazy results, mainly because including Arrhenius, *all the pioneers who worked on greenhouse warming thought about only the radiation balance at Earth's surface.*[84]

In other words, any given parcel of air not only receives radiant energy from incoming sunlight, but also re-radiates energy in all directions. This re-radiated energy is, in turn, absorbed, reflected, or re-radiated by other parcels of air. Therefore, to handle radiation in a nine-level model, Manabe needed his radiation code to work out these radiative transfers *within* the atmosphere, i.e., between model levels.

Furthermore, it was necessary to model the problem Möller had discovered: how changes in $CO_2$ would affect the concentration of water vapor (a problem known as "water vapor feedback"). Möller's 1963 calculation showed that cloud and water vapor feedbacks associated with $CO_2$ doubling could have huge effects, producing "almost arbitrary temperature changes" of up to 9.6°C, for which no existing model could account. He also argued that scientists had focused on $CO_2$ because they could measure its global concentration adequately with relative ease, whereas the extreme variability of clouds and water vapor made those phenomena almost impervious to empirical study at the global scale.[85] A warmer atmosphere at the surface causes greater evaporation, but more water vapor in the atmosphere can translate into more clouds, which may either cool or warm the atmosphere depending on their structure, altitude, and color. Finally, water vapor is implicated in moist convection, another process of vertical heat transfer. Working out the relationships among all these factors would take decades; some remain controversial today.

Throughout the 1960s, Manabe, collaborating with Möller, Smagorinsky, Robert Strickler, Richard Wetherald, Doug Lilly, and Leith Holloway, slowly worked out a new, one-dimensional radiative-convective model. His group then used that model to create radiation code for the three-dimensional, nine-level GFDL GCM.[86] At first, they simply wanted to understand the model's sensitivity to changes in and relationships among key variables involved in vertical radiation transfer, including $CO_2$, water vapor, ozone, and cloudiness. In the first Manabe one-dimensional model, doubling the concentration of carbon dioxide produced an increase of 2.3°C in global average temperature. Revisions to the model later reduced the increase to 1.9°C.[87] In 1975, Manabe and Wetherald became the first investigators to test the climate's sensitivity to carbon dioxide doubling in a GCM. Their results indicated a rise of about 2.9°C at the surface, notably larger than the simple radiative-convective model's predictions.[88]

Table 7.3 lists the most significant papers on carbon dioxide's temperature effects from Arrhenius's seminal work through Manabe and Wetherald's pioneering GCM calculation. Nearly all these papers used carbon dioxide doubling as a benchmark. (Many also calculated the effects of halving the concentration.) In view of the complexity of the problem, these results all fell within a strikingly restricted range. Leaving aside Arrhenius's early result (based on very approximate measures of $CO_2$'s radiative properties) and Möller's 1963 calculation of 9.6°C (by his own admission a model that produced "almost arbitrary temperature changes"), these calculations vary from a low of 0.8°C to a high of 4°C, approximately a factor of 4. Stephen Schneider, reviewing these and other results and comparing his own model with that of Manabe and Wetherald, argued for a likely climate sensitivity of 1.5–3°C.[89] Furthermore, *all the reported values are positive*. In other words, every single result pointed to warming from carbon dioxide doubling; the question was not whether, but how much. This striking consistency would help to place global warming theory on the political agenda, the subject of chapter 14. Thus climate sensitivity to carbon dioxide doubling had already become a paradigmatic modeling experiment by the 1950s. So it was not only natural, but necessary, to try the same experiment with GCMs. The "2×$CO_2$" simulation became a standard modeling exercise.

As Kuhn pointed out long ago, paradigmatic experiments—those that settle an issue and create a widely accepted framework for solving further problems—can unify a scientific field.[90] In this case, however, simulations of $CO_2$ doubling had a paradoxical dual effect. First, because virtually all simulations gave results within a fairly narrow range of positive values, they unified the field in something like the way Kuhn suggested, creating a set of puzzles to be solved within the dominant framework of simulation modeling and establishing a set of techniques for solving those puzzles.

**Table 7.3**
Selected estimates of the effects of carbon dioxide doubling on global average temperature, 1896–1975.

| Investigator(s) | Year | Climate sensitivity to $CO_2$ doubling | Remarks |
|---|---|---|---|
| Arrhenius | 1896 | 5–6°C | 2-D (zonal and vertical) radiative transfer model; hand calculation |
| Hurlburt | 1931 | 4°C | Unnoticed until 1960s owing to general rejection of $CO_2$ theory; Callendar unaware of Hurlburt's work until about 1942 |
| Callendar | 1938 | 1.5°C | 1-D radiative transfer model; $CO_2$ doubling not mentioned in text, but appears in graph; no convection |
| Callendar | 1949 | 2.1°C | Revised version of his 1938 calculations; $CO_2$ doubling explicitly mentioned |
| Plass | 1956 | 3.8°C | 1-D radiative transfer model; no convection or water vapor feedback |
| Möller | 1963 | 1.5–9.6°C | 1-D surface energy balance model; combined $H_2O$ and $CO_2$ absorption reduces overall warming, but water vapor feedback produces "almost arbitrary temperature changes" |
| Conservation Foundation | 1963 | 3.8°C | Consensus statement by Plass, Keeling, and others |
| Manabe and Wetherald | 1967 | 2.4°C | 1-D radiative-convective model; humidity and cloudiness levels strongly influence $CO_2$ effects |
| Manabe | 1970 | 1.9°C | Revised version of Manabe and Wetherald 1967 1-D radiative-convective model; sensitivity is for "average" cloudiness |
| Rasool and Schneider | 1971 | 0.8°C | 1-D radiation balance model with fixed relative humidity and cloudiness |
| Manabe and Wetherald | 1975 | 2.9°C | First use of a GCM to simulate effects of $CO_2$ doubling |

Yet, in contrast with the classic Kuhnian cases, they did not settle the issue. Instead, the range of results remained large enough to generate ongoing controversy for several decades. The "benchmark" aspect of this experiment became more salient as climate change became a policy issue. (See chapter 14.)

At the end of chapter 4, we saw that Callendar's 1938 revival of the carbon dioxide theory foundered on the objections of C. E. P. Brooks, David Brunt, and other climatologists. They argued that Callendar could not connect the apparent warming trend with carbon dioxide levels, because the warming might be caused by circulatory effects that no one yet understood. Though Callendar did not live to see it (he died in 1964), GCMs would permit climatologists to analyze the connection of $CO_2$ and circulation in detail. In a real sense, climate modelers have been working on that problem ever since.

### Data Friction, GCMs, and Climate Change

How do you distinguish a good simulation from a bad one? Of course your simulation should look like the reality it's simulating. But before you can tell whether it is doing that, you need a reliable picture of that reality. You need a data image. The more detailed your simulation gets, the more precise your data image must become if you are going to evaluate your model's quality. In the case of the planetary circulation, that picture of reality proved exceedingly hard to come by.

The long-term data record available at the dawn of the GCM era remained severely fragmented. No single data set covered both the oceans and the land surface. The standard "global" climate data—the Réseau Mondial and the *World Weather Records*—tracked only about 400 surface stations on land. Before 1938, upper-air data—critical to studies of the general circulation—remained extremely sparse. Collecting and publishing climate data still took years, even decades. Large libraries of digitized data had been collected in a few places, but the storage medium (punch cards) was too bulky, heavy, and fragile to be widely or readily shared. Large data voids remained, especially in the southern hemisphere and the extreme polar regions.

An example illustrates the extent of the problem. In 1965, Smagorinsky, Manabe, and their colleagues wanted to evaluate their hemispheric GCM. They could find no single data source containing sufficient information about the entire hemisphere at every altitude represented in the model. Instead, they had to combine observational data from numerous sources

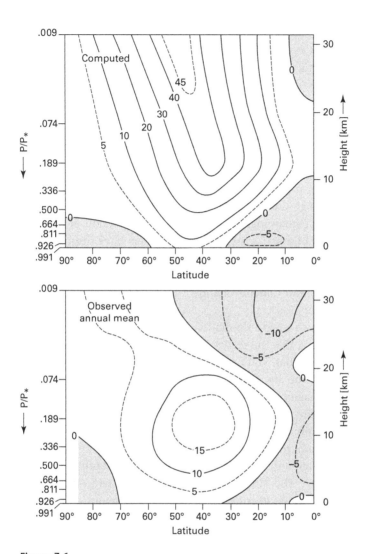

**Figure 7.6**
Zonal mean winds in the northern hemisphere as computed in the GFDL nine-level GCM (top) and as observed (bottom). The diagram represents a cross-section of the atmosphere taken from the equator to the north pole. The observed mean combines tropospheric winds for 1950 with stratospheric winds for 1957–58.
*Source*: J. Smagorinsky et al., "Numerical Results from a Nine-Level General Circulation Model of the Atmosphere," *Monthly Weather Review* 93 (1965), 736.

in dubious ways. For instance, the best wind data for the troposphere came from H. S. Buch's analysis for the year 1950, which was based on 81 northern hemisphere wind stations. But these data did include the stratosphere, so Smagorinsky et al. had to combine them with data for a different year: Abraham Oort's data for the International Geophysical Year (1957–58), from 240 northern hemisphere radiosonde stations at three pressure levels. No traditional climatologist would have accepted combining data from two different years in this way; circulatory patterns varied too much. But all other data on the circulation above the surface exhibited similar limits, so the modelers had no real choice. To top it off, the nine-level GCM contained well over 10,000 gridpoints, a resolution that far exceeded those of all existing observational data sets.[91]

Fortunately, in 1965 no one expected much accuracy. Modelers just wanted to know whether they were in the ballpark. They would be happy if their crude GCMs reproduced even the most basic features of the circulation, such as the trade winds and the vertical temperature distribution. Comparing GCM outputs they expected to be inaccurate with data they knew to be imprecise worked well enough for the time being. But this strategy could not work in the long run. In time, modelers would need better data, especially in the vertical dimension. How else could they tell whether their models were improving?

Furthermore, by 1970 modelers were already becoming embroiled in emerging political controversies over anthropogenic climate change, provoked in part by their own research on carbon dioxide and aerosol effects. This made the demand for accurate global climate data much more urgent. It also changed the nature of the research problems modelers had to solve. As we have just seen, most theoretical models projected a 2–4°C temperature increase from $CO_2$ doubling. But $CO_2$ had not doubled; in fact, as of 1970 it had risen only about 13 percent above pre-industrial levels. Climate scientists therefore faced new questions that could only be answered by comparing simulations with observational data. For example, could they already detect a warming "signal" against the natural "noise"—as Callendar already believed in 1938? GCM simulations might serve as the "control" against which to detect that signal. If they did find a signal, could they prove that greenhouse gases were the cause? Again, simulations would be needed as controls. Could they project how climate would change as $CO_2$ continued to increase? $CO_2$ was not going to double overnight; instead, it would rise slowly. How would this "transient" increase affect the climate's response? In addition, how would particulate aerosols, chlorofluorocarbons, methane, and other anthropogenic pollutants affect the climatic future? What about water vapor? Perhaps the increased evaporation caused by a warmer climate would ultimately paint the planet white with clouds, paradoxically reflecting so much heat back into space that temperatures would plummet, leading to an ice age. Or perhaps the cooling effect of particulate aerosols would cancel out carbon dioxide warming, stabilizing temperatures. How would temperature changes affect ocean circulation? Et cetera, et cetera, in a list that mushroomed along with the new field.

Answering all these questions would require much more, and more precise, climate data. To make their projections credible, modelers would have to show not only that they could simulate the present climate but also that they could reproduce its past. To do that, they would need to know much more about the history of the global atmosphere than anyone had yet attempted to discover. How much information could be wrung from the spotty, mostly two-dimensional historical record? Could new observing systems provide a sufficiently detailed picture of the vertical dimension? How long would it take to accumulate a truly global, three-dimensional climate record? In the following chapters, I trace the evolution of weather and climate data networks since World War II—the sources of the information meteorologists required in order to address these complex questions.

# 8 Making Global Data

As the era of numerical forecasting dawned, atmospheric scientists began to realize that the structure of their discipline had been turned on its head by the computer. In the 1940s the stream of data had already become a flood; both forecasters and climatologists were collecting far more information than either could ever hope to use with the technologies then at their disposal. Yet by the late 1950s, as NWP models reached hemispheric scales and general circulation modeling began, forecasters and climatologists saw that they would soon not have nearly *enough* data—at least not in the right formats (standardized and computer processable), from the right places (uniform gridpoints at many altitudes), and at the right times (observations taken simultaneously around the globe). The computer, which had created the possibility of numerical modeling in the first place, now also became a tool for refining, correcting, and shaping data to fit the models' needs. Meanwhile, the models began to shape data-acquisition strategies, sometimes directly. Over the next three decades, techniques for *making global data* increasingly converged with techniques for *making data global*.

To *make global data*, national weather services developed and refined their observing systems and improved international data communication through a variety of telecommunications networks, old and new. Movement to standardize observations accelerated as a new World Meteorological Organization rose from the ashes of the old International Meteorological Organization. Upper-air observing networks grew, adding the crucial third dimension to the existing surface network. Meteorologists added new stations and new instruments—most important among them satellites, which from their orbits in space could provide more complete and consistent global coverage than any other observing system.

Nonetheless, instrument readings from the emerging global data network did not, in their "raw" form, look anything like the kind of data needed by the regular grids of computer models. In many ways, the proliferation of instruments and observing platforms made the task of integration harder. To *make data global*, scientists developed suites of intermediate computer models that converted heterogeneous, irregularly spaced instrument readings into complete, consistent, gridded global data sets. They also literally created data for areas of the world where no actual observations existed. These intermediate models were turned back upon the data network itself to determine how changes in the location, density, and frequency of observations might affect the quality of forecasts and other simulations. As time went on, these techniques became so tightly intertwined that they transformed the very meaning of the term 'data'. Today, the processes atmospheric scientists invented are ubiquitous not only in geophysics but throughout the sciences. *Virtually everything we now call "global data" is not simply collected; it is checked, filtered, interpreted, and integrated by computer models*. We will explore this second, complementary aspect of global data in chapter 10.

As early as 1950 the outlines of this transformation were already clear to some meteorologists. Speaking on the occasion of the thirtieth anniversary of the founding of the American Meteorological Society, Athelstan Spilhaus noted the spread of the telephone, telegraph, and radio networks, developments in automated instrumentation, and the imminent arrival of computer models. These, he wrote in the *Bulletin of the American Meteorological Society*, would soon combine to comprise "the ultimate weather instrument":

> The complete weather instrument will observe at all suitable points in three dimensions and transmit to a center. There it will not only store data and compile it for study purposes, but will pick off and pre-compute the information needed to feed the electronic computing devices which will prepare predictions and automatically disseminate them to the distribution networks.[1]

The prescient Spilhaus might have been introducing the World Weather Watch project, initiated more than ten years later. Decades before the World Wide Web, this first WWW forged a unified global infrastructure of data, communications, and computing from the tangle of uncoordinated, heterogeneous systems that preceded it. The tensions of the Cold War both spurred and inhibited its development, often in dramatic ways.

This chapter explores the geopolitical and scientific contexts of making global data. The number and the complexity of linked events confound any attempt at a unified chronological narrative. Instead, the chapter considers some of the institutions, political arenas, and projects that most influenced the unification of the weather and climate information infrastructure in the 1950s and the 1960s.

## Meteorology and Cold War Geopolitics

The primary geopolitical context of post–World War II geophysics was the manifest desire of both superpowers for a multi-dimensional form of global reach. In *The Closed World* (1996), I argued that the United States' foreign policy of "containment" conceptualized the Cold War as a global struggle, reading all conflicts everywhere in the world as subsets of the superpowers' contest for military and ideological advantage. Containment strategy materialized in specific technological forms. High-technology weapons— thermonuclear bombs, long-range bombers, nuclear submarines, missiles— would project American power across the globe, while computers, radar, and satellites would enable centralized, real-time surveillance and control. Heavy investment in military equipment would reduce reliance on soldiers. Hence, the extremely rapid improvement of computers between 1945 and 1960 owed much to Cold War technopolitical strategies. At the same time, computers helped to realize America's global ambitions by making possible centralized command and control on an unprecedented scale.[2]

Global reach necessarily involved collecting many kinds of global information. If "we" were to "defend every place"—as General Douglas MacArthur urged the US Congress in 1951—then "we" would need to *know* about every place. The most direct information requirement was, of course, intelligence about military forces, deployments, and strategic intentions. Yet it rapidly became clear that many other kinds of knowledge would also be required, included those of certain geophysical sciences: geodesy (for accurate mapping and missile guidance), oceanography (for submarine and naval warfare), seismology (for detecting nuclear tests), climatology (for anticipating likely conditions at potential sites of conflict), and weather forecasting itself. From the geostrategic perspective, then, meteorology was only one part of a larger project in constructing a global panopticon.

As was noted in chapter 6, weather control became a significant arena of competition in the early years of the Cold War, and therefore it became a source of large amounts of research funding. John von Neumann's unwavering support for weather control projects contributed enormously to their legitimacy. Concerned that the Soviet Union might beat the United States to the discovery of successful techniques, von Neumann "warned that this might ultimately be a more dangerous threat than ICBMs." "Our knowledge of the dynamics in the atmosphere," he told Congress in 1956, "is rapidly approaching a level that will make possible, in a few decades, intervention in atmospheric and climatic matters. It will probably unfold on a scale difficult to imagine at present. There is little doubt [that] one could intervene on any desired scale, and ultimately achieve rather fantastic effects."[3]

By the late 1950s, experiments with cloud seeding and other techniques were still producing equivocal results, yet optimism remained high. Not only the military services, but also the civilian National Science Foundation funded ongoing experiments. By 1959, almost half of the $2.4 million budget of the NSF's Atmospheric Sciences Program went to weather control projects. The prospect of weather control remained very much on the military agendas of the United States and the Soviet Union well into the 1970s, when it was finally abandoned because the results were consistently disappointing.[4] The "rather fantastic effects" of which von Neumann spoke would manifest, instead, in what meteorologists of the 1950s called "inadvertent weather modification": climate change as an unintended side effect of other human activity.

The possibility of weather control was often deployed to justify research on numerical weather prediction, though in actual practice the two remained largely separate tracks. But during the Cold War military forces had other reasons to support weather prediction work. Military technological change increased the superpowers' appetites for global weather data and forecasts. Pilots of high-flying jet aircraft needed information on the jet streams and other high-altitude weather phenomena, which could also affect ballistic missiles. Navigators of long-distance bombers might need weather data from almost anywhere in the world. Tactical nuclear strategy depended on knowing the likely path of fallout clouds and the distances they might travel on the wind. In the 1950s, the US Air Force's Air Weather Service (AWS) grew to be the world's largest weather service, employing an average of 11,500 people. During this period, about 2000 of these AWS personnel were officers with some degree of formal training in meteorology. By the end of the 1950s, military officers accounted for over half of the total enrollment in meteorology programs at American universities.[5] Geostrategy and technological change—mutually reinforcing—thus aligned military interests with the informational globalism of scientists involved in NWP research.

Procurement of global weather data grew into a joint, unified effort of the Weather Bureau and the Navy and Air Force weather services, and (later) of NASA as well.[6] American military weather observations, especially from radiosondes and reconnaissance aircraft flying from remote Arctic and Pacific island airbases such as Alaska, Greenland Hawaii, the Philippines, Midway, and Guam, became important sources of upper-air data in sparsely covered regions. Radio operators at US military bases around the world collected and retransmitted the data broadcast by other countries' weather services. Bases in France, Germany, Korea, and Japan provided coverage of the surrounding regions independent of national weather services. The worldwide forays of American military vessels supplemented coverage of the oceans.

These military observing networks freely shared most of the synoptic data they collected, but they also produced their own separate, classified forecasts and data. There is little evidence, however, that these forecasts and data were any better than those produced by their civilian counterparts. Indeed, military weather services experienced ongoing threats to their survival from commanders who found them redundant or sought to cut costs by relying on civilian forecasts instead. As a co-sponsor of the Joint Numerical Weather Prediction Unit, the Air Force received and tested JNWPU forecast products from the beginning and soon relied heavily upon them.[7] Around 1960 both the US Air Weather Service and the Royal Swedish Air Force discontinued some internally produced forecasts in favor of publicly available results.[8] Even the top-secret Defense Meteorological Satellite Program (DMSP), which operated independent military weather satellites from 1962 until 1994, was nearly identical to its civilian counterpart and was eventually absorbed by it.[9] The principal purpose of the separate military weather network eventually became to provide a backup system in case a war stopped the flow of data through the civilian network.

Cold War politics did, at times, significantly impede the free exchange of data. For example, from 1949 to 1956 the People's Republic of China (PRC) shared no weather data at all. From 1957 to 1963 data from the PRC were generally well distributed, but after 1963 most non-synoptic data, including upper-air data and climatological averages, were closely held until the 1980s, when data-sharing efforts resumed.[10] Figure 8.1 shows synoptic network coverage in 1956. The absence of reports from the PRC is clearly visible as a data void.

The Soviet Union circulated most but not all of its weather data through the international network. For example, it withheld the locations of some weather stations near its northern borders, presumably for military reasons. The US Air Weather Service was able to determine the probable locations

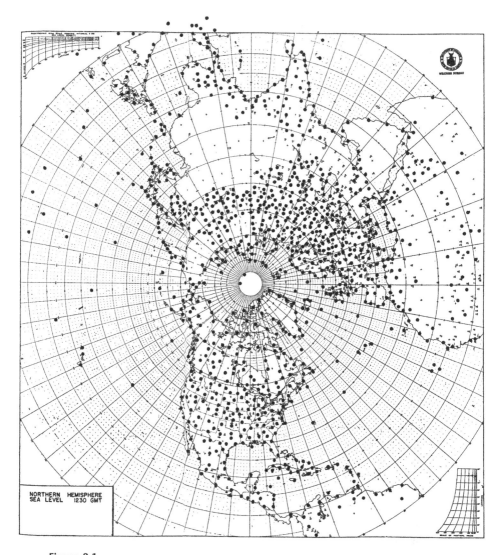

**Figure 8.1**
Average northern-hemisphere sea-level coverage in the 1956 synoptic network, from multiple sources at 1230 GMT. Dots represent surface stations. Note the absence of reports from the People's Republic of China.
*Source*: W. M. McMurray, "Data Collection for the Northern Hemisphere Map Series," *Monthly Weather Review* 84, no. 6 (1956), 227.

of these stations by modeling their fit to several months' worth of weather analyses.[11] Since the PRC and the USSR together accounted for over 20 percent of the world's total land surface, data from these regions were vital.

## The World Meteorological Organization: Infrastructural Globalism during the Cold War

In chapter 1, I argued that infrastructures for collecting and disseminating globalist information play a significant part in producing widely shared understandings of the world as a whole. As we have seen throughout this book, the dream of a global weather data network—a "vast machine," in Ruskin's words—has animated meteorological visionaries since the middle of the nineteenth century. International data exchange was well established in the northern hemisphere by the 1930s, long before numerical weather prediction. Yet the network suffered from enormous data friction because of (among other things) multiple communication and data-storage technologies, data loss in noisy transmission channels such as radio, failure of national weather services to conform to International Meteorological Organization standards, and wide variation in the codes to transmit data.

The disarray reflected the socio-political structure of international meteorology before World War II. National weather services certainly understood the value of standardization and cooperative data exchange, but for structural reasons their principal imperatives came from within. They guarded their independence closely, and regarded IMO standards merely as guidelines. The IMO lacked both political authority and financial support, and no other body existed to organize international weather communication. Data traveled widely through telegraph, telex, radio, and other media. Yet the proliferation of formats, transmission errors, and other data friction prevented most of the potentially available data from ever being used outside the countries that created them.

Thus the global data network circa 1950 had yet to fully transcend the system-building phase (see the section Dynamics of Infrastructure Development in chapter 1). Each national weather service maintained its own standards and systems; these were similar enough to connect to a degree, but different enough to make genuine interoperability impossible. Some regional networks, especially in Europe, had achieved a degree of coherence, while the European colonial empires undergirded weather reporting from far-flung territories elsewhere in the world. But efforts to link these regional and imperial networks into a smoothly functioning planetary internetwork had so far met with little success.

Meteorologists had accepted informational globalism—the principle that standardized data from the whole world should be routinely collected and shared—for many decades. From Maury's ships' logs to the Réseau Mondial, numerous meteorological visionaries had advocated plans best described as *infrastructural globalism*: a permanent, unified, institutional-technical network for global weather data, forecasting, and communication, as discussed in chapter 1. The IMO's institutional weakness, the straitjacket imposed by the sovereignty of national weather services, and the imprecision of pre–World War II forecasting all contributed to the failure to achieve this goal. The end of World War II brought meteorologists' pent-up frustration with this disarray into alignment with technological and institutional opportunity, and transformed meteorology to its core.

During the 1930s, as we saw in chapter 3, the International Meteorological Organization had slowly reached a consensus that it should seek intergovernmental status. That status, the IMO hoped, would commit government authority and financial support to its infrastructural-globalist goals: extending cooperation, improving national conformance to IMO standards, and building out the world data network. By 1939, IMO leaders had drafted a World Meteorological Convention that would implement this plan, but World War II interrupted the effort. Not until 1946 could the IMO meet again. Already primed for major change by its prewar activism, the organization worked at a furious pace, building on the World Meteorological Convention drafted seven years earlier.

Agreement did not come easily. Many participants remained skeptical of the value of intergovernmental status. The perennial issue was whether the change might lead to politicized control of meteorology by governments. Nonetheless, buoyed by the postwar atmosphere of optimism, conferees resolved the major outstanding questions in just over a year. Reassured by negotiators that professional status would remain primary, nations would retain equal rights as members, and that governments would not control its deliberations, the final drafting conference (held in Washington) closed in 1947.

The new organization would become one among many "specialized agencies" of the United Nations, so it would have to conform to the UN's rules of membership. In the final days of the drafting conference, American legal experts advised the conferees that membership in the new organization should be accorded only to "sovereign states" recognized by the UN. This criterion excluded not only divided Germany (the major issue immediately after the war) but also the People's Republic of China, colonial territories, and individual Soviet republics from full membership. As Clark

Miller has observed, for meteorologists this "new vocabulary of 'States' instead of 'countries' superimposed a *geopolitical* imagination of the world over the *geographical* imagination that had previously organized meteorological activities."[12] These debates mirrored others that were occurring simultaneously in the UN General Assembly. In October 1947, representatives of 31 governments signed the World Meteorological Convention. The Convention entered into force in early 1950, and in 1951 the International Meteorological Organization officially became the *World* Meteorological Organization.

The early WMO chipped away at the Herculean task of integrating the unruly complexity of national and regional weather observing and communication systems into a single functional internetwork. It accomplished this by embedding social and scientific norms in worldwide infrastructures in two complementary ways. First, as the process of decolonization unfolded, the WMO sought to align individuals and institutions with world standards by training meteorologists and building national weather services in emerging nations. Second, the WMO worked to link national weather data reporting systems into a global data collection and processing system. Indeed, the new organization's most fundamental and explicit purpose, as outlined in the World Meteorological Convention's opening paragraphs, would be to promote informational globalism through a standardized network infrastructure:

To facilitate world-wide co-operation in the establishment of networks of stations for the making of meteorological observations or other geophysical observations related to meteorology. . . . ;
To promote . . . systems for the rapid exchange of weather information;
To promote standardization of meteorological observations and to ensure the uniform publication of observations and statistics;
To further the application of meteorology to aviation, shipping, agriculture, and other human activities; and
To encourage research and training in meteorology and to assist in co-coordinating the international aspects of such research and training.[13]

In most respects these goals differed little from the expressed ambitions of the IMO. Now, however, meteorologists could call upon the power of governments, as well as the authority (and the finances) of the UN, to implement them.

National sovereignty[14] remained a membership requirement, but the World Meteorological Convention did specify a mechanism by which states not belonging to the UN could join the WMO. Territories (i.e., colonies and protectorates) could also join, under the sponsorship of their governing states. Membership grew quickly. By the mid 1960s most nations were represented.[15] For decades, however, the exceptions to this rule remained extremely significant. Sovereign statehood as a requirement for membership lay in tension with the organization's informational-globalist principles and with the prevailing ideology of scientific internationalism. In 1951 the First World Meteorological Congress immediately moved to soften the rebuff of the PRC's exclusion by inviting that country to participate as an "observer." It soon expanded this decision into a general policy. Any non-member country could send observers to World Meteorological Congresses, and the directors of the country's meteorological services could attend or be represented at technical commission meetings. This uneasy compromise avoided overt conflict with UN policy and the United States, but it did not satisfy the desire of many states for full recognition. Ten of the seventeen non-member nations that were invited to send observers to the Second World Meteorological Congress in 1955 declined.[16] Even many years later, the second-class "observer" status and the exclusion of divided nations provoked anger.

For example, during the Sixth Congress, held in 1971, Rodriguez Ramirez, a delegate from Cuba, insisted on reading into the minutes a statement denouncing the exclusion of "the socialist countries" from full membership. Ramirez accused the WMO of hypocrisy:

The World Weather Watch [see chapter 9] would have more amply fulfilled its objectives had the WMO opened its doors to all countries. . . . The WMO . . . is rejecting UN agreements on the peaceful uses of the World Weather Watch. Viet-Nam, in particular, has suffered the destruction of nearly half of its meteorological stations, loss of the lives of more than 100 scientists and meteorological workers, terrible destruction of its forest wealth by the use of chemical products which have altered its ecology and biology . . . at the hands of the armed invasion forces of the United States and its allies. This declaration, Mr. Chairman, has been supported by the socialist countries of Byelorussia, Bulgaria, Czechoslovakia, Hungary, Mongolia, Poland, Romania, Ukraine and the Soviet Union.[17]

US representative George Cressman responded heatedly that such statements "served no purpose other than to interrupt the proceedings with political propaganda." But he could not resist going on to characterize the US intervention in Vietnam as an invited response to "coercion, organized terror and subversion directed by North Viet-Nam."[18]

Such direct, on-the-record confrontations rarely marred the smooth, courteous surface of WMO meetings. Still, their public eruption marked the subterranean antagonism between the organization's informational globalism and the conflicted, voluntarist internationalism of its predecessor, the IMO. This antagonism mirrored a larger tension that dominated international relations of all sorts in the postwar years, not least in the United Nations itself. Since the middle of the seventeenth century, under the so-called Westphalian model of sovereignty, states had retained absolute control over affairs within their territories, but expected no control whatsoever over the affairs of other states. No state recognized any authority higher than its own. Virtually all international associations were voluntaristic, existing only to promote mutual interests (and only so long as those interests remained mutual). Military alliances were paradigmatic.[19]

The UN system simultaneously perpetuated and eroded voluntarist internationalism. On the one hand, it strengthened the nation-state framework by creating explicit criteria for legitimate sovereignty and codifying the rights of states against one another. On the other hand, the UN system limited sovereignty, since the organization itself had the authority to challenge the legitimacy of governments. Its status as a world organization made withdrawal from the system difficult and costly. Under these circumstances, countries maneuvered carefully to retain a maximum of sovereign rights, while emerging intergovernmental organizations avoided claims to absolute authority that might discourage membership. The language of the WMO Convention therefore hedged carefully, promising only to "promote," "encourage," "facilitate," etc., rather than dictate to its member states. Under Article 8 of the World Meteorological Convention, members were required only to "do their utmost" to implement WMO decisions.

Members could refuse to adopt any WMO recommendation simply by notifying the WMO's Secretary-General and stating their reasons. Such deviations instantly became a principal concern of WMO technical meetings. For example, the Soviet Union and some other countries, "for practical reasons," elected to continue conducting observations every two hours, despite a majority view that every three hours would be sufficient. A compromise "placed emphasis on" the three-hourly observing times. At its first meeting in 1953, the Commission on Synoptic Meteorology expressed confusion about the contradiction between Article 8 of the World Meteorological Convention and Resolution 15(I) of the First World Meteorological Congress (1951), which spoke of *"obligations* to be respected by meteorological administrations."[20] Debate ensued over whether to use 'shall' or 'should' in regulations. Ultimately the commission put off any decision. Nor did the WMO Executive Committee feel ready to impose stronger language. Both bodies instead deferred to the full World Meteorological Congress.

The second World Meteorological Congress, held in 1955, spent considerable time confronting this problem. Finally the Congress decided to issue two separate sets of WMO regulations. All WMO members were expected to conform, "within the sense of Article 8," to the "standard" regulations, while members could implement a second set of "recommended" regulations at their discretion. The criterion dividing these two sets was whether a given practice was considered "necessary" to the collection of a minimal global data set or merely "desirable."[21] Nonetheless, deviations even from "standard" regulations remained common for many years.

The uncomfortable balance between national discretion based on state sovereignty and international governance based on management of a global operational infrastructure slowly tipped in favor of the latter, owing to an increasingly focused combination of technological change and institutional effort. Dramatic advances in information and communication technology during the 1950s made a global real-time data network an increasingly realistic technical possibility; later sections of this chapter address those changes. But other trends during that tumultuous decade, especially decolonization and the Cold War, moved in the opposite direction, disrupting existing meteorological services and even disconnecting substantial areas of the world from the network. For this reason, the most important arenas for infrastructural globalism in the 1950s were institutional and political rather than technological.

The WMO did not replace national weather services as the fundamental unit of organization in meteorology. Yet its UN affiliation and its intergovernmental status conferred both authority and legitimacy on the new organization. Like the IMO before it, the WMO initially served primarily as a central, relatively neutral site for negotiating technical standards. WMO technical commissions worked more vigorously than their predecessors, in part because constant effort was required simply to keep abreast of the many new instruments and techniques arriving in the 1950s. The technical commissions and quadrennial World Meteorological Congresses provided opportunities to resolve differences over standards.

Yet hopes that the organization's new status might rapidly produce general conformity to standards proved vain. Instead, both agreement on standards and implementation of standards took considerable time (often many years). Lacking any sort of police power, the WMO exerted pressure chiefly through meetings and official publications. At first, the central organization in Geneva maintained only a small staff. Except for the

Congresses, most WMO-coordinated efforts took place elsewhere. The WMO Secretariat conducted no research and played no part in managing data networks; all of that was still done by national weather services. The Secretariat's sole functions were to facilitate meetings and to print and distribute WMO publications.

However, the organization's budget grew rapidly in its first two decades. Annual spending increased from about $300,000 in the early 1950s to about $1.3 million by 1965 (See figure 8.2.)

The WMO Secretariat acquired permanent offices in Geneva in 1955, and moved into its own building in 1960. By 1968, its annual budget had reached nearly $4 million. On a symbolic level, the increasingly substantial presence of a central organization mattered enormously.

The WMO's most significant work as an institution took place through its technical assistance program. At the time of the First WMO Congress (1951), the impending independence of Libya, then an Italian colony, created the possibility of a break in meteorological operations, since the existing weather service in Libya was staffed mainly by non-Libyan personnel. The Congress directed the Executive Committee to propose a plan for continuation of service and "to express the willingness of the WMO to provide all possible technical assistance within its available resources."[22]

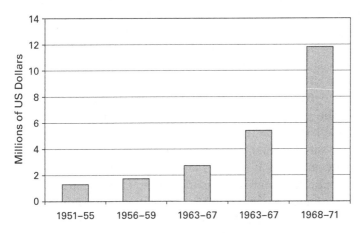

**Figure 8.2**
Budgets of World Meteorological Organization, 1951–1971. Bars indicate totals for each four-year financial period.
*Source*: S. A. Davies, *Forty Years of Progress and Achievement* (World Meteorological Organization, 1990), 162.

From modest beginnings—$23,000 contributed to four countries in 1952—the Voluntary Assistance Program soon became one of the WMO's most significant activities.

Accelerating after 1955, decolonization created some 40 new nations, multiplying the problem posed by Libya. Newly independent, often poor countries typically found few resources and even less attention for meteorology. Throughout this period the United Nations Expanded Program of Technical Assistance for the Economic Development of Under-Developed Countries (abbreviated UNDP) invested in a variety of meteorological assistance projects under WMO guidance.[23] Though hardly the largest UNDP expenditures, neither were these negligible. They typically accounted for 1–3 percent of the UNDP's overall budget.[24]

The WMO hoped to rely entirely on the UNDP for funds, but the UNDP's shifting priorities made it an unreliable ally. Therefore, the WMO established its own Voluntary Assistance Program. Although the majority of funding continued to flow from UNDP, from 1956 to 1959 the WMO's VAP contributed about $430,000 worth of aid to 34 countries. Most of this took the form of expert assistance and fellowships for meteorological training. In the period 1964–1967 the VAP budget reached $1.5 million, with the UNDP contributing another $6.5 million, and by 1972, the WMO and the UNDP together had spent approximately $55 million on meteorological assistance to developing nations, including about 700 expert missions, about 1500 fellowships, and countless seminars and training courses in some 100 nations.[25] Donations of equipment were also a large part of the program.

Who paid for all this? Contributions to the VAP varied from year to year, but as a rule the large majority of the WMO portion came from the United States. The Soviet Union typically offered roughly half as much, almost all in kind rather than in direct financial aid. The United Kingdom and France were the third and fourth largest contributors, each donating amounts roughly one tenth as much as the United States. Sweden led the list of other European countries that provided most of the rest. In all, about 50 nations, including some that also received aid, granted monetary or in-kind contributions to the fund during the 1960s.[26]

The WMO perceived these activities as purely technical. But as Clark Miller has shown, they also contributed to a new politics of expertise.[27] Recipients, particularly those engaged in building new "sovereign states," often understood technical assistance activities as elements of a larger political program. For example, by allowing Israel to provide expert advice to its (mostly immigrant) citizens, WMO assistance to the Israeli weather service simultaneously promoted the legitimacy of the new state.[28] Such initiatives both spurred and grew from a new version of scientific internationalism, which Miller describes as "the idea that cooperation among governments in the areas of science, technology, and other forms of expertise contributes in important ways to the furtherance of broader goals of international peace and prosperity."[29]

It would be absurd to claim some kind of state-building role for the WMO. Yet the organization did help to construct a broad international community of civil servants, science and technology administrators, scientists, and engineers who carried the banner of their native countries. Its Voluntary Assistance Program furthered the representation of weather expertise as a basic *and apolitical* element of the infrastructures which a modern sovereign state should (it implied) furnish to its citizens. Multiplied across many forms of scientific and technical expertise, this representation promoted the integration of expert institutions into emerging liberal states. Additionally, by creating channels and even requirements for an unrestricted flow of scientific information and data, the practice helped reduce the chance—much feared by early Cold Warriors—of being "scooped" in critical areas of science or technology by insular or secret state-sponsored Communist institutions. Ultimately, these and the myriad of similar intergovernmental scientific and technical bodies arising after World War II heralded "a significant shift of foreign policy responsibilities from Departments of State to other government agencies as the participation of experts in international institutions has become central to international affairs."[30]

The technical assistance program also served as a major conduit for the WMO's standardization efforts. Training and expert advisory programs accomplished this not only through their educational content, but also by building human relationships and participatory norms. WMO documents on training often emphasized the importance of communicating to newly trained meteorologists the value of their contribution to the global effort. Efforts were made (and also sometimes resisted) to standardize syllabi for WMO-sponsored training courses.[31] Equipment donated through the VAP also functioned to carry WMO standards, embodied in the machines, from donors to recipients.

The WMO was never a powerful organization in any ordinary sense. With relatively small budgets and a decentralized authority structure, it always remained dependent on the initiative and cooperation of national weather services, and in particular on rich nations with globe-spanning interests. Such interests converged along many fronts in the decades after

World War II. They derived not only from the Cold War but also from cooperative endeavors in air travel, transcontinental shipping, and scientific internationalism. For these reasons, the WMO, though it lacked coercive power, could secure considerable—albeit far from universal—cooperation and involvement. With the World Weather Watch program of the 1960s (discussed in chapter 9), the WMO focused its efforts on building a seamless, permanent, quasi-automatic global weather data network, with inputs from and outputs to every member nation. However, the new organization's first involvement in such an effort began much earlier, during planning for the International Geophysical Year.

### IGY: Data Sharing and Technology on the Cusp of the Computer Age

The technical, institutional, and political issues involved in making global data came together in the International Geophysical Year (IGY). During the "year" between July 1957 and December 1958, scientists from 67 nations conducted global cooperative experiments to learn about world-scale physical systems, including the atmosphere, the oceans, the ionosphere, and the planet's magnetic field and geological structure. Planning for the IGY took place on the cusp of several major, related transitions: from manual to computer methods, from surface-based to space-based observing systems, and from internationalism to globalism. IGY data handling schemes illustrate the powerful tensions between the inertia of the past and the pull of the future at this unique historical moment.

In two previous International Polar Years, 1882–83 and 1932–33, scientists from many nations had collaborated to explore the polar regions. During the first IPY, meteorological observations were taken only at the surface. By the time of the second IPY, the Bergen School had established the concept of the polar front, which governs weather at high latitudes. Hence meteorologists took the opportunity to explore the polar front's characteristics above the surface, using balloons and radiosondes. They also attempted a global program of simultaneous weather observations. With data for only about half of the Earth's surface area, this effort was only nominally successful.[32]

The remarkable fact that shortwave radio signals can be "bounced off" the ionosphere (a layer of ions high in the upper atmosphere) enabled them to travel beyond the horizon, and to serve as the basis for unreliable but worldwide shortwave radio and radiotelegraph transmissions. Hence scientists studying radio waves and the ionosphere "were impelled to think globally."[33] Around 1950, scientists interested in the ionosphere floated the idea of a third International Polar Year. As they elaborated their plans, they began to enlist other sciences, including meteorology.

Official planning for a third Polar Year began in 1952 under the International Council of Scientific Unions, an international nongovernmental organization. Originally founded in 1931, ICSU had reorganized after World War II along new lines that included both national members (usually national academies of science) and international scientific associations. The latter, by the post–World War II period, included the International Meteorological Association and the International Union of Geodesy and Geophysics, among many others. ICSU and its member organizations restricted their focus to "pure" research science. This distinguished ICSU from the WMO and similar intergovernmental organizations, which were oriented largely toward operational activities such as weather forecasting. However, multiply overlapping membership, along with the obvious need to enlist national weather services in any global study of meteorological phenomena, led to early involvement by the WMO in planning for the third Polar Year. Because so many of the interested parties (including many WMO members) wished to explore matters involving the planet as a whole, the WMO Executive Committee suggested changing the name from "Polar" to "Geophysical," and the International Geophysical Year was born. It would include experiments and cooperative synoptic observations on a vast variety of geophysical phenomena, including geodesy, geomagnetism, oceanography, and seismology in addition to meteorology.

The IGY served significant ideological purposes for both superpowers, which used their scientific collaboration to promote their technological prowess and their commitment to peaceful coexistence.[34] Internationalism may have been indigenous to science, but in the IGY it would be deployed to buttress American (and more broadly Western) political interests.[35] Ron Doel has observed that this double purpose implies a kind of cooptation, but in practice it operated more like the "mutual orientation" described in chapter 6 of this volume. Politicians would trade on scientific internationalism to legitimize a particular vision of world order; in exchange, scientists could better promote their own internationalist goals. Scientists' involvement in government and governance would, in the long run, produce pressures to which governments would be forced to respond.

Geopolitics impeded IGY planning in several ways. The Soviet Union (not an ICSU member) had to be persuaded to join the effort by its meteorological representative A. A. Solotoukhine, who was serving on the WMO's Executive Committee.[36] Meanwhile, the People's Republic of China—

representing, at nearly 10 million square kilometers, about a fourteenth of the Earth's land surface—remained unrecognized by the United Nations. Holding only "observer" status at the WMO, its national scientific academy vied with that of the Republic of China (Taiwan) for ICSU recognition.[37] Nevertheless, IGY planners accepted PRC participation in 1955. Though this violated US policy, US scientists were permitted to continue with the tacit approval of the Eisenhower administration.[38] Not long before the IGY began, however, the PRC withdrew in protest. A network of about 200 meteorological stations it had set up for the IGY went ahead as scheduled, but non-Chinese who wished to see the data were required to request them directly from the PRC.

IGY planners settled quickly on an overarching purpose for the venture: to study Earth as a "single physical system." Hence the IGY's meteorological component, with some 100 countries participating, focused most of its attention on observing the global general circulation. Three pole-to-pole chains of observing stations were established along the meridians 10°E (Europe and Africa), 70°–80°W (the Americas), and 140°W (Japan and Australia). Dividing the globe roughly into thirds, these stations coordinated their observations to collect data simultaneously on specially designated "Regular World Days" and "World Meteorological Intervals." In addition to balloons and radiosondes, an atmospheric rocketry program (initially proposed by the Soviet Union) retrieved information from very high altitudes. Planners made an extensive effort to gather data for the southern hemisphere from commercial ships, as well as (for the first time) from the Antarctic continent. In total, the IGY meteorological network claimed about 2100 synoptic surface stations and 650 upper-air stations—far more than the networks for ionospheric and magnetic studies, which counted only about 250 stations each.[39] The program proved so popular that many of its operations were extended through 1959 under the rubric of an "International Geophysical Cooperation" (IGC) year.

Plans called for depositing complete collections of the IGY and IGC meteorological data at three World Data Centers, one in the United States (WDC-A), one in the Soviet Union (WDC-B), and one (WDC-C) at WMO headquarters in Geneva. Each WDC would be funded by the host country.[40] Other IGY sciences created similar but separate WDCs.

Data centers did not undertake to process the data, only to compile and distribute them. National weather services and research groups were responsible for reporting their portions of the IGY data to the WDCs. Merely collecting and compiling these data took several years. Many data sets did not reach the WDCs for many months after their collection; the full set of IGY data was not completed until 1961. In form this resembled the organization of the prewar Réseau Mondial and World Weather Records—but the volume of data for the 18 months of the IGY alone dwarfed that contained in the earlier collections.

Planners recognized that machine methods of data processing would probably soon be widely adopted throughout the sciences. Therefore, in 1955 they recommended that IGY data collections be stored in a machine-readable format. In practical terms this meant punch cards. Magnetic tape and other fast electronic media were, at that time, unreliable, unstandardized, and too expensive to compete with the simple, durable, decades-old IBM card. The WMO, however, chose not to follow this recommendation. Instead, it distributed IGY data on microcards, a miniature photographic reproduction medium similar to microfiche. Each microcard contained reduced copies of up to 96 standard observation forms. The decision to use this medium—a decision made around the time of the advent of operational numerical weather prediction—may seem to indicate a lack of foresight. But the choice was ultimately based on economy and convenience.[41] Though the wealthier weather services had occasionally exchanged decks of punch cards in the past, by the time of the IGY this method of sharing had become impractical. For example, climatologists at WDC-A, housed in the US National Weather Records Center at Asheville, North Carolina, used punch cards to record their IGY observations. The resulting deck amounted to some 10 million cards and weighed approximately 30 tons.[42] Sharing such a deck by reproducing and shipping it would have been very costly—and then the recipient would have to find somewhere to store it. Yet the complete IGY data set could be reproduced on only 16,500 microcards. WMO's WDC-C (Geneva) made this card set available for the relatively affordable price of about $6000.

Whereas most of the other World Data Centers continued to operate without interruption after the IGY, WDC-C for meteorology closed down once it completed the IGY data set in 1961, though WDC-A (Asheville) and WDC-B (Moscow) continued. Rather than take the opportunity to further promote centralized global data collection, the WMO urged each national weather service to publish its own data. The WMO then promised to catalog all the data residing in national repositories around the world. These immense catalogs took years to compile. The IGY data catalog appeared in 1962. Two fat volumes listing various other data sets available at many locations around the world arrived in 1965; a third—listing "meteorological data recorded on media usable by automatic data-processing machines"—did not see print until 1972.[43]

This strategy for acquiring and distributing data reveals two things about the state of meteorology as the tumultuous 1950s drew to a close. First, its philosophy remained internationalist rather than genuinely globalist. WMO members were sovereign states; they might cooperate temporarily, but each behaved as an independent entity. Each contributed data from within its borders; each could purchase a copy of the global data to use however it saw fit. Centralized data collection, opponents feared, might reduce the resources available to national services and duplicate effort.[44] The terms of this debate mirrored an older WMO discussion about a possible International Meteorological Institute, proposed as early as 1946. Though studied repeatedly and recommended politely by various committees, the institute idea finally died. Within the political framework of that era, the WMO could have undertaken a facilitating role for nationally based research, but it could not usurp the center stage by establishing a world research center. Second, planners may have chosen print media for recording and distributing data mainly out of concern for cost and convenience, but their choice also illustrated the tremendous inertia of the existing data infrastructure. For centuries, publishing data in the form of printed tables had been the stable, reliable norm. The IGY microcards simply miniaturized that long-established form. The WMO data catalogs marked the beginning of the end of the print era in meteorology. By the mid 1970s, electronic media such as magnetic tape and disk would largely replace punch cards, microcards, and printed tables as meteorology's primary mode of data storage and exchange.

Thus, although the WMO's founders hoped that the new organization might rapidly ease the data friction inherent in meteorological methods at the middle of the twentieth century, this did not occur. In fact, joining the UN system inhibited the WMO's informational globalism, preventing some countries from gaining membership on equal terms. It involved the WMO in Cold War politics in ways its leaders probably did not anticipate and certainly did not relish. In addition, decolonization created a crisis for data collection as meteorological services once supported by far-flung colonial empires fell under the precarious management of emerging nations much less committed to the project of infrastructural globalism.

Yet Cold War geopolitics and decolonization favored infrastructural globalism in several important ways. First, the superpowers urgently sought global information in numerous arenas, including weather and climate. Cold War intelligence-gathering efforts drove a search for ways to collect globalist information without relying on other countries. (See following section.) Second, two of the Cold War's central technologies—computers

Making Global Data                                                                                                         207

and satellites—would become meteorology's most important tools, producing enormous synergies for meteorology's rapid development. Third, decolonization stimulated VAP-supported efforts to spread meteorological capacity. Finally, international scientific cooperation became an arena for the Cold War contest for moral superiority. Meteorology's image as a benign science made it an ideal showcase for a country's willingness to collaborate toward a common good.

Indeed, the IGY as a scientific project can hardly be separated from its Cold War political milieu. Launched in October 1957, Sputnik dramatically called forth the tension between one-world globalism and Cold War competition that haunted IGY meteorology. Sputnik shocked the American public and set off the "space race," but it came as no surprise to meteorologists, who had eagerly anticipated artificial satellites for over a decade. IGY planning called for satellite experiments to be mounted by both the United States and the Soviet Union. Indeed, the Soviets announced the Sputnik's success during a reception at the Soviet embassy culminating an IGY coordinating committee meeting.[45] In addition to Sputnik, six other satellites were successfully launched during the IGY, though they returned few data of meteorological value.

The IGY marked a dramatic transition. As a concept, with its single-physical-system framework, its emphasis on three-dimensional observing systems, and its satellite data initiative, the IGY's global meteorology represented the cutting edge of science. IGY weather data remained the most precise and comprehensive set of global simultaneous observations until the 1979 Global Weather Experiment. As a showcase for scientific and political cooperation, the IGY went beyond the IMO's voluntarist internationalism to a more permanent infrastructural globalism, creating a much-emulated model. Yet the planned demise of WDC-C—reflecting suspicion that centralized global data collection might erode the sovereign powers of national institutions—was a hiccup in that process. The IGY data strategy looked ultimately to the past, not the future, of information technology. In fact, the IGY observing program can be seen as the last global weather data network whose foundations were not shaped fundamentally by computer modeling and computerized data processing.

## Nuclear Weapons Tests and Global Circulation Tracers

Another critical nexus between meteorology and Cold War geopolitics was the atmospheric testing of atomic bombs. For one thing, nuclear weapons developers needed accurate weather forecasts to plan atmospheric tests.

More important, however, was the need to predict and monitor the spread of nuclear fallout, the vast clouds of radioactive dust produced by nuclear explosions. Research on fallout produced results that proved momentous for studies of anthropogenic climate change.

Awareness that wind might carry fallout very far from the site of an explosion—even around the world—came slowly. Most early weapons tests were conducted at ground level, with relatively low-yield devices whose debris clouds did not penetrate the stratosphere, so most blast products settled to the ground relatively close to the test site. Before 1951, US weapons laboratories conducted only a few tests. The labs did establish a Fallout Prediction Unit in collaboration with the Special Projects Section of the US Weather Bureau, but it initially focused only on predicting fallout patterns within about 200 miles of the detonation.[46]

The early 1950s saw major atmospheric nuclear testing programs begin in the United States and the Soviet Union, along with a smaller program in Australia run by the United Kingdom. By the mid 1950s, dozens of bombs were being detonated each year, some of them thermonuclear. These bombs' debris clouds often penetrated the stratosphere, where winds carried radioactive fallout around the world. In 1951, the United States conducted a high-altitude "shot" at the Nevada test site, the first test at that location. Although the explosion deposited little fallout at the test site, 36 hours later the Kodak manufacturing plant in Rochester, New York—more than 2000 miles away—detected radioactive dust in its air filters. It reported this finding to the Atomic Energy Commission, which promised to inform Kodak of approaching fallout clouds in the future. The AEC directed its Health and Safety Laboratory (HASL) to start a monitoring program to detect "global fallout," defined as radioactive debris injected into the stratosphere and widely dispersed before returning to the ground as particulate ash or in rain and snow.[47]

To track this fallout, the HASL initiated a worldwide network for sampling ground-level fallout deposition using gummed films. The HASL created the network rapidly by cooperating with the US Weather Bureau and the military's Air Weather Service, which simply added these measurements to the observing programs of selected ground stations.[48] By 1962, the HASL's precipitation-monitoring network included about 100 stations in the continental United States and about 100 more around the world (in Canada, Greenland, Antarctica, and Japan, on numerous Pacific islands, and in South America, Eurasia, Australia, and Africa).

By 1962 at least five other programs monitoring global fallout were also underway. Most of them focused on the surface, hunting telltale radioactivity in precipitation, in soil, and in surface air. However, from 1956 to 1983 the AEC ran several high-altitude monitoring programs, using balloons and aircraft to sample air at heights of 6–27 km along several latitudes.[49] The major efforts were based in the US, but the UK organized a 29-station network in 1956, and from 1961 to 1964 a World Meteorological Organization program monitored tritium levels in precipitation at about 100 sites around the world.

In the Cold War context, the fallout-monitoring networks served two main purposes. One was to track and project the probable path of fallout from the nuclear powers' own tests, to reduce danger to military units and other populations downwind.[50] Studying the patterns would also give clues as to how fallout from enemy bombs might spread during a nuclear war. Another purpose was to detect the locations and estimate the yields of other countries' secret nuclear tests. When combined with weather data, fallout patterns could be used to trace flows back in time to an origin point, thus determining the approximate locations of secret test sites.[51] Without this ability, the United States argued, no treaty on limiting or eliminating nuclear weapons could be concluded with the Communist powers.

Tracking global fallout as it traveled through the planetary atmosphere amounted to following the motion of individual parcels of air, something never before possible. This brought unexpected benefits for global circulation studies. Meteorologists could now trace the movement of air around the planet far more precisely.[52] Fallout also served as a tracer for particulate aerosols (dust), and the monitoring programs sampled stratospheric carbon dioxide and carbon 14 for the first time.[53] Thus fallout monitoring produced some of the first three-dimensional studies of global atmospheric chemistry and circulation. According to participants, the fallout-detection networks also "set a precedent for the monitoring of other air qualities on a global as well as local scale."[54]

The issues addressed by fallout monitoring—circulation, aerosols, carbon dioxide, and carbon 14—would also prove central to the study of anthropogenic climate change. Global circulation is, of course, the foundation of modern climatology. Aerosols affect the atmosphere's albedo (reflectivity), so climatologists needed to understand their movement and life cycle in the atmosphere. Stratospheric $CO_2$ measurements provided the first detailed information on the mixing of this greenhouse gas in the upper atmosphere. (It turned out to be very evenly mixed, with concentrations similar to those at ground level. This led to the conclusion that only a few judiciously placed monitoring stations could reliably measure the global $CO_2$ concentration.)

Carbon 14, also known as "radiocarbon," was one of many radioactive isotopes produced by nuclear blasts. Before nuclear testing, all the carbon 14 in the atmosphere had been produced naturally, at high altitudes, by cosmic radiation. Though not as dangerous to human health as other fallout constituents (e.g. strontium 90), carbon 14 holds great interest for science because carbon is the principal non-water constituent of all living things, and therefore also of fossil fuels such as coal, oil, and natural gas.

Tracer studies of carbon 14 in the biosphere began in the late 1940s at the University of Chicago's Institute for Nuclear Studies—a principal site of atomic weapons development—under Willard Libby, who also initiated fallout studies and later became an AEC commissioner.[55] Libby showed that since the atmospheric concentration of natural carbon 14 remained essentially constant, its rate of radioactive decay could be used to calculate the age of fossils. Plants ingest carbon, including carbon 14, from the atmosphere; in turn, animals ingest it from plants or other animals. After animals die and their bodies fossilize, the carbon 14 decays at a known rate. Thus the older the fossil, the less carbon 14 it will contain. This principle applies equally to fossil *fuels*, which are the remains of once-living plants and animals.

By 1950, Willard Libby and Ernest Anderson had created a global "carbon inventory" for the oceans, the biosphere, and the atmosphere, and had estimated the global distribution of natural radiocarbon.[56] Visiting the Institute for Nuclear Studies in 1949, the nuclear physicist Hans Suess met Libby, learned of the radiocarbon research, and decided to pursue the same direction.[57] In 1953, Suess estimated the rate at which the oceans absorb carbon from the atmosphere.[58] Two years later, Suess published a paper showing that recently harvested wood contained less radiocarbon than wood harvested decades earlier. This meant, he hypothesized, that humanity had "diluted" the atmosphere's natural radiocarbon by burning fossil fuels.[59] Suess then revisited the question of fossil fuels first raised by Callendar and Arrhenius. How much of the carbon dioxide from fossil fuel combustion would remain in the atmosphere? For how long?

By 1955, the oceanographer and scientific entrepreneur Roger Revelle had recruited Suess to the Scripps Institution of Oceanography, where Suess organized a carbon 14 research laboratory. Revelle, too, had been influenced by fallout studies; he had worked on the dispersion of radioactive by-products from undersea weapons tests. Revelle estimated that about 80 percent of the carbon from fossil fuels would remain in the atmosphere, rather than dissolve into the oceans (as many had assumed). Writing with Suess, Revelle composed a paragraph calculating that the oceans could absorb only about a tenth as much $CO_2$ as earlier calculations had indicated. Therefore, at current rates of fossil fuel combustion, atmospheric carbon dioxide would continue to increase, making global greenhouse warming a real possibility. Humanity, Revelle and Suess wrote, had engaged in a "great geophysical experiment."[60]

At the time, these observations spurred more curiosity than alarm. Revelle used his considerable influence to help start carbon dioxide monitoring programs in Antarctica and at Mauna Loa, Hawaii (far removed from urban pollution that might skew readings). These measurements became the source of perhaps the sole undisputed fact in debates about global warming: the steady rise in atmospheric carbon dioxide concentrations (figure 8.3).[61]

David Hart and David Victor have argued that anthropogenic climate change became a political issue only when, around 1965, the "carbon cycle

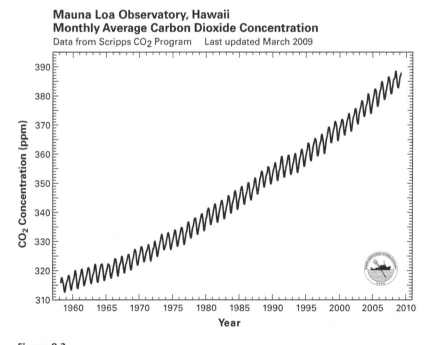

Figure 8.3
Atmospheric carbon dioxide measurements at Mauna Loa, Hawaii, 1957–2009. The regular annual variation reflects the northern hemisphere's seasonal cycle, with carbon uptake during spring and summer growth followed by carbon release during fall and winter as plants die back or shed their leaves.
Courtesy of $CO_2$ Project, Scripps Institution of Oceanography.

discourse" combined with the scientifically distinct "atmospheric modeling discourse" to formulate a new basic research concern known as the "carbon dioxide/climate problem."[62] Yet as we have seen, the link between carbon and climate was first forged in fallout studies of the 1950s. The connection between fallout monitoring, carbon dioxide monitoring, and climate-change studies remained important throughout the Cold War.[63]

As mentioned above, carbon 14 also proved ideal as a tracer for general circulation studies. Unlike the heavier constituents of fallout, which (as the name indicates) gradually "fell out" of the air because of their weight, most carbon 14 became carbon dioxide and remained with its associated air mass. For this reason, following its path allowed meteorologists to determine how long it takes for gas in the stratosphere to mix with lower layers, and to track gas circulation from one "compartment" of the atmosphere (such as the Hadley cells) to another. Figure 8.4 shows the distribution of "excess" (i.e., bomb-produced) carbon 14 in July 1955.

Figure 8.4 illustrates the "remarkable opportunity to study stratospheric motions" provided by carbon 14 and other radioactive tracer studies.[64] In 1955, when these data were collected, virtually all bomb carbon 14 in the stratosphere had been produced by the two atomic test series at the Marshall Islands test site (11°N), one in 1952 and one in 1954. Thus the figure illustrates the poleward spread of fallout, including its spread across the equator, as well as descent of stratospheric carbon 14 into the troposphere (dark lines), over a period of 1–3 years after the stratospheric "injections" of bomb radiocarbon from the Marshall Islands tests. Notably, observed concentrations were higher at the Texas and Minnesota latitudes than at the latitude of the original injections. Observations such as these were used to develop models of stratospheric circulation.[65]

As weapons tests became more frequent and more powerful, public unease about fallout intensified. In 1954, fallout from the Bravo thermonuclear test at Bikini Atoll severely injured the crew of the *Lucky Dragon*, a Japanese fishing vessel. One crew member died. The Bravo test—given a go-ahead despite unfavorable weather reports indicating strong high-altitude winds blowing toward inhabited areas—also exposed many other people to fallout, including hundreds of Marshall Islanders as well as US Navy observers and weathermen. Understandably, the *Lucky Dragon* incident panicked the Japanese population. The Eisenhower administration initially blamed the crew's injuries on vaporized coral from the atoll. The clumsy cover-up only aroused further anger and suspicion, forcing the administration to admit the truth. Ultimately the United States paid Japan $15 million in compensation.

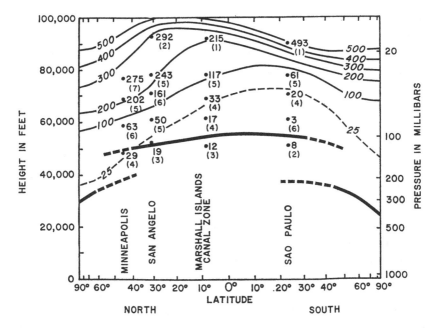

**Figure 8.4**
Original legend: "Cross section of the atmosphere showing the mean excess radiocarbon distribution as of 1 July 1955. Concentrations in $10^5$ atoms per gram of air are indicated near points identifying the altitude. Numbers in parentheses show the number of samples from which the mean concentration was computed. Thin lines are isolines of carbon 14; heavy lines indicate the position of the tropopause." Data are from samples taken by balloons sent up from Minneapolis; San Angelo, Texas; the Panama Canal Zone (on the same latitude as the Marshall Islands nuclear test site); and São Paolo, Brazil.
Source: F. Hagemann et al., "Stratospheric Carbon-14, Carbon Dioxide, and Tritium," *Science* 130, no. 3375 (1959), 550. Reprinted with permission from AAAS.

In the latter half of the 1950s, public dread of fallout rivaled fear of nuclear blasts themselves. In 1957, American scientists initiated a petition to the United Nations opposing atmospheric testing. More than 9000 scientists worldwide eventually signed. Fallout shelters became a centerpiece of civil defense strategies; private homeowners built shelters in basements and backyards. Fallout became a trope not only in science fiction but also in mainstream novels and films. Neville Shute's novel *On the Beach* (1957, film version 1959) portrayed Australians waiting to die as fallout clouds enveloped the world after a nuclear war. The fictional Soviet "doomsday device" in Stanley Kubrick's 1964 film *Dr. Strangelove* contained a thermonuclear weapon coated with "cobalt thorium G"; if detonated, the doomsday device would produce "a lethal cloud of radioactivity which will encircle the earth for ninety-three years. . . . Within ten months, the surface of the earth will be as dead as the moon."

Such fears extended to weather and climatic effects. Rumors spread widely that atomic tests had altered weather downwind of the test sites.[66] The relatively open publication of American fallout studies played a big part in arousing public concern.[67] In 1955 the US House of Representatives held initial hearings on the health, safety, and weather effects of fallout, calling John von Neumann, Weather Bureau research director Harry Wexler, and AEC HASL director Merril Eisenbud as witnesses.[68] More extensive hearings took place in 1957 and 1959.[69]

Americans were hardly the only people terrified of fallout. In 1958, after the Soviet Union declared a moratorium on nuclear testing, experts met in Geneva to debate the feasibility of a total ban on nuclear tests. The main issues all revolved around monitoring. Could a technical infrastructure be built that would be capable of detecting any nuclear test, anywhere in the world, from abroad? Even if such an infrastructure could be constructed, would governments trust it enough to support a test ban? Could technical systems be combined with social mechanisms, such as on-site inspections, to render monitoring acceptably trustworthy? The conference of experts reported that such an infrastructure could indeed be built, by combining the fallout-monitoring systems discussed above with seismic, undersea acoustic, and electromagnetic monitoring, all to be followed up by on-site inspections in case of uncertainty.[70]

Thus nuclear weapons tests spawned multiple infrastructural-globalist projects, connecting cutting-edge geophysics to planetary governance.[71] The fallout-monitoring network was the first successful effort in this direction: the nuclear powers deemed it sufficiently trustworthy to support the Limited Test Ban Treaty, signed in 1963. The network directly generated new knowledge about the global circulation—knowledge that was needed to understand how fallout might spread. Fallout itself made tracing the circulation possible.

Whether seismic monitoring could detect underground tests with equal sensitivity remained intensely controversial in the 1950s. In the end, the push for a comprehensive nuclear test ban foundered on that issue. Neither political proponents of a ban nor technical experts succeeded in making the case that remote sensing could reliably detect underground nuclear tests. Only after the Cold War's end, following decades of improvement in the global seismic monitoring network, did a Comprehensive Test Ban Treaty again reach the negotiating table.

### The Technopolitics of Altitude

For many decades, virtually all data about the atmosphere came from ships and ground stations—points on a two-dimensional surface. But by the late 1930s, many forecast services used upper-air data routinely. Meteorologists now sampled the entire three-dimensional volume of the atmosphere. (See chapter 5.) In the 1950s, when three-dimensional NWP models arrived, upper-air data became vital to the entire meteorological enterprise. One could not run a three-dimensional model without data for the third dimension. Climate GCMs needed no data to generate a simulated circulation, but their results could not be checked against reality without statistical data on the real circulation at altitude.

Exactly how much upper-air data would NWP models and climate GCMs require? How many levels should be sampled, and which ones? With what instruments? How often? Who would collect and transmit these data, and how? Who would pay, and who would benefit? These questions involved meteorologists in what I call the "technopolitics of altitude": concerns and struggles over sovereignty, capabilities, rights, responsibilities, and powers in the upper atmosphere and outer space.

Engaging in technopolitics means designing or using technology strategically to achieve political goals.[72] Symmetrically, it also means using political power strategically to achieve technical or scientific aims. Technopolitics typically involves forming alliances among otherwise distinct actors and communities and aligning their interests.[73] This clearly describes meteorology's situation in the 1950s. To achieve their goal of a global upper-air observing system, meteorologists allied with American military agencies and aligned their interests with those agencies' Cold War goal of global strategic surveillance. Cold War politics also sometimes inhibited upper-air network development, as when the PRC withheld most data from the international synoptic network in the years 1949–1956 and again in 1963–1980.

Figure 8.5 shows the distribution of upper-air observations available to the Joint Numerical Weather Prediction Unit from northern hemisphere radiosonde stations in 1960. Other stations (not shown on the map) generated a limited amount of additional upper-air data from pilot balloons and aircraft reconnaissance flights.

**Figure 8.5**
Northern hemisphere radiosonde stations as received by the JNWP Unit twice daily, circa 1960. Note the small number of stations at lower latitudes (toward the edges of the diagram) and over the oceans.
Source: N. A. Phillips, "Numerical Weather Prediction," in *Advances in Computers*, ed. F. L. Alt (Academic Press, 1960), 48.

Upper-air data networks were considerably more expensive and technically complex than the relatively cheap and simple ground stations. Weather reconnaissance aircraft had to fly on a daily basis. Pilot balloons, rawinsondes, and radiosondes are "throw-away" devices that require frequent replacement.[74] (The balloons eventually burst at high altitude.) Reading their transmissions required radio and radar tracking stations. Early radiosonde data had to be interpreted by skilled human operators. A single radiosonde cost roughly $50 in 1960 (the equivalent of about $300 today). Each station typically launched two new radiosondes a day, one at 00 and one at 12 GMT. US stations alone sent up about 120,000 radiosondes a year.[75]

By 1963, after more than ten years of WMO effort and after the IGY, the global synoptic surface network still reported only about three-fourths of the data requested by the WMO Commission for Synoptic Meteorology. The upper-air network generated less than two-thirds of the requested observations. Data "requested" by the WMO represented a realistic assessment of potential contributions, rather than an ideal network; tables 8.1 and 8.2 provide details. The world's weather network for synoptic forecasting comprised roughly 3000 surface stations. The upper-air network, however, comprised only about 500 stations, almost all of them in the northern hemisphere above 30°N.[76] Particularly noteworthy is the striking discrepancy between the European, North American, and South Pacific

Table 8.1
Percentage of surface stations reporting at synoptic observing hours in 1963

| Regional association | Number of stations requested | Percentage of surface observations implemented at each synoptic hour (GMT) | | | | | | | |
|---|---|---|---|---|---|---|---|---|---|
| | | 00 | 03 | 06 | 09 | 12 | 15 | 18 | 21 |
| I. Africa | 694 | 34 | 63 | 84 | 72 | 84 | 71 | 62 | 24 |
| II. Asia | 963 | 92 | 86 | 93 | 74 | 95 | 71 | 89 | 73 |
| III. South America | 543 | 74 | 26 | 23 | 23 | 84 | 64 | 84 | 37 |
| IV. North and Central America | 469 | 92 | 67 | 75 | 65 | 92 | 70 | 92 | 69 |
| V. South-West Pacific | 286 | 97 | 96 | 98 | 88 | 79 | 55 | 75 | 85 |
| VI. Europe | 831 | 97 | 97 | 99 | 99 | 99 | 99 | 99 | 95 |
| Whole world | 3786 | 81 | 73 | 79 | 70 | 88 | 67 | 83 | 64 |

Source: "The Challenge of the World Network Problem."

## Table 8.2
Percentage of upper-air reports at synoptic observing hours in 1963.

| Regional association | Number of stations requested | Pilot balloon | | | | Radio-sonde | | Radio-wind | |
|---|---|---|---|---|---|---|---|---|---|
| | | 00 | 06 | 12 | 18 | 00 | 12 | 00 | 12 |
| I. Africa | 124 | | | | | 26 | 28 | 27 | 12 |
| II. Asia | 247 | | 38 | | 63 | 77 | 73 | 62 | 60 |
| III. South America | 47 | 36 | 16 | 64 | 37 | 28 | 49 | 28 | 38 |
| IV. North and Central America | 144 | 77 | 77 | 77 | 83 | 90 | 88 | 90 | 88 |
| V. South-West Pacific | 72 | 94 | 90 | 74 | 70 | 74 | 81 | 68 | 69 |
| VI. Europe | 141 | | 76 | | 72 | 91 | 95 | 87 | 90 |
| Whole world | 775 | 69 | 63 | 75 | 65 | 64 | 69 | 60 | 62 |

*Source*: "The Challenge of the World Network Problem." Empty cells represent a complete absence of reports.

upper-air networks and the rest, with Africa and South America the regions of poorest performance.

In 1960 the average distance between upper-air stations on land was about 500 km, but over the oceans and Africa the average spacing was nearly 2000 km. According to Norman Phillips, "if we assume that at least four observation points are necessary to fix the position and amplitude of any wavelike flow pattern, the existing observational network will allow us to seriously consider the forecasting of wavelengths as short as 2000 km over the continental areas, but we will not be able to say much over the oceanic areas about wavelengths shorter than 8000 km."[77] In other words, the existing upper-air data network could resolve only extremely large-scale phenomena, such as the planetary-scale Rossby waves.

The WMO's Commission for Synoptic Meteorology came to grips with this problem immediately, establishing a Working Group on Networks at its first meeting in 1953. Accepting that building an ideally distributed network was not a realistic option, the group soon shifted its attention to the technical question of how much upper-air data was enough. Perhaps some way could be found to dispense with reporting from some places. As we will see in chapter 10, objective analysis techniques developed for NWP would soon provide, for the first time, a quantitative picture of the payoff to be expected from increasing the density of the observing network.

Philip Thompson (a member of the Working Group on Networks) wrote in 1963 that the most fundamental issue was not technical but economic. In principle, evenly spaced global observations would benefit all countries by increasing forecast accuracy. But developing countries, comparing this diffuse benefit to the cost of operating upper-air stations, might (understandably) opt out. In any case, national weather services would naturally concentrate their attention on activities within their borders. Only rich countries with political as well as scientific stakes in global coverage would see good reasons to support the global upper-air network.

Thompson's economic analysis might have been better framed in terms of political economy: only a clear *national* interest in *global* weather observation might motivate most countries to spend scarce resources on the world network. For example, American, Soviet, British, and French military forces operated weather stations outside their national borders in support of potential military operations around the world. Few other countries needed global data for such purposes.

The advent of Earth-orbiting satellites transformed the technopolitics of altitude. Partial photographs of Earth's disk, returned from rockets as early as 1954 (figure 8.6), excited hopes that satellite images might produce dramatic improvements in forecasting. Now meteorologists could literally *see* large-scale weather systems, instead of laboring to construct maps and mental images from instrument readings alone. But whereas the ground-based upper-air network required well-distributed surface stations (and hence international cooperation), a single satellite—operated by a single national owner—could view large portions of the Earth. One satellite in a polar orbit could overfly the entire planet in a matter of hours. This might allow the satellites' owners to reduce their dependency on the ground-based network, if not to ignore it entirely.

Military intelligence agencies looked forward to the satellite era with equal impatience, but for a different reason. For them, satellites promised direct global surveillance. Western intelligence agencies had had great difficulty penetrating the closed Soviet society with human agents. Meanwhile, the Soviet propaganda machine churned out convincing images of rapidly advancing high-tech armed forces. From 1956 on, American knowledge of Soviet military capabilities came largely from secret reconnaissance flights by high-altitude U-2 spy planes. Such flights violated sovereign rights to national airspace, reserved under the 1922 Convention on the Regulation of Aerial Navigation. Satellites offered a

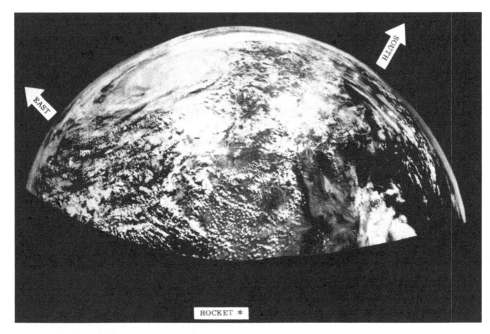

**Figure 8.6**
Photograph taken from a US Navy Aerobee rocket, 1954. A tropical cyclone centered near Del Rio, Texas, is visible at upper left.
Image courtesy of NOAA.

bulletproof alternative to spy planes, more difficult to detect and (at the time) invulnerable to attack.

Here the interests of meteorologists and military intelligence converged. The earliest proposals for military intelligence satellites, in RAND Corporation studies of the late 1940s and the early 1950s, already noted the tight link between satellite reconnaissance and weather forecasting. For intelligence, photographs from space would have to be taken on days when there was little or no cloud cover. Early satellite reconnaissance concepts involved film cameras. Wasting precious film on cloudy skies would be crazy, but this could be avoided by using satellites also to *detect* clear-sky conditions. As a result, both American and Soviet leaders adopted the benign-sounding goal of weather prediction as a cover for military activities. The Explorer VII satellite, launched by the US in 1959, carried both the first meteorological instruments for measuring radiation balance and the first camera from which film was successfully recovered for the secret CORONA spy program (John Cloud, personal communication, 2004).

Figure 8.7
A U-2 spy plane with ersatz NASA insignia on its tail fin, May 6, 1960. Image courtesy of NASA.

When Lockheed defeated RCA in a 1956 Air Force competition for a reconnaissance satellite, RCA salvaged its planning efforts by proposing a satellite-based television weather reconnaissance system to the US Army.[78] Plans for this system, initially designated Janus, evolved as launch vehicles' throw weights increased. Eventually the program—renamed Television Infrared Observation Satellite (TIROS)—was transferred to the fledgling National Aeronautics and Space Administration.[79] On April 1, 1960, the new space agency launched TIROS-I, equipped with television cameras for photographing cloud cover. TIROS-I made headlines nine days later when it photographed a South Pacific cyclone well before instruments detected the storm. Meteorologists rejoiced, though by then their awareness of satellites' potential was already old.

In May 1960, a Soviet antiaircraft missile brought down a U-2 flying over Sverdlovsk. The Eisenhower administration assumed that the pilot, Francis Gary Powers, had been killed, so it denied the Soviets' accusations of spying. Instead, the administration claimed the U-2 had been conducting "weather research" for NASA. To back up the story, officials surreptitiously doctored another U-2 with NASA insignia (figure 8.7) and presented it to the press at Edwards Air Force Base.[80] The hastily devised cover story contained a grain of truth: Powers's U-2 nominally belonged to a weather reconnaissance squadron of the US Air Force Air Weather Service, and it carried meteorological instruments whose data were supposed to be shared with NASA.[81] After the US denial, the USSR produced Powers alive, tried him as a spy, and imprisoned him for nearly two years before releasing him in trade for a spy of their own.

The Powers episode dramatically heightened tensions between the superpowers. U-2 spy flights continued, but their days were clearly numbered. With high-resolution cameras, US intelligence agencies believed, satellites in low orbits might make up for the loss of the U-2 photos. By August 1960, the CIA's top-secret Corona spy satellite program had begun successful photo reconnaissance.[82]

But the U-2 incident underscored the potentially grave implications of violating another country's airspace. Did satellites also violate the sovereignty of nations they overflew? International law did not specify an upper limit to national airspace. Perhaps there *was* no limit. Yet the Soviets, having launched the world's first satellite themselves, could hardly argue that overflight *per se* was illegal.[83] Instead, they used the U-2 incident to emphasize the illegality of espionage, no matter how or from where it was conducted. On this basis the Soviet Union argued that, no matter what its nominal purpose, photography from space—including weather photography such as that provided by TIROS—was illegal.

The Eisenhower administration scrambled to preempt an effective legal challenge to satellite overflight, but escalating Cold War tensions rendered this goal simultaneously more important and more difficult to achieve. During the 1960 presidential campaign, John F. Kennedy ran on an alarmist platform, arguing for a major expansion of the US nuclear arsenal to defend against the putative Soviet threat. Yet well before Kennedy was inaugurated in 1961, Corona satellite reconnaissance had already falsified previous intelligence reports that the Soviet Union possessed as many as 100 ICBMs. The Corona photos showed no ICBMs at all. Kennedy could not risk revealing this intelligence, since that would involve revealing its source, undermining the negotiations on legalizing satellite overflight. Nor could he back down from his campaign promises. So he initiated a major nuclear buildup, even as he proposed cooperation with the USSR in outer space.[84] As president, Kennedy's first major encounter with the Communist world was the disastrous Bay of Pigs invasion of Cuba in April 1961. Days later, Soviet cosmonaut Yuri Gagarin became the first man in space, an event widely reported as the Soviet Union's having "won the space race." In June, Kennedy's first summit meeting with Soviet Premier Nikita Khrushchev did not go well. In August, the Soviet Union closed off Western access to West Berlin. Kennedy responded by calling up 250,000 military reservists.

The best-known political event in the history of meteorology happened in the midst of this crisis, and in response to it. On September 25, 1961, speaking before the UN's General Assembly, Kennedy presented a series of arms-control proposals. He then proclaimed:

... we shall urge proposals extending the UN Charter to the limits of man's exploration in the universe, reserving outer space for peaceful use, prohibiting weapons of mass destruction in space or on celestial bodies, and opening the mysteries and benefits of space to every nation. *We shall propose further cooperative efforts between all nations in weather prediction and eventually in weather control.* We shall propose, finally, a global system of communication satellites linking the whole world in telegraph, telephone, radio and television. The day need not be far distant when such a system will televise the proceedings of this body in every corner of the world for the benefit of peace.[85]

These proposals constituted a comprehensive technopolitics of altitude. Kennedy articulated an infrastructural globalism that joined technical systems to political agreements on the atmosphere and outer space. While sending a conciliatory message to the USSR, Kennedy's proposals simultaneously seized a leadership position for the United States. They implied the project of a satellite-based global weather observing system. They sought to involve countries that had no direct stakes in the space race, creating a potentially worldwide audience for American broadcasting while promising to spread communication technologies with important commercial potential. Finally, they laid out a course of action that, if followed, would render satellite overflight of other nations not only legal but essential to a worldwide communications infrastructure.

As Walter McDougall has shown, Kennedy (like Eisenhower before him) fully understood that a world reaping daily benefits from weather and telecommunications satellites might dismiss objections against satellite photography of airbases and missile silos.[86] Much as John von Neumann had used NWP to promote digital computing in the 1940s, Kennedy exploited weather satellites to justify massive military and civilian space programs. As assistant Secretary of State Harlan Cleveland recalled it,

... we can't even clean up our urban slums, the editorial writers were writing, and this guy wants to go to the Moon! If President Kennedy could, at the UN, hang up a new vision of a space-based system that would move us from chancy three-day weather forecasts to more dependable predictions five days ahead, he could dramatize the prospect that the exciting but expensive US space program would benefit every farmer, every businessman, every picnicker, every citizen who needs to guess, with the help of the atmospheric sciences, what the weather is going to do next.[87]

As it turned out, the dual purposes of the satellite programs served both scientists and politicians. The American and Soviet space programs continued on their dual course, with large secret military programs looming quietly behind the peaceful public space race. In the long run, this mattered a great deal: the relatively reliable, independent intelligence returned by spy satellites helped to keep nervous fingers off the nuclear trigger in the Cold War's final decades.

### Global Data Infrastructures as Cold War Strategy

The international cooperation Kennedy proposed concealed a military agenda, but it was not just a cover story. Kennedy and his successors made good on that promise in many areas, ranging from meteorology to the Peace Corps to the International Cooperation Year proclaimed by President Johnson in 1965. The cooperation was real, yet it also served as a Cold War strategy. As a contest between ideologies and social systems, the Cold War demanded that nations prove their commitments to peace, scientific progress, and improvement of everyday life. Meteorologists and other geophysicists, whether or not they were always aware of the uses to which their ambitions would be put by politicians, had their own globalist goals. Internationalism was a very old element of many sciences, and the sensitivity to international relations promoted by the Cold War spurred a trend toward even greater collaboration across borders. Sometimes, as in the case of the physicist-led "One World" movement against nuclear weapons, political goals were overt.[88] More often, those goals remained in the background; as they saw it, scientists simply wanted to continue their traditional internationalism and openness.

By mid 1961, a large number of interconnected institutions and individuals within the meteorological community had converged on a set of common goals. Three years earlier, the WMO had established a Panel of Experts on Artificial Satellites, with representatives from the United States, the Soviet Union, the United Kingdom, and Australia. At panel meetings, US Weather Bureau chief Harry Wexler had developed the idea of a global observing system he named the "World Weather Watch."[89] The panel reported its conclusions to the WMO in mid 1961. At almost the same time, a US National Academy of Sciences panel also began promoting the World Weather Watch idea, as well as international programs for research, operational weather forecasting, and telecommunications.[90] Meanwhile,

NASA had launched three TIROS satellites, demonstrating the readiness of technical systems.

Kennedy's UN address suddenly and (by all accounts) rather unexpectedly joined these parallel streams. The speech came in September, *before* the NAS report was officially briefed (in October) to Kennedy's Science Advisory Committee, though probably not before its conclusions had reached the ears of his science advisor, Jerome Wiesner. After the speech, Kennedy's national security adviser, McGeorge Bundy, directed the Secretary of State and numerous relevant US government agencies to pursue the stated objectives vigorously.[91]

By the end of 1961, the UN General Assembly pressed further. It directed member states and the WMO to pursue "early and comprehensive study, in the light of developments in outer space, of measures: (a) To advance the state of atmospheric science and technology so as to provide greater knowledge of basic physical forces affecting climate and the possibility of large-scale weather modification; (b) To develop existing weather-forecasting capabilities and to help Members make effective use of such capabilities through regional meteorological centers."[92] The resolution directed the WMO, in consultation with the International Council of Scientific Unions and other relevant bodies, to prepare an organizational plan and financial recommendations for this effort by June 1962. Within a few months, Kennedy and Khrushchev had agreed on the value of global satellite meteorology and the principle of cooperation. On March 7, 1962, Kennedy wrote to Khrushchev:

Perhaps we could render no greater service to mankind through our space programs than by the joint establishment of an early operational weather satellite system. Such a system would be designed to provide global weather data for prompt use by any nation. To initiate this service, I propose that the United States and the Soviet Union each launch a satellite to photograph cloud cover and provide other agreed meteorological services for all nations. The two satellites would be placed in near-polar orbits in planes approximately perpendicular to each other, thus providing regular coverage of all areas.

Khrushchev replied two weeks later:

It is difficult to overestimate the benefit which could be brought to mankind by organizing a world weather observation service with the aid of artificial earth satellites. Precise and timely weather forecasts will be another important step along the way to man's conquering of nature, will help him still more successfully cope with natural calamities and open up new prospects for improving the well-being of mankind. Let us cooperate in this field, too.[93]

A more definitive instance of the technopolitics of altitude can hardly be imagined.

The WMO working group seized the momentum of these collaborative overtures to put forward a dramatic and ambitious plan, building on proposals already outlined by the Panel on Artificial Satellites and the US National Academy of Sciences. The Soviet satellite expert V. A. Bugaev, US Weather Bureau chief Harry Wexler, and a few others met in Geneva to prepare the plan in consultation with a number of other bodies, including ICSU, the International Telecommunications Union, the International Atomic Energy Agency, the International Civil Aviation Organization, and the United Nations Educational, Scientific, and Cultural Organization. In June 1962—exactly on schedule—the WMO group produced its *First Report on the Advancement of Atmospheric Sciences and their Application in the Light of Developments in Outer Space*.[94]

Events soon to intervene would disturb the hopeful atmosphere of these proposals. Within a few months, the Cold War would reach its hottest point with the Cuban missile crisis. The "space race" consumed the attention of both nations, overshadowing work on weather and communications satellites. In the end, Kennedy's proposal for a cooperative two-satellite observing system fell by the wayside; instead, the US and the USSR each launched their own separate series of polar-orbiting weather satellites.

Yet infrastructural globalism remained an effective and important strategy for reducing Cold War tensions, binding the superpowers through common projects, involving their allies, and generating both rhetorics and realities of global data collection. Work toward a global cooperative weather observing system continued, paving the way for the World Weather Watch.

**Box 8.1**
The Limited Test Ban Treaty as Infrastructural Globalism

A profoundly important example of infrastructural globalism in action was the limited nuclear test ban treaty (LTB), concluded in 1963 by the United States, the United Kingdom, and the Soviet Union and later ratified by 110 other countries. The LTB envisaged two major goals: slowing the arms race and eliminating environmental contamination by nuclear fallout. Parties to the treaty, "desiring to put an end to the contamination of man's environment by radioactive substances," agreed not to test nuclear weapons in the atmosphere, in outer space, underwater, or in "any other environment if such explosion causes radioactive debris to be present outside the territorial limits of the State under whose jurisdiction or control such explosion is conducted."

**Box 8.1**
(continued)

> As in the case of the WMO Convention, the language genuflected to a Westphalian conception of national sovereignty: anything states did within their own territories remained entirely outside the treaty's jurisdiction.
>
> At the same time, though, the LTB directly acknowledged that activities conducted within one state might affect not only that state's neighbors but all nations everywhere. In his radio and television address after the conclusion of the accord, an elated President Kennedy announced that the treaty was "a step toward freeing the world from the fears and dangers of radioactive fallout. . . . These tests befoul the air of all men and all nations, the committed and the uncommitted alike, without their knowledge and without their consent. That is why the continuation of atmospheric testing causes so many countries to regard all nuclear powers as equally evil. . . . A ban on nuclear tests . . . requires on-the-spot inspection only for underground tests. *This Nation now possesses a variety of techniques to detect the nuclear tests of other nations which are conducted in the air or under water, for such tests produce unmistakable signs which our modern instruments can pick up.*"[a]
>
> The LTB may thus be seen not only as the first global environmental treaty, but also as the first to recognize the atmosphere as a circulating global commons that could be directly affected on the planetary scale by human activities. At the same time, the LTB was the first treaty to circumvent issues of sovereign territory by means of a global technical monitoring network, exploiting the natural circulation of the atmosphere to detect evidence of events inside national borders. Implicitly, the United States committed itself to maintaining the fallout network in perpetuity. Thus the global fallout-monitoring network—built largely on the back of the global weather data network—became a basic infrastructure of the Cold War.
>
> This was only one of many examples of the powerful boost Cold War infrastructural globalism imparted to the rise of global climatology and climate politics. Without fallout monitoring, concerns about carbon dioxide and climate change probably would have taken much longer to emerge. Without the tracers created by nuclear testing, we would know much less than we do about the stratospheric circulation. Without them we would also know much less about the carbon cycle and its ability to absorb excess carbon from fossil fuels. Even the simple yet critical project of monitoring atmospheric carbon dioxide might not have begun until much later.

a. "Radio and Television Address to the American People on the Nuclear Test Ban Treaty," John F. Kennedy Library and Museum.

# 9  The First WWW

Ruskin dreamed of a "perfect system of methodical and simultaneous observations." Richardson imagined a "forecast-factory." Those were visions of a centrally planned, perfectly uniform observing and forecasting system whose every part and every person works according to a single set of standards and techniques. For the myriad of reasons discussed in earlier chapters, to build such a system in the real world you would need godlike authority and unlimited resources, and even then you would probably fail. Instead, infrastructural globalism succeeded in meteorology because it adopted a particular development strategy—a strategy partly inherited, partly chosen, and partly forced upon it by circumstance. Let us call it the *internetwork strategy*.[1]

Centrally planned systems rarely "scale" well. Instead, real-world infrastructures always build on an installed base. They create standards and gateways to connect existing systems and networks into something larger and more powerful. The goal cannot be absolute compliance or technical perfection, because you will never get that. Instead you seek "rough consensus and running code," to quote the battle cry of Internet architects.[2] If you are lucky, you may get something like the Internet, the telephone network, or the electric power grid: a network of networks that behaves, at least for many relevant purposes, *as if* it were a single unified system. An internetwork gives you reliability and redundancy, checks and balances. It permits some parts to fail without affecting the performance of the whole. It also gives you tensions: friction among the many connected systems; a parliament of instruments, individuals, and institutions, all with their potentials for disagreement, resistance, and revolt.

The World Weather Watch linked existing national weather networks with new, international oceanic and space-based observing systems, data processing, and communications networks. Commonplace today, the internetwork strategy was just emerging in the 1960s. As we will see, the planners of the World Weather Watch confronted all the issues later faced by the designers of the Internet—not to mention that other WWW, the World Wide Web. These included, in addition to the technical problems of internetworking, such political issues as how to balance national sovereignty rights against supranational standards and how to integrate emerging nations in a system dominated by the developed nations and two superpowers.

The World Weather Watch became, and still remains, the World Meteorological Organization's most central activity. At the end of the 1970s, the Global Weather Experiment—the first full-scale test of all World Weather Watch systems together—was the largest scientific experiment ever conducted. In 1997 the US National Academy of Sciences called the World Weather Watch the world's "most formally organized international global observation, communication, processing, and archiving system."[3]

### World Weather Watch Design as Technopolitics

Most accounts of the World Weather Watch portray it as technology driven. For example, a WMO Senior Scientific Officer wrote: "At least initially, the space-based global observing system was the result of technology push. Remote sensing and satellite meteorology were the genesis of the WWW."[4] And indeed, "technology push" does explain a great deal about the development of the World Weather Watch.

Satellites differed from all other instrument platforms. First, instead of sampling the atmosphere through direct contact at discrete points, as most instruments do, satellites observed large areas and atmospheric volumes. Satellite radiometers measured the cumulative radiance of the entire atmospheric column beneath them. Second, satellites could scan the entire Earth with a single instrument. This was a potentially great advantage over the existing system, with its spotty spatial distribution, its data voids, and its hundreds of thousands of instruments, whose intercalibration could never be perfect. It also meshed neatly with the emerging needs of NWP computer models for fully global data. Further, satellites might also serve as communication links. Planners expected that they might pick up transmissions from ships at sea, automated stations, drifting buoys, or sondes circulating over remote areas, then relay those signals to central data-collection stations. For this reason, the *First Report on the Advancement of Atmospheric Sciences and their Application in the Light of Developments in Outer Space* proposed reserving separate, interference-free radio frequencies for data communication between satellites and surface stations. From the outside, this and many other elements of the *First Report*—essentially the manifesto of the World Weather Watch—certainly appeared "driven" by new technology.

Yet the context I described in chapter 8 demonstrates that "technology push" tells only part of the story. In the geopolitical arena, satellites generated a complex mixture of excitement about their potential and trepidation about their consequences for Cold War international relations. Without this anxiety-inducing ambiguity, President Kennedy and the UN General Assembly might never have paid much attention to meteorology. Rhetorics of revolutionary technology in the service of peace and "the betterment of mankind" helped counter fears of technology as the runaway juggernaut of apocalypse.

From meteorologists' point of view, however, the World Weather Watch represented merely the next steps down a path of infrastructural integration first laid out decades earlier. It brought large new symbolic and financial commitments, as well as public visibility. Yet as a concrete scheme it presented little novelty. Indeed, the WMO and its member states were already implementing many of the *First Report's* recommendations at the time of its release. Within the WMO, the WWW was regarded not as a revolutionary step, but simply as the execution of a long-established agenda.

"Technology push" spurred the WWW's start, but a technopolitical strategy of infrastructural globalism governed its design. The project enrolled scientists and weather forecasters in the competitive goals of the superpowers while simultaneously enrolling the superpowers and the United Nations in scientific cooperation. Cold War politico-military competition spurred the United States, Europe, and the Soviet Union to underwrite most of the system's costs and technical burdens. Meanwhile, WMO coordination and support became increasingly crucial, as decolonization created a more complex world, no longer managed by a handful of imperial powers whose far-flung administrative apparatuses might once have facilitated system integration.

Only a few countries possessed the financial means and technical capability to launch and operate their own satellites, and in 1962, as the WWW moved from conception to concrete planning, only the United States and the Soviet Union had actually done so. Satellites' cost and complexity—as well as the political sensitivity of a technology so tightly linked to military intelligence and ballistic missiles—made it virtually inevitable that they would be centrally controlled and their data managed by national owners. This situation threatened the relative parity and independence of national weather services in the existing system. If a small number of countries owned the only genuinely global instruments, how might they alter the time-honored principle of data sharing? What would become of the prerogatives of national weather services in a centralized system? These were as much questions of sovereignty and power as questions of technology and system organization. Furthermore, in the Cold War context no activity in space could proceed without consideration of appearances and strategic implications. In these circumstances, the WMO served as a neutral, legitimizing institution for the project of a global weather infrastructure.

### Hierarchy: Planning for Information Overload

The *First Report* presented a crystalline vision of satellite-based NWP as the inevitable future of meteorology. Satellites would be particularly important for global numerical models because "for this type of prediction any information at all—even intelligent guesses—over areas of no observations will lead to some improvement" (*First Report*, 15). Combining satellite data with conventional observing systems would "make possible the construction of world weather maps—both for the surface and, by differential analysis, for the upper air" (ibid., 16).

In proposing a way to absorb and coordinate huge volumes of new data, the *First Report* went beyond the regional centers envisioned in UN Resolution 1721. It proposed a hierarchically organized, highly centralized *global* system built around three World Meteorological Centers. The WMCs would first process hemispheric and global data, then transmit them to some ten Regional Meteorological Centers. The RMCs would function much like the WMCs, but would produce regional forecasts and data at higher resolutions.

Why three World Meteorological Centers, rather than just one? The *First Report* answered this question as follows:

Theoretically, a single World Centre should suffice, but in actual practice, it is likely that three centres would be required: one in North America, another in the European area and a third in the Southern Hemisphere. In this connection it is pertinent to mention that arrangements are already under way to enlarge the National Meteorological Centre of the US Weather Bureau in Washington into a World Centre in which meteorologists from other countries would be asked to participate. The Hydrometeorological Administration of the USSR has already expressed its intention to establish a World Centre. (29)

Pragmatically, the three-center solution recognized the present and the probable future distribution of the hugely expensive satellite programs, as well as the regional character of existing communication systems. Technopolitically, it offered organizational parity to the superpowers, with the third WMC—located at Melbourne, Australia, in the data-sparse southern hemisphere—providing a small counterweight to the centers controlled by the superpowers.

According to the *First Report*, World Meteorological Centers would require a formidable combination of communication and data processing capabilities:

• Access to conventional meteorological data from as much of the world as possible: this requires location close to a large communications facility.
• Access to world-wide data coming from meteorological satellites. . . .
• Adequate data-processing equipment to convert the conventional meteorological data speedily into objective analyses, numerical weather predictions, etc., and additional specialized equipment for processing and piecing-together [*sic*] of satellite information to produce analyses (such as cloud mosaics, heat radiation charts), covering large areas of hemispheric or world size.
• Staff facilities to blend together the conventional and satellite data. . . .
• . . . Communications facilities to disseminate speedily the analyses and prognostic charts, preferably by facsimile.
• Facilities for training of meteorologists in producing world-wide analyses and prognostic charts using both conventional and satellite information.
• A research facility. . . .
• Close liaison with a World Data and Charting Centre to assist in providing documentation to research groups in all countries interested in hemispheric and world problems. (21–22)

The WWW's primary function would be to serve the needs of weather forecasting. This goal implied near-real-time data analysis and communication. But the volumes of data anticipated from satellites would dwarf the already overwhelming quantities flowing through the network. Separating World Data and Charting Centers from the WMCs would reduce the latter's load by disconnecting data storage and research functions from the more immediate needs of forecast analysis.

To preempt fears that the World Meteorological Centers and Regional Meteorological Centers might displace national weather services, the *First Report* emphasized that the system's main goal would be to *assist* national meteorological centers by reducing redundancy and data friction. At the time this was a very real issue. For example, a 1966 report noted: "A surface analysis for 0600 GMT for north-west Italy is prepared every day by the

Meteorological Services of 15 countries. Some of these Services transmit this analysis by several means (facsimile, radioteletype, CW) and on the whole it is disseminated 25 times."[5] The principal purpose of the RMCs would be "to make available to all countries in the region in a convenient form the vast amount of additional data, in raw and processed form, which meteorological satellites provide. They will in no way replace but will rather assist existing national meteorological services in the preparation of synoptic charts. . . ." (*First Report*, 23)

The *First Report's* catholic approach to satellites noted the three major weather forecast techniques then in use:

*Empirical*—based on present trends, the forecasters' past experience and simplified application of physical reasoning

*By analogy*—with similar past weather patterns

*By mathematical modeling* of the atmosphere and solving of the governing equations. (15)

Satellites, it said, could contribute to all three.

Satellites in fact offered very little to operational forecasting for another decade. Early satellites produced photographs—two-dimensional analog images, principally of cloud patterns—rather than digitized data. Relating these images to synoptic charts proved much more difficult than anticipated. Early predictions of rapid adoption went unrealized; satellite data became central to meteorology only after a complex, decades-long process of "technology reconciliation."[6] (Because they were dramatic, satellite photographs did see widespread use by television forecasters.)

The most important successes of early satellite imagery were political. Starting in 1964, NASA satellites broadcast low-resolution weather photographs, and NASA offered Automatic Picture Transmission receivers (figure 9.1) at cost to any country that desired them; 43 nations took up the offer. Recipients could pick up weather imagery of their own regions whenever the satellite overflew an APT receiver. This popular program muted criticism of the satellite programs' potential to centralize control of weather data in the hands of their national owners.

As for numerical data, operational forecasts first incorporated temperature soundings (vertical profiles) from satellite radiometers toward the end of 1969.[7] But satellite instruments could not replace the radiosonde network, on which they relied for calibration (as they still do today). Indeed, until about 1980 meteorologists disagreed as to whether satellite soundings increased the accuracy of weather forecasts. Authoritative esti-

**Figure 9.1**
A facsimile receiver for NASA's Automatic Picture Transmission system, circa 1964. An APT ground station cost about $30,000.
Image courtesy of NASA.

mates showed that satellite data actually *decreased* forecast accuracy at certain altitudes.[8]

## Growing an Infrastructure

For anyone who followed the technical evolution of the Internet in the 1970s and the 1980s, the numerous World Weather Watch planning reports make fascinating reading. Beginning in the mid 1960s, WWW planners wrestled with what would come to be known, a decade later, as the "internetwork problem": how to link existing, heterogeneous networks so that data could flow seamlessly and automatically from any point to any other point in the global web.[9] The Internet developers who coined the term "internetwork problem" merely needed to link *computer* networks; all the information they were concerned with was "born digital," as bitstreams. The WWW planners, by contrast, confronted the much more difficult problem of blending at least three streams of digital *and analog* information. First came traditional alpha-numeric weather and climate reports, transmitted on a wide range of systems, including telegraph, teletype, and shortwave radio. Second were facsimile (fax) transmissions of weather maps as analog images, such as the APT transmissions discussed above. Fax traffic had grown quickly. Planners projected in 1966 that each RMC would distribute about 110 maps a day by analog fax, requiring a total transmission time of 16½ hours on a dedicated telephone circuit. The report noted that digitizing such maps might cut transmission time by 80 percent. Third, planners anticipated a boom in "grid-point (computer-to-computer) traffic." They expected that the WMCs would transmit processed data (forecasts and climatological charts) to the RMCs as computer files. Using their own computers, RMCs would then produce maps and other products for further distribution to national meteorological centers (NMCs).[10]

As with most telecommunication systems of that era, the proposed structure involved high-bandwidth "trunk" connections, or dedicated telephone circuits, between the WMCs and the RMCs. RMCs would send forecasts and data to NMCs using a variety of "branch" connections, depending on cost and availability. In the future, planners hoped, point-to-point cables—faster and more reliable than broadcast technologies such as radio—would link the entire system. But they recognized that it might be decades before every NMC could acquire such a connection.[11]

Planners were well aware of the potential advantages of direct computer-to-computer data transfer. In the mid 1960s, however, imagining this was far easier than achieving it. None of the available technical means worked very well, and there were a lot of them, including punch cards, punched paper tape, magnetic drums, and magnetic tape. All had advantages and disadvantages. A few large-scale computer networks, such as the American SAGE air defense system, used modems to transfer data directly from one computer to another, but such networks required a degree of central management and technological standardization far beyond the capabilities of international meteorology.[12]

Managing the flow of data through a global point-to-point network would also require resolving a complicated addressing problem. Its nature is evident from this example: Australian contributors proposed that every network message begin with a distribution code indicating the message's destination. The system could utilize "individual codes for each Member, group codes for distribution to fixed groups of Members or centres and multiple addressing for distribution to variable groups of Members or centres. The length of bulletins or messages would be limited, several levels of priority would be indicated by codes, and bulletins or messages would be numbered consecutively on each leg of the system."[13] Soviet meteorologists objected, pointing out that "if the same bulletins have to be sent to several addresses . . . the address [will in some circumstances] consist of as many groups as there are destinations. . . . In reality user requirements . . . are so varied that the proposed system would call for the introduction of extremely unwieldy headings in each bulletin. Also, in this system, the sender of the bulletin would always have to know exactly who wished to receive it, which would lead to intricate operational telecommunication procedures and would necessitate relatively complicated overhaul of the rules and equipment in case of change of user address."[14] This mattered because the International Telegraph Code constrained meteorological messages, requiring address headers to be very short. The Soviets proposed that instead of specifying their destinations, messages should indicate only their origins. The work of determining destinations would then devolve to the regional and national telecommunications hubs.

To those familiar with the routing and addressing issues of the early ARPANET, Internet, and World Wide Web, such debates will sound remarkably familiar. All the problems faced by early computer network designers confronted the planners of the World Weather Watch: error control, routing, specialized telecommunication computers, buffer storage for a "store and forward" system, message discrimination, message priority, message length, start/end codes, and the thorny issue of maintaining "backward compatibility" with existing systems.[15] In the case of the meteorological data network, the need to allow for manual as well as semi-automatic and fully automatic processing complicated these problems. (See box 9.1.)

The idea of a standardized system of gridpoints (to create uniform data files for exchange between forecast centers) ran instantly aground on two objections: that "the number of grid systems in use . . . [is] nearly equal to the number of centres producing numerical analyses and prognoses" and that "new grid or coordinate systems will certainly be introduced with

**Box 9.1**
Japanese meteorological data handling circa 1968

The following long quotation gives a sense of the issues faced by World Weather Watch planners in automating data exchange:

The Japan Meteorological Agency (JMA) makes regular broadcasts such as the JMB (territorial broadcast), JMC (shipping broadcast), JMG (sub-regional broadcast), and JMI (hemisphere broadcast), and various meteorological data exchanges through the Tokyo–Honolulu and Tokyo–New Delhi circuits over the northern hemisphere exchange system. The data necessary for these broadcasts and exchanges are collected by national and international teletype networks and by reception of foreign broadcasts, and incoming meteorological data reach above 500,000 groups per day (one group consists of five letters). From these, selected data are checked, edited, and perforated on paper tape to be fed to the transmitter for broadcasting and exchange on time-schedules. The number of edited bulletins reaches about 4000 a day and more than 20 kinds of format are used. In order to meet this situation an on-line real-time system called the Automatic Weather Data Processing and Communication Control System (APCS) has been projected.

Discrimination by machine meets with the following difficulties:

(a) The countries within JMA's data-collecting area transmit their data in different arrangements (for example, of heading, indication of observation time, and starting and ending symbols).

(b) The meteorological broadcasts made by centres in various countries are the main sources of JMA's data at present. It will be a fairly long time before data can be collected reliably by point-to-point telecommunication channels. Thus incoming data are not free from errors due to noise, deformation of pulses or mishandling of data. The mere reading of headings only, therefore, does not guarantee the correct discrimination of the messages. Further, no request can be made for the retransmission of any missing or unknown part of the message.

(c) The forms now in use for message transmission are designed so as to be discriminated by human beings, and starting and ending symbols are therefore not always used, which causes a great problem in discrimination by machine. Successful machine discrimination is the key to the automatic data-processing, and yet it is the most difficult problem to solve. Discrimination by human beings, backed up by broad experience and knowledge, makes proper treatment of data possible by examination of the contents even if the headings are missing. Any deformation of contents or format by noise can be recognized and amended immediately. The machine, on the other hand, treats data as a series of mechanically composed units and is unable to detect errors that can easily be recognized by men. In this situation, programming of data is not an easy task.[a]

a. World Meteorological Organization, "Global Telecommunication System: Methods and Equipment for Automatic Distribution of Information," World Weather Watch planning report no. 24, 1968.

new numerical models." These objections made agreement on a single gridpoint system extremely unlikely.

Finally, it was unreasonable to expect that most national meteorological centers would be well served by gridded numerical data in the near to medium term. Instead, data "should be in the form most convenient for the users." In view of the wide range of data practices in meteorological services around the world, these would include "gridpoint values for computer-equipped NMCs," "curve co-ordinates to feed plotters," "tabulated form for reception on line printers," "point-to-point facsimile transmissions," and "facsimile broadcasts."[16] System developers thus confronted the fact that a complex combination of analog and digital data would have to be transmitted over a wide variety of analog and digital communication systems.

**An Analog-Digital Internetwork**

The solution reached by the WWW developers was pragmatic. A "stream" of activity would introduce "proven" technology into the existing data network. Meanwhile, a parallel stream would focus on computer models and automated digital data systems, with a goal of moving these to operational status worldwide in the early 1970s.[17] Thus, although computerized, automated data collection and processing would eventually dominate the system, the World Weather Watch began its life as a highly heterogeneous analog-digital internetwork. The pragmatism of this strategy, which built on the installed base (including existing standards and practices), led directly to its success.

In 1967, after four years of intensive planning, the Fifth World Meteorological Congress approved implementation of the World Weather Watch. The Fifth World Meteorological Congress's response to the plan's technopolitical aspects deserves special notice. Planners had anticipated that small and developing nations might balk at network centralization, and indeed at least one delegate to the Fifth Congress did object that the WWW would "make large [national weather] services larger and small services smaller."[18] But planners had already dealt with this objection by linking the WWW directly to an expansion of the existing Voluntary Assistance Program. The document reviewed by delegates noted that "one of the most serious obstacles towards the achievement of [the WWW's] objectives is the lack of sufficient skilled meteorologists . . . in many countries. The successful implementation and operation of the WWW therefore depends on an adequate program of education and training."[19] Clearly, delegates from developing countries saw the WMO's Voluntary Assistance Program as linked to the WWW's success. Minutes of the Fifth Congress are peppered with enthusiasm and thanks offered on behalf of Senegal, Bulgaria, Cuba, Brazil, Argentina, Tunisia, Algeria, and other developing countries.

The strategy for planning and implementing the WWW is best described as lightweight central coordination combined with decentralized design, testing, and implementation. This helps to explain how the WMO could "do" so much with so little. As the political scientist John Ruggie put it in 1975, "the World Weather Watch 'is' national weather bureaus doing what they had always done, doing some things they had never done, and doing some things differently than in the past, all in accordance with a collectively defined and agreed-to plan and implementation program."[20] WMC-A, for example, operated within the US National Meteorological Center and the National Environmental Satellite Service of the National Oceanic and Atmospheric Administration, while the US National Climatic Center provided the archival base for WMC-A.[21] The WMO served mainly as a relatively neutral umbrella and project forum. National weather services provided virtually all the equipment, funding, facilities, and staff for the entire enterprise. Similarly, virtually all the numerous planning documents published under the WMO's aegis were in fact prepared by national weather services and research centers at their own expense.

Nonetheless, the WMO's authority and neutrality were crucial in legitimizing the entire activity. They helped to ensure broad international participation and to minimize at least the appearance of dominance by any single nation. Although the wealthier, more developed weather services (notably those of the United States, the United Kingdom, Germany, Japan, and Australia) carried out the majority of the work, WMO sponsorship helped engage weather services from other countries that might not otherwise have participated.

An odd feature of WWW status reports reflects this decentralized structure: the system's overall cost is literally never discussed. WMO budgets rose substantially after the plan was approved, but they remained minuscule in comparison to the many billions of dollars ultimately spent on the global observing, telecommunication, and data processing systems, including satellites. A very rough estimate of WWW costs can be derived from US reports on the "World Weather Program" in the 1970s. For fiscal year 1971, the United States contributed $1.6 million in "direct" costs "specifically in support of the World Weather Program" (i.e., the WWW). The large majority of these funds supported the WMO Voluntary Assistance

Program. The rest supported atmospheric pollution monitoring stations in Hawaii (Mauna Loa) and in the Arctic. In addition, the United States recorded $41.5 million in "indirect" funding, "primarily for other agency needs but also fulfilling a World Weather Program need."[22] This amount covered mainly satellite observing systems and computer modeling at the World Meteorological Center in Washington. By 1981, direct WWW funding reached $2 million, and "indirect" contributions $122.1 million.[23]

Since the 1960s, the United States has typically borne between one third and two thirds of total world expenditures for meteorology. Using this proportion as a rough guide, we can estimate total annual costs for WWW *operational* systems at approximately $90–135 million in 1971, rising to $250–375 million in 1981. However, the operational World Weather Watch—consisting of the global observing, telecommunication, and data processing systems for weather forecasting—was only one aspect of the overall WWW Program. It also included the Global Atmospheric Research Programme (GARP) and the Systems Design and Technical Development program (SDTD). US annual budgets for these two elements amounted to $64 million in 1971 and $104 million a decade later (including both direct and indirect contributions). Ruggie argued that this decentralized funding and implementation made the WWW a paradigmatic "international regime," defined as "a set of mutual expectations, rules and regulations, plans, organizational energies and financial commitments, which have been accepted by a group of states."[24] Though clearly true, this analysis does not go far enough, since it does not include the vast *technological* networks underlying WWW operations. The expectations, rules, plans, organization, and financial commitments were not the end of the story but the middle of it. They were all aspects of welding together a (relatively) smoothly functioning infrastructure built principally from existing systems and networks.

The final explanation for the WMO's achievement is that, since most of its components were already in place, or were already being built for other reasons, WMO standards and technical guidelines served to orient component systems to each other and to smooth interactions among them. Acting as gateways between pre-existing systems, these standards and guidelines reduced friction in the global data network. Without understanding it in exactly these terms, WWW planners adopted the architecture most likely to result in a robust infrastructure: an internetwork architecture. They strengthened and standardized links between national meteorological services, but left them substantially intact. They centralized some (not all) data reporting and global forecasting, but left finer-grained data and national forecasting to the national services. Finally, they made use of both new incentives and existing norms of scientific culture to draw in new contributors.

The World Weather Watch *could* be understood as many of its planners saw it at the time: simply a series of incremental improvements to an existing global network, driven by an inexorable process of technological change. But to accept this interpretation is to misconstrue the WWW's significance as a technopolitical achievement. It marked the successful transfer of standard-setting and coordinating powers from national weather services to a permanent, globalist intergovernmental organization. Unlike its many predecessors, this global data network has persisted now for over 40 years. It is a genuinely global infrastructure that produces genuinely global information. Virtually all nations contribute data and receive, in turn, WWW data products.

In the following section, we will see how the WWW's research arm, the Global Atmospheric Research Programme (GARP), helped to push the WWW to its full potential, to institutionalize research on the global circulation, and ultimately to create an infrastructure of climate knowledge.

### The World According to GARP

Research was always conceived as an integral activity of the World Weather Watch. Everyone knew that a fully functional space-based observing system and global data network would generate previously unimaginable possibilities. The *First Report* deferred detailed definition of WWW research plans to a panel of experts. Various such panels met during 1965, and in 1966, Planning Report No. 5—written by the D-Day meteorologist Sverre Petterssen—laid out the principal objectives for World Weather Watch research:

(a) To develop a deeper understanding of the global circulations of the atmosphere and the associated system of climates;
(b) To place weather forecasting on a firmer scientific basis . . .
(c) To explore the extent to which weather and climate may be modified through artificial means.

Noting that "research in meteorology and related sciences is presently carried on almost exclusively on the national level," in national weather services and universities, Petterssen described a Global Atmospheric Research Programme.

Petterssen's report placed heavy emphasis on improving understanding of the general circulation. Weather forecasts could best be improved by "considering the atmosphere as a single physical system." This objective would require better knowledge of "the basic atmospheric currents, the behavior of large-scale weather systems, the interactions between atmospheric phenomena of different space-scales and life-spans, the atmosphere's responses to energy sources and sinks and to mechanical and other interactions at the earth-atmosphere interface."[25] Working toward these goals would help to improve the World Weather Watch observing system, because achieving them would require certain kinds of data, such as measurements of radiative fluxes at the top of the atmosphere, that had never been collected before. Initial GARP discussions had already outlined a series of plans. These included a "tropical meteorology project" (to build knowledge about that little-understood, data-poor region) and a "technology development programme" (to create new instrument platforms, such as drifting balloons that would float for months at a specific pressure level, unmanned aircraft, automatic weather buoys, and new satellite and telecommunication techniques).

GARP's "weather and climate modification" objective resonated with certain non-meteorological agendas, especially with military and agricultural interests' dreams of deliberate weather control. "Climate modification" meant something different; by 1966 worries about greenhouse warming had begun to emerge, but so had concerns for aerosol-induced cooling and other anthropogenic climate disruptions, often lumped all together under the heading "inadvertent climate modification." Modelers wanted to provide answers, but just as in Callendar's day those answers hinged on understanding the global circulation.

By 1969, GARP was well underway. Chairing GARP's Joint Organizing Committee was the Swedish meteorologist Bert Bolin, whose had begun his career in the "Scandinavian tag team" that had helped carry out the first numerical weather prediction experiments at the JNWP Unit and had connected the Swedish and American research efforts. Bolin had chaired the IUGG Committee on Atmospheric Sciences that had recommended GARP in the first place, so he was a logical choice. Joseph Smagorinsky also played a major role in the organization. After a 1967 study conference, members of GARP's Joint Organizing Committee developed plans for a series of "global experiments."

"Global experiment" had a very specific meaning in the GARP context. A "global experiment" would include

The implementation of an agreed GARP Global Observing System, with its subsystems . . . integrated into a unified system with internationally agreed characteristics. . . .

The transformation of the signals provided by the various components of the composite observing system into meteorological parameters . . . in accordance with certain agreed standards, and the preparation of global data sets in an adequate form;
The application of the data sets thus obtained to numerical experiments performed by research groups working with adequate models in an internationally coordinated programme
The storage of the data and the application of retrieval procedures for future research.[26]

In the terms laid out in this book's introduction, this is a complete sequence of monitoring, modeling, and memory.

Two aspects of the above description deserve special attention. First, as planners understood it, two different experiments could be conducted at once: a "test-type experiment," testing forecast models' ability to predict the full global circulation, and a "fact-finding type experiment," in which a well-calibrated, long-term data set would provide "a description of the *global* behavior of the atmosphere and thereby [show] in what way existing models fail to describe it." The latter would help scientists find "new hypotheses" with which to modify their models.[27] Earlier in the book, we saw how numerical modeling began to alter scientists' views of the nature of "experiments"; GARP's "global experiments" marked the institutionalization of one such change. You could collect data from the world, then perform your experiments inside the computer. You were both experimenting *with* the model, to find out what it could tell you about the world, and experimenting *on* the model by changing its parameters to make it deliver better results. Second, the "signals" retrieved from instruments were not described as data, though they could *become* data through "transformation . . . into meteorological parameters." This applied primarily to satellite instruments, which measured quantities (such as radiation at various wavelengths) that could be connected with primary meteorological variables (such as temperature) only through data modeling. But it also applied to the blending of data from heterogeneous sources into uniform data sets.

The plan issued in 1969 called for a year-long First GARP Global Experiment (FGGE, pronounced "figgy") to begin around 1975. For the interim, GARP organized a smaller effort, the GARP Atlantic Tropical Experiment (GATE), which took place in the summer of 1974. Forty ships, twelve aircraft, and numerous buoys contributed by twenty countries took part in the experiment. Based in Senegal and co-directed by an American, the experiment lasted 100 days. Ships and aircraft, with about 4000 personnel on board, cruised the tropical Atlantic Ocean between the west coast of Africa and the east coast of South America. NASA's first geostationary weather satellite, SMS-1, became operational on the first day of GATE. Land stations in tropical South America, Central America, and Africa contributed surface, radiosonde, and balloon readings.

The amount of data retrieved from GATE far exceeded the amount normally available for forecasting, and the data-handling procedures required a great deal of post-processing to meet the more stringent quality standards GATE imposed. Data friction was evident in the long (planned) delay between collecting observations and delivering finished data sets. Ten National Data Centers—in Brazil, Canada, France, West Germany, East Germany, Mexico, the Netherlands, the United Kingdom, the United States, and the Soviet Union—validated their parts of the overall data set before delivering them to five Subprogram Data Centers in 1976, well over a year after the observing period. Each subprogram then took an additional year for experiments with the data before delivering finished data sets to the World Data Centers in Washington and Moscow.[28]

Often called "the largest scientific experiment ever conducted," the Global Weather Experiment (as FGGE was known outside the meteorological community) took place in 1978–79. Its scale dwarfed that of GATE. FGGE amounted to the first test of the complete complement of World Weather Watch observing systems. Originally planned for 1974, then 1976, after many delays FGGE finally got underway at the end of 1978. By then full global coverage had just become available from five geostationary satellites in equatorial orbits and four polar orbiters. (One additional satellite failed in orbit just before the FGGE year began.) A "build-up" year before the actual experiment permitted testing of the many new and additional systems added to the existing observing system.

Throughout FGGE's 12-month run, a "special observing system" consisting of 368 drifting buoys, 80 aircraft, and the experimental NIMBUS-7 satellite was added to the WWW's basic observing system. Up to 17 additional aircraft tested an experimental "aircraft-satellite data relay," which reported data to the global telecommunication system in real time by relaying it through the geostationary satellites. (Other aircraft data were reported after their flights, sometimes delaying the data's arrival beyond the 12-hour operational cutoff.) During two 60-day Special Observing Periods, even more systems were added, including up to 43 additional ships and 313 constant-level balloons reporting from tropical latitudes. And during two 30-day Intensive Observing Periods, nine more aircraft joined the special observing system, dropping weather balloons at strategic locations.[29]

An elaborate data management plan described a three-tier structure with "real time" and "delayed" modes. Data designated "Level I" consisted of raw sensor signals, which had to be converted (through models) into meteorological parameters, or Level II data. These included readings from thermometers, barometers, and other primary meteorological instruments, as well as processed Level I sensor signals. In other words, some data, such as thermometer readings, originated as Level II, but *all* Level I data had to be converted first.[30] Level III represented data that had been "assimilated," or processed into gridded global data sets, including model-generated values for all gridpoints.

Planners knew that the existing WWW communication, processing, and forecast systems could not handle the volumes of data FGGE would generate. During its most intensive observing periods, FGGE increased rawinsonde coverage by 30 percent and tripled the amount of data from aircraft and satellites.[31] Aside from the high volume, some of these data had never been collected before. Their effect on forecast quality had to be assessed before they could be included as basic WWW systems. Therefore, FGGE split incoming data into Level IIa and Level IIIa "real-time" information (reported normally to the WWW) and Level IIb and IIIb "delayed mode" data (which either did not arrive before the 12-hour operational cutoff or required further processing before use). "Delayed mode" data had to reach FGGE data centers within 3 months. To reduce data friction, FGGE planners made elaborate efforts to standardize data formats and media and to have "quality control indicators" included. Not all of these standards foresaw every possible problem. For example, satellite data tapes did not include fields specifying which data producer had processed them. When errors were discovered in one set of satellite data after it had been merged with all the others, it was difficult to correct the bad data. Other failures to comply with standards forced data managers to convert formats, eliminate duplicates, apply quality controls, and so on; as a result, it took 75 days to process 10 days' worth of reports.[32] Clearly data friction remained a major force. Nonetheless, the "delayed mode" approach allowed FGGE data managers eventually to accumulate most of the weather observations taken during the FGGE year (see table 9.1).

In the real-time FGGE data stream, national meteorological centers transmitted Level IIa data to the World Meteorological Centers at Washington, Moscow, and Melbourne. From these primary data, the

**Table 9.1**
Global Weather Experiment (FGGE) data levels and data products.

| | Level I<br>Raw signals | Level II<br>Meteorological parameters | | Level III<br>Global analyses | |
|---|---|---|---|---|---|
| | | IIa: Real time | IIb: Delayed | IIIa: Real time | IIIb: Delayed |
| **Observations** | Raw signals: satellite radiances, balloon locations | Real-time operational systems; direct readings plus parameters derived from Level I signals | All available up to cut-off time (up to 6 months after date of observations), including parameters derived from Level I | Level IIa data set | Level IIb data set |
| **Data products** | Signals referenced to Earth coordinates | Global observations referenced to space/time locations | Global observations referenced to space/time locations | Gridded global data (assimilated, including model-generated data for all gridpoints) | Gridded global data (assimilated, including model-generated data for all gridpoints) |
| **Global data sets (stored)** | Numerous | Two complete sets (WMC-A and B), one limited set (WMC-C) | One surface data set (USSR), one Space-Based and Special Observing System (Sweden) | Two complete sets (WMC-A and WMC-B), one limited (WMC-C) | Two complete sets (GFDL and ECMWF) |

WMCs created two principal products: first, complete Level IIa global observations specified by their locations in space and time, and second, the Level IIIa gridded global analyses required by NWP forecast models.[33] This operational mode tested and advanced the capabilities of the still developing World Weather Watch.

The second, "delayed mode" Level IIb data stream included all of the Level IIa data plus other data that arrived after the cutoff, such as some kinds of satellite retrievals, station reports from remote regions, and other reports delayed for various reasons. Level IIb data products specified in Earth coordinates included a complete record of surface observations, collected in the Soviet Union, and a Space-Based and Special Observing System data set collected in Sweden. These collectors then transmitted complete observational data sets to GFDL and the ECMWF, which ran them through 4-D assimilation systems to produce gridded global data, including model-generated data for all model gridpoints. Initially, GFDL required two days of processing time to assimilate this volume of data for each observing day.

The Level IIIb "delayed mode" analyses proved, unsurprisingly, considerably better than the Level IIIa operational ones. Produced many months after observations, these post hoc analyses found no uses in forecasting, of course. Instead, researchers wanted them in order to see exactly how the additional observations affected the analyses, to assess the defects of the operational analysis models, and to test improved operational models. In the early 1980s the NASA Goddard Space Flight Center (GSFC) analyzed the Level IIb data sets repeatedly, applying different data assimilation schemes (see chapter 10) in order especially to determine how the then-novel satellite observations affected them; these represented early reanalyses, though that term did not gain currency until later on. The surprising conclusion was that satellite data, while beneficial in the data-sparse southern hemisphere, had almost no effect on the quality of analyses in the well-instrumented northern hemisphere.

The "delayed mode" data stream existed mainly to serve GARP's second objective: improving understanding of general circulation and climate. That goal became increasingly important during the 1980s, as global warming moved inexorably into the policy arena, bringing both imperatives to quantify climatic change and resources to do so. The FGGE IIIb analyses—the most detailed data image of the global circulation ever created—provided a starting point.

## The Weather Information Infrastructure Comes of Age

GARP and FGGE—the Global Weather Experiment—marked a quantum jump for meteorology in several key respects. The FGGE data set was fully global, including much more coverage of the southern oceans than ever before, and far more extensive and precise than any previous data set. Analyzing the FGGE data occupied legions of meteorologists throughout the following decade. The techniques they developed to do it led not only to improved data assimilation for forecasting, the subject of chapter 10, but also eventually to reanalysis, a major innovation in climate studies that is the subject of chapter 12.

Although weather satellites had already been flying for two decades, only during FGGE did they become fully integrated into the global forecast system. Two important conclusions emerged from the early data analysis. First, including satellite and aircraft data improved Northern Hemisphere forecast skill by more than 1.5 days. Second, satellite and aircraft data together contributed more to forecast skill than did conventional upper-air measurements from weather balloons.[34] These results cemented the role of satellites as the most important instruments in the arsenal of global meteorology. Today, when meteorologists mention the "satellite era," they usually mean the period beginning in 1979, the FGGE year.

Another potent effect of GARP was to tighten the relationship between data and modeling. As we saw in chapter 7, by 1965 GCMs had passed from a pioneering into a proliferation phase. GARP inaugurated regular meetings among the rapidly growing number of modeling groups. Both GATE and FGGE used models to test observing system design. With dummy ("synthetic") data from any planned configuration, one could run forecasts using various models and compare their results. This would tell you something about the advantages and deficiencies of any given system configuration. GATE and FGGE also needed data assimilation models to process the observations (see chapter 10).

Like the WWW itself, FGGE was a deliberate exercise in infrastructural globalism. Already in 1969 GARP planners noted that FGGE might "play a key role, within GARP, as a double test: the test of the technological means [for] . . . a single functional global observing system; and the test of the degree of international cooperation that could be expected to carry out such an ambitious enterprise. . . ." In the event, some 140 nations contributed equipment, satellites, ships, aircraft, data management, personnel, and money to the project. Although the major commitments came from around 20 of these nations, this still marked an amazing level of cooperation, and future programs built on its success.

Finally, GARP planners never lost sight of the climate objective. In the run-up to FGGE, modelers met regularly to compare climate GCMs. At one such meeting, in 1974, fifteen groups from the Soviet Union, Australia, Japan, Denmark, Sweden, England, and the United States presented a total of 19 atmospheric GCMs and one ocean GCM.[35] By 1978, the FGGE "build-up" year, GARP's Joint Organizing Committee (now under Smagorinsky's leadership) sponsored a climate modeling conference to compare climate model simulations at the seasonal time scale. Once they were finally analyzed—a process that took several years—FGGE data provided modelers with a year of relatively well standardized global data against which to evaluate their models. Even as FGGE was winding down, GARP's Climate Dynamics sub-program was reconstituted as the World Climate Research Program, with Smagorinsky as its first chair. Chapter 14 places these events in their political context. Here the significant thing is the role of this long-term, widely international program in building a climate modeling community and in creating methods, opportunities, and data for them to compare and evaluate their models.

Thus the World Weather Watch, GARP, GATE, and FGGE all marked significant steps toward assembling a weather and climate knowledge infrastructure. The internetwork strategy adopted for the WWW continued throughout. The WMO, ICSU, and smaller groups helped organize the infrastructure, but there was no central control, no authoritative system builder at the heart of it all. It was a web of networks of systems, most managed at the national level but some (e.g. satellites) by individual agencies. At the level of data management, this inevitably created breakdowns. At the level of modeling, the internetwork approach relied on a kind of coordinated diversity. Modelers compared their models and shared them, but no single system design took hold. If anything, GARP planners resisted this. Smagorinsky continually reminded modelers that "the problem of modeling the atmosphere is not so straightforward that there is only one path. . . . For many groups to work on the same problem is wasteful duplication only if they make the same mistakes. The cause is moved ahead by the collective activity of different groups when they make different mistakes."[36]

# 10  Making Data Global

Preceding chapters examined the rise of a global weather and climate data internetwork, from the kluged pre–World War II system to the systematic planning of the World Weather Watch for coordinated global observing, communication, and data processing. This is what I have been calling *making global data*: collecting planetary data in standard forms, through interconnected networks, to build data images of global weather and general circulation. Yet difficult as it was, collecting planetary data was only one dimension of the overall effort. This chapter explores the complementary project of *making data global*: building complete, coherent, and consistent global data sets from incomplete, inconsistent, and heterogeneous data sources.

*Making global data* revolved around standardizing data collection and communication. Standards act as lubricants.[1] They reduce friction by reducing variation, and hence complexity, in sociotechnical processes, and they "black box" decisions that would otherwise have to be made over and over again. In both respects, standards are a lot like laws—human laws, that is, not natural ones—as the common distinction between *de jure* ("in law") and *de facto* ("in fact") standards recognizes.[2] Like (human) laws, *de jure* standards are typically negotiated by central bodies and mandated from above. Yet in the end, also like human laws, *standards are always applied or actualized locally, from below*. New standards must therefore always struggle against some combination of the following:

- institutional inertia
- funding constraints
- technical difficulties of application
- problems of integration with other instruments, systems, and standards
- operator training deficits leading to incorrect implementation
- differences among local interpretations of the standard

- passive and/or active resistance from organizations and individual practitioners

For these reasons, few if any standards are ever perfectly implemented. This creates an apparent puzzle: how do standards function at all in the face of widespread differences of implementation? The reason is that in practice people find ways to ignore or work around minor deviations so that common effort can proceed. The preceding chapters have shown numerous instances of this mutual adjustment in global meteorology. *Making global data* begins as a story about grand visions of centralized systems, but it ends as a story about spreading instruments, practices, and knowledge piecemeal across the globe, in the full knowledge that collecting data from the entire planet could only work by accepting deviation, heterogeneity, inconsistency, and incompleteness.

By contrast, *making data global*—building complete, coherent, and consistent global data sets—is a top-down project, strongly shaped by data friction and computational friction. As we have seen, simply assembling global data in one place requires a monumental effort. But gathering the numbers is only the beginning. One must trust them too. Methodological skepticism is the foundation of science, so creating trust is not an easy task, nor should it be.

Precisely because of the deviations and inconsistencies inherent in standards as practice, investigators rarely if ever trust data "out of the box," especially if the data come from multiple sources. To gain confidence, investigators must check the data's quality. This can—and often does—descend to the level of inspecting individual measurements and weather station metadata—a process I described in chapter 1 as "infrastructural inversion." Further, to integrate data from multiple sources, scientists must convert them into common units and formats. How to handle data collected at non-standard observing hours must also be addressed. In effect, *making data global is an ex post facto mode of standardization,* dealing with deviation and inconsistency by containing the entire standardization process in a single place—a "center of calculation," in Bruno Latour's words.[3]

Computer models demanded a degree and a kind of standardization never previously needed in meteorology. NWP models required that values be entered at every gridpoint, both horizontal and vertical, even where no actual observations existed. Missing gridpoint values had to be interpolated from observations, or even (if necessary) filled in with climatological norms. "Objective analysis" computer programs gradually took over this work, replacing human hand-and-eye methods. The most audacious step in this process was to use the previous computer-generated forecast as a "first guess" for empty gridpoints, so that "data" generated by one model (the previous NWP forecast) became input to a second model (the "analysis"), which then became input for the forecast model's next prediction. Ultimately, these techniques transformed the very meaning of the word 'data' in the atmospheric sciences.

Latitude and longitude had long supplied a standard grid for Earth science and navigation, but numerical weather prediction and general circulation models elevated geospatial grids to a qualitatively new status. Adding the third dimension, computer model grids became an ideal Earth: the world in a machine. Data from the real atmosphere had first to pass through this abstract gridspace, which transformed them into new data calculated for the model's demands. Henceforth all predictions would be based on these calculated or simulated data, rather than on actual observations alone.

As numerical weather prediction's techniques of data analysis and data assimilation matured, the data images they created often proved more accurate than many of the observations on which they were based, at least in data-sparse regions. By the 1980s, computerized data assimilation systems routinely generated consistent, complete, uniformly gridded data for the entire Earth. These data became the most accurate available images of the planetary circulation over short periods of time, so climate scientists began to use them for general circulation studies. However, the constant improvement of data analysis and assimilation techniques made it nearly impossible to use those techniques to study climatic change. Frequent revisions of the models, made in the interest of improving their performance on forecasts, rendered any given set of data incommensurable with those generated earlier or later by a different analysis regime. Therefore, by the late 1980s some climatologists began to clamor for reanalysis.

In reanalysis, investigators reprocess decades of original sensor data using a single "frozen" weather analysis and forecasting system. The result is a single complete, uniformly gridded, physically consistent global data set. Reanalysis offered a comprehensive solution to data friction such as that created by heterogeneous data sources, including satellite radiances not easily converted into traditional gridded forms. With reanalysis, many hoped, it would be possible to produce a *dynamic* data image of the planetary atmosphere over 50 years or more—essentially a moving picture that might reveal more precisely how, where, and how much Earth's climate had changed.

Global reanalysis might produce the most accurate, most complete data sets ever assembled. Yet the majority of gridpoint values in these data sets would be generated by the analysis model, not taken directly from observations. Whether or not it eventually leads to better understanding of climate change—a matter about which, at this writing, scientists still disagree—reanalysis represents a kind of ultimate moment in making data global. (See chapter 11.)

**Automating Error Detection and Data Entry**

The principal meteorological benefits of high-speed automatic computing machines during the next few years will lie as much in the processing of large assemblages of data as in numerical forecasting.
—Bert Bolin et al., 1962[4]

In the first decade of NWP research, investigators quickly realized that computers could help them with much more than the weather forecast itself. They could also automate much of the necessary "pre-processing": reading data from teletype receivers, checking them for errors, coding them in machine-readable form, and feeding them into the computerized forecast model. As time went on, these forms of automation contributed as much to meteorology as the forecast models themselves, reducing data friction that might otherwise have slowed computer forecasting to the point of unusability.

Early reports from the US armed services' Joint Numerical Weather Prediction Unit noted that it was "not uncommon to have errors in the reported height of a pressure surface of from 500 to 5000 feet." According to estimates by JNWPU staff, half of these errors stemmed from the manual handling of data prior to analysis. Transmission errors accounted for another 25 percent. The remaining 25 percent came from mistakes in observation or recording at the originating weather stations. Small errors did not matter much, since they were generally eliminated by smoothing "inherent in the analysis process." (See next section.) However, large data errors could cause serious problems, so they programmed the computer to find suspiciously large and small values, as well as values that differed greatly from nearby observations.[5]

The JNWPU then planned "a program for the complete mechanization and de-humanization" of data transmission. Analysts would no longer have to transcribe incoming data from teletype to written lists and/or punch cards. Noting that the "data collection and distribution system was, of course, not designed for and is not particularly suitable for numerical weather prediction purposes," JNWPU staff recommended modifications, such as the use of checksums in data transmission.[6] They also noted the variety of codes in use for international data exchange, including "at least three different upper wind codes and two different radiosonde codes, one of which contains three different possible combinations of units."[7] Message headers identified the reporting station using several different codes; if these headers were not received, as often happened with the teletype transmission system, it could be difficult or impossible to determine a report's origin.[8] Taking data directly from the teletypes, then, would require relatively complex programs and processing. In an era when every byte of computer memory remained a precious resource, streamlined meteorological codes might raise the speed and quality of forecasts significantly.

But numerical forecasters did not wait for this standardization to happen. They made their first attempts to handle incoming teletype data directly in 1957.[9] First they manually edited paper teletype tapes—arriving on ten or more channels simultaneously—to reduce duplicate data. The edited tapes were then punched onto cards by machine. A mechanical card reader then converted the data to magnetic tape and fed it into the computer, which decoded the reports and checked radiosonde data for errors. (See figure 10.1.) This "automatic" processing included a considerable number of manual steps. For example, the computer identification and decoding process kicked out reports with incomplete or missing station identifiers. If the team could work out these reports' station locations from context, they inserted cards manually.

By 1959 the new method had reduced the time required for punching data nearly in half, from ten hours to six—even as data input requirements increased, with NWP models adding more analysis fields and more vertical levels. Paper tape and punch cards remained standard pre-processing media for quite some time. Another ten years passed before electronic transfer from teletype signals directly to magnetic tape and other electronic media became the norm in operational forecasting.[10]

Frederick Shuman neatly summed up both the novelty and the difficulty of the technique, noting that it "introduced a new concept called 'automatic data processing' before that terminology began to have a wider meaning. . . . [The] reading of remotely manually prepared teletype texts into computer-quality databases . . . has many of the qualities of reading natural languages. . . . The input text contains observations in a dozen or so formats, with variations and errors normally found in language that must be recognized in context amid extraneous material."[11] Interpreting the "natural language" of incoming data almost immediately grew closely

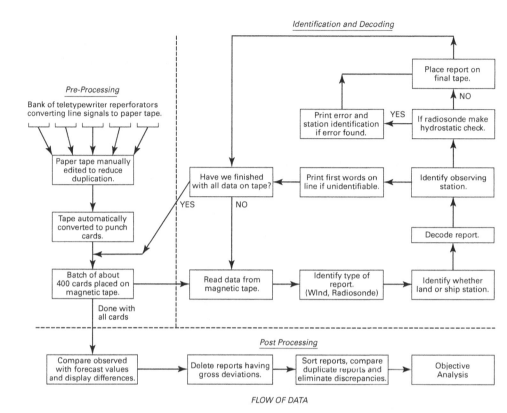

**Figure 10.1**
"Automatic" data processing at the JNWPU in 1957. The entirely computerized part of the procedure appears in the upper right portion of the diagram, separated from the rest by dashed lines. Manual editing occurred at several points during pre-processing and post-processing.
*Source*: H. A. Bedient and G. P. Cressman, "An Experiment in Automatic Data Processing," *Monthly Weather Review* 85, no. 10 (1957), 334.

linked with another major development made possible by high-speed computing: "objective analysis."

**Objective Analysis**

From the dawn of synoptic forecasting, weather forecasting comprised three principal steps: (1) collect the available data, (2) interpret the data to create a picture of the weather situation, and (3) predict how that picture will change during the forecast period. The second step, originally known as "diagnosis," transformed raw data from a relatively few points into a coherent, self-consistent picture of atmospheric structure and motion.[12] As in a medical diagnosis, forecasters combined theory and experiential knowledge to reach a shared understanding of reality from incomplete and potentially ambiguous indications (symptoms). Their diagnosis might or might not be correct, but before they could move on to prognosis (i.e. forecasting) they had to have one. For early NWP, diagnosis or "analysis" proved the most difficult aspect of forecasting. Ultimately, it was also the most rewarding. In the long run, analysis would also connect forecasting with climatology in new, unexpected, and important ways.

To understand how numerical weather prediction changed analysis, we must first review how forecasters worked in the pre-computer era. Before the mid 1950s, human forecasters performed virtually all weather data analysis themselves using a mixed set of techniques. Weather workers at national forecasting centers began with data transmitted using international meteorological codes. As the data arrived—via telegraph, teletype, shortwave radio, or other channels—one person decoded them and read them to a "chart man," who plotted them on blank maps pre-printed with small circles representing each reporting station. One chart depicted the surface. Others showed upper-air data at selected altitudes. Forecasters then studied the charts, rejecting some data as probably erroneous (either because inconsistent with other, nearby data, or because implausible for physical reasons). They also compared the charts with their own previous forecasts. Using both data and their earlier predictions, forecasters then drew maps of the present positions and motions of fronts, zones of high and low pressure, temperature gradients, and other phenomena. These analysis maps were not exactly the same as the weather maps published in newspapers, but they looked very similar, with isolines of pressure and temperature, high-pressure and low-pressure centers, wind speed and direction, and so on. (See figure 10.2.)

Before numerical weather prediction, analysis was an interpretive process that involved a shifting combination of mathematics, graphical techniques, and pattern recognition. Human interpretation played a crucial role in data collection; for example, analysts could sometimes use incomplete station reports if they could deduce the reports' origins or contents from their context. Typically about 2½ hours elapsed from the initial receipt of data, through analysis, to the composition and distribution of forecasts.[13] Analyzing the charts required heuristic principles and experience-based intuition as well as algorithms and calculation: ". . . the manual analyst cannot be expected to use systematic and quantitative

**Figure 10.2**
Weather analysis circa 1960.
Image courtesy of National Centers for Environmental Prediction.

methods in his interpolations and extrapolations. His work is rather a complicated curve-fitting by the eye based on a number of more or less well-established rules. . . . ."[14] As a result, even with identical data, no two analysts produced identical analyses. The forecasting process proceeded in a similar way, with similarly idiosyncratic results.

Like forecasts made by humans, NWP forecasts began with analyzed data. But the analyses NWP required looked much different from the charts used by human forecasters. Instead of isolines, zones of high and low pressure, and so on, NWP models required only precise numerical values at regular gridpoints. But the model grids had nothing to do with the locations of actual weather stations, so the models could not ingest raw data directly. Nor could they generate the values they needed directly from traditional analyzed charts, since isolines rarely intersected gridpoints exactly. Therefore, values at the gridpoints had to be interpolated from the analyzed charts. The IAS meteorology group noticed this immediately, long before completing its first computer model. As a result, in 1948, under von Neumann's influence, the meteorologist Hans Panofsky began research into "objective" (i.e., algorithmic and mathematical) analysis techniques for deriving gridpoint values from observational data.[15] With objective analysis, the digital world of numerical modeling began to assert its fundamental difference from the analog world of pre-computer meteorology. Human pattern recognition, a principal basis of all earlier forecasting techniques, would give way to mathematics. Collectively, these processes began subtly to change the way in which meteorologists understood both the meaning and the purpose of data analysis. In Sweden, at least, "objective analysis seems to have aroused stronger emotions from the meteorological community than the introduction of objective forecasts. . . . Tor Bergeron's reaction to 'Numerical Weather Map' analysis was not positive to say the least. His concern was not so much that the analysis was not accurate enough, but that it was carried out numerically, *by a computer!* . . . The analysis was, for a Bergen School connoisseur, not only a method to determine the 'initial state,' but also *a process whereby the forecasters could familiarize themselves with the weather, creating an inner picture of the synoptic situation.*"[16] Bergeron, a principal developer of Bergen School methods, wrote that "one should not accept the present strict distinction made between 'subjective' and 'objective' methods. . . . In fact, all such methods have a subjective and an objective part, and our endeavor is continually to advance the limit of the objective part as far as possible. . . . Support from intuition will be necessary."[17]

Bergeron's view held true far longer than even he might have expected, and not only for the reason he described. A 1991 textbook on atmospheric data analysis noted that "the objectivity of objective analyses is largely a fiction because analyses of an event produced by two different algorithms may differ to approximately the same extent as the subjective analyses produced by different forecasters. *Every analysis algorithm embodies a mathematical or statistical model of the field structure or process, and the degree of success depends on the artfulness of the model choice.*"[18]

The computer models required values for every gridpoint. Where no observational data existed, one could not simply leave empty points; that would cause the models to produce meaningless results. Figure 10.3 illustrates the severity of the problem.

The dearth of data also posed problems for human forecasters, of course. But unlike the computer models, human forecasters already knew something about typical patterns and atmospheric structures. Usually they could fill voids in data with intelligent guesses based on experience.

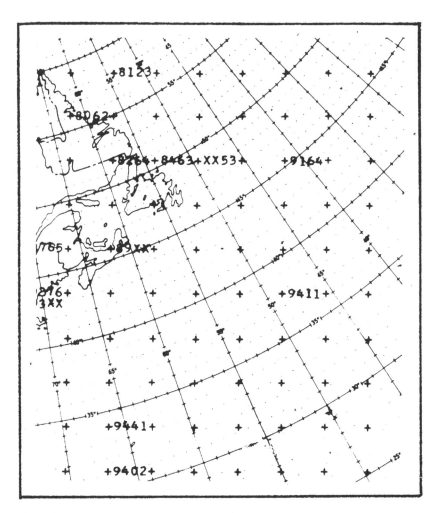

**Figure 10.3**
A section of the JNWPU objective analysis model grid (small crosses) showing observational data (numbers) before analysis. The grid covers part of the North Atlantic Ocean; on the left lie Newfoundland and Nova Scotia. Observational data were available for only twelve of the eighty gridpoints shown.
*Source*: G. P. Cressman, "An Operational Objective Analysis System," *Monthly Weather Review* 87 (1959), 369.

Therefore, in 1955, when the JNWP Unit began operational numerical weather prediction, a team of synoptic analysts gridded the analyzed data by hand. Superimposing the model grid over the analysis map, they interpolated "subjectively" to the gridpoints, using both mathematical methods and interpolation "by eye." The latter technique proved especially important in areas, such as the Pacific Ocean, that had few upper-air stations.

The aforementioned analysis technique soon proved inadequate, largely because the JNWP Unit's forecast model turned out to be considerably more sensitive than existing methods to the quality of input data. While human forecasters typically performed only first-order differentiation during interpolation, the computer model solved second-order and third-order differential equations. In other words, the models performed a first-order differentiation and then differentiated the result. They repeated this process for a third-order equation. Each successive differentiation amplified error. A more painstaking "subjective" analysis could reduce these errors, but it would take too long. For operational purposes, then, the JNWP Unit adopted a sketchier, less accurate routine that could be completed in a few hours. By this point, the forecast model itself took only a few minutes to run. Thus, not only did the "subjective" analysis process produce an unacceptably high error rate; it also consumed much more time than any other part of the forecasting process.

Other features of early models also made them more sensitive than existing practices to certain sources of error. For example, the new models gradually settled on data inputs and forecasts for the 500-mb pressure level, which occurs at about 5 km above sea level, quite a bit higher than the pre–World War II norm of 700 mb, about 3 km above sea level. (The pressure at sea level is 1014 mb.) This shift presented substantial advantages. At 500 mb, where it is less influenced by surface topography, the atmosphere behaves more regularly. Also, theorists now saw 500 mb as a "steering level" that guided the movement of surface weather. And Air Force weather reconnaissance aircraft flew at that altitude, providing crucial input data.[19] However, errors in radiosonde height determination increased with altitude, rendering the 500-mb observations less accurate than those at 700 mb. Deficiencies in the upper-air data network made this problem even worse. Intercalibration of radiosondes, especially those from different manufacturers, remained rudimentary. Slow reporting and communication meant that the full set of upper-air reports did not arrive for at least six hours after the nominal observing time.[20] To complete their task in time, analysts had to settle for an incomplete set of reports.

After about six months of routine NWP using subjective analyses as input data, the JNWPU introduced objective analysis. In fact the earliest models already contained a few interpolation routines. For example, inputs to the first JNWPU prediction model consisted of "subjective analyses" at 1000, 700, and 400 mb. But since the NWP program actually needed values for 900 mb rather than 1000 mb, it interpolated the 900-mb values. Later the program also interpolated its own output, producing forecast charts at 1000 mb and 500 mb from its results at 900 and 400 mb.[21] The 500-mb analysis and forecast became the de facto standard around 1958.

How did objective analysis work? The JNWPU's first, experimental analysis program defined a 1000×1000 km square around each gridpoint. Next, it searched for all available observed data within that square. If it found no data, the program skipped that gridpoint and moved to the next one. If it did find data, the program fitted a quadratic surface to all the data points within the search square. It then interpolated a value on that surface for the gridpoint. The program repeated this process six more times in order to generate values for the gridpoints skipped on the first pass for lack of data. At each pass it progressively enlarged the size of the search square, up to 1800 km. During these additional passes, the program treated both actual observations and the gridpoint values interpolated on earlier passes as data.[22] In this way, the program gradually generated values for every point on the grid. This technique worked well for areas densely covered by observations, but performed poorly in large data-void regions.[23]

Slightly before the JNWPU, the international NWP group in Stockholm pioneered a different approach to objective analysis. Instead of calculating all gridpoint values from the current set of incoming data, as was done in the JNWPU method, it took the most recent previous forecast as a "first guess."[24] Forecasters had long applied this technique to data-sparse areas such as the Pacific Ocean. It worked better in such regions for two reasons. First, sometimes data (perhaps from a ship) *had* been available at the time of the previous forecast, but were not available for the current analysis period (perhaps because the ship had traveled out of the area). Second, the previous forecast represented "the expected positions of high and low pressure systems which moved into the [data-sparse] area from regions in which the density of observing stations *was* sufficient to establish their identity and approximate motion."[25] Substantial pressure systems would not normally dissipate completely during a 12-hour or 24-hour forecast period. If those systems moved into a data-void region, then, the observing system might no longer "see" them, but this did not mean they were gone. The previous forecast gave a reasonable guess as to how they might have moved and developed. Using "prior information" in this way made more sense than running each new forecast only from current data, as if nothing was already known about the atmosphere's recent history. In the case of very large data voids, such as those in the southern hemisphere, climatological averages could provide plausible first guesses, a technique first operationalized in the Soviet Union.[26]

Since they were output from the previous numerical forecast, first-guess "data" gave initial values for the entire grid. The objective analysis program then corrected the first-guess values using current observational data, weighting the observations more than the forecast values. In the forecast-analysis cycle, the forecast model's output became the first-guess input to the current round of objective analysis, which in turn served as input to the forecast model for the next forecast period. The US Weather Bureau soon adopted a modified version of this approach, while the UK Met Office pursued the pure interpolation method. Most subsequent systems of objective analysis descended from these two techniques.

None of the imaginable alternatives to objective analysis were remotely practical. No one even suggested the impossible project of reconstructing the actual observing network on some ideal grid. Nor did anyone contemplate basing models on the actual distribution of observing stations. To make modeling tractable you had to use a grid, and to use a grid you needed data for the gridpoints. So gridpoint values for computer models had to be *created*, using some combination of interpolation and the first-guess method. This did not mean inventing data from nowhere. Instead, an entire subfield of meteorology developed around mathematical interpolation techniques. The following section briefly recounts the evolution of weather analysis and interpolation.[27]

**"Models Almost All the Way Down"**

With ever more sophisticated interpolation algorithms and better methods for adjudicating differences between incoming data and the first-guess field, objective analysis became a modeling process in its own right. Gridded analysis "products," as they are known, constituted *models of data*, in Patrick Suppes's well-known phrase[28]: "structures into which data are embedded that add additional mathematical structure."[29] The philosopher Ronald Giere once put the point as follows:

. . . when testing the fit of a model with the world, one does not compare that model with data but with *another model*, a model of the data. . . .The actual data are processed in various ways so as to fit into a model of the data. It is this latter model, and not the data itself, that is used to judge the similarity between the higher level model and the world. . . . It is models almost all the way down.[30]

Objective analysis fit hand in glove with automated data entry and quality control. The analysis model could compare each observation with both the first-guess "background" field and the values it had interpolated for nearby gridpoints. Any observation deviating greatly from either of these probably contained an error and could safely be rejected. (This assumption did lead, on occasion, to rejection of correct data.) Typically, other quality controls, such as checking for correct code formats and climatological "reasonableness," were also applied during this "pre-analysis" or data selection process. Systematically rejected data could help identify problems with instruments or procedures at the reporting stations. Used recursively in this way, analysis became the cornerstone of a decades-long transformation that automated much of the meteorological workflow while refining the observing system's quality.

Objective analysis also offered meteorologists a precise technique for measuring effects of the global data network's limitations. Exactly how much payoff in predictive skill could they expect from a certain increase in network density? With limited resources, where should they concentrate their efforts to improve the network? Applying objective analysis to idealized observing situations led to the somewhat surprising conclusion that network density mattered less than many believed.[31] In these studies, investigators created bogus data to simulate observing networks of various densities. These "simulated data" were then processed by analysis models, and finally by forecast models. As illustrated in figure 10.4, initially these "observing system simulation experiments" seemed to show that beyond a station separation of about 500 km increased network density would bring little improvement in either analysis or forecast accuracy. Since reducing station separation led to exponential increases in the number of stations required, the costs would be far greater than the benefits.

The increasing recursiveness of simulation-based science—exemplified here by the observing system simulation experiments—caused some members of the meteorological community to suspect flaws in the whole NWP enterprise. For example, these critics noted, as NWP became an unstoppable juggernaut, the quality of 500-mb synoptic-scale forecasts quickly became the benchmark. $S_1$ skill scores, the basic measure of forecast accuracy, did show steady improvement at the 500-mb level.[32] These 500-mb forecasts—large-scale flow patterns at an altitude roughly 5 km

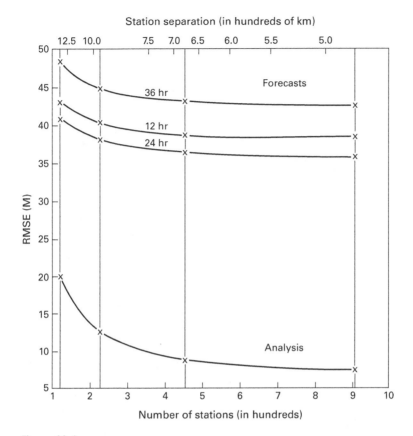

**Figure 10.4**
Effects of observing network density on the root mean square error (RMSE) of weather analysis (bottom curve) and forecasts (top three curves) in the US NMC primitive equation model, circa 1966, from experiments with the models using simulated data.
*Source*: L. S. Gandin et al., *Design of Optimum Networks for Aerological Observing Stations* (World Meteorological Organization, 1967), 18.

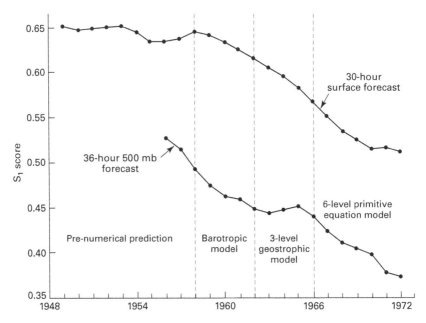

**Figure 10.5**
Changes in $S_1$ skill scores at the US National Meteorological Center, 1948–1972 (based on normalized RMS error of pressure gradient). The surface forecast remained considerably worse, and improved less over time, than the 500-mb forecast.
*Source*: C. E. Jensen, "A Review of Federal Meteorological Programs for Fiscal Years 1965–1975," *Bulletin of the American Meteorological Society* 56, no. 2 (1975), 217.

above the surface—mattered greatly to aviation. Yet the accuracy of surface forecasts lagged far behind, as can be seen in figure 10.5. These meteorologists worried that NWP's large-scale approach would never help much with predicting the small-scale turbulent phenomena, such as rains, tornadoes, and snowstorms, that mattered most to the majority of forecast users. Finally, one critic noted, "the gridpoint data to which the forecasts are compared *are themselves derived from computer analysis* of the actual observations."[33] In other words, the $S_1$ skill score simply measured the difference, at each gridpoint, between the forecast value and the analyzed value, on only one variable (pressure). Thus the $S_1$ score did not directly measure progress in predicting ground-level weather, the only thing that matters to most people. (Indeed, progress on surface forecasting remained meager for many years. Surface precipitation forecasts did not improve at all between 1966 and 1972.) During the 1960s and the early 1970s, much simpler statistical forecast models appeared to deliver roughly equivalent accuracy, raising the question of whether NWP's complexity and expensive computer requirements were really justified.[34]

Simulating data to test simulation models; using one model's output to measure another one's quality: welcome to the mirror world of computational science. These features bothered traditional meteorologists, yet they represented a major trend throughout the natural sciences.[35] Indeed, after about 1960 the recursive application of information technology to its own design became a unique characteristic of IT-based infrastructures of all sorts.[36]

Meteorologists' concerns about the limits of computational methods intertwined with amorphous apprehensions about how automation might affect human skill and professional pride. In part to assuage forecasters' fears of competition or displacement, the US National Weather Service called its global model products "numerical guidance." The service forwarded not only its finalized forecasts (NWP output), but also its objective analysis. Regional and local forecasters used both—rarely without alteration—to create their own forecasts.

Although computerized forecasts have improved steadily, to the point that most now exceed human skill, human intervention remains necessary. In a small percentage of cases involving unusual or extreme weather situations, revisions based on human experience can still improve the quality of a forecast.[37] Since extreme weather situations can be among the most dangerous for human communities, the importance of these experience-based judgments can far exceed their number.

Thus, while forecast models got most of the press, in many ways objective analysis mattered even more in the mid-twentieth-century transformation of meteorology. Because of the time required for fully mathematical interpolation, forecasters could not even have attempted the technique before electronic computers. Further, they probably would not have attempted it in the absence of NWP models' demands for gridded data, or in the absence of their error-amplifying properties. As an added, at least equally important benefit, objective analysis entrained the automation of data handling.

Without objective analysis techniques, subjective analysis would have remained the principal limitation on the speed of forecasting. At first, some weather services kept pace with NWP models' increasing appetite for data using human analysts organized into teams (figure 10.6). But as the teams grew larger, reconciling multiple analysts' contributions—with the variations inherent in manual analysis techniques—grew progressively more difficult. By the early 1960s, dramatic declines in the cost of computer

Figure 10.6
A weather data analysis "assembly line" in the 1960s.
Image courtesy of National Centers for Environmental Prediction.

power had sealed the contest's outcome. Objective analysis simply grew cheaper than human team effort.[38] By speeding the analysis process, the programs soon permitted consideration of more vertical levels than human forecasters had ever tried to handle.

**Everything, All the Time: From Interpolation to 4-D Data Assimilation**

Modelers soon realized that they could carry the logic of objective analysis much further.[39]

First, each forecast already produced a set of values for every model gridpoint. Therefore, to analyze the situation on Tuesday, they could start with Monday's forecast. In other words, rather than zero out the entire grid and start each forecast with nothing but incoming observations, they could use the previous forecast as an intelligent guess at current gridpoint values, then use the data to correct the guess. As weather systems moved

Making Data Global 269

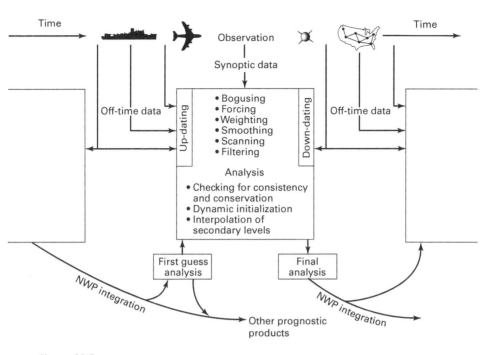

**Figure 10.7**
An early version of data assimilation: the analysis-forecast cycle at the US National Meteorological Center, circa 1971.
*Source*: S. Teweles, "Review of Efforts to Minimize the Data Set Needed for Initialization of a Macroscale Prognosis," *Bulletin of the American Meteorological Society* 53, no. 9 (1972), 855.

from data-rich regions into data-poor ones, this method would "carry" good information from the forecast into an area with few observations. For example, say that Monday's forecast showed a well-defined low-pressure zone, detected by numerous instruments, moving southeast into an un-instrumented ocean area. Tuesday's analysis would then show that same low-pressure zone, now within the un-instrumented area. Even though not verified by observations, this would still be the best information available for that area.

Second, if models could interpolate data in three spatial dimensions, they might also interpolate in the fourth dimension: time. This eventually permitted NWP models to ingest a great deal of data previously left out of forecasts, such as observations taken at times other than the main synoptic hours. Figure 10.7 shows NWP analysis procedures at the US National

Meteorological Center circa 1971, as the analysis procedure began ingesting "off-time" data, i.e. data collected outside the main synoptic hours. In the figure, "up-dating" and "down-dating" indicate that "off-time" data were inserted into the analysis as if they had been recorded at the synoptic hours. "Bogusing" refers to creating estimates, intelligent guesses, or interpolated values for data-void areas.

The analysis process corrected the first-guess field using synoptic data, but it did not simply replace first-guess values with observed ones. Since the observed values did not come from gridpoints, they had to be interpolated to the gridpoints, checked for errors, balanced against nearby observations to keep fields physically consistent, and also balanced against the first guess (which in the case of data-void areas might prove a more valuable source of information than scattered observations far from the center of the void). Forcing, weighting, smoothing, scanning, and filtering were analysis techniques used to balance and combine the observational data with the first-guess field. Meanwhile, other checks ensured that the analyzed data maintained physical consistency. For example, mass, momentum, and vorticity are all conserved in the real atmosphere; if the analyzed data did not reflect this, values had be adjusted to compensate. This mutual adjustment between model-generated gridpoint values and observations came to be known as "data assimilation."

The move from simple interpolation to assimilation marked a profound transformation in meteorology's "models of data." The number of gridpoints in even the simplest forecast models had rapidly far outstripped the number of available observations, especially in data-poor regions, and the disparity became gigantic as model resolution increased. By 2000, global forecast models with a 1° horizontal resolution and 20 vertical levels contained roughly 1.3 million gridpoints. With at least four variables (two wind vectors, temperature, and moisture), about 5 million values would have to be interpolated from the roughly 10,000–100,000 available observations—a procedure unlikely to produce good information at most gridpoints.[40]

Instead, data assimilation typically began *by interpolating the first-guess forecast to the observation locations*—the reverse of the original objective analysis scheme. Next, it compared these interpolated values with the current observations. When they differed, assimilation systems calculated a weighted difference between first-guess values and observations, recognizing that both the observations and the first-guess fields contained errors. Weighting coefficients were based on the estimated statistical error covariance of the two sources (roughly, the degree to which the errors in each source tended to be systematically similar or different). This technique, known as "optimal interpolation" or "statistical interpolation," created a disciplined way to anchor the model-generated first-guess values to the observations, producing an "assimilated" data set whose values all remained constrained by the model physics.[41] Unlike simple interpolation (which could produce physically impossible combinations of gridpoint values), data assimilation ensured consistency with the model physics, if not with the real physics of the atmosphere.

As data assimilation techniques improved, they transformed forecasting's relationship to observational information in numerous ways. Before NWP, forecasters had sought relatively small amounts of data and preferred a few trusted, high-quality sources. Low-quality data, large volumes of data, irregularly reported measurements, continuous measurements (as opposed to discrete ones taken only at synoptic observing hours), and data arriving after forecasters had completed their analysis: under the manual regime, forecasters could do nothing with any of these and simply ignored them. Computer models changed all this.

Now, in areas of the computer grids where few observations existed, almost any information source was better than none: ". . . over the areas where there is no uniform network of rawinsonde stations, aircraft reports are utilized even though an individual report gives information for just one arbitrary level at some arbitrary moment. Similarly, the velocities of clouds derived from their apparent movement between successive satellite scans are used in areas of sparse information even though the velocities and the levels to which they are assigned are gross estimations."[42] Data assimilation made each piece of observational data interdependent with all the other data nearby. This meant that analyzed gridpoint values could sometimes be more accurate than the input observations. Tests of the models using different combinations of available data supported the surprising conclusion that larger quantities of less accurate, more heterogeneous data—data often far below the quality thresholds of previous eras—produced better forecasts than smaller sets of more accurate data.[43] The secret lay in the assimilation system's ability to balance the first-guess data with these lower-quality observations through a physical model that constrained the range of possible values within physically plausible limits, thus reducing the impact of observational errors while still gaining valuable corrections to the first-guess field.

Data assimilation also transformed forecasting's relationship to time. Recall from chapter 2 that decades of effort were required to establish uniform observing hours for simultaneous observations. Analysis maps were the practical products of this work. Each map constituted a sort of snapshot—a data image or mosaic—of weather at a moment in time. Early data assimilation systems continued this tradition, but as techniques grew more sophisticated the old conception of the observing system as something like a still camera receded in favor of "four-dimensional (4-D) data assimilation." Though experiments with 4-D data assimilation began in the late 1960s,[44] the technique did not mature until the early 1980s, with the data assimilation experiments conducted during and after the Global Weather Experiment. More like a movie than a series of snapshots, 4-D data assimilation could absorb data as they arrived, building a continuous, dynamic data image of the global atmosphere, seeing everything, all the time.

Analysis produced through 4-D data assimilation thus represented an extremely complex *model of data*, far removed from the raw observations. With many millions of gridpoint values anchored to fewer than 100,000 observations, one could barely even call the analysis "based" on observations in any ordinary sense. As the data assimilation expert Andrew Lorenc put it, "assimilation is the process of finding the model representation which is most consistent with the observations."[45]

Data assimilation thus reflected a profound integration of data and models, moving forecasting onto a new terrain that has been called "computational science." 4-D data assimilation and analysis models form the core of the modern weather information infrastructure. They link thousands of heterogeneous, imperfect, incomplete data sources, render them commensurable, check them against one another, and interpolate in time and space to produce a uniform global data image. In my terms, they *make data global*. Because of this, by 1988 Lennart Bengtsson and Jagadish Shukla could write: ". . . *a realistic global [analysis] model can be viewed as a unique and independent observing system that can generate information at a scale finer than that of the conventional observing system."* Bengtsson and Shukla offered the following situation—not atypical—as evidence: From August 25 to August 30, 1985, because of a breakdown in the regional telecommunications system, virtually all data for North Africa—a land surface considerably larger than America's 48 contiguous states—were never transmitted to the European Centre for Medium-Range Weather Forecasts. Despite this, on August 29 the ECMWF's analysis model showed a powerful surface vortex in the western Sahara. A Meteosat infrared photograph of the region, taken on the same day, confirmed the reality and the position of this vortex. (See figure 10.8.) With almost no observations available for several days before, most of the data used in the analysis came from the

Making Data Global 273

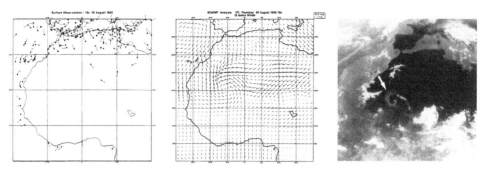

Figure 10.8
Left: Wind observations available at 12:00 UTC, 29 August 1985, five days after last receipt of most data from northern Africa. Middle: ECMWF analysis for winds 30 meters above the surface for the same date and time. Right: Meteosat photograph of the same area, 11:55 UTC, 29 August 1985. The area represented is nearly as large as the 48 contiguous states of the USA.
Source: L. Bengtsson and J. Shukla, "Integration of Space and In Situ Observations to Study Global Climate Change," Bulletin of the American Meteorological Society 69, no. 10 (1988), 1133–36.

forecast for the previous day. Thus, in a sense the analysis model "observed" a phenomenon which the data network itself did not directly register.[46] This was not merely a lucky guess. Analysis models have demonstrated their power as independent observing systems over and over again. In numerous cases, investigating a systematic disagreement between analysis models and instrument readings has uncovered problems with the instruments. Example: Radiosondes sent up from remote Marion Island, 2300 km southeast of Cape Town, South Africa, consistently reported winds that varied by 10–12° in orientation from the winds generated by the ECMWF analysis model. Investigating the discrepancy, the South African Meteorological Service discovered that its software erroneously used magnetic north, rather than true north, to calculate wind direction from the radiosonde signals.[47] The fact that the analysis model proved more accurate than the reported readings, despite the near-total absence of other observations in that region of the world, demonstrates the model's power.

### The Global Instrument: Integrating Satellites

Satellites were the most important new source of measurements since the radiosonde. The most important weather instruments rode on polar-orbiting satellites, which can overfly the entire planet roughly once every 12 hours.[48] These were the first truly global instruments, *making global data* as a matter of course. Yet *making satellite data global* took many years and required unprecedented effort, owing to data friction inherent in the new forms and to the unprecedented quantities of information satellites could provide. For example, NWP required numerical inputs, but early satellites and radar[49] provided mainly analog images—a qualitative form of information not easily converted into numbers. The meteorologist Margaret Courain designated 1957–1987 a period of "technology reconciliation" during which the weather services strove to reconcile their commitments to NWP with the new capabilities of radar and satellites. Instrument makers, satellite agencies, and "data guys" struggled to transform radar and satellite data into forms the NWP community could use. Yet even after numerical data from instruments such as radiometers became available, integrating them into the existing analysis system presented meteorologists with serious difficulties. Radiometer measurements corresponded to large atmospheric *volumes*. These measurements somehow had to be translated into the precise levels demanded by computer model grids. For these reasons, satellites initially helped mainly with local and regional forecasting. Cloud imagery from the polar-orbiting TIROS and geosynchronous ATS satellites located tropical storms at sea and gave local forecasters visual images of weather systems to accompany their forecast maps. The images made for good television, and they rapidly became a staple on the nightly news.

Yet nephanalysis (analysis of cloud types, amounts, water content, etc.) from imagery required extensive interpretation by human beings and did not directly generate any quantitative data.[50] Initially, understanding the images at all proved extremely difficult:

We got the pictures from TIROS-1 in 1960 and spent the first year figuring out what we were seeing. The vidicon TV camera was mounted on a spin axis in the orbital plane; part of the time it looked at the earth and part of the time it didn't. You got oblique views. We struggled to attach latitude and longitude grids to those pictures. The problem was to be able to translate onto flat maps.[51]

New "spin-scan" techniques, invented by Verner Suomi and first used on the ATS satellite series in 1966, scanned a narrow band below on each orbit, as if peeling a single, continuous strip of rind from an orange. These strips could then be reassembled into a single image. Still, despite enormous early optimism, scientists found few effective ways to derive quantitative data from photographs. For example, initially they hoped that cloud photographs would revolutionize nephanalysis. Yet when the US Air Force—for which cloud cover was a principal operational concern—developed a global nephanalysis model, it ignored satellites in favor of conventional surface, radiosonde, and aircraft observations. Later the Air Force incorporated (numerical) satellite soundings measured from visible and infrared radiances. Photographs made their appearance only in a final step, when human analysts "adjusted" model output by comparing it with satellite imagery.[52] "Cloud winds" (wind speed and direction estimated from successive satellite photographs of moving cloud systems, introduced in the late 1970s) became one of the few usable quantitative products of imagery. Although satellite photography improved tropical storm tracking and regional forecasting, global forecast models did not incorporate nephanalysis until the late 1980s. Even then, they restricted nephanalysis inputs to data-sparse regions and, like the Air Force nephanalyses, relied principally on numerical data.[53]

Radiometers or "sounders" became the first satellite instruments to generate data used directly in forecast and analysis models. (See box 10.1.) The first satellite sounder, SIRS-A (Solar Infrared Radiation Station), flew in 1969 on the Nimbus-3 satellite. Since then many other satellite radiometers have observed numerous spectral regions, and have been used to measure temperature, humidity, ozone, and other variables at all altitudes.

The US National Meteorological Center began experimenting with sounder data soon after the SIRS-A satellite reached orbit. Yet for nearly 25 years satellite data could not be shown to improve forecast quality consistently, except in the data-sparse southern hemisphere. Over data-rich areas, radiosondes seemed to provide more accurate data. In the northern hemisphere, satellite temperature retrievals of the 1980s actually had a small *negative* effect on the quality of ECMWF forecasts.[54] Therefore, operational weather analysis models incorporated satellite soundings only over the oceans. Thus the new instruments *made global data*, but meteorologists could not yet *make those data global*.

Not until the mid 1990s did satellite sounding consistently contribute to forecast skill. This counterintuitive result stemmed directly from the specific character of numerical weather models. Analysis required vertically gridded input—i.e., measurements at a precise series of levels, such as those produced by radiosondes. Indeed, "in 1969, the NMC director told a NOAA scientist, referring to satellite soundings, 'If you can make them look like radiosonde data we can use them.'"[55] But the satellite sounders' fields of view typically covered an area 50 km wide, and the radiances they measured could originate in atmospheric layers up to 5 km in depth. Thus they measured large

**Box 10.1**
How Satellite Sounders Work

Satellite sounding instruments measure the atmosphere at various depths (hence the analogy to naval sounding, which also appears in the word 'radiosonde'). Sounders rely on physical principles connecting radiation absorption and emission with gas chemistry. The sun radiates energy at a characteristic set of frequencies. The Earth reflects about 30 percent of this energy back into space. The atmosphere, the oceans, and land surfaces absorb the remaining 70 percent. This absorbed energy is ultimately re-radiated into space at a different set of frequencies, mostly in the infrared portion of the spectrum. Each of the atmosphere's constituent gases absorbs energy in certain spectral bands, and re-radiates in others. Therefore, as viewed by satellites from space, Earth's overall radiance is attenuated in the absorbing bands and amplified in the re-radiating bands. Satellite instruments "see" only the total radiance, not the contribution of each gas or atmospheric layer.

At first, to convert overall radiances into temperature and other meteorological variables at different altitudes, computer models "inverted" the measurements to "retrieve" the contribution of individual gases. For example, the Microwave Sounding Unit (MSU) measured radiation at four frequencies (channels 1–4). Channel 4 received mostly radiation absorbed and re-emitted by oxygen, while Channel 1 received some oxygen radiance, but was dominated by radiances due to surface reflection and cloud water vapor. Inversion applied weighting functions to each channel to determine the radiance contributions of each atmospheric level. (See figure 10.9.) These weighting functions depended, in turn, on other models and measurements such as those contained in the US Standard Atmosphere.[a] Finally, the data thus derived were converted into temperature or other variables and gridded horizontally and vertically.

Since the mid 1990s, new analysis methods have eliminated the need to convert radiances to gridpoint data. Instead, analysis and forecasting models incorporate radiance measurements directly. In this technique, known as "direct radiance assimilation," the analysis models simulate the top-of-atmosphere radiance values—what satellite radiometers actually measure—for the atmospheric state predicted by the previous forecast run. Just as with other first-guess fields, the analysis process then compares the satellite-measured radiances with the modeled radiances and updates the radiance field with measurements wherever the two disagree.[b]

**Box 10.1**
(continued)

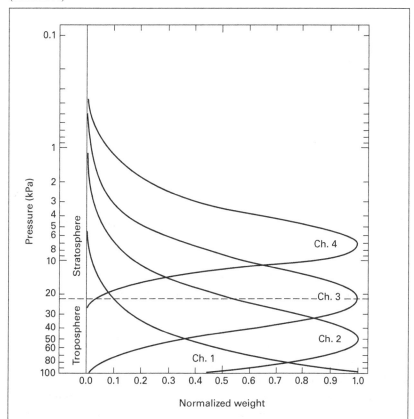

**Figure 10.9**
Weighting functions for Microwave Sounding Unit channels 1–4, used to create a vertical atmospheric temperature profile from top-of-atmosphere radiances in four spectral bands (channels).
*Source*: R. W. Spencer and J. R. Christy, "Precision and Radiosonde Validation of Satellite Gridpoint Temperature Anomalies. Part I: MSU Channel 2," *Journal of Climate* 5, no. 8 (1992), 848.

a. United States Committee on Extension to the Standard Atmosphere, US Standard Atmosphere, NOAA-S/T 76–15672 (National Oceanic and Atmospheric Administration, 1976).
b. UK Met Office, "Satellite Microwave Radiances," 2007, www.metoffice.gov.uk.

volumes at a vertical resolution much lower than most atmospheric models required.[56] Modelers' retrieval algorithms reduced the column-average data to vertically and horizontally gridded temperature or moisture profiles. (See box 10.1.) These mismatched techniques made the gridded retrievals considerably less precise than the original sounder readings.

In the mid 1990s, as greater computing power became available and modeling shifted decisively from gridpoint to spectral techniques, new analysis methods incorporated satellite radiance measurements directly into the forecast models alongside so-called *in situ* data.[57] This produced an immediate improvement in forecast skill. "The reason why the satellite soundings had little impact on the forecast [before 1995] was that the satellite retrievals were treated as poor-quality radiosondes (i.e., point measurements) rather than as high-quality volume measurements (i.e., what the radiances represent). By assimilating the radiances rather than the retrievals, the proper spatial resolution of the data was necessarily represented in the model. . . ."[58] The new methods reversed the original procedure. As they did with other observations, the models now transformed the analysis variables into top-of-atmosphere radiances (the quantity actually measured by the satellite), then updated these first-guess radiances with the satellite measurements, then converted the radiance field back into meteorological variables.

Adding satellite data to 4-D data assimilation systems entailed many layers of increasingly detailed, computationally expensive modeling. For example, assimilating satellite radiances directly required "fast radiative transfer" models, which compute the radiation emitted or absorbed along the satellite's actual viewing path, including surface and cloud reflectance as well as gas emission/absorption. In turn, fast radiative transfer models integrate or depend on other models and measurements, such as the parameterized results of "line-by-line" radiative transfer models.[59] Among many additional complicating factors, satellite orbits slowly decay over time, as they fall ever closer to the Earth. Orbital decay causes the satellite instruments' angle of observation to change; data modeling must compensate. Meanwhile, newly launched radiometers must be calibrated against older ones; continuity of the data record requires an intercalibration factor that adjusts for the slight differences in how individual radiometers respond.[60]

In the early 1990s, the complexity of data modeling precipitated a major debate—discussed at length in chapter 15—between climate scientists and satellite specialists attempting to reconcile global temperature trends derived from microwave sounding units (MSUs) with those measured by radiosondes. Initially, the MSU specialists argued that their data showed a slight cooling in the troposphere (the layer of air closest to the Earth's surface). Meanwhile, radiosonde data showed a steady temperature increase at the same altitudes. Although in public situations this was framed as a controversy about "data," the scientific debate—which continues at this writing in 2008—revolved around whether the models used to invert the MSU radiances had correctly accounted for the complicating factors described above. Yet radiosonde data *also* require model processing, e.g. to intercalibrate the different sensor technologies used by various nations in the approximately 3000 weather balloons launched each day from more than 800 radiosonde stations around the world. Once again, it's models (almost) all the way down.

**From Reductionism to Reproductionism: Data-Laden Models and Simulation Science**

When I discuss 4-D data assimilation in talks, scientists from other fields sometimes assume I must be describing some sophisticated version of interpolation, i.e. a set of transformations on observational inputs leading to data at intermediate points. Yet, although objective analysis began as automated interpolation, 4-D data assimilation systems moved into an entirely different realm long ago. The difference between raw observations and analysis in meteorology is far greater than the difference between raw instrument readings and interpolated data in many other fields, although one can witness various stages of these same transformations taking place throughout the sciences. Assimilation models are full-fledged atmospheric simulations; if run with no observational input at all, they would keep right on going day after day, month after month, generating physically consistent global data images. Where observations are available, they constrain the models, but they do not determine their output in any ordinary sense of 'determine'.[61]

Traditionally, scientists and philosophers alike understood mathematical models as expressions of theory—as constructs that relate dependent and independent variables to one another according to physical laws. On this view, you make a model to test a theory (or one expression of a theory). You take some measurements, fill them in as values for initial conditions in the model, then solve the equations, iterating into the future. Then you compare the model result to the measured result. Do they correspond? If they do, perhaps your theory explains the phenomenon. If they don't, your theory needs some work. This conception of the theory-model relationship jars when applied to weather data assimilation. Yes, analysis models are based on theory, and yes, in refining them modelers constantly compare their output with observations. Yet from the point of view of operational forecasting, the main goal of analysis is not to explain weather but to reproduce it. You are generating a global data image, simulating and observing at the same time, checking and adjusting your simulation and your observations against each other. As the philosopher Eric Winsberg has argued, simulation modeling of this sort doesn't test theory; it *applies* theory. This mode—application, not justification, of theory—is "unfamiliar to most philosophy of science."[62] Analysis models are filled with observational data, but the full global data image they generate is driven by—and the large majority of its data points are created by—applied theory. In data assimilation systems, the theory-ladenness of data reaches a level never imagined by that concept's originators.[63]

At the same time, far from expressing pure theory, analysis models are *data-laden*.[64] And the same can also be said of all forecast models and general circulation models. Stephen Schneider writes:

> ... even our most sophisticated 'first principles' models contain 'empirical statistical' elements within the model structure. . . .We can describe the known physical laws mathematically, at least in principle. In practice, however, solving these equations in full, explicit detail is impossible. First, the possible scales of motion in the atmospheric and oceanic components range from the submolecular to the global. Second are the interactions of energy transfers among the different scales of motion. Finally, many scales of disturbance are inherently unstable; small disturbances, for example, grow rapidly in size if conditions are favorable.[65]

Hence the necessity of parameterization, much of which can be described as the integration of observationally derived approximations into the "model physics." Schneider and others sometimes refer to parameters as "semi-empirical," an apt description that highlights their fuzzy relationship with observational data. For the foreseeable future, all analysis models, forecast models, and climate models will contain many "semi-empirical" elements.

What should we make of all this? Though it looks very little like our idealized image of science, in which pure theory is tested with pure data, that image was arguably always a false one.[66] The reductionist ideals of an earlier age of science sought always to explain large-scale phenomena through smaller-scale component processes. Complementing rather than replacing it, in computational sciences such as meteorology a new ideal has emerged—an ideal we might call *reproductionism*.

Reproductionism seeks to simulate a phenomenon, regardless of scale, using whatever combination of theory, data, and "semi-empirical" parameters may be required. It's a "whatever works" approach—a "Pasteur's Quadrant" method that balances practical, here-and-now needs for something that can count as data, right up alongside rigorous physics.[67] Eric Winsberg begins to get at this when he writes that "since simulations are used to generate representations of systems for which data are conspicuously sparse, the transformations that they use need to be justified internally; that is, according to their own internal form, and not solely according to what they produce."[68] As Sergio Sismondo puts it, "Applied theory isn't simply theory applied, because it instantiates theoretical frameworks using a logic that stands outside of those frameworks." Describing Winsberg's analysis, Sismondo continues:

Simulations and their components are evaluated on a variety of fronts, revolving around *fidelity* to either theory or material; assumptions are evaluated as close enough to the truth, or unimportant enough not to mislead; approximations are judged as not introducing too much error; the computing tools are judged for their transparency; graphics systems and techniques are expected to show salient properties and relationships.[69]

The familiar logics of discovery and justification apply only piecemeal. No single, stable logic can justify the many approximations involved in reproductionist science. Instead, different logics apply to different model elements, and the overall enterprise must be evaluated on outcomes alone.

### Model-Data Symbiosis

The theory-model-data relationship in meteorology is thus exceptionally complex. Models contain "semi-empirical" parameters, heuristic principles derived from observations. Meanwhile, global data sets are produced by simulations, which are constrained but not determined by instrumental observations. In earlier work I described this relationship as "model-data symbiosis," a mutually beneficial but also mutually dependent relationship.[70] This idea aligns with recent work by philosophers of science on "models as mediators"—a semi-autonomous "third force" in science, functioning in the spaces between the real world, instrumentation, and theory.[71] As Margaret Morrison and Mary Morgan argue,

Scientific models have certain features which enable us to treat them as a technology. They provide us with a tool for investigation, giving the user the potential to learn about the world or about theories or both. Because of their characteristics of autonomy and representational power, and their ability to effect a relation between scientific theories and the world, they can act as a powerful agent in the learning process. That is to say, models are both a means to and a source of knowledge.[72]

The concept of model-data symbiosis also supports the claims of the philosophers Stephen Norton and Frederick Suppe, who argue that "to be properly interpreted and deployed, *data must be modeled.*"

Defining scientific methods essentially as ways of controlling for the possibility of artifactual results, Norton and Suppe argue that model-data symbiosis pervades all sciences—even the laboratory sciences, in which data modeling allows investigators to remove or correct for artifactual elements. "Even raw data," they argue, "involve modeling built into the instrumentation." One example is a thermoelectric probe, which derives ambient temperature from the current generated by two dissimilar metals joined inside the probe. Relating these currents to temperature requires parameters for each metal's magnetic permeability. The probe's temperature measurements must be understood as outputs of a physically instantiated mathematical model.[73] If Norton and Suppe are right, seeking purity in either models (as theories) or data (as unmediated points of contact with the world) is not only misguided but impossible. Instead, the question is how well scientists succeed in controlling for the presence of artifactual elements in both theory and observation—and this is exactly how the iterative cycle of improving data assimilation systems (and the observing network) proceeds.

Thus, in global climate science (and perhaps in every model-based science), neither pure data nor pure models exist. Not only are data "theory-laden"; models are "data-laden." Models allow experimental methods to be applied to phenomena that cannot be studied using traditional laboratory techniques. Because they allow investigators to control for artifactual elements not related to the phenomenon under investigation, models also allow the creation of coherent global data sets. In this chapter we have seen how 4-D data assimilation models do this for weather data. In the next chapter, we will explore how they do it for climate data.

### Versions of the Atmosphere: The Changing Meaning of "Data"

In this chapter we have seen how, between about 1955 and 1975, weather forecasters transformed data analysis from an intuitive scientific hermeneutic—an interpretation of data by skilled human beings—to an "objective" product of computer models. Meanwhile, meteorology's arsenal of instrumentation grew to include devices, from Doppler radar to satellites, whose raw signals could not be understood as meteorological information. Until converted—through modeling—into quantities such as temperature, pressure, and precipitation, these signals did not count as data at all.

By the late 1960s, early plans for the Global Weather Experiment assumed that after being converted into meteorological parameters, raw satellite observations had no utility and would normally be discarded.[74] During the same period, general circulation models emerged as primary tools of both weather and climate science. These models simulated the evolution of atmospheric states. In weather prediction, these simulated states rapidly came to be called "forecasts," although even today they are only part of the weather forecasting process (which still involves human judgment). Similarly, climate modelers at first distinguished model outputs from observations using phrases such as "simulated data," but maintaining the distinction proved linguistically awkward. Over time, despite the potential for confusion, climate model results became known simply as "data."[75] Today, data archives at climate laboratories and distributed archives such as the Earth System Grid overflow with petabytes of "data" from simulations, preserved to enable comparisons of models not only with observations, but also with each other.

Today's meteorologists understand the meaning of "data" very differently from meteorologists of earlier generations. The panopticism that ruled meteorology from Ruskin to Teisserenc de Bort and beyond has been slowly but surely replaced by an acceptance of limits to the power of observation. In place of Ruskin's "perfect systems of methodical and simultaneous observations . . . omnipresent over the globe," 4-D data assimilation augments and adjudicates spotty, inconsistent, heterogeneous instrument readings through computer simulation. Modern analysis models blend data and theory to render a smooth, consistent, comprehensive and homogeneous grid of numbers—what I have called in this chapter a *data image*, rather than a data set. Meanwhile, global data images from GCM simulations proliferate.

What, then, is "global" about global data in meteorology? Intuitively, one thinks first of the extent of instrument coverage—what I have called *making global data*. Yet with respect to weather forecasting, at least, instrument readings are literally less than half the story. Rather, it is the *models* that *make data global*. They create a picture—a data image of the world—that is complete and whole, even though the observations are not.

One might assume that making data global also meant making it universal: a single, fixed truth, valid for everyone, everywhere, at all times. Instead, something rather different has happened. Using models to make data global legitimized the possibility of alternative data images. The logic goes as follows: You will never get perfect knowledge of initial conditions. No practical observing mesh will ever be fine enough to do full justice to the atmosphere's huge range of scales of energy and motion, from the molecular to the global. Furthermore, there will always be errors in the instruments, errors in the transmission, and errors in the analysis model. On top of that, the chaotic nature of weather physics means that tiny variations in initial conditions (here, read "analyzed global data") often produce highly divergent outcomes. Therefore, using a single analyzed data set as input to a single deterministic forecast model will always entail a substantial margin of error, especially for periods longer than one or two days.[76] In the early 1990s, forecasters began to turn this apparent defect in their method into an advantage. In a technique known as "ensemble forecasting," for every forecast period they now generate an "ensemble" of slightly different data sets—different global data images, versions of the atmosphere—which collectively reflect the probable range of error. Typically the ensemble contains twelve or more such data sets. Forecasters then run the forecast model on each of these data sets, producing a corresponding ensemble of forecasts.[77] Characterized statistically, the differences among these forecasts represent a *forecast of the forecast error.*

Forecasters can exploit the range of variability represented by the forecast error in numerous ways. One is to create a single forecast by averaging the ensemble of forecast runs. This "ensemble average" forecast generally exhibits greater accuracy than any of the individual forecast model runs used to generate it. Another technique is to calculate confidence levels for individual variables. The forecasts often "cluster" or "tube" into several sets of outcomes, defining the most likely variations. Both the US Weather Service and the European Centre for Medium-Range Weather Forecasts currently provide global ensemble forecasts in addition to their deterministic (single-run) forecasts. Numerous regional and national forecasting centers use ensemble techniques for medium-range forecasts over smaller areas.

An astounding proliferation of data images accompanies this process. At this writing, in 2008, the ECMWF ensemble forecasting system creates 51 different 10-day forecasts—a "51-member ensemble"—every day. The ensemble includes one control forecast based on the unaltered analysis. The Centre then runs its forecasting model on 50 additional data sets.[78] Each of these is created by introducing slight perturbations into the initial global analysis, corresponding to the likely range of error in the analysis. The result is 51 different data images.

Which of these 51 images most closely resembles the state of the real atmosphere? No one can know for sure. Any or all of the 50 perturbed analyses might be more accurate than the initial one. And these 51 daily analyses are merely a few of those weather centers produce each day. Other production ensembles, new analysis models under testing, reanalyses integrating backward as well as forward in time—for any given day in the recent past, hundreds, even thousands of global data analyses, each one slightly different from all the others, exist in the computers and data repositories of the world's weather centers.

# 11  Data Wars

Climate is essentially the history of weather, averaged over time. So one might be forgiven for assuming that climate *data* are simply weather data, averaged over time. After all, we are talking about the same primary variables (temperature, pressure, wind, etc.), measured by the same instruments at the same places, usually by the same people, transmitted through the same communication systems, under standards set by agencies of the same organizations.

Yet in the course of writing this book I was regularly met with blank stares, puzzlement, and hostile comments. "You're talking about *weather* data, not *climate* data," some interlocutors said, apparently offended on some deep level that I would dare to do such a thing. It took me a long time to understand the reason for their discomfort. An outsider, I had walked onto the central battlefield of a long, low-level war. Meteorologists are perhaps the nicest people you will ever meet; they fight quietly and politely. But the war is no less real, and the stakes are high.

Over decades, climate scientists have accumulated a long litany of frustrations with the overwhelming focus of the global observing system on forecasting. Many of them feel that operational agencies mostly ignored climatology's needs until quite recently, managing instruments and handling data in ways detrimental to the long-term climate record. As a result, the colossal quantities of weather data mostly cannot be used by climatologists, who are instead forced to rely on the much smaller "climate quality" data sets they have collected themselves. Since the 1980s, a series of high-level review commissions and WMO programs have repeatedly argued the case for more consistent data practices, better documentation and calibration of changes, and climate-relevant satellite instrumentation in the overall data system.[1] Some weather scientists now think the tables have turned. Burgeoning budgets for climate-change research, they believe, have stripped funding from other priorities, such as the forecasting of severe weather. Since hurricanes, tornadoes, and similar phenomena regularly kill people and cause extreme damage, they bitterly complain, forecasting of severe weather should receive at least as much priority as climate change.

So exactly what *is* the difference between weather data and climate data, and why does it matter so much? Does the difference *make* a difference? How does it affect knowledge of climatic history, validation of climate models, and forecasts of the climatic future? Can computer models make "weather data" into "climate data"?

The distinction stems from important differences between forecasting and climatology. These concern, especially, the following:

- the overriding *purposes* of data collection
- the major *priorities* governing data collection
- which *data sources* fit those priorities and purposes
- how to assess and control *data quality*
- the degree to which *computation* is centralized or distributed
- how each field *preserves data* from the past.

As we have seen throughout this book, observing systems and standards changed often and rapidly over time, creating temporal discontinuities and inconsistencies. These "inhomogeneities," as meteorologists call them, rendered large quantities of weather data unusable for climatological purposes. Meanwhile, until the late 1980s, when climate politics became a major public issue, climatologists' pleas for greater attention to their data needs fell mostly on deaf ears.

Yet in recent decades, with fast, high-quality 4-D data assimilation, forecasting and climatology have begun to converge. Reanalysis of global weather data is producing—for the first time—consistent, gridded data on the planetary circulation, over periods of 50 years or more, at resolutions much higher than those achieved with traditional climatological data sets. Reanalysis may never replace traditional climate data, since serious concerns remain about how assimilation models "bias" data when integrated over very long periods. Nonetheless, the weather and climate data infrastructures are now inextricably linked by the "models of data" each of these infrastructures requires in order to project the atmosphere's future and to know its past.

The next two sections sketch how scientists understand weather data and climate data. The rest of the chapter examines how the historical distinction between these two forms of data is changing, producing a proliferation of data images of the atmosphere's history.

## Weather Data

When scientists talk about "weather data," they mean the information used in forecasting. For this purpose, speed and well-distributed coverage are the highest priorities. In today's World Weather Watch (WWW), synoptic data from the Global Observing System (GOS) are transmitted via the Global Telecommunication System (GTS) to the Global Data Processing and Forecast System (GDFS). GDFS models first analyze the data, filling in values for voids in the observing system, then create forecasts from the analysis. Once analyzed, the original sensor data are mostly archived in vast tape libraries. Since only a few facilities have the connectivity and the computer power to receive and analyze global sensor data in real time, the analyses produced by those facilities matter much more than the raw sensor signals used to produce them.

At this writing, the Global Observing System includes roughly 15 satellites, 100 moored buoys, 600 drifting buoys, 3000 aircraft, 7300 ships, 900 upper-air (radiosonde) stations, and 11,000 surface stations. The GOS's core remains the synoptic network, consisting of six Regional Basic Synoptic Networks (RBSNs) of land-based stations reporting in real time (ideally, eight times a day). This network structure is tightly linked to numerical modeling: "GOS requirements are dictated to a large degree by the needs of numerical [forecasting] techniques."[2] The World Meteorological Organization's goal for the RBSNs has hovered at about 4000 stations ever since the WWW's inception in 1968. However, that goal has been frustrated by the failure of Africa, Central America, South America, and parts of Asia to expand their networks. In 2004, only 2836 RBSN stations met the full reporting goal, a decline of more than 200 stations since 1988.[3] The crucial upper-air network also shows a troubled picture. Station implementation peaked in the early 1990s before dropping off again, with coverage of the southern hemisphere reaching only about 50 percent of the WWW's goal. (Because upper-air stations launch "consumables"—radiosondes and rawinsondes—they cost more than simple ground stations.) Further, more than 30 percent of all weather-station reports are never used in forecasting, having failed to reach forecasting centers within 2 hours. The number of unavailable reports reaches 65 percent in some regions.[4]

Satellites are expensive in absolute terms. Merely launching one into orbit can cost $75 million–$400 million. A typical mission (including the launch, the satellite, insurance, and ground monitoring over the satellite's lifetime of 3–10 years) costs between $500 million and $1 billion. But since a single polar-orbiting satellite can survey the entire planet twice a day, such satellites actually cost little relative to the gigantic quantities of information they provide. Sending up satellites also requires far less *political* effort than expanding the surface network, since it can be done by a single country or by a regional consortium such as the European Space Agency. For these reasons, meteorologists have persistently sought more and better satellites to make up for continuing deficiencies in GOS surface and radiosonde observations. These factors have reduced incentives to build out the surface network, even though many meteorologists would still like to see that happen.

Today satellites provide about 98 percent of the roughly 75 million data items evaluated by the European Centre for Medium-Range Weather Forecasts during every 12-hour forecast period.[5] Not all of these data are actually used by the system, however. In particular, only about 5 percent of satellite radiance data enters the analysis. Even after this filtering, however, satellite data still outnumber—by ten to one—those from all other instruments combined. Nonetheless, recent studies confirm the importance of the terrestrial network (surface plus upper-air). In the ocean-dominated southern hemisphere, where the terrestrial network is small and poorly distributed, satellites improve forecast quality dramatically. But in the extra-tropical northern hemisphere, where the terrestrial network is densest (and where most of the world's people live), satellite data provide only modest improvement over the terrestrial network alone.[6]

Throughout the NWP era, forecasters sought to improve the quality of forecasts as quickly as they could. They introduced new instruments, new mathematical techniques, and new computer models into the observation-analysis-forecast system as soon as they could be shown to work better than their predecessors. As a result, the observing system changed constantly, with new satellite radiometers, radiosondes, precipitation gauges, thermometers, and many other instruments added to the mix. Meanwhile, analysis and forecast models also evolved swiftly. The ECMWF, for example, produced its first operational forecast in 1979. That center soon formalized a continuous revision procedure, testing each new model by running it in parallel with the previous version for several months before replacing the old model with the new one. At this writing, the ECMWF model had already completed 32 revision cycles—on average, more than one revision each year. These kinds of changes in observing systems and forecast models never took place all at once or globally. Instead, national meteorological services managed their own changes and schedules. As we will see in more detail below, the frequent and rather chaotic revisions of observing systems and analysis processes created tremendous difficulties for climate scientists.

The World Meteorological Center system, developed for the World Weather Watch, centralized the computation of global weather forecasts. The weather stations that generate the original sensor data perform virtually no computation; they simply report instrument readings. Indeed, today more than 3000 of the roughly 11,000 surface stations in the Global Observing System are automated. Unstaffed, they broadcast readings of pressure, temperature, winds, and other variables by radio, Internet, satellite uplink, and other means. Under the WWW, stations in the synoptic network transmit reports to the WMCs (Moscow, Melbourne, and Washington) via the Global Telecommunication System (GTS). The Global Data Processing and Forecast System (GDFS),[7] consisting of the WMCs and a handful of specialized centers such as the ECMWF, process the incoming data with their global analysis and forecast models. These centers forward their analyzed global data and their forecasts to national meteorological centers, which may subject them to further processing in order to generate their own forecasts.

Forecasters historically put low priority on preserving raw data. Even in the era of empirical forecasting, forecasters kept vast libraries of analyzed charts, but rarely attended to the fate of the original instrument readings on which the charts were based. Occasionally the forecasting community has adopted some particular set of original sensor data as a benchmark, to test the performance of analysis and forecast models. For example, modelers and forecasting centers mined data from the 1979–80 Global Weather Experiment for well over a decade, using those data to refine analysis and forecast models. Similarly, data about particular hurricanes and other extreme events sometimes acquire the status of benchmarks, used to compare, test, and improve regional forecast models. For the most part, however, once the daily forecast cycle ends, forecasters work only with the *processed* data, i.e. the analysis. Original sensor data may or may not be stored; usually they are never used again. Sensor data that take more than 12 hours to arrive—such as a ship's weather log, data submitted by mail, or even ordinary weather station data not reported on time owing to a power failure or other problems—are never incorporated into the weather data record. For this reason, some have said that four-dimensional data assimilation is actually "three and a half dimensional." In other words, such systems can ingest data that arrive up until the cutoff point for a given analysis period, but they cannot (of course) take account of near-future data that might correct the analysis further. Reanalysis permits true four-dimensional assimilation, in which future observations as well as past observations can influence the state of the analysis at any point in time.[8]

Climate Data

The purposes, priorities, sources, and character of climate data contrast with those of weather data. The purpose of climate data is to characterize and compare patterns and trends. This requires statistics—averages, maxima, minima, etc.—rather than individual observations. And climate scientists care more about measurement quality, station stability, and the completeness and length of station records than they care about the speed of reporting. For example, whereas up to 35 percent of weather station reports are never used in forecasting because they fail to reach forecast centers within 2 hours, climatologists often can obtain much higher reporting ratios.

Climatologists use many of the same data sources as forecasters, but they also use many others. Certain kinds of data, such as precipitation measurements or paleoclimatic proxies, are crucial to climatology yet have little relevance to forecasting. Conversely, some kinds of data useful in forecasting play little or no role in climatology. For example, Doppler radar revolutionized daily precipitation forecasting, but the data it produces are of little interest to climatologists.[9]

Among the best regional data sources is the US Cooperative Observer Network, founded in 1874 under the Army Signal Service. The network comprises more than 5000 "full climatological" stations, and 7000 "B" and "C" stations that record mainly hydrological and other special-purpose data. Cooperative Observer Network stations are run by private individuals and institutions, including university research centers, reservoirs, water treatment plants, and agricultural businesses. The Coop Network provides both weather and climate data to the National Weather Service (supplementing the NWS's official staffed and automated stations) and to many other entities. Of the 5000 "A" stations, about 1200 belong to the Historical Climate Network, which includes only stations that have "provided at least 80 years of high-quality data in a stable environment."[10] The HCN thus includes more than twice as many stations as the 512 in the WMO Regional Basic Synoptic Network for North and Central America. Although at present the total number of US weather stations (including private, educational, and agricultural stations and "mesonets") dwarfs the HCN, these stations cannot provide long-term climatological data, because they do not meet the HCN's quality and stability criteria. Only HCN data meet the standards for detecting long-term regional and global climate change. The point here is that, at least until recently, only some weather-station data were useable for climatological purposes, while many climatological stations did not contribute to the weather forecast system.

When examining pre-twentieth century and paleoclimatic data, climatologists also use numerous "proxy" sources, including data on non-meteorological phenomena that depend strongly on climatic conditions. These data can provide indirect information about past weather conditions. Examples include ice cores, harvest records, tree rings, and species ranges.[11] The precision of proxy data is inherently lower than that of instrument observations, and the number of locations for most proxy data is small relative to the number of climatological stations operating today. Still, before about 1850 proxy data are all we have. Climatologists also make creative use of other non-standard sources, including diaries in which people recorded such seasonal events as when spring flowers first appeared or snow first fell.

Because all knowledge of the climate is based on information about the past, climatology has always prioritized preserving data for the long term. Climatologists prize long-term, stable, homogeneous data sets, accepting transient data sources (such as moving ships or short-lived weather stations) only in regions where no other data are available. To control data quality, climatologists may compare one data set with another for the same area, perhaps taken with different instruments (e.g. radiosonde vs. satellite). Metadata—information about station or instrument history, location, etc.—are crucial to this process.

Table 11.1 summarizes the differences between weather and climate data. Because of these differences, today's Global Climate Observing System (GCOS) bears less relevance to the issue of climate change than its parallel with the Global Observing System (for weather) would suggest. Formally established in 1992, the GCOS coordinates, links, and standardizes elements of existing observing systems to create more consistent global climate data. It is not a separate observing system; instead, it represents "the climate-focused 'system-of-systems' framework, or interface, through which all the global observing systems of WMO and its UN and non-UN system partners work together to meet the totality of national and international needs for climate observations."[12] At present the GCOS network includes about 800 surface stations selected from the roughly 10,000 in the Global Observing System, plus additional stations from other data networks (such as upper-air and ocean stations). GCOS coordination efforts

Table 11.1
The principal differences between weather data and climate data.

|  | Weather data | Climate data |
|---|---|---|
| Purpose | Dynamics: forecasting of atmospheric motion and state<br>Forward-looking, short term (days to weeks) | Statistics: climate characterization, detection of climatic change, evaluation of climate models<br>Dynamics: general circulation<br>Backward-looking, long term (years to centuries) |
| Priorities | Speed<br>Well-distributed coverage<br>Forecast skill | Completeness (spatial and temporal)<br>Stability of data record |
| Principal data sources | Surface synoptic network (land, ocean)<br>Upper air (radiosonde, rawinsonde)<br>Radar<br>Satellites (radiances, visual images, cloud winds)<br>4-D data assimilation (generates additional data) | Climatological stations (some climate-only, some in synoptic network)<br>Upper air (radiosonde, rawinsonde)<br>Satellites: useful but problematic; period of record is short (1979–present); must be correlated with radiosonde record<br>Ships' logs (starting in 1850s)<br>Non-instrument sources: diaries, proxy measures (harvest dates, tree rings, ice cores, etc.), especially prior to 1850<br>Reanalysis (generates additional data) |
| Quality assessment and control | Low quality data sometimes useful in areas of poor instrument coverage<br>Transient data sources OK<br>Automatic consistency and conservation checks during data assimilation<br>Data analysis detects systematic instrument errors | Long-term, stable data sources prized<br>Transient and low-quality data sources rarely used<br>Hand inspection and correction of individual station records for systematic biases and other errors<br>Comparing "duplicate" data sets<br>Comparing models and observations<br>Metadata on station history required for quality control |

Table 11.1
(continued)

|  | Weather data | Climate data |
|---|---|---|
| Computation | Centralized in Global Data Processing and Forecast System, Regional and National Meteorological Centers | Partially distributed: stations calculate monthly means Global statistics calculated by localized collectors (WWR, MCDW, Global Historical Climatology Network, individual dataset projects) |
| Preservation | Low priority Stored data used for model diagnosis and improvement | High priority Stored long-term data are fundamental basis of climate knowledge |

will doubtless improve the global climate record in the long term, but they are still too recent to have much effect on the quality of the long-term climate record.

Meanwhile, climatologists also began to assess the homogeneity of individual weather stations' records over time. Numerous factors can reduce the stability of any time series. Instruments can drift out of calibration. Moving an instrument from a sheltered to an open location, or from the south side of a hill to the north side, can raise or lower recorded values. Over long periods, trees growing around a station can reduce recorded wind speeds. Industrial energy use, automobiles, home heating, and heat-absorbing pavement raise local temperature readings in growing cities, creating the "urban heat island" effect.

In an influential article published in 1953, J. Murray Mitchell dissected the many causes of "long-period" temperature changes in station records, dividing them into two principal types. "Apparent" changes, such as changes in thermometer location or shelters, were purely artifactual, stemming from causes unrelated to the actual temperature of the atmosphere. "Real" changes represented genuine differences in atmospheric conditions. These could be either "directly" or "indirectly" climatic, for example resulting from shifts in the general circulation (direct) or variations in solar output (indirect). But not all "real" temperature changes reflected actual climatic shifts, since some were caused by essentially local conditions (such as urban heat islands, industrial smoke, and local foliage cover) that had nothing to do with the climates of the region or the globe. Figure 11.1 details Mitchell's classification of the issues.[13]

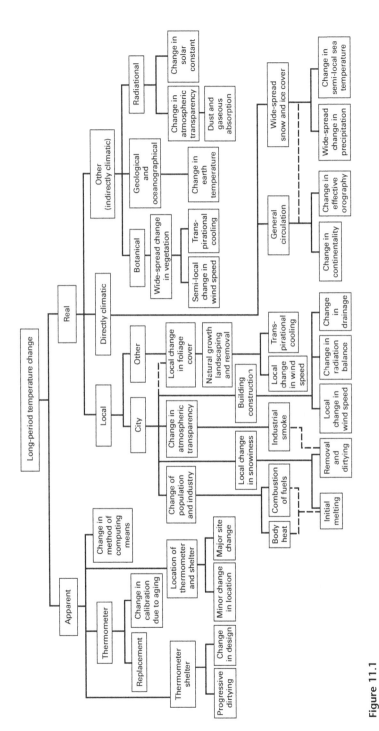

**Figure 11.1**
Apparent and real causes of temperature change at climatological observing stations.
*Source*: J. M. Mitchell, "On the Causes of Instrumentally Observed Secular Temperature Trends," *Journal of the Atmospheric Sciences* 10, no. 4 (1953): 245.

Since the majority of weather stations are located in or near population centers, urban heat islands proved especially problematic for climate data. Stations originally located miles outside a city could be engulfed within it as time went on, slowly creating an apparent climate change. Similarly, at some point virtually every station changed its observing hours, its method of calculating daily means, and/or its method of calculating monthly means. Changes in stations' locations also proved extremely common, accounting for up to 80 percent of the inhomogeneities in station records. Numerous variables related to the siting, the housing, and other local circumstances of instruments can also cause readings to differ substantially, even between sites quite close together.

Imperfect calibration of instruments to a standard is another source of error. Instruments made by different manufacturers often exhibit systematic differences. But even standardized, mass-produced instruments made by a single manufacturer can exhibit slight variations. Over time, exposure to the elements can alter individual instrument behavior. Then there is the problem of *temporal* calibration, as instrument manufacturers change their designs and sensor technologies. For example, between 1951 and the present, the Finnish Väisälä radiosondes used in Hong Kong employed at least five different temperature sensors, including two kinds of bimetal strips, two kinds of bimetal rings, and a capacitive bead. Such changes create discontinuities in the data record.[14] The occasionally dramatic discontinuities caused by adopting new instruments, changing station locations, and other issues are visible in figure 11.2, which illustrates how precipitation readings in ten countries changed over a 100-year period.

Most inhomogeneities in climate data have little political valence, but there are important exceptions. As we saw in chapter 8, during the Cold War the Soviet Union withheld some data, while the People's Republic of China withheld virtually all data. These data were not included in any Western climate data set until the mid 1980s.[15] Another kind of political issue is subtler:

... consider the station Pula, ... now managed by the Croatian Hydrometeorological Service. Its turbulent history started with the K.K. Central-Anstalt für Meteorologie und Erdmagnetismus [Austria]. From 1918 until 1930 it was managed by the Ufficio Centrale in Rome; from 1931 until 1941 it belonged to the Federal Republic of Yugoslavia. During World War II it was occupied by the Germans and after 1945 it belonged to the Socialist Republic of Yugoslavia. Since 1991 it has been part of the network of Croatia.[16]

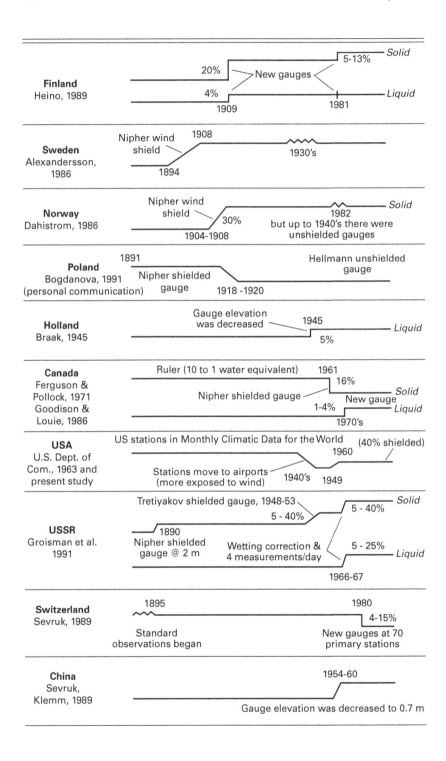

These political changes may have created inhomogeneities due to changes in station management or applicable national standards—inhomogeneities that may be hard to detect if other nearby stations were simultaneously affected by the same changes.

The issues discussed above represent only a small sample of the many inhomogeneities affecting climate data. Many of these, such as recording errors and instrument placement, can be ignored because they are structurally random and thus as likely to be negative as to be positive. Therefore, given a sufficiently large number of stations, random errors of negative sign would approximately cancel errors of positive sign. Indeed, studies confirm that inhomogeneities in the *temperature* record can significantly affect trends on local and regional scales, but on hemispheric scales their effects are minimal. However, it cannot be assumed that all climate variables exhibit this self-cancellation of random errors—in precipitation, for example, the introduction of shielded gauges generally produced higher readings worldwide.[17]

Some recent work suggests that systematic errors may be more widespread than was previously believed. For example, nineteenth-century meteorologists throughout the Alps placed precipitation gauges on rooftops and thermometers in windows; later, they moved precipitation gauges to ground level and mounted thermometers inside screening devices placed in open areas (thus reducing the artifactual effects of buildings and pavement). Although stations modified their instrument placement at different times, precipitation measurements were systematically higher and temperature measurements lower after instrument placements were changed.[18] Similarly, a volunteer survey by surfacestations.org indicates a possible bias toward positive temperature errors at the majority of US climatological stations, due mainly to instrument placement near local heat sources such as air conditioner exhaust and parking lots. (However, the quality of the survey method and the possibly dubious motives of many volunteers leave this result open to question.) Still, on continental to global scales the effects of such changes on temperature trends are likely to cancel out.

A different and much more problematic issue arises with respect to satellite data. As we saw in chapter 10, most raw sensor readings from satellites require some kind of processing to convert them into meteorological

◄ Figure 11.2
Discontinuities in precipitation readings caused by changes in instrumentation, observing practices, and other factors, 1890s–1980s.
*Source*: T. R. Karl et al., "Detecting Climate Variations and Change: New Challenges for Observing and Data Management Systems," *Journal of Climate* 6 (1993), 1483.

information. This can be a complex modeling process, as in the inversion of microwave radiances, but it also can be a much simpler data-reduction process. For example, starting in 1966 the National Oceanic and Atmospheric Administration produced gridded data on snow cover from visual satellite imagery, interpreted by hand. In 1972, a higher-resolution satellite view improved the accuracy of this measurement (still interpreted by hand), resulting in an instrument-related increase in the extent of snow cover. Snow-cover readings were weekly; monthly data counted a grid cell as snow covered if covered with snow for two weeks or longer. The data-reduction process was altered in 1981; now data workers averaged the snow/no snow information from weekly charts, an approach that reduced the calculated monthly snow cover in every month except August. Figure 11.3 shows the effect of applying the newer algorithm (solid line) versus the older algorithm (dashed line) to all data from 1972 on.

Further, the National Aeronautics and Space Administration began producing snow-cover charts from microwave radiances from the Scanning Multichannel Microwave Radiometer (SMMR) in 1978; in 1987 it introduced the Special Sensor Microwave Imager, which included an additional microwave channel. These data required processing by multiple algo-

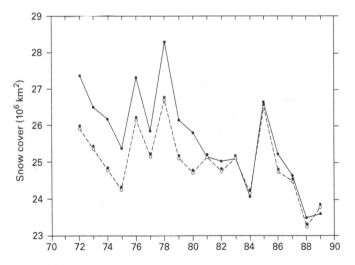

**Figure 11.3**
Northern hemisphere snow cover from NOAA polar orbiting satellites, processed using a consistent data-reduction algorithm (solid line), vs. the same data as processed by earlier algorithms (dashed line).
*Source*: Karl et al., "Detecting Climate Variations and Change," 1487.

rithms, each specific to how a particular region's land surface qualities affect the properties of snow.[19] NASA's microwave data did not agree perfectly with NOAA's visual charts; both were recognized to have virtues and defects. Thus both instrument changes and algorithm changes affected this record dramatically. On satellites, at most only two or three instruments of any given type are usually orbiting at any one time. Therefore, the self-cancellation of random instrument errors that stabilizes the global surface record cannot occur with satellites. Further, the dependence of satellite data on algorithms and modeling makes such data especially vulnerable to an increasingly common problem. Data specialists warn that

> as more and more data become dependent on processing algorithms, problems such as this are likely to grow rapidly unless special care is taken to avoid or at least document changes in processing algorithms. For example, traditional direct measurements of precipitation derived from stick measurements or weighing are being replaced by tipping devices with built-in conversion processing software. Similarly, liquid-in-glass thermometers are being replaced by electronic systems.... Automation will require indirect sensing of all our climate variables. Special procedures are required to archive these measurements in their proper and most basic units so that changes in either external software or internal built-in microprocessors will [enable] homogeneous reprocessing of the data when the inevitable improvements in the system occur.[20]

### Fighting for the Long Term: Building Stability into Change

The assortment of issues described above only scratches the surface of the vast array of data problems scientists face in their ongoing quest to refine and extend the historical climate record. Most, if not all, of those problems stem from the divergence of purpose and focus between the operational systems responsible for weather forecasting and the related but not identical systems for monitoring climate. Historically, climate scientists have often found themselves stuck with damaged goods—leftovers from a much larger forecasting enterprise that ignored consistency and reliability in its constant drive to move forward with newer, better technology. Until climate change became a major public issue, and even afterward, this was often a losing battle.

Concerns about the stability and reliability of data led to calls—repeated over decades in various forms—to improve the climate data infrastructure.[21] The US National Research Council, the Global Climate Observing System, the UN Framework Convention on Climate Change, the Intergovernmental Panel on Climate Change, and other agencies have endorsed principles for climate monitoring similar or identical to the following:

1) *Management of network change.* Assess how and the extent to which a proposed change could influence the existing and future climatology.
2) *Parallel testing.* Operate the old system simultaneously with the replacement system.
3) *Metadata.* Fully document each observing system and its operating procedures.
4) *Data quality and continuity.* Assess data quality and homogeneity as a part of routine operation procedures.
5) *Integrated environmental assessment.* Anticipate the use of data in the development of environmental assessments.
6) *Historical significance.* Maintain operation of observing systems that have provided homogeneous datasets over a period of many decades to a century or more.
7) *Complementary data.* Give the highest priority in the design and implementation of new sites or instrumentation within an observing system to data-poor regions, poorly observed variables, regions sensitive to change, and key measurements with inadequate temporal resolution.
8) *Climate requirements.* Give network designers, operators, and instrument engineers climate monitoring requirements at the outset of network design.
9) *Continuity of purpose.* Maintain a stable, long-term commitment to these observations, and develop a clear transition plan from serving research needs to serving operational purposes.
10) *Data and metadata access.* Develop data management systems that facilitate access, use, and interpretation of data and data products by users.[22]

The ideal is clear: to integrate the weather and climate data networks, forming a genuine, robust, and enduring climate-data infrastructure. Governments and private-sector elements have begun to focus on the need for reliable predictions of seasonal and interannual climate, as well as of long-term climate change. As a result, calls have emerged for an *operational* climate forecasting system that would provide real-time climate analysis and prediction capabilities on a local or a regional scale. The US NOAA/NCEP Climate Test Bed, for example, envisions "'a Seamless Suite of Forecasts' spanning weather, intraseasonal, interannual, and multi-decadal timescales."[23]

Looking backward, climate scientists face the daunting task of refining and reconstructing the historical record. The infrastructural inversion process I have described throughout this book is one major tool for rectifying the climate data record: looking at each station's record, recovering whatever can be learned about the station's history, correcting for some kinds of changes, rejecting anomalous data points as likely errors, and so on. After taking the infrastructure apart, scientists can—sometimes, to some degree—correct the record for errors and systematic biases. From this perspective, the climate data infrastructure looks much more fluid and less stable than the dry and certain-seeming lists of numbers it provides might otherwise suggest.

## Global Climate Data—Plural

We live on just one planet, with just one real climatic history. Yet when climate scientists talk about global climate data, they are talking in the plural. Just as with weather, we are multiplying data images for global climate. Hundreds, even many thousands of variant data images exist, though only a few gain authoritative status, and they have generally (but not always) converged over time.

As an example, let us explore some data sets behind the most important figure derived from global climate data: global average temperature. Usually climate scientists express this figure as a temperature anomaly time series. The "temperature anomaly" is simply the difference (positive or negative) between a given year's temperature and the average temperature of a chosen reference period; in figure 11.4, the reference period is 1961–1990. This technique allows direct comparison of different data sets without regard to their absolute temperature values, which may differ. It also permits comparing records from different types of instruments. For example, temperature anomaly trends for the lower troposphere from radiosonde data correlate well with those from surface thermometer data, but the absolute values of radiosonde and surface readings differ substantially. (Temperature varies as a function of altitude.)

Figure 11.4, from the IPCC Fourth Assessment Report (2007), compares global temperature trends from nine well-known datasets, all expressed as anomalies from a 1961–1990 average. Dozens of other global average temperature datasets (not shown here) have also been created from surface thermometer data.[24] On the decadal scale, all the trend lines show similar tendencies after 1900. Yet clearly they also disagree, sometimes strikingly. The figure reveals a maximum difference between the various trends of about 0.6°C before 1900 and 0.2°C after 1900. This may not sound like a large difference. Yet across the entire period 1840–2005, the total difference between the minimum temperature (Willett's –0.9°C) and the maximum one (Brohan et al.'s +0.45°C) is only 1.35°C. Thus the maximum disagreement among the trends is nearly half as large as the maximum total temperature change one might read from this chart. Further examination reveals other oddities. For example, from 1850 to 1900 the Willett trend line offers values well below all the others; also, its slope rises from 1875

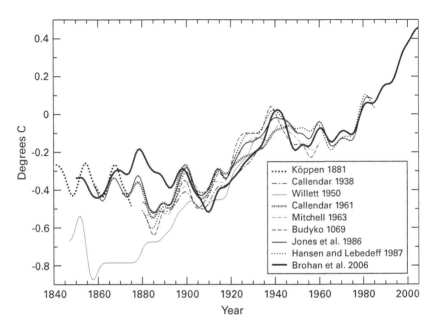

**Figure 11.4**
Original legend: "Published records of surface temperature change over large regions. Köppen (1881) tropics and temperate latitudes using land air temperature. Callendar (1938) global using land stations. Willett (1950) global using land stations. Callendar (1961) 60°N to 60°S using land stations. Mitchell (1963) global using land stations. Budyko (1969) Northern Hemisphere using land stations and ship reports. Jones et al. (1986a,b) global using land stations. Hansen and Lebedeff (1987) global using land stations. Brohan et al. (2006) global using land air temperature and sea surface temperature data is the longest of the currently updated global temperature time series (Section 3.2). All time series were smoothed using a 13-point filter. The Brohan et al. (2006) time series are anomalies from the 1961 to 1990 mean (°C). Each of the other time series was originally presented as anomalies from the mean temperature of a specific and differing base period. To make them comparable, the other time series have been adjusted to have the mean of their last 30 years identical to that same period in the Brohan et al. (2006) anomaly time series."
Source: *Climate Change 2007: The Physical Science Basis* (Cambridge University Press, 2007). Image courtesy of IPCC.

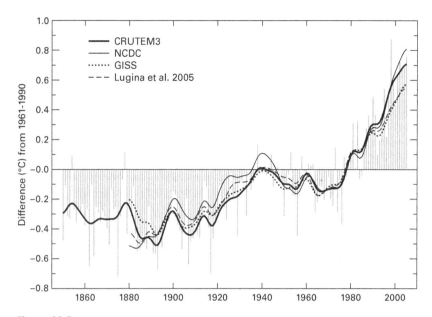

**Figure 11.5**
Four recent global surface temperature anomaly data sets. The thick trend line here represents the same data as the thick trend line in figure 11.4. Original legend: "Annual anomalies of global land-surface air temperature (°C), 1850 to 2005, relative to the 1961 to 1990 mean for CRUTEM3 updated from Brohan et al. (2006). The smooth curves show decadal variations. The [thick] curve from CRUTEM3 is compared with those from the National Climatic Data Center [thin], the Goddard Institute for Space Studies [dotted], and Lugina et al. [dashed]."
*Source: Climate Change 2007: The Physical Science Basis* (Cambridge University Press, 2007). Image courtesy of IPCC. Colors replaced as noted in brackets to allow noncolor reproduction.

to 1895, where several other lines trend downward. Meanwhile, from 1870 to 1895 the Brohan et al. trend shows a considerably higher value than any of the others.

Recently produced global data sets converge more closely, but they still show significant differences. Figure 11.5 compares four data sets produced between 2001 and 2006. The trend lines here exhibit a maximum disagreement of around 0.3°C in 1880–1890 and 0.2°C beyond that. Yet the 1880–1890 trends disagree in both slope and sign. What is going on here? Why do we have multiple climate data sets in the first place? Why do they disagree? How can the "same" data produce different results? The classical scientific approach to these questions tries to narrow the field. Some data sets are better than others; through a process of examination and correction, you hope to approach, asymptotically, a single master data set, a definitive quantitative account of what actually happened in the atmosphere across the period of record. This approach sees data as history; it represents a fundamental premise of scientific understanding. Obviously we would like to know with greater certainty exactly how much the planet has warmed, and numerous scientists have spent their entire working lives trying to do just that. The increased convergence of the trend lines in figure 11.5 (versus those in figure 11.1) represents the fruit of this effort.

Yet just as in the case of ensemble weather forecasting, rather than reducing to a single definitive set, global climate data images have instead proliferated dramatically in the computer age. All the themes I have developed in this book played parts in this proliferation. The path of change closely mirrored that of weather forecasting, but with a delay of 10–20 years. As they proceeded, climate science shifted from the partially distributed computation characteristic of its past toward massively centralized computation in Latourian "centers of calculation."

First, in the 1970s, digitization projects made global climate data widely available for the first time by reducing data friction. Digitization made it possible to transfer formerly inconceivable volumes of data by exchanging data tapes. More recently, data sets mounted on Internet file servers reduced friction even further. Next, computers reduced the computational friction involved in applying complex mathematical techniques. Data centers, using an increasingly sophisticated array of interpolation methods, began producing *gridded* historical data sets. These centers could now apply consistent standards, and techniques could be retrospectively applied in order to *make data global* in the sense of chapter 10. As a result, scientists moved beyond simple monthly and annual averages toward a profusion of statistical methods and modeling techniques. This culminated in the 1990s with reanalysis of 40-year data sets through frozen 4-D data assimilation systems (discussed in chapter 12). Now scientists are working toward reanalyzing the entire 150-year history of instrumental observations.

To unearth the changing "memory practices" of climate science, let us first ask: What exactly *is* a "global climate data set"[25]? As we have seen, "global climate data" are never simply the total collection of relevant instrument readings. Until reanalysis, climatology always required long-term, homogeneous records (see previous section). Box 11.1 shows that, with the exception of Callendar's 1961 effort, the major global climate data sets assembled before 1965 used fewer than 200 station records. Like virtually all hemispheric and global data sets published before 1960, these relied

**Box 11.1**
About the Historical Surface Temperature Series in Figure 11.4

**Köppen 1881** Used fewer than 100 land stations, covering only the tropics and temperate zones.[a]
**Callendar 1938** Discussed in chapter 4; selected about 200 stations worldwide.
**Willett 1950** The most comprehensive global calculation to date, reaching back to 1845. Seeking a geographically representative sample, Willett chose one station to represent each 10° latitude-longitude cell, picking the one with the best available long term record. By this procedure he selected 129 station records of more than 50 years, plus another 54 stations with 20–50 year records. Thus Willett managed to cover 183 of a possible 648 10° × 10° cells. Data for all but two stations were taken from *World Weather Records*. Willett's global average weighted each station record equally.[b]
**Callendar 1961** Examined some 600 station records, "the majority" from World Weather Records and "a few score others from a variety of sources too numerous to mention."[c] Of these, Callendar estimated that about 74 percent were probably reliable; 8 percent showed spurious temperature increases probably related to urban locations (heat islands); and 18 percent were probably unreliable. Therefore he retained only about 450 stations for his study, notable also for promoting (again) the carbon dioxide theory of climatic change.[d]
**Mitchell 1963** A student of Willett, Mitchell updated Willett's 1950 study using a nearly identical method and selection of stations. Unlike his mentor, however, Mitchell weighted stations according to the surface area of the latitude band in which they were located. (Since the surface area of latitude bands is greater near the equator and smaller near the poles, Willett's simple averaging technique gave too much weight to high-latitude stations.) For this reason, Mitchell's series[e] begins in 1882, the first year in which all latitude bands could be adequately represented. The weighting procedure accounts for the considerable difference between Mitchell's calculation and Willett's.
**Budyko 1969** This trend line represents only the northern hemisphere above about 20°N. Budyko's data came from temperature-anomaly analysis maps created at the Main Geophysical Observatory in Leningrad. Although many station records used to create the maps were drawn from *WWR*, they also included Soviet records not previously available to Western climatologists.[f]
**Jones et al. 1986** Produced by the Climatic Research Unit at the University of East Anglia, this global data set used 2194 stations, including hundreds of new records recovered from meteorological archives covering previously undocumented areas of the Soviet Union, the People's Republic of China,

Box 11.1
(continued)

> northern Africa, and northern Europe. Jones et al. also produced a gridded version of the dataset (5° latitude by 10° longitude). Rather than use Willett's and Mitchell's method of choosing a single station to represent a grid cell, Jones et al. interpolated values for each gridpoint from a number of stations surrounding each gridpoint. They further introduced numerous corrections to existing data. This data set was later expanded to include marine temperatures. Continuously updated and corrected, the CRU gridded data set remains one of the most authoritative.[g]
> **Hansen and Lebedeff 1987** Based principally on W. M. L. Spangler and R. L. Jenne, *World Monthly Surface Station Climatology* (National Center for Atmospheric Research, 1984), a digitized, corrected, and updated version of *World Weather Records* and *Monthly Climatic Data for the World*. Like most other investigators, Hansen and Lebedeff introduced corrections of their own. For example, they tested the effect of urban heat islands on the global data by extracting all stations associated with population centers larger than 100,000 people[h]; this reduced the global average change from 0.7°C to 0.6°C across the period 1880–1980.
> **Brohan et al. 2006** The massively revised third release (CRUTEM3) of the Climatic Research Unit 1986 dataset. It includes quality-controlled monthly average temperatures for 4349 land stations. Between 1986 and 2006, numerous previously unreported stations were added, while station records were corrected or homogenized by national meteorological services and others.[i] The trend lines represent only the land-surface component of this dataset. The full dataset, HadCRUT3, combines marine and land surface data to provide complete global surface coverage (Brohan et al., "Uncertainty Estimates in Regional and Global Observed Temperature Changes").

a. W. Köppen, "Über Mehrjährige Perioden der Witterung—III. Mehrjährige Änderungen der Temperatur 1841 bis 1875 in den Tropen der Nördlichen und Südlichen Gemässigten Zone, an den Jahresmitteln Untersucht," *Zeitschrift der Österreichischen Gesellschaft für Meteorologie* (1881): 141–50.
b. H. C. Willett, "Temperature Trends of the Past Century," *Centenary Proceedings of the Royal Meteorological Society*, 1950: 195–206.
c. Callendar, "Temperature Fluctuations and Trends Over the Earth."
d. Ibid.
e. J. M. Mitchell Jr., "On the World-Wide Pattern of Secular Temperature Change," in *Changes of Climate* (1963).
f. A. Robock, "The Russian Surface Temperature Data Set," *Journal of Applied Meteorology* 21, no. 12, 1982.

**Box 11.1**
(continued)

g. P. D. Jones et al., "Global and Hemispheric Temperature Anomalies—Land and Marine Instrumental Records," in *Trends Online*, Carbon Dioxide Information Analysis Center, US Department of Energy, Oak Ridge National Laboratory, 2006; Jones et al., "Southern Hemisphere Surface Air Temperature Variations: 1851–1984," *Journal of Applied Meteorology* 25, no. 9 (1986): 1213–30.

h. J. Hansen and S. Lebedeff, "Global Trends of Measured Surface Air Temperature," *Journal of Geophysical Research* 92, no. D11, 1987: 13,345–72.

i. P. Brohan et al., "Uncertainty Estimates in Regional and Global Observed Temperature Changes: A New Dataset from 1850," *Journal of Geophysical Research* 111, no. D12106, 2006; P. D. Jones and A. Moberg, "Hemispheric and Large-Scale Surface Air Temperature Variations: An Extensive Revision and an Update to 2001," *Journal of Climate* 16, no. 2, 2003: 206–23; P. D. Jones, "Hemispheric Surface Air Temperature Variations: A Reanalysis and an Update to 1993," *Journal of Climate* 7, no. 11, 1994: 1794–802.

almost exclusively on *World Weather Records*. *WWR*, first published in 1927 by the Smithsonian Institution, became the *de facto* standard data source for large-scale climatology.[26] Yet the early *WWR* collection effort was *ad hoc*, based not on a systematic and exhaustive survey but on the personal contacts of the collectors. Therefore, the *WWR* collection omitted vast swathes of data. For example, it captured only about 5 percent of all available mid-nineteenth-century data. Over half of its nineteenth-century records originated in just three countries: the United States, Russia, and India.[27]

Until quite recently, in fact, climatologists studying regional or global climate rarely used raw instrument readings at all. Instead, as we saw in chapter 5, from its earliest days climatology relied on a strategy of partially distributed computation. Each station calculated its own statistics, such as monthly means, and transmitted *only* those results—rather than the instrument readings from which they were calculated—to central collecting entities. Initially, this was the mission of the Réseau Mondial. Later, under sponsorship of the International Meteorological Organization and the World Meteorological Organization, *World Weather Records* assumed responsibility for collecting world climate data. In 1959 the US Weather Bureau took over *WWR*, eventually handing it off to the US National Oceanic and Atmospheric Administration, which also published *Monthly Climatic Data for the World*. Figure 11.6 shows typical data tables from these two publications.

## ZANZIBAR, EAST AFRICA
Lat. 6° 10′ S. Long. 39° 11′ E. H_b = 56 ft.
TEMPERATURE IN DEGREES F.
Means of ½ (daily Max. + daily Min.)

| Date | Jan. | Feb. | Mar. | Apr. | May | June | July | Aug. | Sept. | Oct. | Nov. | Dec. | Year |
|---|---|---|---|---|---|---|---|---|---|---|---|---|---|
| 1891 | ... | ... | ... | ... | *78.3 | 77.3 | 76.3 | †75.9 | 77.7 | 78.9 | ... | 81.1 | ... |
| 1892 | 81.6 | 83.4 | 82.7 | 81.2 | 79.3 | 78.5 | 76.8 | 77.3 | 78.1 | 78.9 | 80.5 | 82.2 | 80.0 |
| 1893 | 81.1 | 82.0 | 81.6 | 79.8 | *77.5 | ‡77.0 | 76.0 | 75.4 | 76.6 | 78.7 | 79.0 | 81.8 | 78.9 |
| 1894 | 82.9 | 83.8 | 82.6 | 81.1 | 78.3 | 76.9 | 76.0 | 77.0 | 78.2 | 79.5 | 79.4 | 81.9 | 79.8 |
| 1895 | 82.9 | 83.8 | 83.2 | 81.0 | 79.3 | 77.6 | 76.5 | 77.3 | 77.6 | 79.5 | 80.2 | 82.9 | 80.2 |

SURFACE        JANUARY 1998

| STATION | LATITUDE | LONGITUDE | ELEVATION | DAYS OBS. | PRESSURE MEAN STATION | PRESSURE MEAN SEA LEVEL | TEMPERATURE MEAN | TEMPERATURE DEPART | VAPOR PRESSURE MEAN | VAPOR PRESSURE DEPART | PRECIPITATION DAYS 1 MM | PRECIPITATION TOTAL | PRECIPITATION DEPART | PRECIPITATION QUINTILE | SUNSHINE TOTAL | SUNSHINE % OF AV. |
|---|---|---|---|---|---|---|---|---|---|---|---|---|---|---|---|---|
|  | o′ | o′ | n | nb | nb | nb | °C | °C | nb | nb | nn | nn | nn |  | hr |  |
| AUSTRIA |  |  |  |  |  |  |  |  |  |  |  |  |  |  |  |  |
| 11028 ST. POELTEN | 4812N | 01537E | 282 |  | 985.0 | 1020.3 | .7 |  | 5.5 | .8 | 6 | 37 | 3 | 4 | 58 | 109 |
| 11035 WIEN/HOHE WARTE | 4815N | 01622E | 209 |  | 994.1 | 1020.1 | 2.2 |  | 5.5 | .7 | 5 | 30 | -8 | 3 | 74 | 132 |
| 11120 INNSBRUCK-FLUGHAFEN | 4716N | 01121E | 593 |  | 949.7 | 1021.9 | .4 |  | 4.9 |  | 9 | 45 |  | 2 | 127 |  |
| 11146 SONNBLICK | 4703N | 01257E | 3109 |  | 689.4 | 2992 Z | -10.8 |  | 1.4 | -.5 | 12 | 83 | -175 | 1 | 148 | 130 |
| 11150 SALZBURG-FLUGHAFEN | 4748N | 01300E | 450 |  | 965.9 | 1021.1 | 1.0 |  | 5.3 | .6 | 7 | 31 | -32 | 1 | 96 | 132 |
| 11155 FEUERKOGEL | 4749N | 01344E | 1621 |  | 833.2 | 1461 Y | -1.8 |  | 3.1 |  | 17 | 94 |  | 1 | 135 |  |
| 11212 VILLACHERALPE | 4636N | 01340E | 2160 |  | 781.2 | 1490 Y | -5.1 |  | 2.6 |  | 5 | 24 |  | 1 | 152 |  |
| 11231 KLAGENFURT-FLUGHAFEN | 4639N | 01420E | 476 |  | 963.5 | 1022.2 | -.9 |  | 4.7 | .6 | 3 | 6 | -30 | 1 | 128 | 178 |
| 11240 GRAZ-THALERHOF-FLUGHAFEN | 4700N | 01526E | 347 |  | 977.6 | 1020.5 | .4 |  | 5.3 | .8 | 0 | 3 | -28 | 1 | 134 | 213 |

Figure 11.6
Sample sections from *World Weather Records* (*WWR*, top) and *Monthly Climatic Data for the World* (bottom). The *WWR* entry includes the station's method of calculating mean monthly temperature. (See table 11.2.)
*Sources*: H. H. Clayton, *World Weather Records* (Smithsonian Institution, 1927); *Monthly Climatic Data for the World: January 1998* (National Oceanic and Atmospheric Administration, 1998).

Starting in 1935, to standardize and facilitate collection of climatic data, the International Meteorological Organization asked all national weather services to transmit "mean monthly values of the main climatological elements" early in the following month. To integrate this reporting into the condensed weather codes sent over the international telegraph network, the IMO (and later the WMO) developed a separate set of CLIMAT codes. These reporting requirements were later extended to ocean stations and upper-air networks. WMO technical regulations initially set the density of the requested climatological network at one station every 300 km, but this proved impractical; it was later relaxed to one station every 500 km.[28]

The distributed computation strategy reduced computational friction, since climatologists could calculate regional or global statistics from the reported monthly averages—a much smaller set of numbers than the individual daily readings. It also reduced data friction, decreasing the volume of data that had to be moved from one place to another. But the strategy also created latent uncertainty about data quality. While assembling and summarizing the incoming data, editors of *World Weather Records* and *Monthly Climatic Data for the World* could (and often did) detect gross errors, such as reversals of sign, as well as missing data. As a quality-control measure, they checked each station's reports against those of neighboring stations. However, the large distances (typically 500–1000 km) between stations made it impossible to detect small or subtle errors with this technique.

As computers and communication systems matured, climatologists began to revisit the entire strategy. In the 1960s, a handful of climate research centers began a painful process of reviewing and recalculating all available global climate data. At first they did this mainly by examining and cross-checking the statistical reports. The next steps were to seek out and include station data missed by *WWR* and other collectors, and to develop new techniques for including data records previously deemed too brief.

The Soviet data sets created in 1969 by Mikhail Budyko and his colleagues represent a special case. In addition to data from *WWR*, Budyko's group included data from Soviet and other sources that had never been incorporated into Western data sets. Whereas most others calculated averages mathematically, the Soviet climatologists deployed a graphical mapping technique. In a project lasting eight years (1960–67), the Main Geophysical Observatory in Leningrad (now St. Petersburg) prepared monthly analysis maps of hemispheric, and later global, surface temperature anomalies starting from 1881.[29] Beginning with 246 stations in 1881, by 1980 the Soviet data set included about 2000 stations. Budyko's group first corrected and "homogenized" the station records. They then plotted these records on maps, creating "hand-drawn, smooth, synoptic-type analyses" much like the subjective analyses of pre-NWP weather forecasting.[30] Finally, they overlaid a 5°×10° grid on the maps and interpolated from the map contours to the gridpoints, releasing the gridded data set on digital tape in 1980.

Until the mid 1980s, many climate scientists regarded this Soviet data set as the best available global source, despite a lack of clarity about exactly what corrections had been applied and how. Alan Robock and other climatologists working with the Soviet authors under a joint US-USSR environmental protection agreement concluded that the Soviet data-homogenization techniques, though mostly undocumented, were probably of high quality. Yet the group's subjective analysis techniques rendered direct comparison with other data sets problematic.[31]

## Reconstructing the Climate Record

In the mid 1970s, as climate models matured, climate data specialists began to digitize historical climate data and grid it for computer processing. These efforts naturally started with the major existing sets of climate data. The US National Climatic Data Center and National Center for Atmospheric Research digitized and combined *World Weather Records* and *Monthly Climatic Data for the World* around 1975, producing a database known as *World Monthly Surface Station Climatology* (WMSSC). NCAR and NCDC made this dataset available to other researchers on magnetic tapes.[32] This became the basis for numerous subsequent datasets. The middle numbers in each cell of figure 11.7 show the number of stations represented in one such dataset produced by NASA's Goddard Institute for Space Studies.[33] The surface temperature trend derived from this dataset appears as the Hansen-Lebedeff line in figure 11.4.

In the early 1980s, in an international effort based at the Climatic Research Unit of the University of East Anglia (with participation from the US Department of Energy's newly formed Carbon Dioxide Research Division), Phil Jones, Raymond Bradley, Tom Wigley, and others embarked on a systematic attempt to recover every available station record from 1850 on. They surveyed numerous meteorological data centers, libraries, and archives to find records not included in *WWR*. The team more than doubled the number of nineteenth-century northern-hemisphere station records. They also increased geographical coverage to include much of northern Asia and a number of Atlantic islands. The group then carefully interrogated the *WWR* data. Among other things, they found that the *WWR* collectors, in their zeal to create a definitive record, had introduced

**Figure 11.7**
Top: Evolution of coverage by surface stations in *World Monthly Surface Station Climatology* (based principally on *World Weather Records* and *Monthly Climatic Data for the World*), with coverage shown as a 1200 km radius around each station. Bottom: Surface stations included in the Goddard Institute for Space Studies version of *World Monthly Surface Station Climatology* as of 1987. Grid cells demarcate regions of equal area. Numbers in each cell represent the date on which coverage began (top), total number of stations in that region (middle), and a grid cell identifier (bottom right). Note that 12 of the 40 cells in the southern hemisphere contain 0–2 stations.
*Source*: J. Hansen and S. Lebedeff, "Global Trends of Measured Surface Air Temperature," *Journal of Geophysical Research* 92, no. D11 (1987), 13,346–47.

Data Wars 313

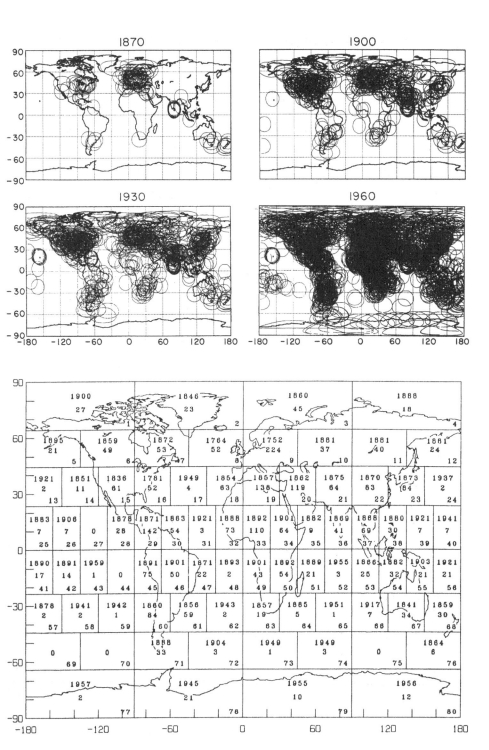

"adjustments" that would create complex difficulties for future investigators. With respect to the US data—"a very mixed bag"—they noted that *WWR* included "records which have had no adjustments, records adjusted for observation time change, records adjusted for location, all of the above, none of the above, or some of the above for some of the time."[34] For example, they found that standard observing times had "varied chaotically from one country to another, and even from one station to another within the same country." Similarly, up until 1950 the *WWR* collectors had applied a stringent standard for mean temperature at US stations: that daily means should be calculated by averaging 24 one-hourly observations. Few stations had ever collected so many readings, so the *WWR* applied an adjustment factor to bring them into line with a "true" 24-hourly mean.[35] The CRU successfully "deconvoluted" the US data by working backward from the *WWR* adjustments. Yet the *WWR* collectors documented many of their other adjustments poorly or not at all, leaving future scientists to puzzle out exactly what they had done, such as how they had compensated for station site changes.[36] The CRU also found that national meteorological services had employed a vast variety of slightly different methods to calculate mean daily temperatures. Table 11.2 shows a sample of the dozens of different methods used in the nineteenth and early twentieth centuries. Only fifteen countries maintained a consistent method from the nineteenth century onward.

Changes in observing hours and methods of calculation could have a considerable effect on temperature averages. Figure 11.8 illustrates the problem. Since most of the stations in the US Cooperative Observer Network are staffed by amateurs paid (if at all) only tiny sums for their work, the network originally accommodated their sleep schedules by setting a "climatological day" ending at sunset, typically between 5 p.m. and 8 p.m. local time, when observers record the maximum and minimum for the preceding 24 hours. Later on, the climatological day was shifted to end at 7 a.m., introducing a "time of observation bias" into the climate record, first noted by Mitchell.[37] For example, "observers who report the minimum temperature ending at 0700 local standard time can have twice as many days with temperatures below freezing under certain climate regimes than if they were to observe the 24-hour minimum at 1700 local standard time."[38] This and other changes in standard observing hours caused an artifactual reduction in the average annual temperature of the United States of 0.16°C across the period 1931–1985 (figure 11.8). Similarly, neither the 0700 nor the 1700 climatological-day regime syncs with the midnight-to-midnight calendar day. Yet readings from both are *attributed* to the calendar

**Table 11.2**
Different methods used to calculate mean daily temperatures in selected countries, primarily during the nineteenth century. 00, 03, etc. refer to observing hours. Reproduced from J. P. Palutikof and C. M. Goddess, "The Design and Use of Climatological Data Banks, with Emphasis on the Preparation and Homogenization of Surface Monthly Records," *Climatic Change* 9, no. 1, 1986, 139).

| Country | Methods used to calculate mean daily temperature |
|---|---|
| Egypt | ½(max + min); means of 3-hourly observations, ⅛(00 + 03 + ··· 21); ¼(09 + 21 + max + min); ¼(06Z + 12Z + 18Z + min); means of 24 hourly values (exact hours unknown) |
| France | ½(max + min); ⅓(06 + 13 + 21); ⅓(06 + 14 + 22); means of 24 hours, ¹⁄₂₄(01Z + 02Z + ··· 24Z); means of eight 3-hourly observations |
| Ghana | ⅛(03 + 06 + ··· 24); ½(max + min) |
| Guyana | ½(max + min); ¹⁄₁₂(00 + 02 + ··· 22); ⅓(07 + 13 + 18) local time; ½(12Z + 18Z). |
| Tunisia | Means of 24 hours (exact hours unknown); ½(max + min); ¼(07 + 13 + 19 + (19 + min)/2) |
| USSR | ¼(01 + 07 + 13 + 19); ⅓(07 + 13 + 21); ¼(01 + 07 + 13 + 21); ¼(07 + 14 + 21 + 21) 105°E meridian time; means of 2–4 daily observations in 53 different combinations |

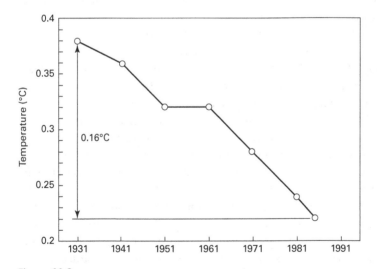

**Figure 11.8**
Estimated changes in US national average temperature caused by changes (at circled points) to standard observing hours.
*Source*: Karl et al., "Detecting Climate Variations and Change," 1486.

day on which they end. The use of climatological days can thus shift minima from the first or the last calendar day of a month into the preceding or the following month, altering the calculated monthly average.

Before digitization, most efforts to calculate global temperature had selected a few stations to represent a given area, such as a 10° latitude-longitude cell.[39] The Soviet graphical effort described above was an exception, but the lack of documentation for adjustments and interpolation techniques made it unreplicable and caused an enduring debate over its quality. After digitization, data specialists began to apply mathematical interpolation techniques similar to those used in weather forecasting to produce gridded global data sets. Here, however, the purpose of gridding was not to feed numerical models. Instead, gridding offered a principled way to integrate data from multiple stations to generate a single value for each grid cell. This, investigators hoped, would reduce the effects of minor inhomogeneities and "locally unrepresentative individual stations." Various interpolation strategies and gridding techniques could then be compared. An early CRU effort produced the gridded Northern Hemisphere data set shown in figure 11.9. The criterion used required at least six stations within 300 nautical miles of each gridpoint, each with at least ten years' data. Temperature anomalies calculated from these data correlated very closely with the Soviet data set originally produced by Budyko's group and updated by Vinnikov et al.[40]

Another project took on the 70 percent of Earth's surface area occupied by oceans. Beginning in 1981, the International Comprehensive Ocean-Atmosphere Data Set (ICOADS, originally COADS) project prepared a digitized version of marine atmosphere and sea surface temperatures, using marine logbooks dating back to Maury's pioneering effort in 1854. Digital versions of some of these records had already been created as early as the 1890s, when the US Hydrographic Office introduced Hollerith punch cards. By the 1920s, punch-card recording had become routine in the United States and in Europe. During and after World War II, collectors in the United States and elsewhere began accumulating and combining decks of cards, recoding them into a single format and eventually transferring them to magnetic tape. By 1981, various digitized sources included more than 140 million records, about half of them duplicates. These data suffered from most of the homogeneity problems described earlier in this section, and from many other problems unique to marine data. For example, marine observers originally measured sea surface temperatures by throwing a canvas or wooden bucket over the side of a ship, hauling up some water, and inserting a thermometer. Some of these buckets were

Data Wars 317

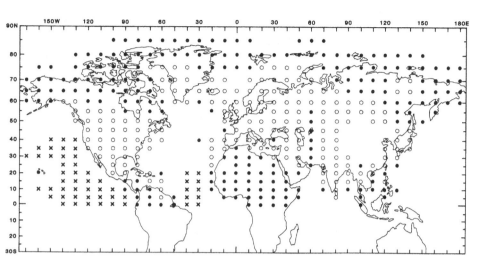

Figure 11.9
Northern hemisphere climatological data coverage, 1881–1980. To count as "covered," a gridpoint needed at least six stations within 300 nautical miles, each with at least 10 years' data. Open circles: gridpoints with data available starting in 1900. Full circles: gridpoints with data coverage starting in the 1950s. Crosses: gridpoints without sufficient data even after a "relaxed interpolation/extrapolation procedure." Full circles plus open circles represent usable gridpoints. Note the absence of data points across most of the People's Republic of China.
Source: P. D. Jones et al., "Variations in Surface Air Temperatures: Part 1. Northern Hemisphere, 1881–1980," *Monthly Weather Review* 110, no. 2 (1982), 67.

insulated; others were not. With the advent of powered vessels, many fleets switched to a different technique, recording temperatures with a sensor placed in the ship's engine-cooling-water intake. Heat from the engine can raise the detected value if the sensor is nearby, biasing these readings warmer than the sea outside the vessel. The exact depth of the engine intake also affects readings. These three methods produce systematically different results, and variations within each method create further inhomogeneities.[41] The more scientists inverted the infrastructure and recovered metadata, the more they could use algorithms to render data collected by these various methods comparable.[42]

## Metadata Friction

One might imagine that by now every conceivable source of error and possible improvement would have been found, but this is certainly not the case. Take an example: In the 1990s climatologists began mining WMO Publication No. 47, *International List of Selected, Supplementary and Auxiliary Ships* (published annually starting in 1955, with numerous irregularly issued supplements). These documents described numerous features of ships and their onboard observing systems, such as what type of thermometer, anemometer, and barometer they used to measure atmospheric variables and their height above the sea surface. Publication No. 47's original purpose was to help weather forecasters identify and interpret data reported by Voluntary Observing Ships. Once a new version arrived, these operational users had no need for the old one, so they discarded it. When climate scientists began building the Comprehensive Ocean-Atmosphere Data Set, they used these manuals as metadata. But most of the year-by-year manuals and supplements had been discarded long before.

The effort involved in finding existing metadata, digitizing them, and combining them with whatever metadata you already have might be termed "metadata friction." Starting in the latter half of the 1990s, investigators gradually recovered older copies of Publication No. 47, its supplements, and other similar manuals, such as the 1963 UK *Marine Observer's Handbook* and *Lloyd's Register of Shipping*. They then digitized them and added them to existing marine metadata. Where early investigators had applied corrections on a fleet-wide basis, these metadata permitted an increasingly fine-grained application of corrections, down to the level of individual ships.[43] Recently these metadata recovery efforts have led to the detection and explanation of a large, sudden drop (−0.3°C) in sea surface temperature (SST) starting in 1945, probably related to a sudden switch in the COADS database from dominance by data from US-based vessels using engine-intake measurements (1942–1945) to data from UK-registered vessels using bucket measurements (starting in 1945).[44] This spurious drop, if confirmed, will be corrected in future versions of SST data.

Unlike the surface station reports in *WMSSC* and its predecessors, most marine records were not inherently climatological. Instead, because ships, weather buoys, and most other marine observing platforms are always moving, these logs were simply weather records, each for a particular location on a particular day. Despite the large number of individual records, outside the major shipping lanes the amount of data for any given grid cell in any given month is small, especially in the little-traveled southern hemisphere. The ICOADS project gridded these data, initially at 2°×2°. Then it calculated climatological statistics for each grid cell. In effect, ICOADS treated each cell as a single weather station, integrating all measurements from all platforms within that cell for each analysis period (monthly, yearly, etc.) to create a climatological data set. Combining data from such a large area (up to 200×200 km, depending on latitude) makes more sense at sea than it would on land, where topography and vegetation create large local and regional variations.

The aforementioned studies have brought increasing attention to sea surface temperature, rather than land air temperature, as a potential marker of global climate change:

> Surface air temperature over land is much more variable than the SST. SSTs change slowly and are highly correlated in space; but the land air temperature at a given station has a lower correlation with regional and global temperatures than a point SST measurement, because land air temperature (LAT) anomalies can change rapidly in both time and space. *This means that one SST measurement is more informative about large-scale temperature averages than one LAT measurement.*[45]

Furthermore, marine air temperature near the surface generally correlates closely with sea surface temperature. Therefore, SST can serve as a reasonable surrogate for marine air temperatures.[46]

**Proliferation within Convergence: Climate Data Today**

By 1986, the Climatic Research Unit had combined the COADS marine data with its land surface data to produce the first comprehensive global surface temperature dataset.[47] Since then, this dataset—now in its third release, known as HadCRUT3—has been co-produced by the UK's Hadley Centre, which manages marine records, and the CRU, which handles land surface data.[48] In the decades since these pioneering efforts, more and more land and marine records have been retrieved and digitized. The third release of the CRU land surface temperature dataset CRUTEM3 contained records for 4349 land stations—more than twice the number contained in the 1986 release. While many of the additional records represent new stations added since 1986, many others were historical records omitted from *WWR*, *MCDW*, and *WMSSC* and recovered through an ongoing search.

Figure 11.10 shows temperature anomaly trends for land surface (top) and SST (middle) components of the HadCRUT3 data set. The much larger uncertainties in the land data stem from the much greater variability of land surface temperatures, as noted above. The combined whole-global trend (bottom) reduces the overall uncertainty because the SST (at 70 percent of total surface area) dominates. A notable feature of these charts is that while the best-estimate land and SST trends are essentially identical

320                                                                                                          Chapter 11

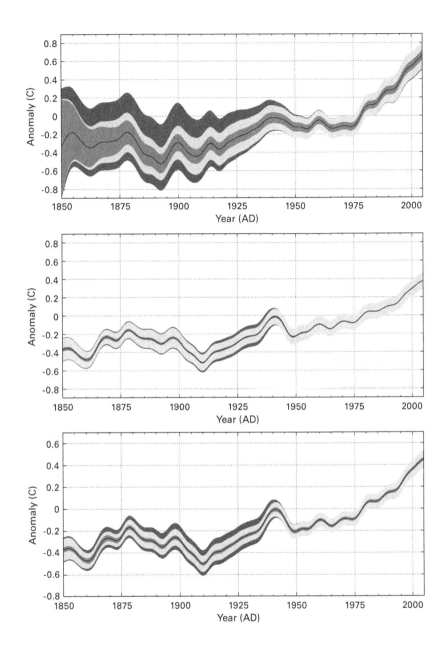

until about 1980, they diverge after that, with land temperatures rising faster than SST. The reasons for this remain unclear. As Brohan et al. explain, this "could be a real effect, the land warming faster than the ocean (this is an expected response to increasing greenhouse gas concentrations in the atmosphere), but it could also indicate a change in the atmospheric circulation, it could indicate an uncorrected bias in one or both data sources, or it could be a combination of these effects."[49] Hence, the work of infrastructural inversion remains unfinished, as it probably always will.

The trend lines in figure 11.10 also represent model/data symbiosis (discussed in the previous chapter) in action. The data behind these trend lines were first adjusted to reduce homogeneities, then gridded using an interpolation procedure, then adjusted for variance (differences in the number of observations available within each grid cell). Without these adjustments, the data could not be joined into a (relatively) uniform and usable whole.

What you learn from all this is that if you want global data, you have to make them. You do that by inverting the infrastructure, recovering metadata, and using models and algorithms to blend and smooth out diverse, heterogeneous data that are unevenly distributed in space and time. There are many ways to do this, and as a result there have been many versions of global data: global climate, plural, a moving target that continues to shift (albeit within a restricted range). Since the 1980s, a series of projects have vastly expanded the number and type of measurements available to do this, as well as the available tools. These began with digitized traditional climatological data from *WWR* and *MCDW*, but they have expanded to incorporate other sources such as marine weather records

◄ **Figure 11.10**
Global averages of land (top), marine (middle), and combined land-marine (bottom) atmospheric temperatures from HadCRUT3. The combined land-marine data have lower uncertainty than the land surface data alone because 70 percent of Earth's surface is ocean. From original legend: "The black [center] line is the best estimate value; the [medium gray] band gives the 95% uncertainty range caused by station, sampling, and measurement errors; the [light gray] band adds the 95% error range due to limited coverage; and the [dark gray] band adds the 95% error range due to bias errors."
*Source*: P. Brohan et al., "Uncertainty Estimates in Regional and Global Observed Temperature Changes: A New Dataset From 1850," *Journal of Geophysical Research* 111, no. D12106 (2006). Colors replaced as noted in brackets to allow non-color reproduction.

(COADS). The sustained effort and expertise required has had the effect of concentrating analysis in a few "centers of calculation" (in Bruno Latour's phrase)—principally the Climatic Research Unit at the University of East Anglia, the UK Hadley Centre, NASA's Goddard Institute for Space Studies, and the US National Climatic Data Center. These efforts represent the *ex post facto* standardization process I described at the beginning of chapter 10. They have narrowed the range of global temperature trend estimates considerably since 30 years ago, when digitization began. Yet each of these efforts continues to produce new versions of its global data. Furthermore, as we are about to see, this is not the only way scientists have found to re-create the history of climate.

## 12 Reanalysis: The Do-Over

From the earliest national and global networks through the 1980s, every empirical study of global climate derived from the separate stream of "climate data." Climatological stations calculated their own averages, maxima, minima, and other figures. Central collectors later inverted the climate data infrastructure, scanning for both isolated and systematic errors and working out ways to adjust for them, seeking to "homogenize" the record. All of these efforts presumed—for the very good reasons discussed above—that only traditional "climate data" could form the basis of that record.

But as numerical weather prediction skill advanced and computer power grew, a new idea emerged: What about a do-over? What if you could rebuild climate statistics "from scratch," from daily weather data? And what if you could do this not simply by recalculating individual station averages, but by feeding every available scrap of weather data into a state-of-the-art 4-D assimilation system, as if taking a moving data image *with a single camera*? What if you could recreate the entire history of weather, at every altitude, every gridpoint, every place on the planet—and then calculate climate statistics from that?

### Gathering Resources and Assembling Data

The roots of reanalysis lay in the Global Weather Experiment's parallel data streams. Recall from chapter 9 that "delayed mode" data from the First GARP Global Experiment (FGGE) took years to analyze. In fact those data were analyzed repeatedly—re-analyzed—in order to compare different analysis methods and improve the final data quality. Yet although the FGGE data gave climate scientists good circulation statistics for that one year, World Weather Watch analysis systems kept changing, creating inhomogeneities that built up over time, rendering weather analyses unusable for climate studies.

But the FGGE experience sparked an idea. In the early 1980s, while it was still underway, Lennart Bengtsson, director of the European Centre for Medium-Range Weather Forecasts, conceived a project to reanalyze ten years' observations with a "frozen" data assimilation system (one that would not change during the reanalysis).[1] If such a project were to begin in the mid 1980s, Bengtsson argued, it could start with data from the FGGE period and continue through 1987. Conveniently, ECMWF's tape archive already contained most of the data it would need (or so Bengtsson thought at the time). Reanalysis would create a single internally consistent set of analyses that could be used to diagnose problems with forecast models as well as to study the general circulation. Though a ten-year reanalysis would not be long enough to show meaningful climate trends, perhaps future reanalyses could reach further back.

In 1982, thinking along the same lines, Jagadish Shukla, head of the climate section at NASA's Goddard Space Flight Center in Maryland, proposed reanalysis leading to a "description of the state of the atmosphere, ocean and land surfaces for a ten-year period, 1979–1988."[2] This reanalysis would have supported a proposed Global Habitability Program at Goddard, aimed at characterizing long-term environmental trends affecting food production, air and water quality, and the habitability of the planet, all essentially based on what would later be called an "Earth system model." In the early 1980s, this idea went nowhere, Shukla recalled, and he left Goddard in 1983. But in 1985 he received encouragement to propose a shorter reanalysis as part of the Tropical Ocean/Global Atmosphere (TOGA) subprogram of the Global Atmospheric Research Program. "A plan must be developed," Shukla told a TOGA panel in 1986, "so that all data are merged through a global reanalysis scheme, at least once and perhaps several times, as analysis and initialization schemes improve." In this way, data received after the operational cutoff time could be integrated into the analysis, just as in the FGGE Level IIIb analyses. Meanwhile, Kevin Trenberth and Jerry Olson of the National Center for Atmospheric Research were comparing the operational analyses produced by the US National Meteorological Center and the ECMWF between 1980 and 1986 to determine whether they might be reliable enough for use in climatology. Trenberth and Olson found substantial differences between the two analyses, especially in the tropics and the southern hemisphere. Although these differences diminished between 1980 and 1986, Trenberth and Olson noted that reduced differences between the analyses do not necessarily mean improved accuracy. This is because changes that improve NWP model forecast performance at one center are likely to be adopted at the other center, with resulting parallel changes in the analyses. Interannual variability and climate change can only be explored with these data sets if proper account is taken of the impacts of changes in analysis procedures. Consequently a strong case can be made for a reanalysis of the original data, preferably enhanced with further observations (either delayed or non-real time) that did not arrive before the operationally imposed cut-off, using the same state-of-the-art analysis system.[3]

Trenberth argued that the name "four-dimensional data assimilation" misstated the nature of operational analysis, which was actually "three and a half dimensional." In other words, operational analyses looked backward in time, integrating data from the recent past (up to the observational cutoff), but they did not look forward in time, correcting the analysis with data arriving in the first few hours after the cutoff. But data assimilation systems purpose-built for reanalysis could potentially offer this capability, leading (in principle) to more accurate, more smoothly varying analyses.[4]

Despite the very high anticipated cost in time, money, and effort, the TOGA panel not only endorsed Shukla's proposal but noted that "many elements of the TOGA Program cannot be justified unless there is the prospect of reanalysis."[5] Trenberth and Roy Jenne, head of NCAR's Data Support section, began examining what data might be available for reanalysis from both NCAR and ECMWF. Jenne's personal notes from this period reflect not only the building ambition, but also the many reservations expressed by more cautious participants and as the institutional difficulties of bringing operational forecast centers into the climate business:

Good idea, but who is crazy enough to do it?

Will cost huge amount of money!

Basic variables will be better, but no guarantee that derived quantities will be better (heating, fluxes, etc.)

Can freeze the model, but what about the observing system?

Why should the forecast centers, with a mission to improve forecasts for the future, use their resources to analyze past data?

If two separate reanalyses are different, how do you know which one is correct?

Let us wait till we have developed methods to include "future" data in the assimilation system.[6]

Additional questions about which institutions would do the work, how they would divide the labor, and how the effort would be funded remained to be resolved. But apparently the 1986 TOGA meeting convinced administrators such as Jay Fein, director of the NSF's Climate Dynamics Program, to begin marshalling the necessary resources.

Bengtsson and Shukla joined forces to compose a landmark paper, "Integration of Space and In Situ Observations to Study Global Climate Change," published in 1988. Having demonstrated the power of the analysis model as a "unique and independent observing system" (see chapter 10 above), they proposed to begin a long-term reanalysis (at least ten years) no later than 1990. A twenty-year reanalysis could provide decadal climate averages, which, Bengtsson and Shukla noted, had "never been calculated from a homogeneous data set." It could also help in specifying atmospheric forcings for ocean circulation models, for analyzing the global hydrological cycle (using model-calculated rainfall estimates), and for numerous other climatological purposes.[7] Bengtsson and Shukla urged the community to get moving right away, but they also recognized that data assimilation systems would continue to improve, so that any reanalysis probably would have to be repeated later.

**Production: Data Friction, Again**

Three major reanalysis efforts took shape between 1986 and 1990. First, the Data Assimilation Office (DAO) within NASA'S Goddard Laboratory for Atmospheres employed a "frozen" version of the Goddard Earth Observing System assimilation system known as GEOS-1. The DAO completed a 5-year GEOS-1 reanalysis in 1994 and continued through the 1990s.[8] Meanwhile, the ECMWF planned a 15-year reanalysis, known as ERA-15, to cover the years from 1979 (the FGGE year) through 1994. ERA-15 ran from 1994 to 1996.[9] The third and most ambitious effort was a cooperative project of the US National Centers for Environmental Prediction (then called the National Meteorological Center)[10] and NCAR. The NCEP-NCAR project initially targeted a 35-year period, 1958–1992. Other, smaller reanalysis projects also began in the early 1990s.[11]

The three major reanalyses eventually extended their temporal coverage. The GEOS-1 reanalysis covered 1980–1995. At this writing the NCEP-NCAR reanalysis covers the 60-year period 1948–2008 and is ongoing. These have been followed by "second-generation" reanalyses using newer, higher-resolution assimilation systems. ERA-40, produced with an updated version of the ECMWF model, began in 2000 and covered 1958–2002 (actually 45 years, not 40). Driven by model intercomparison efforts in support of the Intergovernmental Panel on Climate Change assessment reports, NCEP collaborated with the US Department of Energy on "Reanalysis-2," for 1979 on, correcting known problems with the NCEP-NCAR reanalysis.[12] Reanalysis-2 continues at this writing.

Great increases in computer power and related improvements in data-sharing techniques—especially the possibility of data transfers over the Internet— made reanalysis possible, but these projects still faced enormous data friction and computational friction. Remember that we are talking not about traditional "climate data," pre-digested using the partially distributed computation strategy, but about the daily and hourly readings putatively produced by the forecast data network—*all of them*. Not only surface-station data, but also data from satellites, radiosondes, rawinsondes, pilot balloons, aircraft, ships, and other sources would have to be assembled in one place. No one had ever done this before. None of the world's weather centers—not even the World Meteorological Centers—maintained a complete library of original weather records.

Reconstructing a quasi-complete global weather data record required creative, painstaking, long-term efforts by data managers. In 1989, the ECMWF sent Daniel Söderman on a mission to scout out possible data sources. Söderman visited fourteen organizations in the United States and Europe in June of that year and corresponded with another twenty agencies around the world, collecting information about the location, nature, and quality of data that ECMWF did not yet possess.[13] Meanwhile, NCEP and NCAR cemented their reanalysis plans.

Early in 1991, NCAR's Data Support Section, led by Roy Jenne, embarked on an exhaustive search to supplement its already extensive data archive with collections from other agencies and national weather services around the world. "I feel a little scared at the magnitude," Jenne admitted in an early paper on the project. "We are talking about preparing data sets with nearly all of the world's synoptic scale observations."[14] Numerous individuals and institutions around the world (including NOAA, NASA, GFDL, ECMWF, the UK Met Office, and the Japan Meteorological Agency) contributed data. Jenne's notes toward a history of reanalysis include numerous recollections of the complexity involved in this tedious process. For example:

The USAF had worked during about 1955–74 to gather data from many nations and key-enter the data onto hundreds of millions of punched cards. This resulted in a surface dataset (TD13) with 100 million observations and a large dataset of rawinsondes (in TD54). In 1990 the tapes arrived at NCAR. When these arrived, I felt that we would have enough observations from at least 1948 for a good reanalysis. Many other data sources would also be used. NCDC had worked on a "Northern Hemisphere Data tabulation" project that resulted in the C-Cards dataset that had punched card upper air data from GTS (NH) for 1949–1965. . . . Plus we got NCDC data for US raobs [radiosondes] and it had data from some other countries in early years. And NCAR also got data directly from a number of countries.

NCAR also had been gathering data from GTS world telecom that could be obtained on magnetic tapes. About 1973 we obtained NCEP upper air data from NCEP for 1962–1972 before these 2000 tapes were purged. And we have collections from NCEP, USAF, and Navy for all of the recent years starting 1970 or 1973. . . .

We had to do a lot more work on the observations to prepare them for reanalysis. The location for a lot of observing stations was wrong. One set of aircraft recon data had all longitudes wrong by 10°. We found date/time problems. We found data assigned to the wrong levels in rawinsondes. We found data biased by truncation vs. the proper rounding of units. And much more.

In conference papers and reports, Jenne and others detailed a myriad of other issues with individual data sets.[15] Figure 12.1 shows another sample from Jenne's personal notes, intimating the frustrations experienced during nearly a decade of work.[16]

In 1998 NCAR finished assembling the full 50-year data set (1948–1998). The process had consumed about 30 person-years of dedicated effort at NCAR alone.[17] Concurrently, the NCEP group led by Eugenia Kalnay had prepared the "frozen" 4-D data assimilation system, a modified, reduced-resolution version of the operational forecast assimilation model. Production began in 1994, concurrently with data assembly, in a series of stages. The project's first major publication appeared in a special issue of the *Bulletin of the American Meteorological Society* in 1996. The special issue included (for the first time ever) a CD-ROM with model documentation and monthly climatologies produced from some of the reanalyzed data. This rapidly became one of the most cited articles in the history of meteorology.[18] By 1998, after numerous stops, starts, and reruns of problematic periods, NCEP had completed reanalysis of the entire 50-year period.[19]

ECMWF's long-term reanalysis, ERA-40, began in 2000 and finished in 2003, profiting from the ERA-15 experience and from subsequent ECMWF

D. Produce analyses for 1968-72 (5 years).  *[handwritten: Update: All of VTPR was used in the ECMWF ERA-40 (1957-on)]*
- Oct 29, 1996: NCEP ready to start
- Oct 31: NCEP "dead in the water." Needs type of rawinsonde instruments for radiation corrections. Not available. Invent other methods.
  - Handle raob radiation humidity error (old US instruments).
  - And much trouble with surface snow input.
- Dec 24: Start production on 1968-72.

- 31 Mar 1997: Finished 1968-1972 (includes 2.9 years of reruns).

29 years are done, 1968-1996

E. April-June 1997: Do cleanup work, receive data from NCAR.
- Did FGGE (1979) again. The earlier runs had lost most raobs over Antarctica. These reruns were done with and without satellite data.

- Did TOGA COARE again (temporary fix). For special raobs in TOGA COARE, the soundings were very long and this caused trouble at NCEP. The raob temperatures were used as if they were virtual temps (so analyses were too cold). This was discovered because of too much ocean to air surface flux. So TOGA COARE (4 mo) was done again (and the special TOGA COARE data was not used).

**Figure 12.1**
A portion of Roy Jenne's 1998 progress notes on the NCEP-NCAR reanalysis. Reproduced with permission.

assimilation model improvements to introduce a "second-generation" reanalysis model. ERA-40 also benefited substantially from data-preparation work done at NCEP and at NCAR. Since ECMWF's archive extended back only to 1979 (when the Centre began operational forecasting), NCAR provided most of the pre-1979 data used in ERA-40 in exchange for free access to all ERA-40 data products. For the post-1979 period, ECMWF relied mainly on its own extensive data archives, but also collected additional data from Australian, Japanese, Russian, and other sources. Like the NCEP-NCAR project, ERA-40 encountered data problems, including "a coding error in surface-level data as received, . . . incorrect times . . . assigned to radiosonde reports, and . . . assimilation of erroneous ERS-1

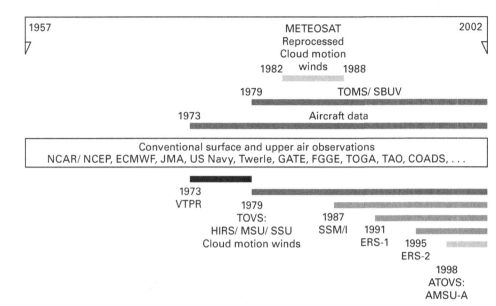

**Figure 12.2**
Observing systems used in the ERA-40 reanalysis. Except for aircraft data, shaded bars represent satellite instruments.
*Source*: S. Uppala et al., "ERA-40: ECMWF 45-Year Reanalysis of the Global Atmosphere and Surface Conditions 1957–2002," *ECMWF Newsletter* 101 (2004), 4. Graphic courtesy of European Centre for Medium-Range Weather Forecasts.

altimeter ocean-wave-height data."[20] Figure 12.2 lists the numerous data sources included in ERA-40, illustrating some of the discontinuities in the data stream as new satellite instruments were launched and old ones taken offline. Reanalysis models had to adjust for the different instrument characteristics and different variables measured.

Reanalysis efforts also had to cope with vast increases in input data volumes, especially from satellite radiances. Not only inputs, but also outputs grew dramatically. The model used for ERA-15 resolved to a surface grid of about 250 km, with 31 vertical levels; its output files occupied about 130 gigabytes per year. Subsequently, the ERA-40 model used a 125-km horizontal grid and 60 vertical levels, increasing the size of output data files by more than an order of magnitude to about 1500 GB per year. For comparison, the ECMWF's original FGGE IIIb reanalysis contained just 10 GB of data.

Reanalysis provoked enormous excitement. By the early 2000s, other institutions, including the Japan Meteorological Agency, had launched major reanalysis projects, and numerous smaller, experimental projects had been started.[21] Investigators at NOAA's Earth System Research Laboratory used surface pressure data from the pre-radiosonde era to extend reanalysis back to 1908, complementing existing studies to create a full century of reanalysis data, and they have begun to consider reaching even further back, into the late nineteenth century.[22] By 2007, publications concerned with reanalysis for climate studies were appearing at a rate of 250 per year.[23]

At the same time, reanalysis revealed the depth and intractability of problems in both the observing system and the models. The observing system's highly heterogeneous instrument types, models, and manufacturers create numerous ongoing disparities in the input data. For example, about two dozen radiosonde models are currently in use around the world. Each model's measurements differ slightly from those of other models, not just as a result of varying materials and construction but also because of fundamental differences in how the various radiosondes work: some contain temperature and altitude sensors and derive pressure readings from those, whereas others measure temperature and pressure directly and derive altitude. Various tracking techniques and radiation-correction techniques also affect their readings. Combined, these variations create differences in measurements of geopotential height. In regions dominated by a single radiosonde model, that model's signature can appear in weather analyses as fictitious circulation features.[24]

All the assimilation models used in reanalysis to date exhibit biases of various kinds, due mainly to imperfect physical parameterizations. (See chapter 13.) Though the injection of observational data continually corrects the analysis, the models continually "pull" in the direction of their inherent biases. These small biases do not much affect skill in the short time periods addressed by forecasters, but in reanalysis over decades model bias asserts itself. For example, for the first years of satellite observations the ECMWF ERA-40 assimilation model exhibits a cold bias in the lower stratosphere, and other discrepancies in the southern hemisphere below 45°S.[25]

Figure 12.3 illustrates how model bias can affect reanalyses. In the lower diagram, a biased model systematically reaches toward a higher value. Each time observations are injected, they pull the result back toward the atmosphere's actual state, but over time the model bias produces a markedly higher trend—more so when observations are less frequent. In the upper diagram, both model and observations are unbiased. The model still drifts, but now its fluctuations are more randomly distributed around

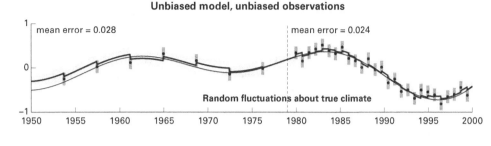

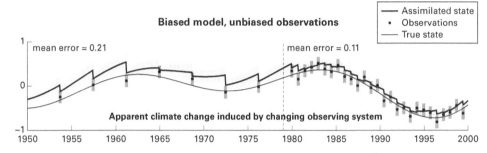

Figure 12.3
A schematic illustration of the effects of reanalysis model bias on climate trends. The lower diagram shows how model bias can create an apparent climate trend even when corrected by unbiased observations. (This conceptual diagram does not reflect a real model or model variable.)
Courtesy of Dick Dee.

the real state of the atmosphere, producing a more accurate trend calculation over time. Statistical and empirical corrections can make reanalysis more accurate, but the underlying problem is not likely to be fully resolved.[26] Furthermore, actual observations are not unbiased; instead, they are subject to the various errors discussed above, some of which are systematic.

### Fingerprint: Reanalysis Data and Climate Change

How well has reanalysis worked? Reanalyses and traditional climate data agree well—though not perfectly—for variables constrained directly by observations, such as temperature. But derived variables generated mainly by the model still show considerable differences.[27] For example, reanalysis models do not yet correctly balance precipitation and evaporation over land and oceans, whose total quantity should be conserved.[28] This affects

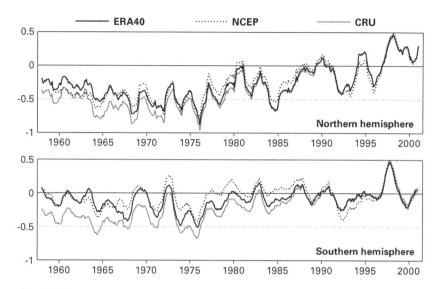

**Figure 12.4**
Comparison of hemispheric surface temperature anomalies, 1957–2001, from ERA-40 and NCEP/NCAR reanalyses and CRU traditional climate data. Anomalies for each data set are calculated against their 1987–2001 averages.
*Source*: A. J. Simmons et al., "Comparison of Trends and Low-Frequency Variability in CRU, ERA-40, and NCEP/NCAR Analyses of Surface Air Temperature," *Journal of Geophysics Research* 109 (2004), 3.

their calculations of rainfall distribution, a climate variable that is extremely important to human populations and to natural ecosystems.

Significant differences remain between the major reanalyses, as well as between the reanalyses and major traditional climate data sets. Figure 12.4 compares the ERA-40 and NCEP-NCAR reanalyses with the Climatic Research Unit surface temperature data set. Notably, the CRU data set systematically shows lower temperatures than either of the reanalyses in the period before the advent of satellite observations in 1979, and especially before 1967. For ERA-40, investigators attribute this difference to "limited availability of surface observations . . . combined with a near-surface warm bias in the background forecasts" used in the reanalysis.[29] After 1979, however, the reanalyses—especially the ERA-40 second-generation reanalysis—track the CRU surface temperature data set quite closely.

As a replacement for traditional climate data, then, reanalysis has so far been disappointing. At this writing, even reanalysis' most ardent proponents believe that their highest hopes for the do-over—a definitive, higher-resolution climate data set that might supersede traditional climate data—remain unrealized. For the long-term record, most investigators prefer to trust traditional data, especially for single variables such as temperature or precipitation. As a result, while the 2007 IPCC report referred to the ERA-40 and NCEP-NCAR reanalyses frequently, it treated them as supplementary evidence rather than primary data, warning that "long-timescale trends over the full length of the reanalyses [remain] potentially unreliable."[30]

Yet the do-over has proved at least modestly successful for certain specific purposes. Reanalysis offers something that traditional climate data will never achieve: physically consistent data across all climate variables. Traditional climate data are "single variable": you get a set of averages for temperature, another one for pressure, a third for precipitation, a fourth for sunshine, and so on. Each type of observation is independent of the others, but in the real atmosphere these quantities (and many others) are interdependent. Reanalysis models simulate that interdependence, permitting a large degree of cross-correction, and they generate all variables for every gridpoint. This allows scientists to study structural features of the atmosphere and the circulation not directly measured by instruments. For this reason, the greatest successes of reanalysis data for the post-1979 period have come in so-called fingerprint studies.

The "fingerprint" metaphor suggests that climate forcings from human activity, such as greenhouse gases and aerosols from fossil fuels, may produce different patterns from those caused by natural forcings such as volcanic eruptions and sunspots. Such fingerprints would be multivariate—that is, they would appear in the relationships among multiple variables, rather than in single-variable climate statistics such as global average temperature. Also, they would be four-dimensional: part of the pattern would be temporal, related directly to the increases or decreases in forcings. Once you identify a likely fingerprint, you will need a data set that includes all the fingerprint variables, and you will need these data to be temporally consistent. That sounds like reanalysis.

Predicting a fingerprint requires forcing climate models with a realistic array of relevant factors on a time-evolving or "transient" basis.[31] For example, rather than simply run the model once at pre-industrial $CO_2$ concentrations and again for $CO_2$ doubling, you increase the $CO_2$ concentration year by year, corresponding to its actual historical rate of change. You do the same for other factors, such as aerosols and other greenhouse gases. Now you analyze the changing circulation patterns and climatological averages statistically and temporally, comparing them with control runs that contain only natural forcings. You also calculate the amount of "noise" produced by natural variability in the control runs. If your fingerprint signal remains after subtracting this noise, you've found a candidate for a *unique* anthropogenic effect, one that could not be caused by any known combination of natural events. Then you check the observational data. If you see the same fingerprint there, you've found some evidence of anthropogenic change.

In the early 1990s, climate modelers put forward a number of fingerprint candidates, but could not strongly confirm any of them.[32] By the second half of the 1990s, numerous studies had detected a variety of probable fingerprints, but the level of uncertainty in the observational climate record still made confirmation difficult.

But in 2003 and 2004, a group led by Ben Santer at Lawrence Livermore Laboratory used ERA-15 and then ERA-40 reanalysis data to clearly identify one model-predicted fingerprint: a significant change in the height of the tropopause.[33] Heating in the troposphere and cooling in the stratosphere would cause the height of the tropopause to rise. Both are GCM-predicted responses to anthropogenic greenhouse gases and stratospheric ozone depletion. GCM runs including only natural forcing factors, such as solar and volcanic activity, do not produce this result. Though still not fully confirmed, this and similar results reached from reanalysis data have created cautious optimism that the tropopause height changes seen in the ERA-40 reanalysis are real, and that they represent a significant fingerprint of anthropogenic climate change.

**The Future of the Past**

Where is reanalysis going? It is hugely expensive, requiring many weeks of supercomputer time as well as ongoing data cleanup like that described above. And the do-over has already been done over several times. Investigators have proposed new techniques to improve reanalysis products, such as restricting reanalysis inputs to "pre-validated" data that have been temporally and spatially calibrated to reduce observing system heterogeneities. Longer-term reanalyses have also been proposed, some beginning as far back as 1850.[34]

Detecting and attributing climate change and diagnosing model biases are no longer the only reasons to reanalyze weather data over and over again. NCEP has committed to ongoing reanalysis—at least of the period since 1979—on a regular cycle, because it needs reanalyses at current model resolutions to calibrate its operational climate forecasting system (Louis Uccellini, personal communication). So reanalysis has been integrated into the infrastructure of climate knowledge. Each round will bring new revisions to the history of climate. Well into the future, we will keep right on reanalyzing the past: more global data images, more versions of the atmosphere, all shimmering within a relatively narrow band yet never settling on a single definitive line.

# 13 Parametrics and the Limits of Knowledge

In chapters 8–12, I explored how scientists use models to process, interpret, and reconstruct observations, to *make data global*. Models are how we know what we know about the history of climate. They are also why we can keep learning more about the history of climate, without ever securing a single definitive version of that history once and for all. The past shimmers. What about the future?

If we need models to make sense of data about the past, we need them all the more to predict whether, and how, the climate will change. Chapter 7 described how climate models work. Recall the main difference between general circulation models of weather and climate: Climate GCMs don't start from observations. Instead, they simulate their own climates. Then we compare their simulated climates with the real one—or, rather, with the data about the climate that we have (re)constructed, *made global*, in all the ways the last five chapters have described.

Climate models don't use data, I said—and this is true, in the sense that they don't ingest observations of the real atmosphere and use them to correct their simulations, as weather forecast models or data assimilation systems do. Yet climate models use data in many other, very important ways. This chapter is about a fuzzy boundary between data, theory, and algorithms, a place deep within the practical, everyday work of general circulation modeling where modelers combine measured quantities with code to calculate the effects of processes too small, too complex, or too poorly understood to be modeled directly. These are the "semi-empirical" parameter schemes by which modelers handle sub-grid-scale processes, known as the "model physics." (See chapter 7.) Later in the chapter, I will talk about how modelers "validate," or "verify," or "evaluate" climate models, a perilous choice among seemingly similar terms that amounts to the $100 trillion question for climate modeling: Can we trust model predictions of climate change?

'Parameter' is an interesting word, meaning literally a "para-metric" or "para-measure." The prefix 'para', from Greek, can mean many things, including beside, near, along with, beyond, resembling, faulty, undesirable, assistant, and auxiliary. According to a glossary maintained by the American Meteorological Society, 'parameter' refers specifically to "any quantity of a problem that is not an independent variable. More specifically, the term is often used to distinguish, from dependent variables, *quantities that may be more or less arbitrarily assigned values for purposes of the problem at hand*" (emphasis added). So a parameter is a kind of proxy—a stand-in for something that cannot be modeled directly but can still be estimated, or at least guessed.

Parameterization illustrates the interaction of computational friction with the limits of human knowledge. In an ideal climate model, the only fixed conditions would be the distribution and the altitude of continental surfaces. Virtually all other variables—sea-surface temperature, land-surface albedo (reflectance), cloud formation, etc.—would be generated internally by the model itself from the lower-level physical properties of air, water, and other basic elements of the climate system. Instead, *most* physical processes operating in the atmosphere require some degree of parameterization; these parameterized processes are known as the "model physics." Generating these parameters (or "parameterizations," as scientists somewhat confusingly call them) is a large part of any climate modeler's work.[1]

## Types of Parameterization

Parameters may be fixed quantities, such as coefficients, or they may be mathematical functions containing both coefficients and dependent variables. The simplest fixed parameters are specified directly from empirical measurements. These include such quantities as solar radiation, the size and location of continents, the gravitational constant, and the concentration of carbon dioxide in the atmosphere.

Much more commonly, parameters represent a variable physical process rather than a fixed quantity. An easy way to think about this is to imagine modeling rainfall. In principle, you could try to model individual raindrops, but it's hard to see how that will help you if you're interested in rain over an area the size of Ohio (about the size of today's climate model grid cells). You just want to know how much water fell on the whole state, and when, so your model of rainfall need not include individual drops.

A major parameterization in all climate models is radiative transfer. The atmosphere contains both gases ($CO_2$, methane, nitrogen, ozone, oxygen, water vapor, etc.) and solids (particulate aerosols, ice clouds, etc.). Each one of these materials absorbs solar energy at particular frequencies. Each also emits radiation at other frequencies. Those emissions are then absorbed and re-radiated by other gases and solids. These radiative transfers play a huge role in governing the atmosphere's temperature. Thus, models must somehow estimate how much radiation the atmosphere in a given grid box absorbs, reflects, and transmits, at every level and horizontal location. This is the problem discussed at the end of chapter 7, which Callendar, Plass, Möller, Manabe, Rasool, and Schneider all tried to tackle, each adding a new level of understanding of the problem's complexity. "Line-by-line models," which combine databases of spectrographic measurements for the various gases with physical models, can carry out this summing.[2] Today's best line-by-line models achieve error margins better than 0.5 percent. Yet the computational cost of line-by-line models precludes using them directly in climate GCMs. Therefore, 1980s GCMs used fixed parameters derived from line-by-line model results to approximate actual radiation absorption in a given grid box on the basis of its specific mixture of gases, the angle of the sun's rays at that location, and other factors. Today's GCMs go further, incorporating "fast radiative transfer models" that can calculate radiative interactions among various gases and radiative interactions with aerosols, clouds, ice clouds, and other features of the atmosphere. Modelers have systematically compared these fast radiative transfer codes (i.e., parameterizations) against the results of line-by-line models, over time bringing them ever closer together. At this writing, the best fast radiative transfer models contained in climate GCMs track the results of line-by-line models to within about 10 percent.[3] The nesting of complex subprocess models within larger GCMs exemplifies the intense interplay among models that has become a hallmark of computational science.

Let's take another example: cloud physics. Clouds play crucial roles in climate. Convection processes responsible for towering cumulus clouds are a primary mechanism of vertical heat transport in the atmosphere. Clouds also strongly affect surface temperature; they can raise it or lower it, depending on conditions. For example, clouds can form a heat-trapping "ceiling" at night, yet their white tops reflect incoming solar energy back into space during the day, cooling the surface. Clouds' climatic effects depend on many factors, including their altitude, density, and color. They interact with aerosols, which can provide nuclei around which cloud droplets form, and which can also affect clouds' color, changing their reflective characteristics.

Because clouds affect climate so greatly, and because they are still so poorly understood, cloud parameterization remains among the most controversial areas of climate modeling. Clouds develop from "complicated interactions between large-scale circulations, moist convective transport, small-scale turbulent mixing, radiation, and microphysical processes," and they remain notoriously difficult to study empirically.[4] Many kinds of clouds form on scales of a kilometer or less, two orders of magnitude below climate GCM resolutions. Therefore, rather than represent cloud formation directly—in terms of convection columns, condensation nuclei, and their other physical causes—a GCM might calculate the amount of cloud cover within a grid box as some function of the independent variables temperature and humidity. To deal with the issue of scale, climate GCMs typically calculate a "cloud fraction" (the percentage of each grid box covered by clouds). This raises the question of what to do when clouds appear at two or more levels in the same atmospheric column. If a grid box on one level is 50 percent cloudy and another grid box two levels higher is also 50 percent cloudy, how much of the sky is cloudy at ground level: all of it? 60 percent? 75 percent? GCMs must be provided with functions that determine how much overlap to specify; this can depend on other variables.[5]

Some cloud parameterization schemes in the earliest GCMs resulted in cloud "blinking," an oscillation between the presence and absence of cloud cover in a given grid box at each time step when certain variables lay just at the critical threshold. (Real clouds do not, of course, behave this way.) Cloud parameterizations remain problematic even in relatively recent models. The Rosinski-Williamson study of round-off error, described in chapter 7, discovered that such errors affected cloud parameterization in NCAR's CCM-2 far more than they affected the model's dynamical core, owing to interaction of the equation specifying cloud behavior with the fixed maximum value specified for relative humidity in the parameterization.[6]

The list of model parameters and their interactions is very long, and many of these are just as problematic as those described above. Other phenomena not well captured by parameterization include the planetary boundary layer (the turbulent layer of air nearest the Earth's surface) and such characteristics of land surfaces as roughness, elevation, and heat retention. For example, climate models of the early 1990s, operating at a resolution of 500 km, represented the entire region between the Sierra Nevada and the Rocky Mountains as a single plateau of uniform elevation. Even today's 200-km GCMs can scarcely separate California's Central Valley from the mountain ranges on either side of it. Much less can they resolve the 50×120 km bowl of the San Francisco Bay, which also is surrounded by mountains.

Modelers develop parameterizations by reviewing observational data and the meteorological literature. They identify the range of observed values and try to find relationships between small-scale processes and the large-scale independent variables in their models. When they succeed in finding such relations, they call the resulting parameters "physically based." Often, however, they do not find direct links to large-scale physical variables. In this quite common case, modelers invent *ad hoc* schemes that provide the models with the necessary connections.

An example of an ad hoc parameter is "flux adjustment" in coupled atmosphere-ocean circulation models (AOGCMs). The interface between the atmospheric model and the ocean model must represent exchanges of heat, momentum (wind and surface resistance), and water (precipitation, evaporation) between the atmosphere and the ocean. These fluxes—flows of energy and matter between atmosphere and ocean—are very difficult to measure empirically. Yet they profoundly affect model behavior. From the 1970s until the mid 1990s virtually all AOGCMs included *ad hoc* parameters, known as "flux adjustments," that artificially altered the modeled fluxes to keep them in line with observed values.[7] Modelers spoke of flux adjustments as "non-physical" parameterizations—i.e., ones not based on physical theory—but also sometimes characterized them as "empirically determined."[8] In 1997, version 2 of NCAR's Community Climate System Model became the first AOGCM to function realistically and stably during long runs (centuries) without flux adjustment.[9] Since then, many other major climate models have also been weaned from flux adjustment, but a small minority of models included in the 2007 IPCC assessment still used the technique.[10]

Any given GCM's model physics contains hundreds or even thousands of parameterizations. The model's overall outputs depend on the interactions among all of these parameterizations, as well as on how they interact with the model's dynamical core (equations of motion). The extreme complexity created by all these interactions can make it difficult to determine exactly why models do what they do. An entire subfield—climate model diagnosis—works out ways to isolate the origin of particular problems to specific parameterizations and their interactions.

## Tuning

Another important part of modelers' work is known as "tuning" the parameters. "Tuning" means adjusting the values of coefficients and even, sometimes, reconstructing equations in order to produce a better overall model result. "Better" may mean that the result agrees more closely with observations, or that it corresponds more closely to the modeler's expert judgment about what one modeler I interviewed called the "physical plausibility" of the change. In some cases, parameters are well constrained by observational data. In such cases, you know the range of values found in the real world precisely, so if your parameterization takes you outside that range, you know that you must fix it. In other cases, empirical data don't constrain the range so much. A variety of values and ranges may all be plausible within the limits of existing knowledge, so you have to make some choices, guided mainly by scientific intuition—and by how your choices affect model behavior. Such parameterizations, which include clouds and aerosol effects, are said to be "highly tunable." Since many parameters interact strongly with others, changing one parameterization may push the behavior of others outside an acceptable range, requiring further tuning.

Since the mid 1990s, modelers have increasingly focused attention on parameterization and tuning of aerosol effects in GCMs. Aerosols are microscopic airborne particles, such as dust, soot, and sea salt. Some are blown into the air. Others are formed in the atmosphere by interactions among gases; their origins are both natural (dust, salt) and human (automobile exhaust, coal soot). Aerosols influence climate in numerous ways, most importantly by altering cloud albedo (reflectance) and by directly reflecting or absorbing heat. As with clouds, much of their behavior derives from very small-scale interactions, and they are notoriously difficult to study *in situ*. High-resolution aerosol models now exist that can treat these processes explicitly on scales up to a few kilometers, but their computational demands are too great for use in GCMs. Hence parameterizing aerosols in regional and global models remains a work in progress.

Though each particular aerosol's properties are different, overall aerosols seem to exert a cooling influence on climate (figure 13.1). Yet the range of uncertainty regarding aerosols' climatic effects is large. At one extreme of that range, aerosols could exert an influence nearly as large as those of all anthropogenic greenhouse gases together, *but opposite in sign*.[11] Climate scientists say that aerosols "mask" the warming effects of greenhouse gases, because their lifetime in the atmosphere is short (days to months). If

# Parametrics and the Limits of Knowledge

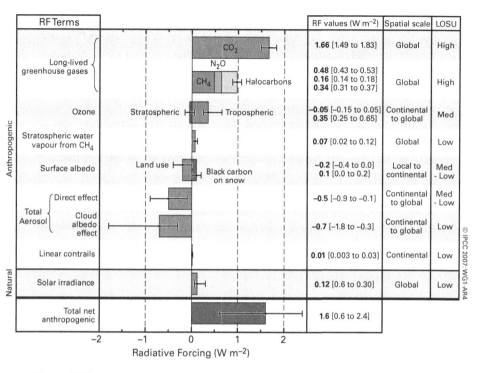

**Figure 13.1**
The major global radiative forcing factors, with uncertainties indicated by black bars. At the extremes of their uncertainty range, the combined negative forcing from direct and indirect aerosol effects could be at least as large as the positive forcing caused by long-lived greenhouse gases. LOSU (rightmost column) represents the IPCC-assessed "level of scientific understanding."
*Source: Climate Change 2007: The Physical Science Basis* (Cambridge University Press, 2007). Image courtesy of IPCC.

aerosol production were to fall dramatically—for example, because coal-fired power plants adopted exhaust scrubbing systems—aerosol levels in the atmosphere would drop quickly and the warming effects of greenhouse gases would assert rapidly themselves. Further, as the atmosphere warms, it becomes more humid. This tends to increase aerosols' dropout rate, reducing their lifetime in the atmosphere enough to decrease aerosol levels even without emissions reductions.

This strong sensitivity of the climate system means that getting aerosol parameters right matters greatly to modelers' ability to project future climate change. The crude aerosol parameterizations used in models for the earliest reports of the Intergovernmental Panel on Climate Change resulted in an untenably high projected warming. When Mount Pinatubo erupted in 1991, ejecting a huge dust cloud that reduced global temperatures by about 0.5°C for a number of months, attention to aerosols and their parameterization increased dramatically. Yet our understanding of aerosol chemistry and physics remains much more limited than our understanding of greenhouse gases and other important climate forcings, as reflected in figure 13.1's long error bars for aerosol effects. Because of this, Jeff Kiehl and others have recently raised concerns about the fact that the overall outputs of GCMs seem to agree more closely than their aerosol parameterizations should permit. The (unconfirmed) suspicion is that modelers are adjusting these "highly tunable" parameters, perhaps more than they can justify from empirical knowledge, in order to keep overall model outputs within a plausible range.[12] Nothing is likely to improve current climate models more than better, less highly tuned parameterizations of aerosol effects.

### Living in the Semi-Empirical World

Parameterization and tuning remain, in effect, scientific art forms whose connection to physical theory and observational data varies widely from case to case. As Myanna Lahsen, Simon Shackley, and other ethnographers of climate modeling have observed, modelers can become so deeply involved in adding parameters and tweaking them to adjust model output that they lose track of the physical justifications for these practices.[13] One of Lahsen's interviewees, an empirical meteorologist, told her:

> Building a model is a full-time job and it takes a lot of expertise just to do that, and I think that sometimes it's easy to lose track of what the model is being built for. It's not enough simply to build a sophisticated model that includes a lot of physics, physical parameterization, etc., and to integrate the model in time and to take just sort of a general view of it, saying, "yeah it looks like it's simulating something that appears to be like the real atmosphere." . . . It's just as important to develop a model but, at the same time, to fully realize the limitations of the model and to always take a very critical look at the capabilities of the model and what kind of questions one may be able to answer and pursue using the model. The climate system is so complex, with all of the feedback loops, that it's very difficult to actually model all of those loops.[14]

Or, as one climate modeler told me: "Sure, all the time you find things that you realize are ambiguous or at least arguable, and you arbitrarily change them. I've actually put in arguable things, and you do that all the time. You just can't afford to model all processes at the level of detail where there'd be no argument. So you have to parameterize, and lump in the whole result as a crude parameter." Another Lahsen interviewee described the psychological effect of doing model-based research as making it hard to differentiate the model from reality:

> You spend a lot of time working on something, and you are really trying to do the best job you can of simulating what happens in the real world. It is easy to get caught up in it; you start to believe that what happens in your model must be what happens in the real world. And often that is not true. . . . The danger is that you begin to lose some objectivity on the response of the model [and] begin to believe that the model really works like the real world. . . .[15]

Another ethnographer, Simon Shackley, described two "epistemic lifestyles" in climate modeling. In his terms, *climate seers* use models to "understand and explore the climate system, with particular emphasis on its sensitivity to changing variables and processes," seeing the models as tools for this purpose. Meanwhile, *climate model constructors* see models as an end in themselves. They seek to "capture the full complexity of the climate system [in models] which can then be used for various applications." Climate model constructors are more likely to focus on increased "realism," an adjective referring not to accuracy but to the inclusion in the model of all physical processes that influence the climate. Climate seers, by contrast, tend to focus on modeling the most fundamental and best understood processes, and to use a variety of different models, including simpler zero-, one-, and two-dimensional models. "For the seer," Shackley writes, "*which model is 'state-of-the-art' depends on the model's intended application. . . . For the model constructor, by contrast, a single state-of-the-art model exists irrespective of its application.* It is defined as one whose control is close to the observed climate, which uses the most up-to-date observations as inputs and for verification, and which contains the most detailed, most physically realistic parameterizations."[16] This distinction, Shackley points out, does not necessarily divide good modeling practices from bad ones. Rather, it represents two legitimate strategies aimed at producing different kinds of knowledge. Climate seers may be more prone to tune their models according to intuitive judgments of how model results can be "improved," but they may also insist on simpler parameterizations whose physical basis and effects on model output can be more readily understood than the more "realistic" but more complex parameterizations preferred by model constructors. By contrast, model constructors' efforts at realism attempt to capture the full complexity of the atmosphere, seeking advances in knowledge from the potentially unexpected results of adding more processes to the models. The model constructors' approach also has a political dimension, since those who challenge GCM results often argue that they do not account for the effects of some unincluded process, such as cosmic rays.[17] Adding more processes reduces modelers' vulnerability to this line of attack, though at the same time it increases the opportunities to question the accuracy of parameterizations.

Controversies about tuning rage both inside and outside the climate modeling community. The philosopher-physicist Arthur Petersen notes that "simulationists hold divergent views on the norm of not adding *ad hoc* corrections to models."[18] Some accept these corrections as necessary; others view them almost as morally suspect and seek to eliminate them. David Randall and Bruce Wielicki argue that tuning "artificially prevents a model from producing a bad result." Noting that some modelers refer to tuning as "calibration"—exploiting that term's positive connotations—Randall and Wielicki write: "Tuning is bad empiricism. Calibration is bad empiricism with a bag over its head." Yet Randall and Wielicki also acknowledge that, in the case of physical processes that are known to be important but are not well understood, there may be no choice.[19]

In general, modelers view tuning as a necessary evil. Most try to observe certain constraints. Tuning should not, for example, bring the tuned variable outside its known range of observed behavior. In addition, models should be evaluated on observational data for which they have not been explicitly tuned. In other words, tuning lower-level variables may be justified if it produces better higher-level behavior, for example in reproducing multi-dimensional phenomena such as the North Atlantic oscillation or regional monsoons. Ultimately it may be possible to use techniques borrowed from data assimilation to tune models automatically, reducing the potential for subjective effects to influence the process.[20]

**Models vs. Data: Validation, Verification, or Evaluation?**

In the 1960s and the 1970s, climate models were still too primitive for anyone to worry about getting the details right. Furthermore, as we saw in chapter 7, data on the full global circulation (especially the vertical dimension) remained sketchy until the 1978–79 Global Weather Experiment. In those days, you couldn't be completely sure which side problems were coming from. How much of the disagreement between your model and the data was due to problems in the model, and how much to noise in the data?

In the 1980s, as both models and data improved and climate change became a high-stakes political issue, modelers began to look for more than ballpark resemblances. They now wanted to show that their models could capture—reliably and with some degree of precision—the real climate's behavior. Until the end of that decade, there were essentially two ways to do this. The simplest was to set your model parameter values to present-day conditions, run the simulation for 20–100 years to get a climate, then compare your results with statistics for the actual circulation. The other way was to set your parameters to a *different* set of conditions, such as those of the last ice age or the hot Cretaceous. How would your model respond to changing the distributions of continents, the Earth's orbit, solar input, carbon dioxide concentrations, and other factors characteristic of the distant past? If your model's results reasonably resembled paleontological estimates of past climate conditions, that could boost your confidence in its skill. If setting your model's parameters to present-day conditions also produced a reasonable result, this made it more likely that your simulation was accurate. From there, you could argue that setting its parameters to some future state—such as carbon dioxide doubling—could give you a reasonably accurate picture of how the climate would respond.

As they gradually became available in the early 1980s, data from the First GARP Global Experiment provided a much more detailed and reliable data image of the global circulation against which to compare GCM simulations. The FGGE data assimilation (see chapter 9) produced three-dimensional gridded data sets from which circulation statistics for the FGGE year could readily be derived. Modelers soon began to use these statistics as checks on the realism of GCM outputs. Owing to the short length of the record, however, *these were not climate data*. You needed a longer record to get a true climatic picture. Meanwhile, the Climatic Research Unit, the Goddard Institute for Space Studies, and other groups discussed in chapter 11 were assembling an increasingly detailed and widely trusted set of climate data for the surface.

The increasingly precise data emboldened modelers to talk, by the end of the 1980s, about "verifying" or "validating" climate models.[21] A "verified" or "validated" model would be one that reliably reproduced observed climate. In the political context of that period, these terms soon became hotly contested. In 1994, Naomi Oreskes, Kristin Shrader-Frechette, and Kenneth Belitz argued in the pages of *Science* that talk of "verification" or "validation" of models was bad epistemology.[22] The word 'verification', they wrote, normally implies definitive proof. But models, Oreskes et al. argued, are essentially intricate inductive arguments. Since no inductive proposition can be proved with perfect certainty, models can never be verified in this strict sense. The fact that a model agrees—even perfectly—with observations does not guarantee that all of the principles it embodies are true. Furthermore, as was discussed in chapter 10, the logic of simulation modeling does not require, or even permit, definitive proof. For example, parameterizations by definition do not derive from exact physical principles; no one expects them to prove perfectly accurate.

Oreskes et al. saw "validation" as a somewhat less stringent standard. Strictly defined, it refers to demonstrated internal consistency and an absence of detectable flaws. On this reading of the term, a model might be "valid" without being an entirely accurate or complete explanation. Nevertheless, Oreskes et al. pointed out, scientists commonly used 'validation' and 'verification' as synonyms. If scientists did not see a difference between them, neither would policymakers or ordinary citizens.

Oreskes et al. argued that models can at best be "confirmed." This implies only that model results agree with observations. This agreement, by itself, tells us nothing about whether the model reached its results for the right reasons. Thus, "confirming" a model does no more than make it a viable candidate for a true explanation. In other words, confirmation raises the likelihood that the model embodies true principles, but does not make this certain. This view is consistent with Popper's well-known doctrine of falsificationism, which holds that scientific hypotheses (or models) can be proved false by observations, but cannot be proved true.[23]

This critique had a considerable impact on the Intergovernmental Panel on Climate Change, which dropped the words 'validation' and 'verification' altogether. (However, those words still appear regularly in the scientific literature reviewed in IPCC assessments.) The agency substituted 'evaluation', defined as an assessment of "the degree of correspondence between models and the real world they represent"[24]: a claim of relevance and trustworthiness, rather than truth.

By the late 1980s, increased computer power permitted the first efforts to simulate the actual climate of the twentieth century. This marked a major change in the role of GCMs. Previously, most studies simply compared a control run under pre-industrial conditions with a second run under $CO_2$ doubling. The product was a calculation of the "climate sensitivity" (a phrase which invariably refers to $CO_2$ doubling), the figure shown in tables 7.3 and 14.1. But in the real atmosphere, of course, $CO_2$ does not double overnight, and many other climate-related events occur at various times. Simulating the year-by-year course of twentieth-century climate would demand "transient forcing" runs, with $CO_2$ levels specified at their actual time-varying amounts and, perhaps, with major volcanic eruptions and other consequential events included in the appropriate years. Since climate is an average state, no one expected such transient runs to match the course of change exactly, but averaged over periods of 5–10 years or more a good simulation should roughly track the basic trends shown in figure 11.5. The first such transient runs were attempted, with some success, in the second half of the 1980s. (See figure 14.3.)

### Model Intercomparisons as Standardized Gateways

After the establishment of the Intergovernmental Panel on Climate Change in 1988, modelers initiated a number of exercises in model evaluation. The most important of these exercises—and one that became crucial to integrating the climate knowledge infrastructure—was model intercomparison. The Atmospheric Model Intercomparison Project, mentioned in chapter 7, was among the first of these projects. It was established at the Lawrence Livermore International Laboratory in 1989, when the US nuclear weapons laboratories began seeking non-military uses for their supercomputer facilities and modeling expertise. The AMIP strategy required each modeling group to run its model using a specific set of "boundary conditions," or parameters (specifically, monthly average sea surface temperature and sea ice distribution for the years 1979–1988). All model runs had to provide specific output variables in a standard format. Whereas previously each modeling group had chosen boundary conditions and output variables for itself, these common specifications created a basis for intercomparing model performance and diagnosing the sources of differences in model behavior.

After the Atmospheric Model Intercomparison Project, intercomparison evolved into a series of more elaborate projects. Many of these participated in what became a regular cycle of model evaluation linked to the Intergovernmental Panel on Climate Change assessment process. Today, most intercomparison projects are organized under the umbrella Program for Climate Model Diagnosis and Intercomparison.[25] Some 40 different projects evaluate different aspects of all types of climate models. The current phase of the Coupled Model Intercomparison Project (CMIP), for example, compares model simulations of the following:

pre-industrial control run
present-day control run
climate of the twentieth century experiment (realistic greenhouse gas increases)
1 percent per year $CO_2$ increase to doubling and quadrupling

stabilization at 550 ppm $CO_2$
stabilization at 720 ppm $CO_2$

and numerous others, prescribing well over 100 output fields the models should generate to permit direct comparison.

In terms of the theory of infrastructure development developed in chapter 1, intercomparisons function as standardized gateways linking the various models and modeling groups. By permitting regular, direct, and meaningful comparisons of the models with one another and with standardized data sets, they have helped to transform climate modeling from a craft activity of individual laboratories into a more modular and standardized collective activity involving virtually all of the world's climate modeling groups; in theoretical terms, they linked a set of isolated systems and created a network. Technological change, of course, facilitated this transformation. With the rise of the Internet during the 1980s, modelers could exchange data far more easily.[26] The broad international participation and frequent interactions that characterize model intercomparisons today would not have been possible under previous information regimes. These projects can thus be seen as early versions of what would today be called "cyberinfrastructure" or "e-science."

How do current GCM results compare with each other? How well do they track the twentieth century climate record? What do they tell us about the relative roles of natural and anthropogenic factors? Figure 13.2 provides a glimpse of all these comparisons simultaneously. In both top and bottom charts, the thick black line represents the Climatic Research Unit surface temperature data set discussed in chapter 11. The year-by-year outputs from numerous simulations fluctuate, as expected, around the actual climate record. (No climate model would *or should* reproduce the record exactly, due to climate's natural variability.) The average trend from all the simulations together is also shown.

Figure 13.2 shows that when both natural and anthropogenic forcings are included in the models, their average trend tracks the twentieth-century record reasonably well. However, when anthropogenic forcings (greenhouse gases, aerosols, and other human activities that affect the atmosphere) are removed, the model trend begins to fall below the observed one around 1960, departing from it altogether by 1980. This is true not just of the models' average; it is true of every individual model run too. *Not one* of the models compared here could reproduce the most recent period of warming without including human activities.

Figure 13.2 shows global averages, but since 2000 model resolution has increased to the point that modelers can calculate similar figures for climate

Parametrics and the Limits of Knowledge 351

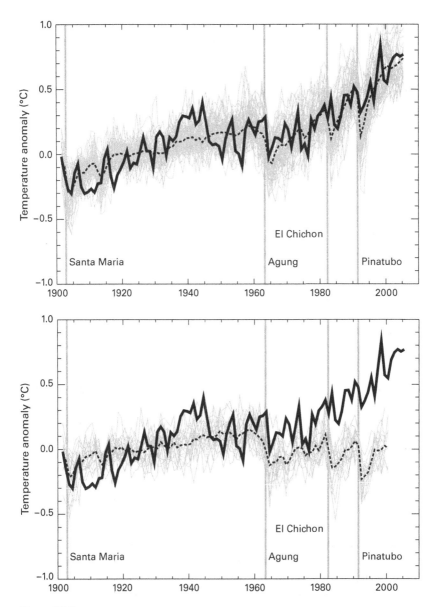

**Figure 13.2**
AOGCM simulations (thin gray lines) of twentieth-century global surface temperature anomaly with (top) and without (bottom) anthropogenic forcing. Black line in both graphs represents observations (HadCRUT3). Dashed lines represent the trend across all model simulations. Vertical gray lines mark major volcanic eruptions.
*Source: Climate Change 2007: The Physical Science Basis* (Cambridge University Press, 2007). Image courtesy of IPCC. Colors replaced to allow non-color reproduction.

on a continental scale. Figure 13.3 shows the same comparison as in figure 13.2, broken down by continent and also by land surface and ocean components. In all cases, model-calculated trends track observations (black trend lines) so long as anthropogenic factors are included, but diverge substantially by the last decades of the twentieth century if only natural forcings are used.

Do the results shown in figures 13.2 and 13.3 "verify" or "validate" climate models? No, because Oreskes et al.'s critique of those terms still applies. Under that analysis, it is conceivable (though unlikely) that somehow all the various modeling groups have independently parameterized and/or tuned their models to bring them into close convergence with the twentieth century trend. If not, it is still conceivable (though again unlikely) that for reasons we cannot anticipate, it will turn out that none of these independently parameterized models predict the climate accurately. Ultimately, neither comparing climate models with observations nor comparing them with each other can prove that they are correct. Nor would any climate modeler ever make such a claim: models are, after all, simulations.

Yet as a principled and detailed mode of evaluation, the comparisons can help us *trust* the models, even if they do not provide proof in any strong sense. Here the epistemological undercurrents of this book's argument have profound practical implications. Distinguishing evaluation and confirmation from validation or verification helps to clarify the proper role of models in forecasting climatic change: not as absolute truth claims or predictions, but as heuristically valuable simulations or projections.[27]

The picture that I hope is emerging here is that *all* knowledge about climate change depends fundamentally on modeling. It's not that there is no such thing as an observation separate from modeling. It's that putting together a trustworthy and detailed data image of the global climate— getting enough observations, over a long enough time span—requires you to model the data, to *make* them global. It's not that climate simulation models are perfectly reliable, any more than weather forecast models always get it right. Instead, it's that the simulations already include a lot of data in their parameters; they are precisely *not* pure theories, but theories already partially adjusted to the conditions we observe in the real world. That's model-data symbiosis. So whether we are looking at data images or model projections of climate futures, we will always experience them as probabilistic, as shimmering rather than fixed.

At the end of chapter 10, I talked about ensemble forecasting—the deliberate use of multiple, slightly perturbed data analyses to create a set

Parametrics and the Limits of Knowledge 353

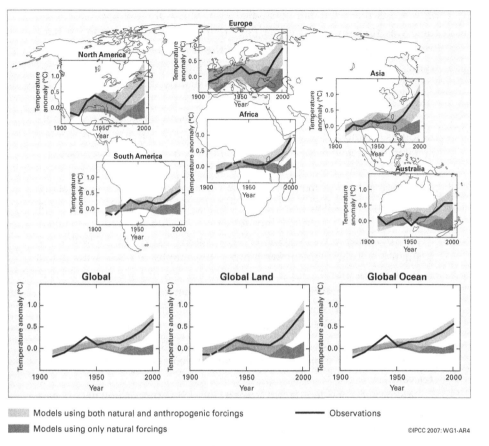

**Figure 13.3**
Original legend: "Comparison of observed continental- and global-scale changes in surface temperature with results simulated by climate models using natural and anthropogenic forcings. Decadal averages of observations are shown for the period 1906 to 2005 (black line) plotted against the centre of the decade and relative to the corresponding average for 1901–1950. Lines are dashed where spatial coverage is less than 50%. [Dark] shaded bands show the 5–95% range for 19 simulations from five climate models using only the natural forcings due to solar activity and volcanoes. [Light] shaded bands show the 5–95% range for 58 simulations from 14 climate models using both natural and anthropogenic forcings."
Source: *Climate Change 2007: The Physical Science Basis* (Cambridge University Press, 2007). Image courtesy of IPCC. Colors replaced as noted in brackets to allow non-color reproduction.

of forecasts that then can be studied as a statistical ensemble. The point there was that once you understand that no observations and no data analysis will ever be a perfect image of the real atmosphere, you can stop trying to make them so. Instead, you can exploit the certainty of imperfection by making your forecast from many slightly different data images: shimmering data.

Complicating this picture even further, in terms relevant to parameterization, is the most recent development in ensemble prediction: perturbed model physics, now being explored by both weather and climate modelers. This is the logical counterpart to perturbed data analysis. Its premise is that models are at least as likely as observations to contain errors, particularly in parameterized quantities. Often the physically plausible range of a parameter is well established, but not its exact value. To exploit this fact, perturbed model physics systematically varies the values of important parameters within their known ranges. In complex models with many parameters, the number of possible combinations of parameter values is large; each combination constitutes a variation of the model. The perturbed-physics technique thus generates an ensemble that can include dozens to thousands of model variations (or more). It then runs each variation and compares the outputs of all the runs statistically. In effect, this procedure replaces subjective tuning with systematic variation of parameter values.

One project developing this technique, climate*prediction*.net, began around 2000 at Oxford University and other institutions in the United Kingdom, including the UK Met Office and the University of Reading. Climate*prediction*.net links thousands of personal computers into a gigantic "grid computer," functionally equivalent to a supercomputer, over the Internet. At this writing, its computing grid claims more than 100,000 individual host computers, whose owners volunteer idle computer time for climateprediction.net to run its climate models. The project has already made tens of thousands of runs of various Hadley Centre models on a variety of modeling experiments using perturbed physics, and has begun to issue publications.

Originally climate*prediction*.net hoped to narrow the predicted range of climate sensitivity by using the perturbed-physics technique. Instead, initial results *widened* the range rather dramatically. An early study using more than 2000 runs produced a climate sensitivity range of 1.9°C to *11.5°C* (an upper bound almost 2½ times higher than that of the IPCC).[28] However, a later, better-constrained study from the same group reduced that range to 2.2–6.8°C.[29]

More than 2000 variations on the same model, 2000 possible versions of atmospheric physics, 2000 potential planetary futures—what do these shimmering images tell us? It is too early to evaluate the perturbed-physics technique definitively. For one thing, despite the thousands of variations, these runs used only one of the available GCMs; perturbed-physics runs on other GCMs might reach different outcomes. Still, the relatively greater agreement among climate models used in the IPCC reports could conceivably be due to questionable parameterization and tuning practices—"bad empiricism with a bag over its head," as Randall and Wielicki put it. Yet even if true, this argument cuts against climate-change skepticism. Most skeptical arguments implicitly or explicitly promote the idea that warming is minimal or nonexistent and that greenhouse warming is a false threat. Yet *not one* of the perturbed-physics model runs predicted warming of less than 1.9°C. No combination of parameters produced anything like a flat or downward trend. Instead, the main effect of these runs was to raise the *upper* end of the sensitivity range. In other words, if these runs show anything, they show that the IPCC sensitivity estimate—currently 2–4.5°C—may be much too conservative.

Like data about the climate's past, model predictions of its future shimmer. Climate knowledge is probabilistic. You will never get a single definitive picture, either of exactly how much the climate has already changed or of how much it will change in the future. What you will get, instead, is a range. What the range tells you is that "no change at all" is simply not in the cards, and that something closer to the high end of the range—a climate catastrophe—looks all the more likely as time goes on.

# 14 Simulation Models and Atmospheric Politics, 1960–1992

Between 1960 and 1980, as World Weather Watch systems came online and the Global Atmospheric Research Program built toward the crescendo of the Global Weather Experiment, concerns about anthropogenic greenhouse warming slowly began to travel outward from a small scientific elite into policy circles.

Policymakers care much less about the past than about the future. The overriding questions for them take the form "What will happen if . . . ?" For most policymakers or policy institutions even to notice any given issue among the thousands that confront them daily generally requires at least three things: a crisis, a constituency that cares about the crisis, and a theory of why the crisis is occurring and how to resolve it.

In the 1960s, $CO_2$-induced global warming met none of these criteria. Rising concentrations of $CO_2$ were evident and essentially uncontested —but did they constitute a crisis? Few contended that dangerous levels of warming had already occurred. At most, some believed they had detected an increase of about 0.2°C since the pre-industrial era, a signal that was very hard to detect against the background of natural variability. Furthermore, the data seemed to show that after five decades of steady warming (1890–1940), global temperatures had fallen somewhat between 1940 and 1970.[1] All this made it very difficult to argue that rising CO2 levels called for immediate action.

What about a constituency committed to the issue? Here again, no. Among those aware of the carbon dioxide issue, in those years only scientists commanded sufficient social and political authority to raise the issue. And few scientists, even those who did see a crisis in the offing, were prepared to play canary in the coal mine—at the possible cost of their reputations—without a well-accepted theory of climate change to back them up. In the 1960s the causes, the potential extent, and even the sign (positive or negative) of global temperature change remained uncertain.

Debates about the balance between greenhouse-gas effects, aerosol effects, other anthropogenic factors, and the many natural causes of climate variability and change were still unresolved. Meanwhile, simulation models remained too primitive for use in forecasting climate change.

Another reason why global warming achieved little serious policy attention in the 1960s, however, has been less often remarked: computer modeling, the tool on which warming theories depended for their credibility, had yet to acquire full scientific legitimacy. Today, digital simulation modeling is virtually a knee-jerk scientific response, the first and most effective tool for analyzing almost any problem. One can scarcely imagine a scientific life without it. Yet before about 1970 most sciences had barely begun to think about simulation modeling, let alone to accept it as a fundamental method of discovery. The idea that computer simulations could inform a policy discussion of anything at all remained to be established. As it happened, climate models would figure significantly in that process.

This chapter sketches the story of how computer models became policy tools, how interaction between climate modelers and policy bodies boosted the legitimacy of modeling, and how global warming "arrived" as a crisis with a constituency first within policy agencies and then, around 1988, among the general public. Climate modeling first engaged with politics during the controversy surrounding supersonic transport (SST) aircraft in the late 1960s. The SST controversy resurrected global atmospheric politics—in decline since the 1963 Limited Test Ban Treaty—by providing a crisis that made climate change a focus of international concern. The SST controversy would be followed, in quick succession, by a series of new global atmospheric issues, including ozone depletion and acid rain (mid 1970s), "nuclear winter" (early 1980s), and global warming (late 1980s). Together and separately, these concerns manifested the theme that human activity could affect Earth's atmosphere not only locally and regionally—as "pollution," the typical frame for environmental issues of the 1960s—but on a planetary scale. By 1992, global warming would reach the floor of the United Nations Conference on Environment and Development, which culminated in the UN Framework Convention on Climate Change.

My goal in this chapter is not to trace the entire path of global warming as a political issue. Many excellent studies of that topic already exist, with more appearing almost daily. Nor will I attempt an exhaustive account of computer modeling's role at the science-policy interface across the last five decades. Instead, I am going to characterize several periods in global atmospheric politics as a way to illustrate the significance of the model-data relationships I have explored throughout this book. My argument is that climate scientists, through repeated involvement in the series of issues just mentioned, helped establish simulation modeling as a legitimate source of policy-relevant knowledge. Because of this, climate science gained a permanent foothold in policy circles around 1980, when scientists began to stabilize a set of model predictions, principles of greenhouse warming, and data trends. *Considered as heuristics*, these figures, principles, and trends have remained essentially the same ever since.

This relative stability did not silence political debate, of course. Nothing that so challenged the energy economy of the entire planet could have avoided vociferous controversy, nor should it have done so. In chapter 15, I will examine the sources of stability in climate knowledge and some of the attempts to challenge it. Almost invariably, the latter have revolved around the very model-data relationships I have explored throughout this book.

### Climate Change as Weather Modification: The 1960s

$CO_2$-induced global warming first made its way into policy discussions in a desultory way during the late 1950s.[2] Then, and throughout the 1960s, most studies framed the issue as a speculative aspect of the post-World War II weather-modification agenda. (See chapter 6.) Aviation, military, and agricultural concerns were the chief clients for weather-modification ideas, which reached a zenith during the Vietnam War. For example, between 1967 and 1972, a US Air Force effort called Operation Popeye attempted to wash out parts of the North Vietnamese resupply route known as the Ho Chi Minh Trail by seeding clouds over Laos during monsoons. Among the largest efforts was Project Stormfury, which tested techniques for steering or extinguishing tropical cyclones before they made landfall. A joint operation of the US Department of Commerce and the US Navy, it ran from 1962 through 1983. The exact amount of money expended on these and other weather-modification projects remains unknown, since many of the military projects were classified, but a conservative estimate would place the figure at well over $150 million in today's dollars.[3]

Aviation, the armed forces, and agriculture were powerful interests, but their goals mainly involved deliberate, short-term changes to local or regional weather and climate. None of them wanted much to do with *global* climatology, or with the slow, steady rise in $CO_2$ concentrations. But especially in the context of Vietnam, the military, at least, had the ingredients mentioned above: a crisis, a constituency, and a plan. Also, all of them needed better forecast models. The better the models, the better they could predict the best moments for cloud seeding. But they also needed better models in order to confirm the efficacy of weather modification: to know whether your weather-modification scheme had worked, you had to know how much rain you could have expected, or what the cyclone's speed and direction would have been, without your intervention.

Seeking relevance (and funding), climate scientists sought to connect climate change with the weather-modification agenda by framing it as "inadvertent weather modification." GCM builders at UCLA proposed research on "the adaptation of mathematical models for studying possible modification techniques of global circulation patterns."[4] Studies of $CO_2$-induced climate change began appearing as subsections in larger reports addressing cloud seeding, hail suppression, airport fog control, and weather or climate as a weapon of war.[5] The US National Academy of Sciences raised "inadvertent climate modification" as a potential policy concern.[6] The National Science Foundation appointed a Special Commission on Weather Modification.[7] These study groups argued for further research into fundamental processes related to this "inadvertent modification" of climate—specifically, more and better climate modeling. In the United States, the primary result was expanded support for research at the National Center for Atmospheric Research and the Geophysical Fluid Dynamics Laboratory, scientific organizations that were relatively distant from policy processes.

Another major 1960s framework for environmental concerns was "pollution," also usually understood as a local, national, or regional issue. In 1965, an Environmental Pollution Panel appointed by the President's Science Advisory Committee included a long appendix on carbon dioxide and global warming. A number of major figures in climate research—Roger Revelle, Wallace Broecker, Joseph Smagorinsky, and Charles Keeling—joined in the report. Noting the steady rise in $CO_2$ concentrations, the PSAC report pointed to the potential for global warming, but hedged on its likely extent. PSAC recommended extended monitoring of $CO_2$ and global temperature, and further work on GCMs. It also outlined possible geo-engineering measures for reducing warming, such as spreading reflective particulates across vast ocean areas to increase Earth's albedo.[8] (In the weather-modification era, such discussions of geo-engineering were commonplace.) Though the PSAC report made no policy recommendations

Simulation Models and Atmospheric Politics            361

and produced no major policy outcomes, it did name carbon dioxide as a major "pollutant."

### SCEP, SMIC, and the SST

The rise of the environmental movement in the latter half of the 1960s, which reached its climax in the first Earth Day in 1970, prepared the ground for a renewal of global atmospheric politics. According to David Hart and David Victor, "by 1968 the notion that pollution could modify the climate was a commonplace"—one the popular press often treated as a truism.[9] Yet as I have argued, the *specific* predictions and confirming data necessary to turn this "notion" into an actionable crisis did not yet exist.

General circulation modeling, along with studies of aerosols and carbon dioxide effects using simpler one- and two-dimensional models, first entered the larger discourse of environmentalism around 1970, when two important scientific working groups raised concern about anthropogenic climate change to the level of an international issue. These groups, the Study of Critical Environmental Problems (SCEP, 1970) and the Study of Man's Impact on Climate (SMIC, 1971), presented GCMs and other models as the principal sources of climate knowledge.[10]

Both SCEP and SMIC were organized by Carroll Wilson, an entrepreneurial professor at the Massachusetts Institute of Technology. Wilson—who had begun his career as Vannevar Bush's assistant during World War II, served as general manager of the Atomic Energy Commission, and represented the United States on the UN Advisory Committee on Science and Technology—was uniquely positioned to make seminal contributions at the intersection of environmental science, technology, and policy. He timed SCEP and SMIC to contribute to the planned major UN Conference on the Human Environment, scheduled for Stockholm in June 1972. He organized month-long summer conferences for SCEP and SMIC participants, making them into bonding experiences as well as opportunities for scientific exchange.

SCEP and SMIC focused on *global* environmental problems and policies, marking a significant departure from the regional, national, and local issues that dominated the early environmental movement. The conference reports—*Man's Impact on the Global Environment* (SCEP, 1970) and *Inadvertent Climate Modification* (SMIC, 1971)—were published as elegant trade books, bound between sleek black covers, rather than as academic volumes. Scientists and policymakers alike often cite SCEP and SMIC as watershed events in the emergence of anthropogenic climate change as a political issue.

In the background of both working groups, especially SCEP, lay the then-controversial plans for a commercial supersonic transport (SST): a jet that would fly high in the stratosphere, well above the typical altitude for jet aircraft. Though noise from sonic booms probably concerned the public more, scientists feared that water vapor, aerosols, and nitrogen oxides in SST exhaust, released at stratospheric altitudes, might destroy the ozone layer and/or cause climatic changes on a global scale. The *New York Times* reported the SST controversy closely and editorialized against the planes. In 1970, the *Times* trumpeted the SCEP report's conclusions about possible climatic effects on its front page.[11] The following year, bowing to public outcry, Congress eliminated federal funding for the American SST program, but the Anglo-French Concorde project proceeded apace. Starting in 1976, the twenty Concorde jets flew only transatlantic routes between London, Paris, New York, and Washington. All Concorde service ceased in 2003.

The SST controversy had two important outcomes for climate science. First, as a direct result, in 1971 the US Department of Transportation established a Climate Impact Assessment Program (CIAP), with an annual budget of about $7 million, to research the SST's possible atmospheric effects.[12] CIAP has been called "the first major . . . integrated assessment of an environmental issue," involving hundreds of atmospheric scientists in an intensive three-year study.[13] However, the demise of the American SST program minimized CIAP's impact by deleting the immediate cause for concern.

A second, perhaps more enduring effect was to focus scientific attention on the issue of anthropogenic aerosols, not only from the SST but also from power plants, factories, automobiles, and other sources. (In Los Angeles and other large cities, smog had reached nearly intolerable levels, becoming a close-to-home problem that anyone could see.) In a 1971 study using a one-dimensional radiation-balance model, S. Ichtiaque Rasool and Stephen Schneider found that the climate risk from aerosol cooling substantially exceeded any realistic risk from $CO_2$-induced warming: "An increase by a factor of 4 in the equilibrium dust concentration in the global atmosphere, *which cannot be ruled out as a possibility within the next century*, could decrease the mean surface temperature by as much as 3.5°K. If sustained over a period of several years, such a temperature decrease could be sufficient to trigger an ice age!"[14] Though entirely speculative, this result had the effect of toning down the SMIC discussion of $CO_2$ warming the following year. However, it was soon overturned by models that included more feedbacks (as we saw in chapter 7), and few other studies of this period ventured predictions of near-term global cooling. (See box 14.1.)

SCEP focused on world-scale environmental problems, conceived as "the effects of pollution on man through changes in climate, ocean ecology, or in large terrestrial ecosystems."[15] Among other things, it introduced the concept of "environmental services" (now usually called "ecosystem services"), including under that heading climatic self-regulation as well as pollination, pollution filtering, and many other activities of the natural environment that directly benefit human beings. The National Center for Atmospheric Research and the Scripps Institution of Oceanography—important centers of atmospheric and ocean modeling respectively—were heavily represented among the invitees. Joseph Smagorinsky attended on a part-time basis, and Roger Revelle consulted, ensuring a prominent place for the $CO_2$ issue in both reports.

**Box 14.1**
Global Cooling? A New Ice Age?

> In 1972, a Brown University paleoclimate conference on "The Present Interglacial: When Will It End?" produced a consensus statement, reported in *Science*, that "global cooling and related rapid changes of environment, substantially exceeding the fluctuations experienced by man in historical times, must be expected within the next few millennia *or even centuries*. In man's quest to utilize global resources, and to produce an adequate supply of food, global climate change constitutes a first-order environmental hazard which must be thoroughly understood well in advance of the first global indications of deteriorating climate."[a] Two CIA reports on climate, released to the President's Domestic Council in 1974, claimed that scientists were "quite positive" that a new ice age would arrive "within the next 2500 years." One reported that "the western world's leading climatologists have confirmed recent reports of a detrimental global climatic change. . . . [The] earth's climate is returning to that of the neoboreal era (1600–1850)—an era of drought, famine, and political unrest in the western world."[b] Popular media reports and grandstanding Congressmen pumped up these cooling speculations into fears of an imminent ice age.[c]
>
> Yet despite the often-repeated claim that scientists stoked fears of cooling, few if any studies of the 1970s predicted a *near-term* return of the glaciers. An oft-cited National Academy of Sciences report did mention the likelihood that the present interglacial period would end within the next 10,000 years, but

**Box 14.1**
(continued)

> it made no near-term predictions. Mainly it called—like virtually all studies of its era—for further research to reduce the large uncertainties.[d] Similarly, the body text of the 1974 CIA reports described very large uncertainties, and mainly recommended more climate research leading to better climate forecasting. A recent survey of the peer-reviewed climate science literature between 1965 and 1979 showed that while a handful of papers did describe possible cooling scenarios, even in those years most scientific articles either articulated warming scenarios (the majority), or else were neutral between the two possibilities.[e]

a. G. J. Kukla and R. K. Matthews, "When Will the Present Interglacial End?," *Science* 178(4057), no. 4057 (1972), 191, emphasis added.
b. "A Study of Climatological Research as It Pertains to Intelligence Problems," US Central Intelligence Agency, reprinted in *The Weather Conspiracy* (Ballantine Books, 1977).
c. L. Ponte, *The Cooling* (Prentice-Hall, 1976); N. Calder, *The Weather Machine* (Viking, 1975); P. Gwynne, "The Cooling World," *Newsweek*, April 28, 1975: 64; A. J. Bray, "The Ice Age Cometh: Remembering the Scare of Global Cooling," *Policy Review* 58, fall 1991: 82–84.
d. US Committee for the Global Atmospheric Research Program, National Research Council, *Understanding Climatic Change: A Program for Action* (National Academy of Sciences, 1975); J. H. Douglas, "Climate Change: Chilling Possibilities," *Science News* 107, no. 9 (1975): 138–40; W. Connolley, "Was an Imminent Ice Age Predicted in the '70's? No," 2005, www.wmconnolley.org.
e. T. C. Peterson et al., "The Myth of the 1970s Global Cooling Scientific Consensus," American Meteorological Society, 20th Conference on Climate Variability and Change, 2008.

SCEP and SMIC both presented general circulation models as "indispensable" for research on anthropogenic climate change, citing GCMs as "the only way that we now conceive of exploring the tangle of relations" involved in climate. SCEP pointed to two uses for GCMs: as "laboratory-type experiments on the atmosphere-ocean system which are impossible to conduct on the actual system" and as a way of producing "longer-term forecasts of global atmospheric conditions." While noting their many deficiencies, the report argued that models were "the only way that we now conceive of exploring the tangle of relations" involved in climate. It recommended an expanded program of climate, ocean, and weather modeling research.[16] Both SCEP and SMIC also militated for new or expanded global monitoring programs, such as a 100-station network to sample air and precipitation chemistry, measure solar radiation, and gather other climatological data.

SCEP's recommendations focused on the need for uniform global data. Letting forth a scientific *cri de coeur*, the report hammered on the point that "critically needed data were fragmentary, contradictory, and in some cases completely unavailable. *This was true for all types of data—scientific, technical, economic, industrial, and social.*" It recommended three initiatives: (1) new methods for global information gathering that would integrate economic and environmental statistics, along with "uniform data-collection standards," (2) "international physical, chemical, and ecological measurement standards," administered through a "monitoring standards center," and (3) integration of existing monitoring programs to produce an optimal global monitoring system.[17]

Whereas the majority of the participants in SCEP were Americans, SMIC included a much more international cast of characters. Held less than a year before the Stockholm UN Conference on the Human Environment, the summer workshop was hosted in Stockholm by the Royal Swedish Academy of Sciences. SMIC provided a detailed technical discussion of GCMs and other, simpler climate models. Its conclusions with respect to human impacts on climate, however, were decidedly muted. William Kellogg, one of the SMIC organizers, recalled: "After most of our conclusions were out on the table, I was intrigued by the apparent impasse in the assembled group between two opposing schools of thought: the climate 'coolers' and the climate 'warmers,' if you will. It depended on whether you came from the atmospheric particle or aerosol camp, or the carbon dioxide and infrared-absorbing gases camp." This split stemmed, in part, from the recent Rasool-Schneider study indicating the potential for aerosol cooling to overwhelm carbon dioxide's warming effects.[18] Further, Kellogg noted, "there was a clear reluctance on the part of many [SMIC participants] to make any predictions at all about the future."[19] As a result, the SMIC report warned of possible "serious consequences," but took no consistent position on exactly what those might be. SMIC recommended a 100-station global monitoring program specially designed for climate, more elaborate than the one proposed the previous year by SCEP. SMIC estimated that an adequate program could be established for a yearly budget of about $17.5 million.

SCEP and SMIC had at least one direct outcome: the UN Conference on the Human Environment approved their calls for a global climate monitoring network, appearing as Recommendation 79 of the UNCHE Action Plan:

It is recommended:

(a) That approximately 10 baseline stations be set up, with the consent of the States involved, in areas remote from all sources of pollution in order to monitor long-term global trends in atmospheric constituents and properties which may cause changes in meteorological properties, including climatic changes;
(b) That a much larger network of not less than 100 stations be set up, with the consent of the States involved, for monitoring properties and constituents of the atmosphere on a regional basis and especially changes in the distribution and concentration of contaminants;
(c) That these programmes be guided and coordinated by the World Meteorological Organization;
(d) That the World Meteorological Organization, in cooperation with the International Council of Scientific Unions (ICSU), continue to carry out the Global Atmospheric Research Programme (GARP), and if necessary establish new programmes to understand better the general circulation of the atmosphere and the causes of climatic changes whether these causes are natural or the result of man's activities.[20]

The UNCHE led immediately to the establishment of the UN Environment Programme, thus in some sense institutionalizing this goal. However, the recommendation punted implementation to the World Meteorological Organization, the International Council of Scientific Unions, and the Global Atmospheric Research Program, raising awareness and adding legitimacy to these existing efforts but ultimately contributing little of substance.

*The Limits to Growth*

At the same time he was organizing SCEP and SMIC, Carrol Wilson also catalyzed another project connected with global simulation modeling: the "world dynamics" models used in the famous report *The Limits to Growth*, published in 1972. Though not directly related to climate modeling, the history of world dynamics bears attention here for two reasons. First, it connects strongly with model-data relationships I have discussed throughout this book. Second, *The Limits to Growth* was the first document based on simulation modeling to gain the attention of a mass audience as well as of political leaders outside the restricted circles of science advisory bodies.

Wilson was an early member of the Club of Rome, a small international group of prominent businessmen, scientists, and politicians founded in 1968 by the Italian industrialist Aurelio Peccei. Among the Club's earliest projects was a proposal to model what it called the "world *problématique*"— global, systemic problems—using techniques borrowed from cybernetics, but the Volkswagen Foundation had rejected that proposal for methodological vagueness.[21] At this point, Wilson invited his Sloan School of Management colleague Jay Forrester to attend the Club's first general meeting, to be held in June of 1970. Forrester, a major computer pioneer in the 1940s and the early 1950s, had begun developing simulation models of industrial and urban dynamics in the late 1950s.[22] His modeling techniques, essentially complex series of weighted feedback relationships, strongly resembled cybernetic approaches (although Forrester usually denied any influence from that quarter). Wilson saw the potential for a fit. At the Club of Rome meeting, Forrester suggested that his approach might work for modeling the "world *problématique*," and he invited Club members to attend a workshop on industrial dynamics.

When the Club of Rome's executive committee arrived at MIT three weeks later, Forrester had worked up and programmed on the computer a rough model, "World 1," based on earlier ideas. This very rapid work-up of the world model reflected one of Forrester's fundamental beliefs: that system structure and dynamics mattered far more than precise inputs. In *Urban Dynamics*, published in 1969, Forrester had argued that systems as complex as cities are "counterintuitive," in the sense that policies developed to fix problems often end up making them worse. In a classic example, citizens and downtown businesses complain about a lack of available parking. The city responds by building parking structures, but the result is gridlock on the streets when the structures empty simultaneously at the end of the workday. In effect, policymakers tend to see and treat symptoms rather than causes. This occurs because "a complex system is not a simple feedback loop where one system state dominates the behavior, [but] a multiplicity of interacting feedback loops . . . controlled by nonlinear relationships."[23]

No matter what they were simulating, Forrester's models tended to be insensitive to changes in most parameters, even changes of several orders of magnitude. To Forrester, the models offered a way to discover the few parameters and structural changes that *could* produce desirable effects in systems that would otherwise escape anyone's attempt to control them. Furthermore, in his simulations many systems' short-term responses to change were of opposite sign from their long-term responses; in other words, effective policies usually followed a "worse before better" pattern. This was not a happy message for policymakers to hear, much less to deliver, so Forester also advocated using simple simulation models to educate a broad public about nonlinear systems and the power of models as aids in understanding.

The World 1 model divided world systems into five major subsystems (natural resources, population, pollution, capital, and agriculture) and incorporated some sketchy data and guesses about relationships among these variables. Perhaps not surprisingly, World 1's characteristic modes resembled those of the industrial and urban models Forrester had already developed, especially the phenomenon of "overshoot and collapse." Eduard Pestel recalled that the Club of Rome's founder, Aurelio Peccei, was tremendously impressed "by the fact that all computer runs exhibited—sooner or later at some point in time during the next century—a collapse mode regardless of any 'technological fixes' employed," and that Peccei "obviously saw his fears confirmed."[24] The Club of Rome returned to the Volkswagen Foundation with a new proposal. This time, with Forrester's methodology in hand, the foundation approved the application for an 18-month modeling project at MIT. Forrester's former student Dennis Meadows led the System Dynamics Group, with Forrester acting as consultant.[25] The team went on to develop two successor models, World 2 and World 3.[26]

The fundamental conclusion of *The Limits to Growth* was that a finite planet could not sustain the exponential growth rates exhibited by many important trends, including natural resource consumption, pollution, and population growth. Carbon dioxide was explicitly included under the heading "pollution." The world dynamics models continued to show, after refinement and even on the most optimistic assumptions, that natural resources would be rapidly exhausted, that pollution would rapidly increase to life-threatening levels, and that catastrophic collapse, including massive famine, would follow around the year 2050. *The Limits to Growth* sold more than 7 million copies worldwide, in some thirty languages.

The System Dynamics Group's self-described "bias" followed Jay Forrester in favoring model structure and dynamics over precise data. The group did attempt to calibrate the models by starting model runs in the year 1900 and adjusting parameters until the model results roughly matched historical trends. But the data used to estimate those trends, as well as the data used to parameterize the world models, were generally poor in quality. In many cases, they were simply guessed. Reviewers attacked *Limits* savagely for this apparent sin.[27] But the situation was more complicated than it appeared.

Like the weather and climate modelers before them, the System Dynamics Group encountered great difficulty in its search for high-quality information about the real-world systems they simulated. Since they generated very long runs (up to 150 years) to project long-term future trends, the group wanted to validate the models against equally long historical time series. As Dennis Meadows put it in an interview, "it was hard in those days to find the kind of comprehensive, cross-national time series data on the issues we wanted to see, except on population. So we were looking where we could, with the United Nations and the World Bank as principal sources of information."[28] But in 1970 the UN and the World Bank were not even a quarter-century old. Where were Meadows and his colleagues to find complete, accurate *world-scale* information on pollution, agriculture, trade, and other macro variables before 1950?

Thus the System Dynamics Group faced data-collection problems much like those that faced climatology: no central authority or source, widely differing standards, many data inaccurate, large gaps in temporal and spatial coverage, and so on. The scant available information, such as pollution measurements and resource-consumption figures, was mostly either in a highly aggregated form that prevented analysis using the model categories, or else so disaggregated and incommensurate that the small modeling group could not make use of it. In the years to come, these data problems would emerge over and over again, in arena after arena. Whether in natural science, social science, economics, or policymaking, every attempt to study global change of any kind had to deal with collecting inconsistent, incomplete data from a vast variety of heterogeneous national and local sources, then correcting, filtering, and combining them to create global data sets. Whether explicit or implicit, simple or complex, *modeled global data have been the norm rather than the exception in every field.*

The world dynamics models drew heavy fire from the scientific community, which saw them as simplistic. The World 1 model contained only 120 lines of equations; all the models lumped numerous complex problems into a relatively small set of single, dimensionless variables. Though later versions disaggregated some of these variables, the world dynamics models certainly represented a much lower order of sophistication and completeness than climate models of the same era. Other critiques took aim at the very global-ness of the models (their attempt to pull together so many heterogeneous factors in a single frame) and their confident century-scale predictions, pointing out that few predictions made in 1872 would have been accurate for 1972.[29] Disaggregating the world models into regions, adding complexity and heterogeneity, might produce quite different results—and indeed, in 1974 the second Club of Rome report did just that.[30] Another criticism went even deeper. The "overshoot and collapse" tendency of all Forrester's models drew well-founded suspicions that the modeling techniques themselves—not the phenomena supposedly modeled—generated the characteristic behavior. Within a couple of years, most scientists regarded world dynamics with indifference or even contempt, and attention faded.

Nevertheless, during the rest of the 1970s the Club of Rome commanded considerable international respect. It convened a series of meetings among senior politicians to discuss global resource concerns. Some of the meetings, held in major world capitals, were attended by presidents and prime ministers.[31] *The Limits to Growth* and the Club of Rome had few if any *direct* effects on policy, but that was never their intent. Instead, through its models, popular books, meetings, and person-to-person canvassing of politicians, the Club succeeded in communicating, to both a broad public and a policy elite, its two basic conclusions: that ever-increasing population, pollution, and consumption levels would eventually break the world system, and that attempts to control problems piecemeal, without taking into account the interconnected and often global nature of socio-technical-environmental systems, would almost certainly miscarry. Thus, although world dynamics failed as a scientific enterprise, it succeeded as an intervention in political culture. *The Limits to Growth* also succeeded in legitimizing and promoting global simulations as a policy-relevant method of analysis—at least in some circles, especially the popular press and the environmental movement.

In many scientific circles, however, the system-dynamics modelers' hubris sparked suspicion and even derision. A telling example regards a joint East-West international project, the International Institute for Applied Systems Analysis (IIASA). Early briefs for the Institute, first conceived in 1966, had nothing to do with modeling. By 1968 the project's working name was "International Center for the Study of Problems Common to Advanced Industrialized Societies." Between 1968 and 1972, the project homed in on applying cybernetics, operations research, and other modeling techniques to social, technical, and environmental concerns. Howard Raiffa, the IIASA's first director, recalled that, as a direct result of the controversy surrounding *The Limits to Growth*, "the issue of global modeling was very intense" by 1972, when the Institute opened near Vienna:

> Some people thought [global modeling] was the main purpose of IIASA. Aurelio Peccei, who was president of the Club of Rome, was a strong advocate. So was the Canadian representative. But Lord Zuckerman insisted that there be nothing about global modeling and he threatened to pull out the [British] Royal Society. The enmity between Sir Solly [Zuckerman] and Peccei was very severe. The compromise was that IIASA itself would not do any work on global modeling, but would host a series of conferences to review contributions to global modeling and document the results.[32]

Despite being shunned by "establishment" scientists, members of the original *Limits* modeling group continued to work and publish, attracting a devoted following with a countercultural flavor.[33]

On a different, more infrastructural level, system dynamics did achieve wide currency. Software based on the system dynamics approach—such as DYNAMO, an early programming language for system dynamics simulations developed by Forrester's group at MIT, and Stella, a widely used graphical modeling package developed for the Apple Macintosh in 1984—facilitated simple simulation exercises incorporating numerous feedback loops. This software grew popular especially for ecological modeling and in college and secondary education. Forrester went to considerable lengths to deploy system dynamics in MIT Sloan School pedagogy, and supported numerous other educational uses of modeling throughout his long career.

Forrester-style system dynamics never became a mainstream scientific approach, but it did inaugurate an important minority tradition that has gradually returned to center stage. Related simulations such as energy economy models would rise to prominence in the wake of the 1973–74 oil crisis. *The Limits to Growth* also spawned a continuing series of comprehensive global simulation models. These included a United Nations World Model and a US Government Global Model.[34] By the late 1980s, these had morphed into "integrated assessment models" and gained real currency in policy circles.

In short, despite their poor scientific quality, the *Limits* world dynamics models played a central role in moving global simulation modeling into the public sphere. The Club of Rome played a major part in building public awareness that the complex, multiply interlocking relationships of natural resources and environmental pollution with human economies and societies had to be considered as a global whole. Taken together, SCEP, SMIC, and *The Limits to Growth* mark the public debut of global simulation modeling as a primary technology of environmental knowledge.

At the same time, SCEP, SMIC, and *Limits* also revealed the weaknesses of global knowledge infrastructures. SCEP and SMIC limited their knowledge claims, arguing for a human impact on climate without specifying the nature or degree of change. They succeeded in framing climate change as a global policy problem, and in setting an agenda for further data collection and more modeling, but—recognizing the limits of their predictive powers—they made no attempt to push further. Much more ambitiously, *Limits* sought to demonstrate an actual if non-specific crisis of overconsumption and pollution. It foundered precisely on the absence of high-quality, long-term, commensurable global data against which to validate its models. It tried to draw together a vast number of threads through computer simulation, but failed to demonstrate their realism; it made many claims it could not cash out, especially when pitted against more detailed, sectoral simulations by economists. Nonetheless, *The Limits to Growth* succeeded in establishing its eponymous heuristic as a basic tenet of 1970s environmentalism.

**Climate Politics in the 1970s**

Throughout the 1970s, scientist "issue entrepreneurs," including Roger Revelle, Joseph Smagorinsky, William Kellogg, and many others, continued to articulate reasons for policy concern with climate issues, especially greenhouse warming and aerosol-driven cooling. In general, however, they rarely stepped beyond recommending further modeling and monitoring.[35] Once climate change became associated with the "energy crisis" of the 1970s, the issue gained constituencies within the US policy apparatus, arousing the sustained concern of the Carter administration, the Department of Energy, and the Environmental Protection Agency. Similar constituencies emerged in several European nations, notably Sweden, but this section will focus principally on the United States.

The executive branch of the US government first became concerned with climate in 1973, when a long drought in the Sahel region of Africa caused widespread famine, Peruvian anchovy fisheries collapsed, and the Soviet Union's wheat crop failed disastrously, resulting in the controversial sale of American wheat to the USSR.[36] The Nixon administration framed these issues as threats to geopolitical stability. Two Central Intelligence Agency reports on the security implications of global cooling scenarios, supposedly related to a long list of recent climate-related crop failures around the world, bolstered this framing. One argued that a substantial climate change had begun in 1960, and that an "agricultural optimum" was being "replaced by a normal climate of the neo-boreal era . . . which has existed over the last 400 years. . . . Climate is now a critical factor. The politics of food will become the central issue of every government."[37] The combination of an apparent climate-related crisis and forward-looking geopolitical concerns provoked executive-branch interest. However, the glaring lack of credible short- and medium-term predictive tools limited the focus to improving climate monitoring and prediction, the same goals outlined by SCEP and SMIC. The White House Domestic Council proposed a "United States Climate Program" in 1974; two administrations later, this resulted in the National Climate Program Act of 1978.

The National Climate Program Act authorized $50 million annually for climate monitoring, climate forecasting, and basic research. Its chief concerns involved short-term climatic effects "on agriculture, energy supply and demand, land and water resources, transportation, human health, and national security."[38] In addition to supporting new research, the National Climate Program was to serve a coordinating function for the many existing federal programs related to climate. Emphasizing research and reporting (rather than action) on climatic change, it sought better understanding of local and regional climate trends; long-term, global climate change was a relatively minor focus of the program. In addition, the NCP supported the Global Atmospheric Research Program and the World Climate Program, as well as US-based global data collection.[39]

In retrospect, one of the NCPA's most striking features was its *national* framing of the climate issue. Despite some support for international activities, the NCPA focused on forecasting only the North American climate. Scientists knew, of course, that such forecasting would require global models and global data networks, but on the political level the act recognized only a national interest in climate. Over the next decade, this would evolve toward a recognition of global concerns as a legitimate focus of climate policy.

Although it did support a considerable amount of greenhouse research, the National Climate Program's major accomplishments were to strengthen user-oriented "climate services" and to develop impact-assessment methods. In a letter to Secretary of Commerce Juanita Kreps, the bill's sponsor, Representative George E. Brown Jr., wrote: "I see climate services and impact assessments . . . as the real areas of pioneering in this Program. It is a chance to use technology to increase the socioeconomic resilience of our society."[40] The NCP thus marked a transition point between the short-term weather-modification frame that dominated US climate policy during the 1960s and the 1970s and the concern with long-term, quasi-permanent climate change that took the stage in the 1980s.

In another Carter-era modeling effort, the Council on Environmental Quality (the Carter administration's principal environmental policy unit) spent three years conducting a world modeling exercise charged with projecting "the long-term implications of present world trends in population, natural resources, and the environment," including climate. But like the Club of Rome modelers before them (to whom they compared themselves directly), the Council on Environmental Quality ran into the brick wall of incommensurable models and data:

> Several [federal] agencies have extensive, richly detailed data bases and highly elaborate sectoral models. Collectively, the agencies' sectoral models and data constitute the Nation's present foundation for long-term planning and analysis. Currently, the principal limitation in the Government's long-term global analytical capability is that the models for various sectors were not designed to be used together in a consistent and interactive manner. The agencies' models were created at different times, using different methods, to meet different objectives. Little thought has been given to how the various sectoral models—and the institutions of which they are a part—can be related to each other to project a comprehensive, consistent image of the world.[41]

Therefore, rather than develop a single integrated model, the CEQ sequenced the sectoral simulations, using one as input to the next.

The lack of genuine integration had serious consequences when it came to including climate in the Global 2000 model. The CEQ report included five global "climate scenarios" derived from a survey of experts. They ranged from a –0.9°C cooling to a warming of 2.2°C by the year 2000 (relative to the 1880–1884 average). The experts' opinions as to the most likely scenario centered on a small but significant warming (0.5°C). Yet "unfortunately, after the climate scenarios were developed, it was discovered that *none of the projection models used by the Government were designed to accept climate as an input variable,* and as a result the Global 2000 climate scenarios could not be used."[42] This strange episode highlights the low level of concern about global climate change within US policy agencies in the late 1970s, even those presumably tuned in to environmental and world dynamics issues. Yet it also illustrates the extent to which global simulation modeling had already penetrated political culture, at least at the level of agency analysis. Global modeling practices were emerging in parallel, within individual sectors. The integrated modeling pioneered by *The Limits to Growth*, combining social, technological, and environmental elements, would not flower fully until the latter half of the 1980s.

The first major greenhouse study program developed in the Department of Energy (DOE). When the Carter administration created a multibillion-dollar synthetic fuels program to reduce US reliance on imported oil, scientists pressed the case for concern about greenhouse warming from the resulting emissions.[43] Responding in 1977, the DOE initiated a Carbon Dioxide Research and Assessment Program. The nascent National Climate Program then assigned "lead agency" responsibility for the carbon dioxide question to the DOE. With an annual budget of $10–15 million, the DOE program became the largest greenhouse research effort in the US.[44]

These policy-related programs increased and focused funding for climate modeling, but scientific developments were already pushing a widespread trend in that direction. The Global Atmospheric Research Program was building to its crescendo, the 1978–79 Global Weather Experiment (also known as the First GARP Global Experiment, or FGGE). Even as FGGE began, GARP's focus began to shift from its first objective (improving the global observing and forecast systems) to its second (improving understanding of circulation and climate). In April 1978, GARP's Joint Organizing Committee (JOC) held a major conference on climate modeling, attended by representatives from most of the world's major modeling groups.[45] In his keynote address, JOC Chairman Joseph Smagorinsky discussed "climate predictability of the second kind," i.e., studies of climate sensitivity to the forcing of major parameters such as carbon dioxide. He argued that, despite vast improvement in modeling techniques over the previous decade, "fundamental data are lacking to guide the formulation of improved parameterizations or for providing sufficiently long global records of key climatic parameters." Satellites offered the only realistic hope for global data, though space sensor technology remained "inadequate to many of the already identifiable tasks." Smagorinsky pointed to data from the upcoming FGGE year as "the most immediate opportunity for a definitive global data set."[46]

By the second half of the 1970s, the number of institutions conducting serious climate research had grown dramatically. In addition to the major labs at NCAR, GFDL, UCLA, and RAND, these now included several NASA laboratories, the National Climate Program (led by the National Oceanic and Atmospheric Administration), and the Department of Energy's $CO_2$ program. In 1975 the Advanced Research Projects Agency's climate program was transferred to the National Science Foundation, where it became the Climate Dynamics Program, extending NSF support to a considerable number of university-based research programs.[47] As we saw in chapter 7, during this period climate models and modeling groups spread rapidly around the world.

A watershed of sorts was reached in 1979. By then the White House Office of Science and Technology Policy (OSTP) had become concerned about environmental problems related to the Carter administration's synfuels program, including acid rain as well as climate change. OSTP requested a "fast-turnaround" study of the climate question from the National Research Council. Convened by the MIT weather modeling pioneer Jule Charney, the group included the Swedish climate scientist and GARP chairman Bert Bolin. It also consulted British and Australian scientists as well as numerous Americans. The study group examined six GCM simulations by Syukuro Manabe's group at the Geophysical Fluid Dynamics Laboratory and James Hansen's group at NASA's Goddard Institute for Space Studies, as well as various simpler models. The six GCM runs gave results ranging from 2°C to 3.5°C for $CO_2$ doubling. To this narrow range of results, the Charney group added an additional margin, concluding that "the most probable warming for a doubling of $CO_2$" would be "near 3°C with a probable error of ±1.5°C."[48]

The "Charney report," as it was known, is often cited as the first policy-oriented assessment to claim a concrete, quantitative estimate of likely global warming. Heavily hedged with expressions of uncertainty, the report nonetheless declared that despite a diligent search it had been unable to find any legitimate reasons to discount the probability of global warming. On the other hand, it made no prediction as to exactly when this would occur, arguing that the heat capacity of the oceans might delay its manifestation "by several decades." Nor did it predict how the warming would affect any particular place.

Similarly, the first World Climate Conference, held in Geneva earlier in 1979, issued a "declaration" containing an "appeal to nations" for "urgently necessary" attention to climate issues. Yet while it mentioned the steady rise in carbon dioxide concentrations—along with land-use change, deforestation, urban heat, and other human activities affecting climate—it offered no quantitative estimates of change. It pressed the case for further research and better data. On the other hand, it also expressed a need for concerted study of the possible impacts of climate change.

The World Climate Conference sparked the formation of the World Climate Programme (WCP) under the joint sponsorship of the WMO, ICSU, and the UN Environment Programme. The WCP itself focused principally on applied climate issues, such as agricultural meteorology, but it also established a World Climate Research Programme (WCRP, initially known as the Climate Change and Variability Research Programme). The WCRP, initially chaired by Smagorinsky himself, essentially took over GARP's Climate Dynamics sub-program and comprised most of the same members. It developed a series of increasingly ambitious programs, beginning with the international Satellite Cloud Climatology Project (started in 1982) and the first research initiative on ocean-atmosphere interactions coupling observational and model-based studies: the Tropical Ocean and Global Atmosphere (TOGA) program, begun in 1984.

At the first meeting of the WCRP's Joint Scientific Committee, in early 1980, WMO Secretary-General Aksel Wiin-Nielsen noted increasing awareness of "the $CO_2$ problem" and a rising tide of $CO_2$-related research in many nations. "However," he told the committee, "there is a real need for some machinery to maintain regular critical scientific appraisal of . . . the research in a form which also renders possible definitive and authoritative statements from time to time interpreting the results in terms meaningful to those responsible for policy. . . . These requirements could best be met by some form of international board for the assessment of all scientific aspects of the $CO_2$ question."[49] This was among the first articulations of what would become, eight years later, the Intergovernmental Panel on Climate Change.

**Sticky Numbers: Stabilizing Projections of Global Warming**

After the Charney report and the World Climate Conference, the "carbon dioxide problem" became the subject of focused, sustained international attention. The climate science community recognized the human significance of its scientific work and a responsibility to speak out about it. The notion of ongoing "international assessments" of the state of scientific knowledge would take another decade to gain a permanent institutional foothold in the form of the IPCC. In the interim, climate science began to concentrate its resources on extending and stabilizing both its knowledge of the past (as described in chapter 11) and its projections of a future under $CO_2$ doubling.

Not only did the Charney report make a definite call (unlike most of its predecessors); even more important, its numbers *stuck*. Its 1.5–4.5°C climate sensitivity range, and its 3°C "best guess" warming estimate, marked the stabilization of these crucial figures, which continues in a nearly unbroken line into the present. (See table 14.1.) The Charney group's reasoning in regard to these numbers bears extended quotation and commentary:

We believe that the snow-ice albedo feedback has been overestimated in the H [Hansen] series and underestimated in the M [Manabe] series. For the above reasons [including others not quoted here], we take the global or hemispheric surface warmings to approximate an upper bound in the H series and a lower bound in the M series. . . . These are at best informed guesses, but they do enable us to give rough estimates of the probable bounds for the global warming. As we have not been able

Table 14.1
Range of GCM results, estimate of equilibrium climate sensitivity, and "best guess" global temperature increase for $CO_2$ doubling from successive assessment reports. Corrected and updated from J. P. van der Sluijs, *Anchoring Amid Uncertainty* [*Houvast Zoeken in Onzekerheid*] (University of Utrecht, 1997), 43. (Compare with table 7.3, which shows climate-sensitivity results from simpler models published between 1896 and 1975.)

| Assessment | Range of GCM results (°C) | Equilibrium climate sensitivity | "Best guess" (°C) |
| --- | --- | --- | --- |
| NAS 1979 | 2–3.5 | 1.5–4.5 | 3 |
| NAS 1983 | 2–3.5 | 1.5–4.5 | 3 |
| Villach 1985 | 1.5–5.5 | 1.5–4.5 | 3 |
| IPCC 1990 | 1.9–5.2 | 1.5–4.5 | 2.5 |
| IPCC 1992 | 1.7–5.4 | 1.5–4.5 | 2.5 |
| IPCC 1994 | not given | 1.5–4.5 | 2.5 |
| Bolin 1995 | not given | 1.5–4.5 | 2.5 |
| IPCC 1995 | 1.9–5.2 | 1.5–4.5 | 2.5 |
| IPCC 2001 | 2.0–5.1 | 1.5–4.5 | 2.5 |
| IPCC 2007 | 2.1–4.4 | 2–4.5 | 3 |

to find evidence for an appreciable negative feedback due to changes in low- and middle-cloud albedos or other causes, we allow only 0.5°C as an additional margin for error on the low side, whereas, because of uncertainties in high-cloud effects, 1°C appears to be more reasonable on the high side.[50]

Jeroen van der Sluijs has perceptively argued that, despite the self-admittedly "rough" way in which this range was determined, it became an anchoring device—a "highly aggregated and multivalent consensus knowledge construct" bridging climate science and policy.

Across the next 28 years, successive scientific assessments reported varying ranges of GCM results. The latter changed considerably, especially at the top end (from 3.5°C in the Charney report to 5.5°C in the 1985 Villach report; see table 14.1). Yet *the next eight successive assessments all retained the 1.5–4.5°C range* for climate sensitivity, which changed only in 2007—and only slightly—with the fourth IPCC report. Further, while the Charney report *added* 1°C to the top of the climate sensitivity range as an expression of known uncertainties, most of the later reports *reduced* the top end of their climate sensitivity estimates by up to 1°C (relative to the range of GCM results). "Best guess" figures for global warming also varied very little, changing only twice in ten assessments over 28 years.

The remarkable stability of the climate sensitivity range, and the fact that there have been just two best-guess estimates of warming (3°C and 2.5°C), seems curious in a field marked by substantial uncertainties.[51] Scientists explained that since their judgment as experts was based on a wide range of knowledge, not just on general circulation model outputs, changes in the range of GCM results did not by themselves provide a sufficient reason to alter their assessment of the climate sensitivity range. The 1990 IPCC report noted that the full set of 22 model runs gave estimates ranging from 1.9°C to 5.2°C. Yet, it went on, "most results are close to 4.0°C but recent studies using a more detailed but not necessarily more accurate representation of cloud processes give results in the lower half of this range. Hence the model results do not justify altering the previously accepted range of 1.5–4.5°C."[52] Similarly, Lawrence Gates told Jeroen van der Sluijs that when it reconsidered the climate sensitivity range in 1992, the IPCC decided against revising it because "in the absence of a comprehensive exploration of parameter space in the doubled $CO_2$ context, there appeared to be no compelling scientific evidence to change the earlier estimated 1.5–4.5°C range (which was in itself an educated guess) since such a step would have given greater credibility to any new values than was justified."[53] By itself, however, this is not a compelling account. Another element was the belief—not quantified, but widely shared—that if the highest GCM results did reflect the actual climate sensitivity, more warming would already have appeared in the data record. Other factors included the usually understated importance of simpler, one- and two-dimensional climate models as a counterbalance to GCMs in expert assessments, as well as the role of experts' subjective judgments in shaping final conclusions. For example, Bob Dickinson recalled that at the 1985 Villach conference (discussed below), "My 5.5°C . . . was inferred by showing you would get at least that [much warming] if you took the current GCMs with the strongest ice albedo feedback and combined it with the model with the strongest cloud feedback, so that both strong feedbacks were in the same model. At the meeting Suki Manabe was personally skeptical that such a large number could be achieved, and I recall that led the meeting to adopt the previous range."[54] These kinds of qualitative expert judgments were later brought explicitly into the IPCC's assessment process.

Climate sensitivity to carbon dioxide doubling is a key figure for climate modeling because it provides a simple, direct way to compare the very complex models. Yet other zero-dimensional global averages (such as precipitation and evaporation) and one-dimensional zonal averages (such as sea-level pressure, ocean temperature, specific humidity, and global overturning streamfunction) can also be compared, as well as even more complex two- and three-dimensional output variables. The Program for Climate Model Diagnosis and Intercomparison, which organizes the comparison of climate models used in IPCC assessments, lists hundreds of such variables. Modelers value these more complex comparisons because they reveal much about the various models' strengths and weaknesses. But for communicating the meaning of model results to broader audiences, simple global averages are far more effective, and this is what began to happen with the Charney report and its successors.

The stable 1.5–4.5°C range and 2.5–3°C "best guess," van der Sluijs argues, represented "means of managing uncertainty" as climate change moved from science into a broader policy arena. Like a ship's anchor, they limited "drifting of the primary scientific case." Thus they helped to focus an emerging climate politics on aspects of the issue *other than* uncertainty about the underlying projections. Though they did not stop GCM projections from becoming a central focus of public debate, these sticky numbers helped move the greenhouse issue once and for all into the arena of public policy.[55]

**Nuclear Winter and Ozone Depletion**

During the 1980s, global warming became the central concern of climate politics, rather than just one among several possible forms of "inadvertent climate modification." It also emerged from the shadows of science-based policy advice to become a full-blown issue in mass politics. The climax of this period came in 1988 with the formation of the IPCC and, in the United States, with dramatic congressional hearings during an intense heat wave that placed the issue on the front pages of major newspapers.[56]

Several developments prepared the ground for the surge of public awareness and political activity. First, climate science was maturing, with a narrowing range of model-based projections that eliminated cooling scenarios. Second, three other major atmospheric issues of the late 1970s and the 1980s—acid rain (which I will not discuss), "nuclear winter," and ozone depletion—generated enormous public concern, even more in Europe than in the United States.[57] As genuinely global atmospheric problems, these paved the way for widespread awareness of the greenhouse problem. Meanwhile, new "green" political parties, particularly in Germany and a few other European nations, kept environmental issues of all sorts in the foreground. Relatively successful national and regional pacts on acid rain in the 1970s and the 1980s, followed in 1987–1990 by an unprecedented global ban on ozone-depleting chemicals, bolstered confidence in the possibility of concerted international action. The third major context for global climate politics emerged quite suddenly when the Cold War ended "not with a bang but a whimper." Fizzling out in less than two years, from the fall of the Berlin Wall in 1989 to the collapse of the Soviet Union in 1991, the Cold War left in its wake a yawning "apocalypse gap" that was readily filled, in political discourse, by environmental doomsday scenarios.

The "nuclear winter" hypothesis resulted from model-based investigations into the atmospheric effects of nuclear war. Despite a vast stream of (largely classified) research into the blast effects of nuclear weapons, for more than three decades Pentagon scientists essentially ignored the massive urban and forest fires that would inevitably occur after a nuclear war.[58] Even when they did begin to incorporate post-nuclear firestorms in their macabre calculus, they paid no attention to the staggering quantities of smoke those fires would generate. Instead, interest in that subject came from two other sources: ozone-depletion concerns and studies of volcanic eruptions' effects on climate. Both were related to the SST controversy.

The SST debate had raised the issue of stratospheric ozone depletion by nitrogen oxides in jet exhaust. As the SST controversy wound down, the same hypothesis re-emerged around the nitrogen oxides that would be created in the stratosphere by nuclear blasts. During the same period, interest in aerosols prompted the first modeling of volcanic dust as a factor in climate change.[59] (This interest also had origins in NASA observations of dust storms and climate on Mars.) A 1975 National Research Council study, *Long-Term Worldwide Effects of Multiple Nuclear-Weapon Detonations*, found that serious ozone depletion—up to 50 percent of the global total, 70 percent in the northern hemisphere—would result from a nuclear war, but did not confirm a major climatic effect from aerosols, stating that the dust cloud from a nuclear war would be similar to a major volcanic eruption.[60] However, the focus at that time was on dust rather than smoke.[61] This changed in the early 1980s.

In 1980, Luis Alvarez, his son Walter, and two colleagues published their dramatic discovery of a thin iridium layer at the Cretaceous-Tertiary (K-T) boundary, a stratum of bedrock laid down approximately 65 million years ago. That boundary marks a sudden mass extinction of species, including all non-avian dinosaurs. The K-T extinction presented a paleontological puzzle that no one could explain. The iridium layer indicated the cause: a gigantic meteor, which Alvarez et al. calculated to have been about 10 km in diameter. The colossal collision, they hypothesized, would have created a dust cloud sufficient to darken the atmosphere for several years, suppressing photosynthesis. The explosion would have produced far more dust than any known volcanic eruption. It might also have set off sweeping firestorms in forests and grasslands around the world, generating vast amounts of smoke.[62] This discovery—confirmed when other investigators located the meteor crater near the Yucatan Peninsula—revolutionized paleontology.

Although the gigantic meteor explosion would have dwarfed even a full-scale nuclear exchange, the obvious analogy played a part in galvanizing studies of the smoke effects of nuclear war. In 1982, the atmospheric chemists Paul Crutzen and John Birks published an article titled "The Atmosphere After a Nuclear War: Twilight at Noon" in the journal *Ambio* (figure 14.1). On the basis of results from a two-dimensional simulation of atmospheric chemistry and transport, Crutzen and Birks argued that smoke and photochemical smog would darken the Earth for "several months" after a nuclear war, drastically reducing sunlight and potentially causing massive crop failures and other environmental damage.[63] A group known as TTAPS (the initials of the five co-authors), led by Richard Turco and including the astronomer and science popularizer Carl Sagan, then followed up on this work. TTAPS sequenced three models—a nuclear war scenarios model, a particle microphysics model, and a one-dimensional radiative-convective climate model—to further explore the climatic effects of the massive smoke clouds such fires might produce. Their simulations predicted that for "exchanges of several thousand megatons, . . . average light levels [could] be reduced to a few percent of ambient and land temperatures [could] reach $-15°$ to $-25°C$."[64] Such effects would endure at least for several weeks and possibly much longer. They might further extend to altering the global circulation.[65] TTAPS coined the phrase "nuclear winter" to describe these effects.[66]

The metaphor had extraordinary political import in the context of early Reagan administration claims, already controversial, that an atomic war could be meaningfully "won." Deputy Undersecretary of Defense Thomas K. Jones famously told a journalist that civilians could safely hide from nuclear blasts in shallow holes. "If there are enough shovels to go around," he claimed, "everybody's going to make it."[67] Immediately taken up by the anti-nuclear-weapons movement, nuclear winter made its way into the mass media and the popular imagination, playing a significant role in the widely viewed 1983 TV movie *The Day After*. In the journal *Foreign Affairs*, Sagan argued that nuclear winter might unleash a cascade of famine and environmental disasters sufficient to kill everyone on Earth.[68] The Stanford

**Figure 14.1**
The first page of the *Ambio* article on the atmospheric consequences of nuclear war.
Reprinted with permission of Allen Press Publishing Services.

University scientists Paul Ehrlich and Donald Kennedy collaborated with Sagan and atmospheric scientist Walt Roberts (founding director of NCAR) on a popular book, *The Cold and the Dark: The World After Nuclear War*.[69] Congress held hearings on nuclear winter in 1985.

When other climate scientists tested the TTAPS claims using three-dimensional GCMs, they found the likely outcomes to be considerably less dramatic.[70] Starley Thompson and Stephen Schneider, countering Sagan in *Foreign Affairs*, argued that, though extremely serious, such effects would be more like "nuclear fall" than nuclear winter, and would vary considerably by latitude and region as the pall of smoke circulated around the globe.[71] Thus, although they might indeed cause major agricultural failures, especially if the exchange occurred in springtime, they would probably not extinguish all human life, as Sagan had suggested. By then, however, the "nuclear winter" metaphor had gained a permanent place in the popular political imagination. Like nuclear fallout in the 1960s and the SST controversy in the 1970s, nuclear winter was an issue of global atmospheric politics.

Around the same time, the dramatic discovery of an "ozone hole" over Antarctica (figure 14.2) also drew attention to the vulnerability of the global atmosphere. The full story behind that event lies beyond the scope of this book, but I raise it here because it involved a data-modeling issue strikingly similar to the ones we have encountered in the analysis of historical climate records. As noted above, the ozone-depletion issue began with studies of the supersonic transport's effects on the stratosphere, but these were found to be relatively small and to be manageable by means of emission controls on nitrogen oxides.[72] In 1974, the chemists Mario Molina and Sherwood Rowland hypothesized that chlorofluorocarbons (CFCs)— artificial chemicals used in refrigeration and air conditioning and long thought to be completely inert—could break down in the stratosphere, enchaining chemical reactions that could destroy the ozone layer.[73] A political debate ensued, and some uses of CFCs were banned in 1976. But although the CFC-ozone reaction could readily be demonstrated under laboratory conditions, confirming that the reaction also occurred in the stratosphere proved extremely difficult. Without data to demonstrate real damage, further policy action was not forthcoming.

Until the late 1970s, ultraviolet readings taken by Dobson spectrometers at surface stations made up the bulk of the data record relevant to the ozone layer. A WMO-sponsored World Ozone Data Center had cataloged ozone measurements since 1960. However, many ozone scientists regarded the center's procedures for collecting and correcting data as so poor that

Simulation Models and Atmospheric Politics 385

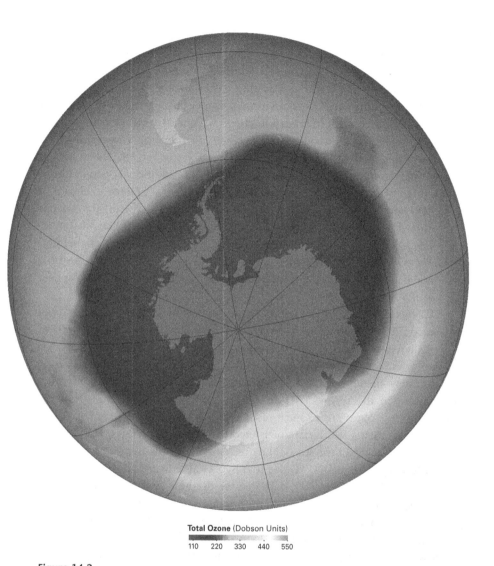

**Figure 14.2**
The largest Antarctic ozone hole ever recorded, imaged using data from NASA's Aura satellite in September 2006. Image courtesy of NASA.

they stopped reporting station data, leaving the global record spotty and inconsistent. Just as in the case of the surface thermometer records discussed in chapter 11, this made it difficult to detect or confirm a global trend. Meanwhile, various modeling exercises proceeded. Also as in the global warming case, one-dimensional models, though easier to calculate, could not account for circulation effects, so scientists developed two- and three-dimensional models. Because reactions involving numerous chemicals other than CFCs affect ozone concentrations, modelers kept adding further complexity. Yet until the mid 1980s, "the additional computational burden of even a 2-D model required using highly simplified chemistry, a disadvantage that persisted for years because rapidly increasing [knowledge of] chemical complexity was a moving target for 2-D modelers. Most modelers judged it more important to include a fuller, current set of chemical processes than to represent two-dimensional movement."[74] The first 2-D model representing most of the major chemical interactions did not appear until 1985.

Thus, the atmospheric chemistry and transport models available in the early 1980s remained in a relatively primitive state. By then, some of them predicted that stratospheric ozone loss from CFCs should be greater at high latitudes. Yet none of these models forecast the very large seasonal ozone declines—up to 40 percent—first detected at British Antarctic observing stations in the springs of 1982, 1983, and 1984.[75] Satellite instruments capable of volumetric ozone measurements first flew in 1978, but although the instruments had in fact registered the Antarctic declines, NASA scientists did not detect the seasonal pattern until the British group was already writing up its results.[76] The idea that NASA's algorithms for controlling the quality of data "filtered out" the ozone declines as outside the range of possible values has been widely propagated.[77] But this is not exactly what happened. According to Richard McPeters of NASA, "there has never been a filter applied as described. . . . Our software had flags for ozone that was lower than 180 DU [Dobson units], a value lower than had ever been reliably reported prior to 1983. In 1984, before publication of the Farman paper, we noticed a sudden increase in low value from October of 1983." The detected low value lay at the 180 DU threshold, not below it. "We had decided that the values were real and submitted a paper to the conference the following summer when [Farman's] paper came out, showing the same thing. . . . It makes a great story to talk about how NASA 'missed' the ozone hole, but it isn't quite true."[78]

The terrifying metaphor of a "hole" in the protective ozone "shield" constituted a crisis sufficient to galvanize the international community. It led in short order to the Vienna Convention (1985) and the Montreal Protocol on Substances That Deplete the Ozone Layer (1987, strengthened in 1990). These treaties rapidly phased out all production of most CFCs. Since then, the *rate* of ozone depletion has slowed, but the stratosphere is still losing ozone; the Antarctic ozone "hole" continues to grow. The turnaround point, when the ozone layer will actually begin to regenerate, will not be reached until around 2025. At this writing, models project full recovery of the ozone layer around 2065.[79]

The account I have given so far may make it sound as if all controversy ceased following revelation of the ozone hole, but this was not the case. Few disputed the hole's existence, but its causes remained highly debatable, *especially in the absence of good long-term data*. For example, the flawed WODC data showed a similar dramatic decline in global total ozone values in 1960–1962, including very low values at the north pole (but not in Antarctica), well before the buildup of CFCs.[80] In 1987 one ozone researcher described the ozone hole as "a mystery" and reported that "none of the models proposed to date seems capable of explaining the . . . spectacular local depletion."[81] The policy historian Edward Parson notes that even after the influential International Ozone Trends Panel report of 1988, widely regarded as having settled the issue, "important uncertainties and gaps in knowledge remained. . . . [These] included quantitative models of Antarctic depletion, good satellite records of global ozone, an explanation for the observed temperate-latitude depletion, and any basis for making quantitative projections of future depletion."[82] Similarly, the political scientist Peter Haas has argued that "consensual knowledge does not tell the full story" of the rapid move to a global treaty on CFCs. The broad scientific consensus marked by the Ozone Trends Panel report occurred *after* the Montreal Protocol had already been signed. Haas explains this as the action of a strong "epistemic community." That term refers to a knowledge-based professional group that shares a set of beliefs about cause-and-effect relationships and a set of practices for testing and confirming them. Crucially, an epistemic community also shares a set of values and an interpretive framework; these guide the group in drawing policy conclusions from their knowledge. Its ability to stake an authoritative claim to knowledge is what gives an epistemic community its power.[83] By 1985, the collection of scientists, scientific organizations, policymakers, and environmental NGOs concentrating on ozone depletion since the mid 1970s had come together as such a community.

For Haas, the epistemic community's strength accounts for the rapid move to a treaty in the absence of a full consensus on the outstanding scientific questions. The infrastructure perspective this book provides would extend that explanation considerably, pointing to the material basis of that community's work and its knowledge. Meteorological interest in stratospheric ozone has an eighty-year history, with some stations monitoring ozone levels by 1930. The global ozone monitoring network dates to the 1957 International Geophysical Year, long before human effects on the ozone layer were of concern to anyone. A full description of ozone modeling and data networks lies beyond the scope of this book, but clearly that infrastructure underlies the more recent epistemic community surrounding the ozone-depletion issue.

Long after the Montreal Protocol and the London Amendments, skeptics continued to argue that, although CFCs probably did affect the ozone layer, the balance between anthropogenic effects and natural variability remained unknown. Furthermore, they contended, data did not support a claimed increase in ultraviolet-B radiation reaching the Earth's surface, the main cause of human health effects attributed to ozone depletion. Finally, the ozone hole was a local effect, created by unique meteorological conditions occurring only during the Antarctic spring. Therefore it was not actually a global problem.[84] Notably, the principal advocates of these arguments included many of the same individuals skeptical of carbon dioxide warming. Few of their findings were published in peer-reviewed journals, and their criticisms were eventually dismissed by the larger scientific community. However, at this writing a new finding in chlorine chemistry may have revived the controversy by casting doubt on existing understanding of fundamental chemical interactions involved in stratospheric ozone destruction.[85]

### Global Warming as Mass Politics

Nuclear winter and ozone depletion drew attention to the vulnerability of the global atmosphere and provoked enormous public anxiety. $CO_2$-induced climate change emerged as a mass political concern during the same period—and it was tightly connected to those issues in both scientific and popular contexts. An era of global atmospheric politics had dawned. In the early 1980s, debates about these issues took place in the context of the Reagan administration's militant anti-environmentalism, especially during Reagan's first term (1981–84). Among other things, the new administration cut the budget of the Department of Energy's carbon dioxide program and dismissed its leader as part of a systematic attack on the department itself (which it attempted to dismantle) and on other government bodies concerned with climate change, including the National Oceanic and Atmospheric Administration and the Environmental Protection Agency. One result was to suppress a 1980 project for "integrated scenario analysis," sponsored by the American Association for the Advancement of Science and the Department of Energy, that would have brought climate scientists together with agronomists, hydrologists, health specialists, economists, and others to project the potential effects of climate change. This cancellation probably delayed the first such comprehensive studies by about five years.[86] As a result of the increased politicization of their activities, one participant wrote, "hostility, sometimes quite open and extreme, became a feature of the agency scene that only began to abate in the late 1980s."[87]

Reagan's efforts failed to stem concerns about climate change within the federal agencies responsible for climate-related policy. Nor did they stop Representative Al Gore—a former student of Roger Revelle—from organizing House hearings on global warming in 1981 and 1982, though little of substance resulted. In 1983, a National Research Council report expressed measured concern about greenhouse warming. It called, like most studies before it, for further research rather than policy action.[88] But the NRC report was overshadowed by a more radical EPA study of warming-related sea-level rise—released on the same day—that drew frightening conclusions about threats to coastal cities, island nations, and low-lying lands around the world.[89] This was the kind of human-impacts information most likely to arouse concern among policymakers. The former National Academy staffer Jesse Ausubel has noted the irony of the situation: "In a surprising turnabout for the [Reagan] administration, its 'own' report, the EPA report, based on the Hansen [climate] model and involving several assumptions from the 'high' end of plausible scenarios, was much more alarming in tone than the report to the government from the NRC."[90] Intense debate over the conflicting NRC and EPA reports brought the greenhouse debate into the legislative arena in its own right, detached for the first time from the "issue handles" of weather modification and energy politics. Gore organized more House hearings in 1984. The Senate took up the issue the following year, after a major international conference on $CO_2$-related climate change, held at Villach, Austria in 1985.

In a public statement issued after the meeting, the Villach conference delegates expressed their concerns much more forcefully than had any previous scientific group. Based largely on GCM projections, the scientists contended that "in the first half of the next century a rise of global mean temperature could occur which is greater than any in man's history."[91] The following year, in further Senate hearings on global warming, NASA scientist Robert Watson testified that he believed global warming to be "inevitable." "It is only a question of the magnitude and the timing," Watson asserted.[92] The increasingly definitive tone of these and other scientific assessments, widely reported in the media, began to raise the issue's public status from a hazy background concern to something much more immediate and clear-cut. Non-governmental organizations engaged the issue for the first time during this "agenda-setting" phase in global warming's transition from science to policy.[93] The first exclusively climate-oriented nongovernmental organization, the Climate Institute, emerged in 1986. Some other NGOs, especially the Natural Resources Defense Council, developed credible independent expertise. Yet polls taken around the same time still showed that 55 percent of Americans had never heard of the greenhouse effect.[94]

In 1987, an alliance among the NGOs, the Environmental Protection Agency, and concerned federal legislators succeeded in passing a Global Climate Protection Act (GCPA). The tone of the GCPA was vastly different from that of its predecessor, the National Climate Program Act, nine years earlier. According to a summary in the Congressional Record, the GCPA

Expresses certain congressional findings regarding global climate protection, including [that] there is evidence that manmade pollution may be producing a long-term and substantial increase in the average temperature on the surface of the Earth, a phenomenon known as the "greenhouse" effect. . . .

Provides that US policy should seek to: (1) increase worldwide understanding of the greenhouse effect and its consequences; (2) foster cooperation among nations to coordinate research efforts with respect to such effect; and (3) identify technologies and activities that limit mankind's adverse effect on the global climate.

Directs the President, through the Environmental Protection Agency, to develop and propose to the Congress a coordinated national policy on global climate change.

Directs the Secretary of State to coordinate such US policy in the international arena.[95]

Now climate policy was firmly framed as a global rather than a merely national concern. However, it was not until 1988 that global warming achieved the status of mass politics.

In June 1988, during testimony before the House Energy Committee, NASA climate modeler James Hansen asserted "99 percent" certainty that the unusually warm global temperatures recorded in the 1980s were due to a buildup of greenhouse gases. The House hearings on global warming had been deliberately scheduled for the hot summer months. Coincidentally, their impact was strongly enhanced by a devastating US drought. In addition to widespread water shortages and crop failures, the summer featured huge forest fires throughout the western states—most notably in Yellowstone National Park, a treasured icon of unspoiled nature for US citizens. "The Earth is warmer in 1988 than at any time in the history of instrumental measurements," Hansen told the sweating representatives on an intensely hot June day. "There is only a 1 percent chance of an accidental warming of this magnitude. . . . *The greenhouse effect has been detected, and it is changing our climate now.*"[96]

Figure 14.3 shows one of the charts Hansen used in his testimony, projecting temperatures reaching into and beyond the peak temperatures of two previous interglacial periods by 2020. This was among the first studies to use "transient" forcing: instead of simply comparing a control run with another run at doubled $CO_2$, transient forcing raises the levels of greenhouse gases in the simulation gradually, at a rate consistent with the actual or projected increase. All the simulations in Hansen's graph run from 1958 to 1985 using observed increases of various greenhouse gases, then diverge into high (A), medium (B), and low (C) scenarios. Scenario C freezes all greenhouse gases at year 2000 levels after 2000 (i.e., no new emissions). Notably, all three scenarios approximately reproduce observed temperatures from 1958 to 1987. (One would not expect an exact reproduction.)

Though many climate scientists disagreed with Hansen's "99 percent" level of certainty, his testimony is widely regarded as the moment when global warming entered mass politics. From that point forward, the issue received substantial, high-profile coverage in the American press. By 1990, awareness of the issue among the American public had shot up to around 75 percent—a high figure for almost any non-economic issue in the United States.[97]

### The Road to Rio: Climate Change as World Politics

Meanwhile, pressure had mounted around the world for policy action on global warming. The UN World Commission on Environment and Development initiated a major review of environmental issues and sustainable development strategies in 1983. In 1987, after two years of public hearings and written testimony, it issued a comprehensive call to action, *Our Common Future*, also known as the Brundtland Report. The Brundtland Report's short section on climate change endorsed the Villach

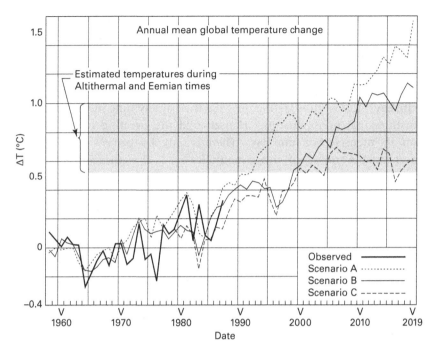

**Figure 14.3**
A published version of a chart shown during James Hansen's 1988 testimony to the House of Representatives and representing annual mean surface temperatures for three greenhouse-gas-emission scenarios vs. observations. Shaded area represents "an estimate of global temperature during the peak of the current and previous interglacial periods, about 6000 and 120,000 years before present."
Source: J. Hansen et al., "Global Climate Changes as Forecast by Goddard Institute for Space Studies Three-Dimensional Model," *Journal of Geophysical Research—Atmospheres* 93, no. D8 (1988), 9347.

conference's call for a move toward global policies.[98] The following year, foreseeing an eventual UN climate treaty, the UN Environment Programme and the World Meteorological Organization established the Intergovernmental Panel on Climate Change to conduct an authoritative scientific assessment.

Not long after the IPCC's formation—and in the same month as Hansen's congressional testimony—the government of Canada convened the Toronto Conference on the Changing Atmosphere. That conference issued another strong call for action, promoted the IPCC, and initiated the process that would lead to the 1992 Framework Convention on Climate

Change. Drawing upon political contexts described earlier in this chapter—the nuclear winter idea and the "apocalypse gap" opened by the Cold War's end—the Toronto Conference Statement asserted that "humanity is conducting an unintended, uncontrolled, globally pervasive experiment whose ultimate consequences could be second only to a global nuclear war."[99]

By the end of 1988, the UN General Assembly had adopted Resolution 45/43, "Protection of the Atmosphere for Present and Future Generations of Mankind," its first resolution on climate change. The resolution assigned the role of scientific support for the negotiation of climate treaties to the Intergovernmental Panel on Climate Change, a function it has served ever since.[100] Climate negotiations were rolled into the UN Conference on Environment and Development (UNCED), planned for Rio de Janeiro in 1992. Originally aimed principally at sustainable development and biodiversity, the UNCED ended up being dominated by the climate-change issue.

As table 14.1 shows, the IPCC's First Assessment Report, completed in 1990 as input to the UNCED, reported the same climate sensitivity range (1.5–4.5°C) as all of its major predecessors since the Charney report a decade earlier. But by 1990 more models were available, and their sophistication had increased considerably. The FAR reported 2×$CO_2$ results for 13 coupled ocean-atmosphere models in 22 variations—for example, the same model using several different parameterizations for albedo, sea ice, and clouds. Modeling groups from the United Kingdom, Australia, Japan, Germany, the Soviet Union, and Canada were represented, as well as four groups from the United States. Model resolutions remained quite coarse. Most had grid scales from 5°×7.5° to 8°×10°. Three higher-resolution (2.5°×3.75°) models were also included.

The First Assessment Report concluded that global temperatures had increased by 0.3–0.6°C over the past 100 years. Crucially, it did not claim that this warming could be definitively attributed to anthropogenic causes. The increase "could be largely due to this natural variability; alternatively this variability and other human factors could have offset a still larger human-induced greenhouse warming. The unequivocal detection of the enhanced [i.e., anthropogenic] greenhouse effect is not likely for a decade or more."

The body of the report discussed the weaknesses of simulation models at some length, noting that "we can only expect current simulations of climate change to be broadly accurate and the results obtained by existing models may become obsolete as more complete models of the climate system are used." Nonetheless, the Policymakers' Summary argued, GCMs were "the most highly developed tool we have to predict future climate." The report predicted an increase of 2°C above the pre-industrial era's average (1°C above 1990 values) by 2025 under its best estimate of a "business as usual" scenario.[101] A 1992 supplemental report confirmed the same values for climate sensitivity, while noting that anthropogenic aerosols and stratospheric ozone depletion might produce an opposite cooling effect and were not yet included in GCMs. Notably, the 1992 report included GCM results from transient studies similar to those James Hansen had used in his 1988 congressional testimony. Because transient studies better simulate the gradual way in which human activities add greenhouse gases to the atmosphere, by the time of the 1995 Second Assessment Report such studies would overtake instantaneous $CO_2$ doubling as the benchmark for model projections.[102]

American scientists were heavily involved in all of this, but leadership also came from the United Kingdom, Canada, and Sweden, as well as from international bodies including the UN Environment Programme, the Group of Seven (an economic forum comprising Canada, France, Germany, Italy, Japan, the UK, and the US), and the WMO. By the end of 1990, the European Community had adopted an explicit climate policy that committed member states to stabilizing $CO_2$ emissions at 1990 levels by the year 2000. Soon afterward, the EC began considering a carbon tax. Also in 1990, formal negotiations opened on the UN Framework Convention on Climate Change (FCCC).

The desire of the Reagan and George H. W. Bush administrations to stem environmental regulation and prevent a possible carbon tax—balanced against their own agencies' and staffers' growing concern over global warming—led those administrations to a strategy of avoiding policy decisions by arguing for further basic research.[103] Though this strategy succeeded in holding the line against policy action, it required expanding research budgets, leading ultimately to the world's largest climate-research effort: the Global Change Research Program, established in 1990.[104]

Since 1990, the GCRP has coordinated most US-government-sponsored research related to global environmental change, broadly defined. It has supported a wide range of research and operational programs, including thousands of individual projects as well as satellite and other observing systems. The organization's annual budget has typically ranged between $1 billion and $1.5 billion, marking it as one of the largest science programs in history. Although some projects traveling under the GCRP umbrella predated the program (notably NASA's Earth-observing satellite systems), much of the budget was "new money"—an unprecedented level of investment in environmental science. Ever since, the GCRP has supported major portions of the global climate knowledge infrastructure, especially (very expensive) satellites.

The Framework Convention on Climate Change opened for signature at the UN Conference on Environment and Development at Rio de Janeiro in 1992. Signed by 165 nations, the FCCC entered into force in early 1994. It sets voluntary goals for stabilizing greenhouse-gas emissions. More important, it requires signatories to prepare national inventories of greenhouse-gas emissions and commits them to ongoing negotiations toward international treaties on climate change. In 1997 these negotiations culminated in the Kyoto Protocol, which sought to reduce greenhouse-gas emissions through a "cap and trade" system and differentiated the responsibilities of developed and developing countries. Though many countries adopted its framework sooner, the Kyoto Protocol did not formally enter into force until 2005. Many commentators have regarded it as weak, ineffective, and unenforceable. At this writing, a successor agreement is under negotiation.

### Global Models as Policy Tools

This chapter has described a series of phases in the rise of global simulation modeling as a forecasting tool for policymakers. In the 1960s there were sporadic attempts by a few issue entrepreneurs to place the climate-change issue on the table, but at that time neither models nor data provided substantial guidance. Owing to the long-term character of the concern and the lack of details about when and how much change to expect, the climate issue gained little attention and these efforts resulted in little more than continued funding for research. Starting around 1970, global simulation modeling gained a central role for the first time as the environmental movement became international.

The earliest gestures toward GCMs as policy tools—in the Study of Critical Environmental Problems, the Study of Man's Impact on Climate, and the Climatic Impact Assessment Program stimulated by the SST controversy—produced no major policy outcomes (SCEP and SMIC because they made few specific predictions, CIAP mainly because limits on the SST program had already extinguished the controversy by the time it completed its work). Nonetheless, along with the much-maligned Club of Rome report *The Limits to Growth*, these efforts succeeded in establishing global modeling as a tool that might be used—at least at some future point—to generate policy-relevant forecasts. SCEP and SMIC also helped expand global atmospheric monitoring programs. At the same time, SST-related studies, particularly those related to nitrogen oxides and stratospheric aerosols, created uncertainty about the balance of cooling vs. warming factors in climate change.

In the 1970s, the rise of energy politics, environmentalism, and concerns about ozone depletion and regional acid rain created a series of slow-motion crises. By 1979, atmospheric models had grown more sophisticated, and scientists felt ready to make firm predictions based on their results. The 1979 Charney report inaugurated a series of assessments, based centrally but not entirely on GCMs, that forecast global warming of 1.5°–4.5°C with carbon dioxide doubling; meanwhile, carbon dioxide concentrations rose steadily. Atmospheric modeling also featured in the nuclear winter and ozone hole debates of the mid 1980s. In 1988, James Hansen's dramatic congressional testimony and the Toronto Conference on the Changing Atmosphere—both dominated by simulation model results—propelled the international community from sporadic handwaving to sustained policy action. They spurred the creation of the Intergovernmental Panel on Climate Change and led to the signing of the Framework Convention on Climate Change at Rio de Janeiro in 1992. An era of global climate politics had dawned, and the knowledge infrastructure that made it possible had computer models at its very heart.

# 15 Signal and Noise: Consensus, Controversy, and Climate Change

If you make a widget and you want people all over the world to use it, you need to do three things. First, you need to make the widget work—not just where you are, under your local conditions, but everywhere else too. It should be bulletproof, stable, and reliable. It should also be simple, accessible, and cheap, so that most people can afford it and will be capable of using it. Second, you need to distribute it around the world. Making it isn't enough; you have to get it out there, everywhere, to everyone. Finally, your widget needs to serve an important purpose. Think clothing, shoes, or 1.6 billion mobile phones.

If your widget is expert knowledge, you need to do these same things, but not in the same way. First, you have to make the knowledge work. It should be stable and reliable, not just for you but also for most other people, wherever they are. Here "working" means that they have to see it for themselves, subject it to whatever tests they can devise, and conclude that they agree. In the case of expert knowledge, most people will *not* be able to do this, because they lack the highly specialized tools and training. So making knowledge work means getting people to trust it—to buy it on credit, as it were, where the credit belongs to an authority they are willing to believe. For that you need representatives: experts from all over, whose presence provides a symbolic guarantee that the knowledge works from many local perspectives, not just from the perspectives of the centers of power. You may also need another kind of representative: not an expert, but a political ambassador to serve as a watchdog against corruption of the knowledge-creation process by partisan interests. Second, you have to distribute the knowledge you create around the world. Here your two types of representatives can serve you well, connecting with local experts and spreading your knowledge to a broader local public. Finally, your knowledge must serve an important purpose, something almost everyone needs to do. In the case of climate knowledge, that purpose is to predict where, how, and how much the climate will change, so that societies can begin to work toward mitigating the worst impacts of global warming. Such impacts may well include severe damage not only to agricultural and natural ecosystems, but also to human life, livelihood, and property.

Most of this book has been about making the widget: building stable, reliable knowledge of climate change. I have argued that this knowledge-production process works through infrastructural inversion: constantly unpacking, re-examining, and revising both historical evidence and predictive models. In a knowledge-production process that involves continuous contestation, you are never going to get a single universal data image, or a single uniformly agreed-upon projection. Instead you will get shimmering data, shimmering futures, and convergence rather than certainty.

In this chapter, I discuss the consensus on climate knowledge and how it is maintained through the work of the Intergovernmental Panel on Climate Change. I then explore how climate controversies develop as they travel beyond the realm of expertise into full public view, where experts no longer can control interpretations completely. Finally, I examine the remarkable opening of expert knowledge processes currently taking place on the World Wide Web. This opening, which is too recent to evaluate in any definitive way, has both benefits and drawbacks. At first glance we might conclude that weblogs, "citizen science" projects, and other online forums must improve the quality of knowledge production because they involve more people, more eyes examining the data, more witnesses to find mistakes and root out corruption, and more creative ideas for new approaches. On a second look, however, we see that increased transparency may also entail heightened suspicion, confusion, and above all *friction*, potentially slowing the production of stable climate knowledge and damaging its credibility.

### The Intergovernmental Panel on Climate Change

The Intergovernmental Panel on Climate Change is a remarkable institution, unique in the history of science. Run almost entirely by scientists, essentially on a volunteer basis, it is an intergovernmental agency under the United Nations Environment Programme and the World Meteorological Organization. With scientists from most nations and government representatives from 193 member nations, the IPCC is a genuinely global organization. It marks the institutional achievement of infrastructural globalism in climate science, the organizational backbone of today's climate knowledge infrastructure.

# Signal and Noise

The IPCC does not *conduct* scientific research; instead, its purpose is to *assess*—collect, synthesize, and evaluate—knowledge about climate change, its impacts on people and ecosystems, and the options for mitigating its extent and adapting to its effects. To make this assessment, the IPCC solicits and compares virtually all of the most current research in climate-related fields. Thousands of scientists are involved, either directly (in composing the assessments) or indirectly (as reviewers, or simply as researchers whose work is considered during the assessments). Large teams of contributing authors, organized by smaller teams of lead authors, work to prepare each chapter of an IPCC report. IPCC rules specify that these teams of authors "should reflect a range of views, expertise and geographical representation," and potential authors from developing nations are recruited aggressively. The 2007 IPCC report involved more than 500 lead authors and thousands of contributing authors from around the world.

The IPCC's full technical reports, each issued in three volumes, typically run to more than 2000 pages. Once these are largely completed, a "synthesis report" for each working group[1] and an overall "summary for policymakers" are prepared, drawing together the main conclusions and presenting them in relatively accessible language. Finally, in a series of plenary sessions, scientists and government representatives conduct a detailed review of the full report and the summary for policymakers. Only after acceptance and approval at these plenary sessions is the official report issued. The completed assessments serve as expert advice to the periodic Conferences of Parties to the Framework Convention on Climate Change. Figure 15.1 illustrates the IPCC's report preparation process.

The IPCC's goals are to represent fairly the full range of credible scientific opinion and to identify likely climatic impacts for several scenarios of future greenhouse-gas emissions. When consensus cannot be reached, the IPCC's charge is to summarize the major viewpoints and the reasons for disagreement. IPCC reports undergo intensive and repeated peer review by scientists. Unusually for a scientific report, drafts are also reviewed by national governments and by stakeholders, including environmental groups and corporate organizations, all of whom may submit comments. Chapter authors are required to respond to all comments, even if they make no changes as a result. IPCC draft reports undergo more scrutiny than any other documents in the history of science.

As a hybrid science-policy body, the IPCC faces the delicate task of maintaining credibility and trust in two different communities: the scientists who make up its primary membership, and the governments and international bodies (the Conferences of Parties to the Framework

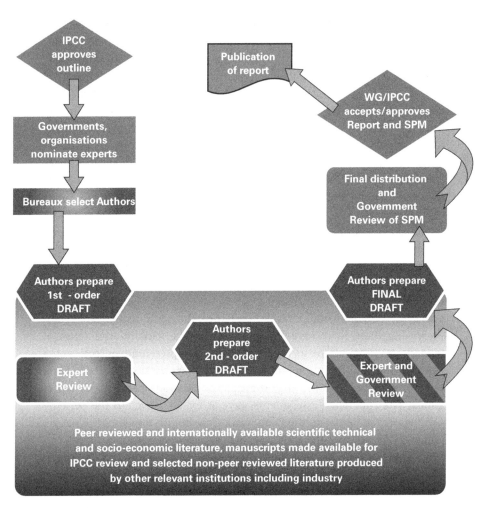

**Figure 15.1**
The IPCC assessment report process.
Graphic courtesy of IPCC.

Convention on Climate Change) to which it provides input. Independent self-governance is one of the primary mechanisms by which the IPCC achieves this goal. Its procedural rules spell out a variety of methods, chief among them peer review, designed to ensure that its reports include the best available scientific knowledge and that they represent this knowledge fairly and accurately.

At first the IPCC's peer-review practices were organized rather informally, mirroring the everyday reviewing practices of its constituent scientific communities. But in a situation in which almost any scientific finding can have political implications, the agency quickly found (like any other organization in a politically charged situation) that, without clear procedures to ensure openness and full rights of participation, dissenters would find their voices ignored—or that they would believe, or claim, that they had been ignored. This happened in 1995, when a loud and highly public controversy erupted over chapter 8 of the Second Assessment Report.

That chapter moved the IPCC statement over the line from merely *detecting* climate change (which might still be due solely to natural variability) to *attributing* it to human causes. Critics attacked the revision procedure, claiming that chapter 8's lead author, Ben Santer, had overstepped his authority in making final revisions to the report's language. As a result, they argued, the statement on attribution was stronger than the evidence could justify. In fact Santer had followed IPCC procedures precisely, but the loud and punishing controversy that ensued led the IPCC to tighten and formalize its guidelines in order to ensure that future reports could not be attacked on procedural grounds.[2] In 1999 the IPCC finalized a major revision to its rules of procedure that introduced "review editors." These editors remain outside the drafting process, coming in at the end to ensure responsiveness to the comments and criticisms collected during review.[3]

The highly inclusive, heavily reviewed, and responsive IPCC process has gained the IPCC enormous stature as representing the consensus of scientific opinion. Most scientists and policy leaders regard it as the most authoritative source of information on climate change and its potential impacts. At this writing, the IPCC has produced four scientific assessments: one in 1990, one in 1995, one in 2001, and one in 2007. Preparations for a fifth report, due in 2013, are already well underway. The agency has become a durable, resilient, and trusted knowledge institution.

By the time of its Second Assessment Report, released in 1995, the IPCC was prepared to assert that "the balance of evidence suggests that there is a discernible human influence on global climate."[4] Its 2001 report went further, claiming "new and stronger evidence that most of the warming observed over the last 50 years is attributable to human activities."[5] In its 2007 assessment, the IPCC strengthened these conclusions, expressing a level of confidence approaching certainty: "Warming of the climate system is *unequivocal*, as is now evident from observations of increases in global average air and ocean temperatures, widespread melting of snow and ice and rising global average sea level. . . . Most of the observed increase in global average temperatures since the mid-20th century is *very likely* due to the observed increase in anthropogenic greenhouse-gas concentrations."[6] These conclusions were endorsed not only by thousands of IPCC scientists, but also by most IPCC member governments—nearly every nation on Earth.

Thus the tentative scientific consensus achieved by the early 1990s has grown steadily stronger over time, reaching very high confidence levels since 2000. In 2006, former US senator and vice president Al Gore's film *An Inconvenient Truth*—based in part on his 1992 book *Earth in the Balance: Ecology and the Human Spirit*—expressed and explained that consensus for a popular audience, to wide acclaim and an Academy Award. In 2007, Gore and the IPCC shared the Nobel Peace Prize for their "efforts to build up and disseminate greater knowledge about man-made climate change, and to lay the foundations for the measures that are needed to counteract such change."[7]

It is important to be crystal clear about the exact nature of the knowledge consensus I am describing. As of 2007, most scientists agreed on at least the following points, illustrated graphically in figure 15.2:

• Global warming of between 0.5 and 0.9°C (best guess: 0.74°C) has occurred since the end of the nineteenth century.
• The global average temperature will continue to rise at a rate of about 0.2°C per decade unless emissions of greenhouse gases are reduced drastically.
• The climate will continue to warm for several decades as the climate system comes into equilibrium, even if drastic emissions reductions are achieved.

In addition, scientists agree that most of the observed temperature increase is probably attributable to anthropogenic greenhouse-gas emissions.[8] (See figures 13.2 and 13.2). These statements are made not with absolute certainty, but with very high confidence.

The consensus view is held by the large majority of scientists publishing in peer-reviewed journals, though not by all. The historian of science Naomi Oreskes surveyed 928 scientific articles containing the keywords "global climate change" and published in peer-reviewed journals between

Signal and Noise

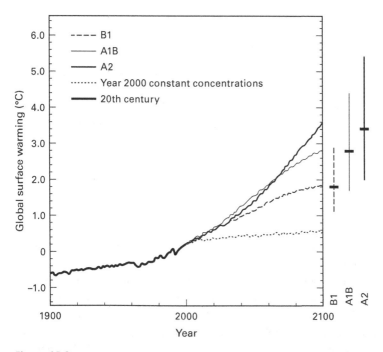

**Figure 15.2**
Historical climate change (left half of plot) and possible climate futures (right half of plot). Original legend: "Solid lines are multi-model global averages of surface warming for scenarios A2, A1B and B1, shown as continuations of the 20th-century simulations. These projections also take into account emissions of short-lived GHGs and aerosols. [Lowest line in right half of plot] is not a scenario, but is for Atmosphere-Ocean General Circulation Model (AOGCM) simulations where atmospheric concentrations are held constant at year 2000 values. The bars at the right of the figure indicate the best estimate (solid line within each bar) and the likely range assessed for [three] SRES (Special Report on Emissions Scenarios) marker scenarios at 2090–2099. All temperatures are relative to the period 1980–1999."
*Source*: *Climate Change 2007: Synthesis Report* (Cambridge University Press, 2007). Image courtesy of IPCC. Redrafted for non-color reproduction by MIT Press.

1993 and 2003. Across the entire period, Oreskes found not a single article that refuted the statement "global climate change is occurring, and human activities are at least part of the reason why."[9] In the same sample, well over 200 articles explicitly endorsed the same claim. The rest of the articles concerned impacts of climate change (the majority), methods, historical climate data, and mitigation. Oreskes concluded that the scientific community had reached a strong consensus on global warming by 1993.

The Oreskes survey suggested unanimity, but this is not quite accurate. Later in this chapter I discuss the controversy surrounding temperature data from the microwave sounding unit (MSU). Those data seemed, for a time in the 1990s, to indicate a slight global cooling in the lower troposphere (not the whole global atmosphere, but a particularly significant part of it). During that controversy, the MSU data were revised repeatedly, and since the early 2000s all parties have agreed that they, too, indicate a warming trend.

"Consensus" does not mean that every issue is settled. In particular, the exact magnitude and effects of the warming trend remain open to debate, as do the relative contributions of human and natural causes to the warming. But no credible scientist now asserts either that no warming has occurred or that the planet is cooling. And no credible scientist now asserts that the entirety of the warming trend can be explained by natural variability. The world is getting warmer, and people are making it happen.

### The Structure of Climate Controversies

Despite the knowledge consensus, the period since 1990 has been marked by extraordinary political controversy. Once the climate problem had been recognized and a global treaty negotiation set in motion, a seismic shift in the world's energy economy became a genuine possibility. This threatened numerous entrenched and extremely powerful interests, including fossil-energy corporations, automobile manufacturers, and oil-exporting countries. Meanwhile, many developing countries feared the potential for restrictions on their ability to expand their economies on the fossil-energy-intensive basis already exploited by now-developed countries in the nineteenth and twentieth centuries. Thus, as in any political process, interest groups began to form.

These interest groups quickly found themselves in strange-bedfellows alignments with and against one another. Many environmental organizations took up the cause, which fit with some of their existing values (sustainability, energy efficiency) and confirmed their belief in ultimate limits to consumption-based growth. Yet anti-nuclear-power elements within the environmental movement suddenly found themselves opposing a newly "greened" nuclear power industry, which now pitched itself as a competitive, low- or no-emissions alternative to fossil energy. Meanwhile, the Organization of Small Island States—led by the Republic of Maldives, 80 percent of whose 1200 islands lie no higher than a meter above sea level—lobbied desperately for quick action in the face of predicted sea-level rise. Members of this organization, many of them developing nations, found themselves pitted against another group of developing nations (notably China) that adopted the stance that developed nations should bear the brunt of the reckoning since they had created the problem in the first place. Alignments, alliances, and conflicts among these various interests remain in flux today, as the world and its individual nations struggle to reach some sort of accommodation to what is now generally seen as an inevitable warming, perhaps with consequences even more serious than most foresaw. As I composed the final pages of this book, a conference of more than 2000 climate scientists was hearing reports that warming could reach 6°C and sea levels could rise a meter by 2100—worse than the direst outcomes predicted by the IPCC.[10]

I will not chronicle in detail the various interest groups, their organizations, or their political impacts. Many excellent discussions of these topics can be found elsewhere.[11] Instead, in the rest of this chapter I will focus on the *structure* of climate controversy as it has occurred in the United States. The few vignettes I present as examples could be multiplied almost *ad infinitum*. Similar debates have also occurred in other countries, but mostly these have been more muted and less concerned with the underlying science; in no case of which I am aware have they involved the degree of acrimony, distortion, and deliberate disinformation encountered in the United States. I conclude the chapter with the case of the MSU temperature data, the major unsettled controversy in climate science during the 1990s and the early 2000s and the only one to present a serious scientific challenge to the documented warming trend.

Policy controversies in the United States often take the form of debates about the scientific basis of the policy. This happens for two reasons. First, American government has an exceptionally open constitutional structure. The bulk of legislative activity is a matter of public record, and most administrative decisions must have an explicit basis that is subject to specific procedural (and sometimes substantive) standards. These procedural standards often require public hearings and the solicitation of expert advice and stakeholder opinion. This open, pluralist system creates many points at which policy choices can be challenged; indeed, that is the system's fundamental purpose. Second, science too values open communication of methods and results, including critical challenges by peers, as basic mechanisms of knowledge production. In the case of climate science, there is an additional element. As I have argued throughout this book, climate science works by infrastructural inversion. Examining and re-examining the past and constantly looking for possible sources of error and new ways to correct them are fundamental.

Thus, in the case of policies based on science, *two systems specifically designed to promote challenge-response-revision cycles multiply each other's effects*. Such cycles can become very difficult to close. Especially when knowledge claims are being bounced back and forth between multiple scientific and policy communities in multiple venues, it becomes possible to go on raising questions about whether there is enough evidence, how much agreement constitutes consensus, and whether all credible views have been fully aired and answered—and to keep on doing so, almost no matter what happens. In such situations, it becomes tricky to distinguish legitimate questions aimed at improving the quality of knowledge from illegitimate questions aimed only at prolonging debate. Such issues are difficult enough within science, where controversies often linger for decades as minority views long after the mainstream considers them closed.[12] Environmental controversies are particularly likely to exhibit this structure when they concern preventing problems that have not yet fully manifested, as in the case of climate change before about 1995, when warming was definitively detected. Such issues often involve incomplete data and theoretical projections of possible effects, rather than after-the-fact analysis of verifiable damage. This structure of controversy characterizes case after case of environmental regulation, as well as controversies in other arenas (for example, concerning the Food and Drug Administration's review of new drugs).[13]

Another factor shaping the structure of climate controversies is the ambiguous role of science in US-style pluralist democracy. On the one hand, science—considered as an abstract force, a voice from nowhere—is widely recognized and respected as an unbiased, apolitical source of knowledge. On an idealized view, high-quality scientific knowledge should and will automatically command policy choices, limiting dispute by partisans to issues of implementation. Yet the implication of this profound authority is that credible science can be translated directly into political power.[14] This is certainly not always the reality (especially regarding issues with religious overtones, such as evolution), yet science still commands exceptional respect.

As a result, in controversies in which scientific knowledge is relevant, all sides seek to enroll experts who can promote or cast doubt on scientific claims relevant to particularist interests. *Science* is held up as universal and impartial, but *scientists* are treated as proxies for interest groups. Stakeholders—especially economically powerful groups seeking to avoid taxes or regulations, such as (in this case) the fossil-energy industry—are well aware that they can delay the closure of debate by challenging key scientific results or, sometimes even more effectively, simply by raising the level of certainty expected of scientific knowledge. Since scientists themselves exhibit a spectrum of opinion over the amount and quality of evidence needed to confirm a claim, a wedge can be driven between "high-proof" scientists, who demand conclusive evidence, and "frontier" scientists, for whom theories or models may dictate accepting a conclusion based on partial or provisional evidence.[15] This is exactly what happened in the case of climate change.

Delaying closure by stirring scientific controversy—or even creating an appearance of controversy where none actually exists—first became a deliberate strategy in the 1950s, when the tobacco industry created organizations (the Tobacco Industry Research Council, the Tobacco Institute, the Center for Indoor Air Research, and others) to fund and promote studies (and authors) that challenged the connection between cigarette smoke and lung cancer.[16] Ever since, this same strategy has reappeared, often by deliberate emulation, in numerous other controversies, including those over the health effects of ground-level ozone and the protection of endangered species.[17] Some scientists have moved serially from controversy to controversy, in each case fighting a rear-guard action against the emerging consensus by claiming a lack of evidence, an overreliance on imperfect models, or both. Among them are S. Fred Singer (tobacco, acid rain, ozone depletion, climate change), Frederick Seitz (tobacco, climate change), and Sallie Baliunas (ozone depletion, climate change).[18]

In the case of climate, organizations such as the Global Climate Coalition (sponsored by the US National Association of Manufacturers, including oil and coal companies) and the Information Council for the Environment (sponsored by coal suppliers and coal-fired power companies) formed around 1990 with the express intent of slowing momentum toward regulation of greenhouse-gas emissions. Conservative think tanks such as the George C. Marshall Institute and the Cato Institute also took up the cause on ideological grounds, principally their opposition to government regulation. These organizations expressly sought out and supported scientists willing to argue against the consensus view.

In 1991, for example, the Information Council for the Environment adopted a strategy to "reposition global warming as theory rather than fact." Funded by the coal industry's Western Fuels Association, the ICE produced a videotape, *The Greening of Planet Earth*, at a cost of $250,000. In the film, eleven scientists discussed how increased levels of carbon dioxide might benefit humanity by improving the productivity of both agriculture and wild plants. That film was said to be influential in the George H. W. Bush White House and among OPEC governments.[19] Western Fuels' 1993 annual report outlined the strategy directly: ". . . there has been a close to universal impulse in the trade association community here in Washington to concede the scientific premise of global warming (or as the lawyers put it—concede liability) while arguing over policy prescriptions that would be the least disruptive to our economy (or as the lawyers put it—arguing damages). We have disagreed, and do disagree, with this strategy." In a section titled "Balancing the Argument Over Global Climate Change," the report describes Western Fuels' decision to "take a stand" by successfully seeking out scientists "who are skeptical about much of what seemed generally accepted about the potential for climate change."[20] And in 1998 the American Petroleum Institute circulated a "Communications Action Plan" for a public-relations effort designed to undermine the case for global warming. "Victory," the API told its members, "will be achieved when . . . average citizens 'understand' uncertainties in climate science . . . [and] recognition of uncertainties becomes part of the 'conventional wisdom'."[21]

On the day I completed the manuscript of this book, in 2009, the *New York Times* reported on internal Global Climate Coalition (GCC) documents brought to light by a recent lawsuit. An "approval draft" of an internal GCC primer on climate change, prepared by the GCC's Science and Technical Advisory Committee, stated unequivocally that "the scientific basis for the Greenhouse Effect and the potential impact of human emissions of greenhouse gases such as $CO_2$ on climate is well established and cannot be denied." It went on to note that "the current balance between greenhouse-gas emissions and the emissions of particulates and particulate-formers is such that essentially all of today's concern is about greenhouse warming." (It also noted uncertainty regarding the likely extent of change and whether it had already been detected.) A section of the report outlined *and debunked* various "contrarian" theories of climate change (as the report itself called them) often put forward as reasons to doubt the IPCC consensus. Yet when the GCC distributed a "backgrounder" to lawmakers and journalists later that year, the public document denied the conclusions of the group's own advisory scientists, claiming that "the role of greenhouse gases in climate change is not well understood."[22]

This strategy of deliberate confusion, disinformation, and denial continues. In 2002 the Environmental Working Group obtained a copy of a briefing book composed by the Republican pollster and opinion architect Frank Luntz. Its guidelines for how Republican candidates should speak about environmental issues included tips on "winning the global warming debate." Luntz stressed that "voters believe that there is *no consensus* on global warming within the scientific community. Should the public come to believe that the scientific issues are settled, their views about global warming will change accordingly. Therefore, *you need to continue to make the lack of scientific certainty a primary issue in the debate*. . . ."[23]

Rhetorical excesses and twisting of evidence to support extreme views have not, of course, been confined to conservative organizations. In a popular book published in 1990, Michael Oppenheimer, for a number of years the chief scientist at the Environmental Defense Fund, wrote: "What is needed is a knockout punch—warming must be understood to threaten the continuation of life on Earth or no one will pay attention until things get out of hand."[24] James Hansen, Stephen Schneider, and other outspoken climate scientists were also frequently accused of pushing beyond the limits of scientific knowledge in their public statements on global warming.

These political stratagems must to some extent be forgiven on all sides; however deceptive, they belong to a standard arsenal of persuasion techniques on which political interest groups have come to rely. At the same time, they point to the real character of the supposed controversy over global warming. Since 1995, at least, that controversy has been primarily partisan rather than scientific; it has been deliberately prolonged by powerful interests seeking to generate uncertainty and doubt.[25]

During the George W. Bush administration (2001–2008), even as the scientific consensus grew ever stronger, political appointees carried the manufacture of controversy to the point of criminal corruption. They censored scientists at federal agencies, successfully blocked the reappointment of Robert Watson as IPCC chair, and altered numerous government scientific reports in an attempt to conceal their most alarming conclusions.[26] In 2007 hearings, the House of Representatives' Committee on Oversight and Government Reform found that for several years the White House Council on Environmental Quality (CEQ) had required federal scientists to seek approval before speaking with reporters. The CEQ often denied such approval to scientists whose views might conflict with the Bush administration's official line that the science remained uncertain and more research was needed. In testimony, former CEQ officials admitted to extensive editing of documents from the Environmental Protection Agency, the Centers for Disease Control, the Climate Change Science Program, and other agencies "to exaggerate or emphasize scientific uncertainties or to deemphasize or diminish the importance of the human role in global warming." Similarly, an internal investigation by NASA's Inspector General into alleged attempts to muzzle James Hansen found that "during the fall of 2004 through early 2006, the NASA Headquarters Office of Public Affairs managed the topic of climate change in a manner that reduced, marginalized, or mischaracterized climate-change science made available to the general public."[27] This "management" included denying Hansen opportunities to speak with the press.

The structure of climate controversies is also shaped by the framing chosen for the issue in the popular media. American journalists, driven both by their professional training and by market demand, routinely seek to represent all opinion poles in their reporting. Professional norms of "balanced" reporting assume (probably correctly) that most journalists are not competent to judge the credibility of scientists *qua* experts. Even where they do feel competent to evaluate credibility, reporters are trained—and constrained by editors who may know even less—to offer readers multiple viewpoints and let them judge for themselves. Meanwhile, anything that sharpens controversy appears both more "newsworthy" and more likely to attract revenue-generating readership. Together, these features create a near-imperative to report scientific conclusions as controversial. Consensus just does not work as news.

"Balanced" journalism can perpetuate an impression of active controversy long after a matter has essentially been settled. In a survey of 340 news articles on climate change published in the *New York Times*, the *Washington Post*, the *Los Angeles Times*, and the *Wall Street Journal* in the period 1988–2002, Maxwell Boykoff and Jules Boykoff found that around 53 percent were "balanced," representing the positions of both proponents and skeptics of global warming as plausible.[28] On television, "balanced" reports made up 70 percent of the total in the period 1995–2004.[29]

Boykoff and Boykoff concluded that the "balanced" journalistic coverage of global warming—the majority of all 1988–2002 articles in the "prestige press" —in fact represented *biased* coverage, because it presented skeptical views lacking scientific support on an equal footing with more credible scientific opinion. They argued that journalists' imperatives to develop stories filled with strong characters, drama, and novelty had biased their reporting toward including skeptics and their arguments regardless of their scientific credibility. This led to "informationally deficient mass-media coverage" of the climate-change issue.[30]

The result of all these factors acting together is that, especially when stakes are very high, as in the case of climate change, small minorities can retain a disproportionate grip on public debates over very long periods. These features of the American science-policy interface transformed climate change into a matter of ideology. Contests between "warming hawks" or "global warmers" and "skeptics" or "contrarians," as they dragged on over many years, became aligned with generic positions on environmental policy. Whether climate science was credible became a murky question that much of the general public felt incompetent to resolve. "Balanced" journalism regenerated controversy at every turn, long after the scientific consensus had been thoroughly achieved. Climate change became a religion, something in which one either believed or didn't believe; for some skeptics, it became an environmentalist conspiracy. This distinguished the American scene from those of most European nations, which accepted the consensus much earlier. Oddly, it aligned the US scene more closely with the scene in some developing countries, where many saw climate concerns as a cynical plot by developed countries to slow their growth, rather than as a genuine global problem requiring a global solution.

### The Scientific Integrity Hearings

An excellent example of the structure of climate controversy occurred in 1995, when the House of Representatives' Subcommittee on Energy and Environment convened a series of hearings on "Scientific Integrity and the Public Trust." Chaired by Representative Dana Rohrabacher, the hearings were part of a sweeping attack on established federal environmental-policy procedures by the 104th Congress' new Republican majority. The three hearings addressed three environmental policy controversies in which "abuse of science" was alleged to have occurred: climate change, ozone depletion, and dioxin regulation.

In each hearing, scientific witnesses of the "high-proof" school were called. Some, including Patrick Michaels and S. Fred Singer, testified that empirical observations failed to bear out the theoretical predictions of the science "establishment"—predictions embodied, at least in the cases of climate change and ozone depletion, in computer models. These skeptical scientists went on to claim that observational data failed to confirm the models. Many, including Michaels, Singer, and Sallie Baliunas, also claimed that their own interpretations of observational data, and/or their own alternative theories or models, had been systematically ignored by the science establishment (e.g., in the case of climate change, by the IPCC). The science establishment's self-interest in maintaining government funding for its research was alleged to be among the corrupting influences leading to supposedly deliberate suppression of "sound science."

"Sound science" was a phrase used by Republican representatives to promote new, high-proof standards for scientific results used to justify policy. "Science programs must seek and be guided by empirically sound data," they contended, rather than by theory or models. Representative John Doolittle articulated the Republican version of "sound science": "I think we need a clear scientific conclusion that there is a definite cause for the problem and that so-called problem is producing definite effects. Theories or speculation about it are not sufficient. We need science, not pseudo-science. I think we've been in an era of pseudo-science where these dire consequences are portrayed in order to achieve a certain political objective." The use of general circulation models to project future climate change received particularly heavy criticism.

Other groups, including the left-leaning Union of Concerned Scientists, immediately adopted the phrase "sound science" in an attempt to seize control of its meaning.[31] In a report on the hearings, Representative George W. Brown Jr., the Science Committee's ranking Democrat, accused the Republican majority of a "totally unrealistic view both of science's present capabilities and of the relationship between data and theory in the scientific method." Its approach to science, he warned, could paralyze policymaking by raising the bar of acceptable evidence impossibly high. "Uncertainty is not the hallmark of bad science; it is the hallmark of honest science," Brown wrote.[32]

The symbiotic relationship between models and data that I described in chapter 10 shows why this "impossible standard" would be fatally flawed with respect to global climate change even if it were not motivated primarily by ideology. The distinction between data and theories or models on which the Republican "sound science" crusade relied does not survive close scrutiny. As I have shown, all modern climate data are modeled in a variety of ways to correct systematic errors, interpolate readings to grids, and render readings from various instrument types commensurable. *If we cannot trust models without evidence, neither can we trust evidence without models.*

## Tilting the Line: The Microwave Sounding Unit

Perhaps the most important data war related to global climate change concerned temperature readings from the microwave sounding units (MSU), which have been carried on satellites since 1978. An MSU scans the atmosphere beneath its flight path as a volume, rather than sampling it at points (as surface stations do) or along a line (as radiosondes do). Recall from chapter 10 that an MSU (like most other satellite instruments) reads radiances at the top of the atmosphere. From those radiances, it is possible to derive the temperature structure of the atmosphere below the instrument. This derivation involves extremely complex data analysis, taking into account the atmosphere's chemical composition (which varies with altitude), the satellite's flight path, the sun's position relative to both the planet and the satellite, and many other factors. Since MSU radiance measurements are fully global, scientists regard them as an extremely important source, even the most important source, of global temperature data.

Like all satellite instruments, MSUs wear out in space and must be replaced every few years. Nine MSUs have been orbited since 1978, usually with periods of overlap to permit calibrating each new instrument against its predecessor. The MSU was originally intended to provide weather data, not climate data. Yet its global coverage—and the absence, at the time, of instruments better suited to climate studies—made treating the long-term MSU record as climate data irresistible. For several years, the MSU data were interpreted principally by a group at the University of Alabama at Huntsville led by Roy Spencer and John Christy. This group is known in the meteorological literature simply as "UAH."

The MSU measures radiances on several channels (frequency bands). Channel 2 readings reflect the "bulk" temperature of the middle troposphere, a thick layer extending from the surface to 15 km. But the region of greatest concern for climate studies is the *lower* troposphere, the thinner layer from the surface to about 8 km. The MSU was not originally designed to observe this region specifically. However, in the early 1990s UAH developed an ingenious way to modify and combine readings from Channel 2's side-looking and downward-looking components to create a "synthetic" channel, known as 2LT, that could resolve readings in the lower troposphere.[33]

The UAH analyses of channel 2LT seemed to show a slight *cooling* in that region. This mattered greatly to the case for global warming, since models predicted that the lower troposphere should warm slightly more than the surface, where records showed warming at the rate of 0.07–0.1°C in the period 1979–1993.[34] In 1995, Christy reported, "with high confidence," that "the global tropospheric temperature has experienced a decline since 1979 of –0.07°C ± 0.02°C per decade."[35] This trend conflicted with longer-term radiosonde data, which at that time showed a lower-troposphere warming trend of nearly 0.1°C per decade since the late 1950s. Christy himself concluded that, even after adjusting the radiosonde data for a potential cool bias early in the record, the 1958–1995 global trend from a "hybrid" radiosonde-MSU dataset showed warming of +0.07–0.11°C per decade.[36] Still, Christy argued, the MSU lower-troposphere cooling trend presented a serious problem for climate models.

In political debates, skeptics seized on the UAH data as proof positive that global warming was a myth. Congressional committees frequently called UAH members as expert witnesses in hearings on climate change. In 1997, for example, Christy testified before the Senate's Committee on Environment and Public Works. He reported the (then) –0.05°C downward trend. Then, after some caveats about ambiguities in the data and its relatively short temporal coverage, he argued: "Because of its precision and true global coverage, we believe that the MSU dataset is the most robust *measurement* we have of the Earth's bulk atmospheric temperature."[37] Similarly, responding to questions raised by other scientists about potential defects in the MSU data, Spencer, Christy, and colleagues wrote: ". . . Hurrell and Trenberth *estimate* the temperature of the atmosphere through a simple linear regression model based only on the sea surface temperatures, and a global climate model simulation with the same sea surface temperatures but no stratospheric volcanic aerosols, [but] *the MSU data actually measure the temperature of the free atmosphere.*"[38]

The main point I want to make here concerns how, in the political sphere, this controversy became a contest between models and data. In the UAH version of the story, models offered only unreliable, purely theoretical estimates, while the MSU supplied reliable, empirical evidence. This structure fit neatly into the Republican "sound science" ideology. Not only less credentialed skeptics, but Christy and Spencer themselves repeatedly presented the MSU data as direct measurements of tropospheric temperature. Yet these data, like all global data (as I have been saying), were in fact heavily processed by data models.

The MSU record begins in 1979, a relatively brief period by climatological standards. Especially in the mid 1990s, when the MSU record covered only about 15 years, a single exceptionally hot or cool year could strongly affect the trend. And indeed, year by year, the UAH trend line changed.

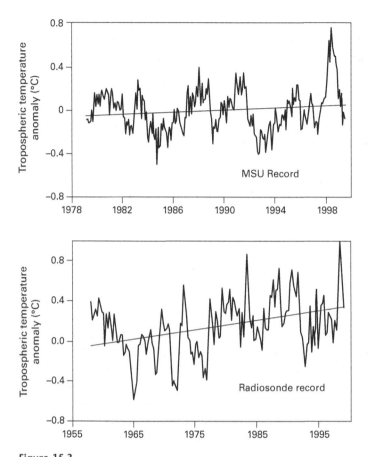

**Figure 15.3**
Top: Lower-troposphere trend 1979–1999 from MSU data, indicating warming of 0.05°C per decade (a reversal of the 1979–1997 trend of –0.05°C per decade). Though the trend reversal is due mainly to the 1998 spike, it also stems partly from adjustments to the UAH data-analysis model, made in 1998. Bottom: 1958–1999 lower-troposphere trend from radiosondes, indicating warming of 0.095°C per decade.
*Source*: W. Soon et al., "Environmental Effects of Increased Atmospheric Carbon Dioxide," *Climate Research* 13, no. 2 (1999): 153.

Figure 15.3 shows UAH data from 1979 to 1999. Here the slope of the trend has shifted from downward to upward, indicating a slight warming (+0.05°C per decade). Clearly the sharp spike in 1998 strongly affected the slope.

What is not obvious from the graph in figure 15.3, however, is that in the interim UAH had adjusted the underlying MSU data. Satellites' orbits decay over time, and as a satellite comes closer to the surface its angle of observation changes. In 1998, UAH adjusted its analysis algorithms to account for this factor, revising the entire data set accordingly.[39] Though this particular adjustment had only a slight effect on the overall trend, it was neither the first nor the last of many such changes to the UAH data model.

Remote Sensing Systems (RSS), a private research firm led by Frank Wentz, began to provide an independently constructed MSU data set (constructed from the same instrument readings). The RSS data analysis showed a positive and slightly higher trend than the UAH version. A number of other groups also contributed to the controversy, offering various new corrections, adjustments, comparisons to other data, and analysis techniques. By 2003 the UAH data had undergone five major revisions (and numerous minor ones) to remove "nonclimatic influences" from the 2LT trend. In addition to orbital decay, these included corrections for bias between different satellite instruments, multiple corrections for diurnal drift, and adjustments to remove stratospheric radiances that overlapped the 2LT frequency channel. In 2005, a correction to the algorithms contributed by RSS raised the UAH trend by +0.035°C per decade (a 40 percent change), tilting the slope of the UAH trend line considerably further upward.

Meanwhile, RSS was also revising its analysis algorithms. As of early 2009, the RSS data set was on version 3.2. Rather than try to resolve the differences for themselves, other scientists who use the MSU data now typically show both UAH and RSS trends. The controversy continues at this writing, but (as figure 15.4 illustrates) as of early 2009 UAH and RSS agree much more closely, and both agree roughly with the surface trend. RSS currently calculates the trend at +0.155 °C per decade, while UAH shows +0.13°C per decade.[40] All these numbers were produced from the same set of instrument readings; the differences between the UAH and RSS data sets *are differences between their data models*. The same thing can be said, of course, for trends calculated from radiosondes, which suffer from sampling errors, location bias, and numerous other shortcomings, and in fact the MSU controversy led to considerable adjustments in that record as well.

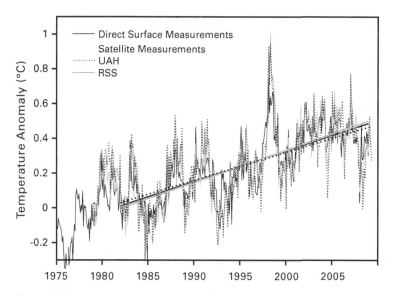

**Figure 15.4**
Global temperature anomaly in surface data vs. lower-troposphere MSU data from UAH V5.2 and RSS V3.2.
Source of surface data: HadCRUT3 (P. Brohan et al., "Uncertainty Estimates in Regional and Global Observed Temperature Changes: A New Dataset from 1850," *Journal of Geophysical Research* 111, D12106). Source of RSS data: TLT V3.2 (C. Mears and F. Wentz, "Construction of the Remote Sensing Systems V3.2 atmospheric temperature records from the MSU and AMSU microwave sounders," *Journal of Atmospheric and Oceanic Technology*, in press). Source of UAH data: TLT GLHMAM 5.2 (J. Christy, "Error Estimates of Version 5.0 of MSU/AMSU Bulk Atmospheric Temperatures," *Journal of Atmospheric and Oceanic Technology* 20 (2003), 613–29). All data updated through 2008. Graphic by Robert A. Rohde.

My point here is that the legitimate scientific debate about the MSU data became translated, in the public politics of climate change, into a generic contest between weak, unconstrained, speculative "models" and strong, firm, empirical "data." Without impugning anyone's motives, one can say that the UAH group contributed substantially to this construction of the issue by insisting that the MSU data were "measurements" and by insisting that its data were inherently superior to data produced by radiosondes and to "estimates" produced by climate models.

As I have shown, the reality is much murkier. Global data cannot exist without analysis models, all the more so in the case of the MSU. I am not suggesting that the MSU doesn't measure the atmosphere; of course it does.

What matters is how one transforms those signals into meteorological parameters, and *that* is a modeling process. The distinction between modeling and measuring was strongly overstated in this case, and that overstatement contributed to an ill-formed political debate based on a fundamental misrepresentation of the nature of climate knowledge.

The MSU controversy exemplifies the infrastructural inversion that characterizes so much of climate science. To find out more about the past, you keep digging into exactly how the data were made. You reanalyze them; you make new versions of the atmosphere. As a result, the past shimmers. In climate science, at least, models and data are symbiotic. "Raw" data are noisy, shapeless, and uninterpretable. Models give them a definite form. Neither models nor data alone can support a living understanding of physical phenomena.

**Models as Gateways in the Climate Knowledge Infrastructure**

Associated with the political arrival of the climate-change issue in the late 1980s was a trend toward increasingly comprehensive global models. This came from two directions. On the science side, Earth system models (ESMs) couple atmosphere and ocean circulation models to models of other climate-related systems, including the land surface, the cryosphere (glaciers, sea ice, and snow cover), hydrology (lakes, rivers, evaporation and rainfall), and vegetation.

Entire scientific fields, often including multiple disciplines, lie behind these component models. Coupling the models also means joining these communities. In many cases, bringing their knowledge to bear on global climate models has required dramatic changes in focus, in instrumentation, and in collective practice. Many of these changes have to do with scale. Consider GENESIS, a pioneering Earth system model created at the National Center for Atmospheric Research in the early 1990s (and later abandoned). One component of GENESIS was an Equilibrium Vegetation Ecology model (EVE) that generated global maps of plant communities at a resolution of 1°×1°. These maps, in turn, became input to the Leaf model, which generated a "leaf area index" (surface area covered by vegetation) within each grid square. The leaf area index fed, in turn, into an atmospheric general circulation model, where it affected the GCM's calculations of water flux, albedo, and other factors. A soil hydrology model determined soil moisture retention, interacting with both EVE and the atmospheric GCM.[41]

At the time, global ecosystem models were mainly based on the biome classification scheme, which divides ecosystems into from five to twenty basic types (temperate forest, tropical rainforest, desert, tundra, etc.)—a very high level of aggregation. At the other end of the scale, forest succession models used species as basic units, down to the level of individual trees—far too fine a granularity to use in a global model. EVE sought a middle ground, relying on the medium-granularity "life form" concept: aggregates of species with similar physiognomic and morphological characteristics—that is, plants that look and function in similar ways with respect to their ecological context, such as broad-leaf and boreal needle-leaf trees. Using the atmospheric GCM's climate statistics, EVE calculated the fraction of each grid box covered by each of 110 life forms.

Just as with climate, and for similar reasons, finding data against which to evaluate these model results proved extremely difficult. In the mid 1990s, EVE vegetation maps agreed with maps made from observations only to about 75 percent—but the various extant observation-based maps agreed *with each other* only to about 75 percent, which put EVE's results within the limits of observational uncertainty. The reason is that field ecology has historically worked chiefly at the level of plots—boxes whose edge length is defined, roughly, by the height of the tallest tree within the study area. Indeed, in a 1992 overview one ecologist wrote that "in no other pairing of disciplines are the temporal and spatial scales at which research is conducted so tremendously mismatched. Current GCMs operate at time steps of minutes to hours while dividing the Earth's surface into blocks which are hundreds of kilometers on a side. In contrast, most terrestrial ecosystem models operate at monthly to annual time steps on a spatial scale measured in tens of meters."[42] In experimental ecology, the scale discrepancy was even greater: a survey of articles published between 1980 and 1986 found that half of the studies used plots a meter or less in diameter, and that very few of the studies lasted more than 5 years.[43] Against this background, simulation modeling offered a way to generate a consistent global data image—which might be as accurate as any available empirical map.

Since that time, a great deal of effort has gone into building global empirical data sets and into simulating ecological processes at larger scale. Meanwhile, ecological models are using atmosphere-ocean general circulation model (AOGCM) projections to project possible environmental consequences of climate change. This is just one example among many of how the climate knowledge infrastructure has brought formerly distant scientific disciplines together through global simulation modeling, forcing each discipline to examine the methods and data of the others. Since the 1990s, global modeling projects and practices have sprung up in nearly every climate-related field.

At first, coupling models involved *ad hoc* programming. The one-off nature of the coupling process made it difficult to compare the outputs of different coupled models, since the particular computational methods by which the component models exchanged information significantly affected the overall output. The proliferation of component models led to an increasing desire to be able to "mix and match" them, for instance by combining an ocean model from one lab with a sea-ice model from another lab and an atmospheric model from a third. As a result, in 2002 a large, open-source, collaborative project—spearheaded by NCAR but also involving NOAA, NASA, and other institutions—began developing an Earth System Modeling Framework. ESMF is essentially a set of standardized software gateways that allow models to exchange information. With ESMF, component models of individual physical systems become modular, with standard interfaces. Scientists can readily assemble new combinations of models, reuse model codes, re-grid, and perform other modeling tasks with much less custom coding. A similar European project, the Programme for Integrated Earth System Modeling (PRISM), promotes "shared software infrastructure tools" and collaborates with ESMF. Both ESMF and PRISM conceive a new stage of climate modeling. These projects are moving coupled models beyond the characteristic lab-based craft processes that still dominate the field into a new era of open-source, more transparent and portable modeling tools and components. Both ESMF and PRISM promote their products explicitly as infrastructure.

The Earth System Modeling Framework has spread rapidly, moving beyond climate modeling to weather forecasting. The US Navy, the US Air Force, and the National Oceanic and Atmospheric Administration's operational weather services are revising their weather-prediction models to meet the ESMF standard. Recent collaborations have taken ESMF beyond the domain of Earth system science to include applications in emergency response management and military battlespace simulations.

A second approach to joining knowledge domains by coupling models emerged from environmental impact assessment. Like their predecessors in the 1970s and the 1980s, integrated assessment models (IAMs) sought to provide a way to compare policy scenarios and to forecast future trends. An example is the Integrated Model to Assess the Greenhouse Effect (IMAGE), begun in the Netherlands in the late 1980s. At the time, IMAGE's principal developer, Jan Rotmans, wrote: "Today, more than twenty years after the foundation of the Club of Rome, it is now widely accepted that ... global, empirical scenario models are a powerful tool for analyzing long-term decision problems." In fact, Rotmans's book on IMAGE bore a significant resemblance to *The Limits to Growth*, presenting a qualitative analysis based in quantitative methods and building its conclusions from somewhat arbitrary assumptions, limited historical data, and highly parameterized relationships. Even the book's appearance, with numerous graphs illustrating exponential growth curves, seemed to borrow *Limits'* visual style.[44] Similar climate-related assessment models of the late 1980s included the US Environmental Protection Agency's Atmospheric Stabilization Framework model and the World Resources Institute's Model of Warming Commitment. These three models provided input to the climate scenarios used in the IPCC's first assessment.[45] Owing to the looser, more speculative nature of IAMs (relative to ESMs), the process of generating scenarios is among the most controversial elements of the IPCC process. Social, behavioral, economic, and policy sciences all feature in this mix.

These increasingly integrated models—both ESMs and IAMs—serve a central organizing function for large and growing epistemic communities, both within and beyond science. These communities might be said to share the scientific macro-paradigm I have called "reproductionism." Reproductionism accepts computer simulation as a substitute for experiments that are not feasible on a global scale. It also accepts the use of data modeling as a control on heterogeneity in space and time. Once again, it's "models almost all the way down." In this very important sense, comprehensive model building is a central practice of global knowledge infrastructures.

### Climate Science on the World Wide Web

Since the early 1990s, with the advent of the World Wide Web, "citizen science" projects of all descriptions have proliferated, making using of the Web's unique capabilities to engage amateur scientists as observers. The earliest such projects, such as Project Feederwatch (organized by Canada's Long Point Bird Observatory and Cornell University's Laboratory of Ornithology), began as simple translations of existing citizen science projects to the Web medium. Project Feederwatch provides simple instructions for amateur birdwatchers to count and report the various species they observe. About 40,000 people have participated in the online count.

Many other citizen science projects were new. Weather Underground, run by the University of Michigan, began in 1992 as a simple telnet-based Internet weather reporting service. After its incorporation in 1995, Weather Underground began allowing individuals to link their personal weather stations into its network, providing site visitors with local readings from about 10,000 stations in the United States and 3000 elsewhere in the world. At the end of chapter 13, I discussed climate*prediction*.net, which runs perturbed-physics climate models on a gigantic computing grid consisting of personal computers linked by means of the Berkeley Open Infrastructure for Network Computing (BOINC). Individuals who donate processor time can watch the model runs progress on their own computers.

After the year 2000, as weblogs ("blogs") came into vogue for communication and discussion on the Web, climate science moved swiftly into the new arena. In 2004, the self-described "mainstream skeptic" newsletter *World Climate Report* moved onto the Web as a blog. Another site is RealClimate, subtitled "climate science from climate scientists." These sites offer lively debates, discussions, and explanations at various levels of technical sophistication. The content ranges from meticulously argued claims closely linked to the meteorological literature and to authoritative data sources, at one end of the spectrum, to rambling diatribes, dark conspiracy theories, and name-calling ("deniers," "global warmers," and much worse). Andrew Revkin, an environment reporter for the *New York Times*, opened his DotEarth blog in 2007. In the same year, the prestigious scientific journal *Nature* mounted Climate Feedback ("the climate change blog"). Today a complete list of relevant blogs would occupy several pages. Meanwhile, more debates—and extensive "edit wars," in which rival contributors alter or delete each other's text—raged behind the pages of Wikipedia entries related to climate science and climate change. (To see these, click the "discussion" and "history" tabs at the top of any Wikipedia entry.)

Blogs and web-based citizen science herald a new stage in the science and the politics of climate change, and they are certain to have considerable impact over the long term. Here I will discuss just three of many possible examples. All of them represent projects in infrastructural inversion.

Climate Audit emerged from a controversy over the "hockey stick" graph (figure 15.5), originally published in 1998 by a group led by the University of Virginia climatologist Michael Mann. A version of the same graph featured prominently in the 2001 IPCC Second Assessment Report.[46] This graph combined data from thermometers with proxy measures of

Signal and Noise

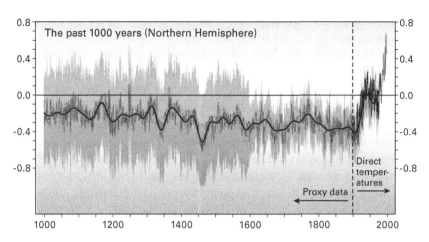

**Figure 15.5**
The "hockey stick" graph from the IPCC's Third Assessment Report (2001). The nickname derives from the trend line's shape, with its "blade" at left and its "stick" at far right. Original legend: "[This chart] merges proxy data (year-by-year [dark gray] line with very likely ranges as [light] gray band, 50-year-average [black] line) and the direct temperature measurements [black line after 1850] for the Northern Hemisphere. The proxy data consist of tree rings, corals, ice cores, and historical records that have been calibrated against thermometer data. Insufficient data are available to assess such changes in the Southern Hemisphere."
*Source: Climate Change 2001: Synthesis Report* (Cambridge University Press, 2001). Graphic courtesy of IPCC. Colors replaced as noted in brackets to allow non-color reproduction.

temperature from tree rings, ice cores, corals, and historical records to chart temperature changes over the past 1000 years. Stephen McIntyre, a former mining industry executive and government policy analyst with a background in mathematics and economics, and Ross McKitrick, an economist, challenged one of the statistical techniques Mann's group had used to analyze the data. This challenge led to a long and bitter controversy involving Mann, the US Congress, the Intergovernmental Panel on Climate Change, the National Science Foundation, the American Association for the Advancement of Science, the National Academy of Sciences, the National Research Council, and numerous other entities and individuals. Across this period, McIntyre, McKitrick, Mann, and various other parties published a series of exchanges in peer-reviewed journals and in more partisan venues.[47]

I will not discuss the details of the scientific issue, which involved arcane mathematics and proxy data sources (rather than the historical instrument records on which most of this book has focused). Instead, here I am interested in the aspect of the controversy that involved access to data and models. Essentially, Stephen McIntyre requested the original data that had been used to construct the graph. Michael Mann provided most of the data, but not all. McIntyre pursued the missing data, but Mann rebuffed him. McIntyre pressed the case—now requesting access to Mann's data-analysis source code and methods as well as the data themselves. Mann at first resisted, then eventually provided these, but by then a congressional investigation had begun. The House Subcommittee on Oversight and Investigations asked Mann to deliver his curriculum vitae, a list of all his grants and other financial support, all the data for all his published work, the source code used to produce his results, and an "explanation" of all his work for the IPCC.[48] Mann reluctantly did so, and the controversy resolved into a more ordinary scientific one about techniques for analyzing data under large uncertainties.

In 2005, during the "hockey stick" controversy, McIntyre began to use his Climate Audit blog to promote the idea of "auditing" climate data and even climate models. He sought to audit, among other data, the Goddard Institute for Space Studies' surface-temperature data set. He began to request, politely but insistently, that the institute release both the raw observations and its data-analysis model. Because McIntyre chronicled all of his requests and GISS' responses on the Climate Audit blog, his efforts gained a degree of publicity they probably would never have received otherwise. GISS resisted McIntyre's requests at first, but after some negative press coverage it complied. In 2007, McIntyre's "audit" discovered an anomaly in the GISS data set involving corrections GISS had applied to records from the US Historical Climatology Network. The blog provided an unprecedented forum for any interested party to signal audit-worthy issues, and Climate Audit and other blogs uncovered further errors made by GISS in an early release of October 2008 data. GISS thanked McIntyre publicly for these contributions. In response to the calls of Climate Audit and other blogs for greater transparency, many climate centers have begun mounting data and even some climate models on public web servers.

SurfaceStations.org represents another sort of "audit," this one based on a "citizen science" model. As we have seen, climate scientists have long expressed frustration with the inadequate and in some cases deteriorating state of the climate observing system. During 2006, a widely read blog (Climate Science: Roger Pielke Sr. Research Group News) conducted a long discussion of problems in the US surface station records. The following year, Anthony Watts, a TV and radio meteorologist, opened the website SurfaceStations.org to organize a survey of all 1221 stations in the US Historical Climatology Network. Through Watts's site, individuals can volunteer to visit a USHCN station and document the placement of its instruments. Their particular mission is to note and photograph how the station's surroundings might be influencing its instrument readings. For example, thermometers placed near air conditioner exhausts, heating vents, or asphalt parking lots, or anemometers placed near tree lines or large buildings, might give inaccurate readings. At this writing, SurfaceStations.org claims to have surveyed more than 80 percent of all USHCN stations, and Watts plans to extend the survey to cover climatological stations outside the United States.

The volunteer observers fill out a simple form describing conditions at the station, and they are encouraged to post photographs. From these site surveys, SurfaceStations.org estimates each station's temperature bias, using the official NOAA Climate Reference Network Site Handbook. In an interim report on its survey,[49] Watts claimed that 58 percent of the 865 USHCN stations surveyed to date by SurfaceStations.org exhibited "expected" biases of +2°C or more, and that another 11 percent displayed an "expected" bias of +5°C or more—an enormous warm bias, if confirmed. However, since the SurfaceStations.org site survey has not undergone peer review, the accuracy of its estimates cannot be adequately determined. Regardless, SurfaceStations.org's photographic documentation of USHCN sites alone represents a significant public service. In a sign of the linking processes the Web medium encourages, Watts thanks Stephen McIntyre and Roger Pielke Sr. for assistance with data analysis.

Another citizen project, Clear Climate Code, "writes and maintains software for climate modeling and analysis, with an emphasis on clarity and correctness." Its first effort involves rewriting GISTEMP (the software used by the Goddard Institute for Space Studies to analyze surface temperature) in the programming language Python, to make the code easier to understand. (GISTEMP now is written in the legacy scientific language FORTRAN.) In 2008, Clear Climate Code detected two minor bugs in the GISTEMP code, which GISS quickly corrected without significant effect on the data.[50]

These blogs and citizen science websites, and others like them, represent remarkable new possibilities for open access and citizen involvement in climate monitoring and modeling. On its face this seems salutary, and perhaps it will be. Yet such projects pull in multiple directions, not all of which lead to improvement in the quality of climate science. For example, while the National Science Foundation and numerous other agencies promote and even require data sharing, the paradigmatic case of such sharing is re-use of data by other scientists—not auditing by amateurs, no matter how knowledgeable and well educated they may be. The more open you make your science, the more effort you must expend to provide your data and to assist people in interpreting it. In the case of climate change, that effort can become onerous, even overwhelming; it can stop you from doing science at all. The "hockey stick" case dragged Michael Mann and his co-authors into a years-long morass of hearings, letters, and public defenses of their data and methods, during which they could have been doing research. Yet if you close your science up, excluding outsiders by refusing them access to data and methods, not only will you raise suspicions and open yourself to accusations of elitism; you also may miss real scientific benefits from unusual critiques and creative ideas "outside the box" of your field's traditions.

Have the projects described here provided such a benefit? Here a decidedly mixed pattern emerges. Climate Audit's detection of errors in the GISS temperature data and Clear Climate Code's discovery of bugs in GISTEMP are clearly beneficial, as GISS has acknowledged. But SurfaceStations.org went beyond surveying stations. It analyzed the survey results, then posted graphics and published a report indicating a large warm bias in the US Historical Climatology Network. Perhaps the survey is accurate, but in the absence of peer review this conclusion remains highly uncertain, and the rationale for posting those results on a public website is highly questionable. Similarly, the value of citizens' interventions in the "hockey stick" controversy is not clear. The National Research Council concluded that the critique by McIntyre and McKitrick helped to improve temperature-reconstruction methods. But it also noted that, in practice, the "principal-component analysis" method used by Mann et al. "does not appear to unduly influence reconstructions of hemispheric mean temperature; reconstructions performed without using principal component analysis are qualitatively similar to the original curves presented by Mann et al."[51] Furthermore, other scientists had noted most of the issues raised by McIntyre and McKitrick; had scientific review processes been allowed to proceed normally, without public uproar, the ultimate outcome would probably have been much the same.

The cases I have described here represent relatively responsible efforts by outsiders to participate in the scientific process. Many other climate blogs and citizen science projects display poorer judgment, ranging from misinformed good-faith opinions to rants, religious screeds, and deliberate disinformation. It is hard to divine where this new openness will lead.

Blogs and citizen science initially appear to increase the transparency of climate knowledge. On their face, they look like another mode of infrastructural inversion. They can certainly contribute to extending "ownership" of the knowledge-production process, which can broaden consensus. But on closer examination, their effects so far are decidedly mixed. Some have contributed new insight, helping to improve the scientific infrastructure by inverting it. At least as often, however, they promote confusion, suspicion, false information, and received ideas.

To the extent that these new forms can be brought within some framework of credentialing and peer review, they may contribute substantially to climate knowledge. To the extent that they undermine those processes—and the danger that they will do so, at least in the near term, is great—they represent ideological and political strategies rather than knowledge projects. In all cases, they represent a new form of friction—perhaps we could call it "inclusion friction"—that slows the knowledge-production process, generating heat and damage as well as, sometimes, light. Ever-greater numbers of ordinary people are learning to create and use Web resources like the ones discussed here. As we move into this future, balancing the public interest in transparency and open access with the legitimate needs of researchers to limit participation in review processes will be difficult indeed.

**Controversy within Consensus**

In principle, scientific knowledge is always provisional, always open to revision. The long history of "scientific revolutions" demonstrates that even universally held paradigms can occasionally be revised or completely overturned.[52] This line of thinking, originally the province of obscure academic analysis in history and philosophy of science, gained a new role as political strategy in climate controversies.

For example, during the Scientific Integrity hearings discussed earlier in this chapter, skeptical scientists and conservative members of Congress argued that a self-interested science "establishment" enforces the acceptance of false theories. Invoking Galileo and other scientific revolutionaries directly,

some Members and witnesses suggested that scientific "truth" is usually more likely to be found at the scientific fringes than in the conventional center. As the

Subcommittee chair stated, "I am not swayed by arguments that 'here's a big list of scientists that are on my side and you only have a smaller group of scientists on your side.' I'm just not swayed by that at all." A similar sentiment was echoed by the Chairman of the Science Committee: "My experience in 18 years of watching science policy being made is it is [sic] often those small groups of scientists, though, who differ with conventional wisdom that in fact are always producing the scientific judgments of the future."[53]

Could the dominant paradigm of climate science, the one behind general circulation models and climate theories, not one day be overturned by a heroic minority? Could not some climate Galileo, Copernicus, or Darwin emerge to reveal that everything we thought we knew was wrong?

Never say never. Yet this contingency, while not beyond imagination, remains extremely unlikely in the climate case, for at least two reasons. First, and more important, the climate-change consensus is supported not by one discipline, but by many relatively independent disciplines. Not only many different models but many different lines of evidence from various Earth and environmental sciences—each with its own methods, measurements, models, and modes of thinking—converge on similar results. This fact alone renders the likelihood of a new Copernican revolution in climate knowledge vanishingly small. Second, in the ongoing IPCC process every alternative theory and every alternative line of evidence is routinely considered and reconsidered by a large community of experts, as well as by governments and non-governmental organizations representing stakeholders of every stripe.

Perhaps, some argue, this IPCC review process is inherently biased in favor of a "scientific establishment," rejecting even legitimate ideas that run against the mainstream view. Indeed, some research on peer review does support the idea of systematic biases.[54] Yet several other studies have reached positive verdicts on peer review's ability to improve the quality of publications.[55]

Peer review is not a truth machine, automatically separating good science from bad. But the peer review of IPCC assessments differs in several ways from the processes used by scientific journals. First, most of the literature considered during IPCC assessments has already undergone peer review once (at the time of publication). Second, IPCC reports are assessments, not primary science; as a result, IPCC peer review is designed to capture both agreement and disagreement. IPCC rules of procedure specifically direct authors to call legitimate controversies to readers' attention: "In preparing the first draft, and at subsequent stages of revision after review, Lead Authors should clearly identify disparate views for which there is significant scientific or technical support, together with the relevant arguments.... It is important that Reports describe different (possibly controversial) scientific, technical, and socio-economic views on a subject, particularly if they are relevant to the policy debate."[56] Finally, IPCC review reaches well beyond the scientific community. Unlike peer review for journals, where only expert opinions are solicited, here partisan, non-expert views are deliberately solicited and their concerns addressed (to the extent the scientific framework permits). Despite its imperfections, this exhaustive, multiple-level, highly transparent review process remains the best approach we have for evaluating climate knowledge. It brings controversy within consensus, it limits bias, and it connects the world's far-flung climate science communities in an ongoing process. This extraordinary process distinguishes climate science from nearly all other scientific arenas, and it warrants my concept of a "climate knowledge infrastructure."

It could be that all the global models contain some error of understanding, as yet undetected, that accounts for the warming they all predict. Maybe their warming forecasts are just groupthink, with scientists unconsciously influencing one another to parameterize and tune their models in similar but unrealistic ways. Perhaps systematic errors in the data make global warming only apparent, not real. Perhaps Climate Audit, SurfaceStations.org, or some successor project will one day prove that so many station records have suffered from so much artifactual heating that really there is no warming at all, or at least none outside the range of natural variability. It could be that the slope of the lower-troposphere trend from the MSU data—entirely independent of the surface record—will flatten, or even fall again, under the influence of some new adjustment to the data-analysis model. And it could be that the IPCC is just an elite club, just another interest group, representing a self-interested scientific establishment bent on defending its research empire.

It could be, but it's not. There are too many models, there are too many controls on the data, too much scrutiny of every possibility, and there is too much integrity in the IPCC process for any of those things to be remotely likely. Knowledge once meant absolute certainty, but science long ago gave up that standard. Probabilities are all we have, and the probability that the skeptics' claims are true is vanishingly small. The facts of global warming are unequivocally supported by the climate knowledge infrastructure.

Does this mean we should pay no attention to alternative explanations, or stop checking the data? As a matter of science, no. If you are doing real science, you keep on testing every new possibility; in climate science, you keep on inverting the infrastructure. As a matter of policy, *yes*. You bring the controversy within the consensus. You get the best knowledge you can. And then you move, or try to move, against the enormous momentum of the fossil-energy infrastructure on which the world depends.

# Conclusion

If engineers are sociologists, as Michel Callon and Bruno Latour have taught us, then climate scientists are historians.[1] Their work is never done. Their discipline compels every generation of climate scientists to revisit the same data, the same events—digging through the archives to ferret out new evidence, correct some previous interpretation, or find some new way to deduce the story behind the numbers. Just as with human history, we will never get a single, unshakeable narrative of the global climate's past. Instead we get versions of the atmosphere, a shimmering mass of proliferating data images, convergent yet never identical.[2]

Let me retrace the arc of this book's narrative—its version of the past—once more, not merely to rehearse its main points, but also to point the way toward other places it might lead. In 1839, the young British critic John Ruskin dreamed of "a vast machine . . . systems of methodical and simultaneous observations . . . omnipresent over the globe, so that [meteorology] may be able to know, at any given instant, the state of the atmosphere on every point on its surface." Today that "vast machine" is largely complete, constructed from components—satellites, instantaneous telecommunications, and computers—that Ruskin could scarcely have imagined. Computer models assimilate observations in near-real time from a far-flung network of sensors on land, at sea, in the upper air, and in outer space. Global weather forecasts and analyzed data zoom around the world in minutes. Refashioned and interpreted by national weather centers and commercial forecast services, they serve countless human ends, from agriculture, shipping, insurance, and war to whether you are going to need an umbrella in the morning.

Weather forecasting today is woven tightly into the fabric of everyday life. It's an infrastructure: ubiquitous, reliable (within limits), widely shared, and transparent. You can get a forecast for any place on the planet. You can get one almost anywhere you happen to be, from any radio, television, or newspaper, or on the Internet. You can pull up radar images, satellite photographs, and webcams. Forecasts aren't perfect, of course, and never will be. But you have learned how, and how much, to depend on them. You understand that they are probabilities and ranges, not certainties, and you act accordingly. As with most things infrastructural, you attend to their mechanics mainly when they fail.

Weather knowledge works like that because its "vast machine" is old and robust. Through countless rounds of revision, countless generations of technology, forecasters fought data friction and computational friction. They lubricated the machine's many moving parts with standards, institutions, computers, and digital media. Much friction remains, of course, but relative to nineteenth-century weather telegraphy, forecasting's data flows, computing speeds, and predictive skill have grown by many orders of magnitude.

The climate knowledge infrastructure is built around and on top of weather information systems. It also, and increasingly, possesses information systems of its own. It too is old and robust; it too has passed through many rounds of revision. Yet unlike weather forecasting, climate knowledge—so far—remains very much *present*, obstinately failing to recede noiselessly into the background. Instead, climate controversies constantly lead down into the guts of the infrastructure, inverting it and reviving, over and over again, debates about the origins of numbers.

Beyond the obvious partisan motives for stoking controversy, beyond disinformation and the (very real) "war on science," these debates regenerate for a more fundamental reason. In climate science you are stuck with the data you already have: numbers collected decades or even centuries ago. The men and women who gathered those numbers are gone forever. Their memories are dust. Yet you want to learn new things from what they left behind, and you want the maximum possible precision. You face not only data friction (the struggle to assemble records scattered across the world) but also metadata friction (the labor of recovering data's context of creation, restoring the memory of how those numbers were made). The climate knowledge infrastructure never disappears from view, because it functions by *infrastructural inversion*: continual self-interrogation, examining and reexamining its own past. The black box of climate history is never closed. Scientists are always opening it up again, rummaging around in there to find out more about how old numbers were made. New metadata beget new data models; those data models, in turn, generate new pictures of the past.

Another of this book's arguments regards the mutually constitutive character of models and data: *model-data symbiosis*. Since the 1950s, computer models have played four complementary roles in the infrastructures of weather and climate knowledge. First, the demands of forecast models provided powerful incentives to build planetary data networks—to *make global data*. Without the models, forecasters could never have used huge volumes of information, so they would never have tried to collect them. Second, data-assimilation models gave scientists the ability to *make data global*: to process heterogeneous, spotty, inconsistent, error-ridden signals into homogeneous, complete, physically consistent data images. Modern 4-D assimilation systems literally synthesize global data, constrained but not determined by observations. Third, general circulation models—based on theory, yet constrained by data used in their parameterizations—let scientists forecast the whole world's weather, simulate climate dynamics, and perform climate experiments. Finally, in the 1980s the reanalysis of historical weather data for climate studies, using 4-D data-assimilation models, reunited forecasting with climate science.

If you think about "thinking globally" from the perspective this book offers, you'll soon see that it has implications far beyond climate knowledge. How do we know what we know about *any* world-scale process or system—especially when, as in climate science, we want to know about change over any long period?[3] Take just one example. As I write these words, in the gloomy spring of 2009, a lot of people are very worried about something called "the global economy." What is that? How do we measure it, track it, calculate its swells and contractions, and predict what will happen to it? What do we know about, say, how the collapse presently engulfing our jobs, homes, banks, and corporations compares with the *global* effects of the Great Depression in the 1930s? And how do we know it?

I can't give you a detailed answer, but I can tell you what such an answer would look like. Just as in weather forecasting, the agencies that collect economic data are mostly run by national governments. Therefore, to get a data image of the global economy you first have to assemble all these national data in one place. How far back can you do that? The concept of national income was first articulated in 1690. Yet until the 1920s, in general only estimates were available, and only from a few governments. By the time of World War II, statistical agencies in the United Kingdom, France, the Soviet Union, the United States, and a few other countries had begun collecting economic data more systematically. But the communist nations used a "material product system" of reporting whose structure differed fundamentally from other national accounting systems, making it nearly impossible to compare them accurately with Western economies.

In 1953 the United Nations created the System of National Accounts, the first uniform international standard for economic reporting. This was another deliberate move in the direction of what I have called *infrastructural globalism*. But that standard took hold slowly and piecemeal, and implementation at the national level varied. Some countries still don't report. Other countries' data are known to be defective. Further, the UN standard and most national accounts deliberately ignore, as "nonproductive," a great deal of economic activity, such as black markets in drugs, weapons, and sex. Housework, unpaid child labor, barter, and other phenomena that are arguably economic in nature never show up as money transactions. Such issues make the very definition of a "global economy" subject to debate.

For all these reasons, even today it is difficult to construct an accurate picture of the world economy directly from data. Even if you stick with the officially reported figures, differences in national income accounting render cross-national comparisons complex. For example, the growth rates of gross domestic product in the United States and the European Union have diverged since 1997. But this divergence may be more related to price-deflation methods introduced into US national income accounting around that time than to a genuine difference in growth rates.[4] If you try to look back before 1955, things get much worse. Tracking the world economy before that date requires an archival search for usable records, or proxy measures where good records don't exist. That's a lot of data friction. An entire subdiscipline, historical economics, exists for precisely this reason. Whether you are looking at the present or the past, you have to *make global data*.

Next, to transform your assembled records into a consistent and accurate picture, you will have to *make data global* in both time and space. To do that, you will have to invert the infrastructure, find out how the data were made, and adjust them (to whatever extent you can) by means of computer models. You will face *metadata friction*, the struggle to learn exactly when, how, and how much your sources revised their accounting standards and recording practices. Once you get these metadata, you'll want to adjust data from each period and place to render them commensurable with data from other periods and places. To do that, you'll need a *data model*. If other people get interested, they may create a different data model. Then you will see *data wars*: rather than one definitive global data set, multiple, competing data sets will emerge. As data models evolve, global *data images* will proliferate.

As for projecting the global economy's future, economists do that, of course, with models. Just as in climate modeling, if you invert this part of the economics infrastructure, you'll find profound issues of scale. For example, it's possible to envision an economy as the sum of all individual transactions. But there are trillions upon trillions of transactions in a year, and just as you don't model statewide rainfall by simulating individual raindrops, you don't model the world economy by simulating every single transaction. Instead, macroeconomic models *parameterize*, using a combination of empirically derived numbers and (say) models of corporate behavior.

Nobody thinks economic models are perfect. Yet despite the notorious imprecision of economic forecasts, firms, banks, and governments place considerable trust in them. They act on shimmering data, shimmering knowledge, because they *must* act—and models give them the best information they are likely to get.

My analogy here is imperfect, of course. Economies differ dramatically from the physical systems involved in weather and climate. But the truth of the analogy does extend to this: anywhere you look, in any field that tries to track and understand any global phenomenon, you will see these same structures of monitoring, modeling, and memory. And increasingly, you will see models supplying data that cannot be obtained from monitoring or memory, as they already have for several decades in weather forecasting and climate science.

Since we are speaking of history here, notice that we need climate models not only to predict possible futures but also to reveal *possible pasts*. Look again at figures 13.2 and 13.3, which show how climate models simulate the twentieth century with and without anthropogenic forcings. Such simulations give us the only information we can ever get about what *might* have happened without our influence. Similarly, during the rest of the twenty-first century, models will be the only way we can witness the effects of whatever climate policies we may adopt, by simulating what would have happened in their absence.

Will climate forecasting, like weather forecasting, one day disappear beneath the surface of everyday life, taken for granted as a smooth-functioning, reliable background system? Will we ever get beyond the "models versus data" structure of climate controversies? Perhaps not. Yet I see two paths by which these things could occur.

One path—perhaps the most likely one—follows a sociotechnical systems trajectory much like that of weather forecasting. The Global Climate Observing System, the Group on Earth Observations System of Systems, and similar projects might make progress toward stabilizing metadata standards, intercalibrating instruments, coupling component models, and improving cooperation among their institutional constituents. Such steps could reduce temporal noise in the observing system and build a reliable climate record laced with comprehensive metadata. Meanwhile, new climate-specialized satellites—such as Japan's GOSAT (Greenhouse Gas Observing satellite), launched in 2008—will provide information that weather satellites cannot. Such changes should eventually generate more precise, more trustworthy climate data, which should in turn permit further refinement of climate models. Yet it will be at least ten years, and probably much longer, before this sociotechnical systems path can place the quality of climate prediction on a par with that of today's weather forecasts. Also, the necessary changes depend on a degree of institutional and national coordination that in the past has invariably proven difficult to achieve. Finally, as long as partisan interests still perceive benefits in contesting climate knowledge, no amount of precision will push climate prediction entirely below the surface of awareness.

The other path I can envision—less likely, but equally desirable—would involve a sea change in how people see climate knowledge and what they think it is good for. If you understand why climate data shimmer, now and always, and why climate predictions too will always shimmer, you may come to accept *proliferation within convergence.* Today, an Enlightenment ideal of knowledge as perfect certainty still holds us back from this acceptance. Oddly enough, so too does a widespread relativism—promoted not least by some of my colleagues in science and technology studies (STS)— that elevates virtually any skeptical view to the same status as the expert consensus.

Let me pause briefly on this point, since STS scholars (among whom I proudly count myself) will be among this book's principal audiences. During the 1960s and the 1970s, when STS was emerging as an interdisciplinary social science, we attacked a technocratic elitism that elevated science above other ways of knowing and seemed to place scientists beyond the reach of moral values and democratic ideals. We argued that scientific authority had reached a dangerous apogee by eliding the human dimensions of scientific practice, claiming unique methods that made nature speak for itself. We sought to overthrow an internalist historiography of science that ignored larger contexts and questions of power. To emphasize the truth that science is a human thing, we talked a lot about what we called the "social construction of knowledge."

Early on, that notion served a useful critical purpose. Social constructivism correctly asserted that, whatever the power of scientific methods, scientific knowledge *also* depends on norms, values, aesthetic principles, and mechanisms of persuasion, challenge, agreement, and evidentiary standards. Such things can't be reduced to mechanically applied methods or technical apparatus. They are inherently and deeply social. To understand how they work in science, as in any other human endeavor, you need historical, sociological, and ethnographic approaches as well.

So far, the "social construction of knowledge" idea makes almost literal sense. If you want to build a skyscraper, you start with natural materials: iron ore, trees, gypsum deposits, and so on. To turn those into a building, though, you have to process them quite a bit. Making iron ore, trees, and gypsum into girders, floorboards, and drywall—not to mention designing, financing, insuring, and erecting your skyscraper—requires not only technology but also social organization, coordinated action, persuasion, standards, and norms. Thus any building is made as much from labor relations, design discussions, banking, politics, and other social processes as from metal, wood, or wallboard. In exactly this sense, science constructs knowledge from natural materials through a combination of technical, social, and political processes. This much of the social constructivist argument seems incontestable—and it is exactly how I have approached the climate knowledge infrastructure in this book.

Yet as STS pushed these ideas further, the notion that social processes are *necessary* to knowledge production sometimes blurred into the far more dubious claim that social agreement is *sufficient* for knowledge production. "Socially constructed" stopped meaning "built by people from natural materials" and started meaning something more like "negotiated collectively by social groups," full stop. It was as if people thought we could stop bothering about the iron ore, the trees, and the gypsum and just make skyscrapers directly from blueprints, mortgages, and contracts. At an extreme, this view regarded physical reality as unknowable or unimportant, and the history of science as purely contingent. Science became little more than ideology or groupthink, within which any belief at all might come to count as "knowledge." (I have actually heard some of my colleagues utter such phrases.) As a result, all too often STS scholars characterized all sides in a scientific controversy as equally plausible, and saw knowledge simply as the outcome of struggles for dominance among social groups. Any outcome, any knowledge could always, one day, be overturned.

This strange social version of Cartesian dualism soon commits epistemological suicide. It depicts physical reality as inaccessible and insignificant even while taking social realities—people's views and their ways of influencing each other—as transparently and directly knowable, not to mention all-powerful. It entails a relativism that soon becomes, if not entirely incoherent, at least useless for practical analysis. Probably few STS scholars ever really held these views in their extremes. But some did, and others, too enamored of overstated claims, expended less effort than they might have done to clarify the meaning of social constructivism. The result, for many, was a corrosive suspicion of all scientific knowledge.

Recently, there have been signs of a return to sanity. Even some once-ardent proponents of radical constructivism have reconsidered its wisdom. Writing in *Nature*, Harry Collins recently proposed that "the prospect of a society that entirely rejects the values of science and expertise is too awful to contemplate. What is needed is a third wave of science studies to counter the skepticism that threatens to swamp us all." Collins's "third wave" would recognize and respect the value of scientific evidence, the tacit knowledge gained from disciplinary experience, and the wisdom of expert communities. At the same time, Collins's "third wave" would require of scientists that they "teach fallibility, not absolute truth"—recognizing the provisional character of all knowledge.[5]

The climate knowledge infrastructure, I have argued, not only accepts the provisional character of knowledge but constructs its most basic practices around that principle. This is the meaning of infrastructural inversion (with respect to the past) and model intercomparisons (with respect to the future). The Intergovernmental Panel on Climate Change explicitly recognizes the provisional nature of climate knowledge by bringing *controversy within consensus,* and by articulating the climate's past and its future as ranges and likelihoods, not bright lines. Indeed, since the mid 1990s the IPCC has reduced its use of quantitative expressions of uncertainty (a 25 percent chance, etc.) in favor of qualitative language ("likely," "very likely," "with high confidence," and so on), especially in its synthesis reports intended for a largely non-scientific audience. Such language communicates appropriate levels of trust rather than measurable "uncertainties"—a massively overused term that naturally invites a negative evaluation of knowledge quality.

Do we really need to know more than we know now about how much the Earth will warm? *Can* we know more? From about 275 ppm in the pre-industrial era, the carbon dioxide concentration reached 387 ppm in 2008—its highest level in 650,000 years. And the *rate* of $CO_2$ increase is rising: from about 1.5 ppm per year between 1970 and 2000, recent measurements put the growth rate at over 2.1 ppm per year since 2004. It is now virtually certain that $CO_2$ concentrations will reach 550 ppm (the doubling point) sometime in the middle of this century. By 2100, they could shoot as high as triple or even quadruple pre-industrial levels, even under optimistic emissions scenarios.

Climate*prediction*.net has run thousands of "perturbed physics" simulations, varying model parameters to find the full range of possible climate futures that models predict. (See chapter 13.) From the results of these large ensembles, leaders of that project have concluded that the actual climate sensitivity might be considerably higher than IPCC estimates—perhaps greater than 6°C. And that's just for starters, since the planet will almost certainly overshoot $CO_2$ doubling.

Even more important, these scientists speculate that *we will probably never get a more exact estimate than we already have*, because all of today's analyses are based on the climate we have experienced in historical time. "Once the world has warmed by 4°C," Myles Allen and David Frame write, "conditions will be so different from anything we can observe today (and still more different from the last ice age) that it is inherently hard to say when the warming will stop." If that is true, the search for more precise knowledge has little hope of success. Worse, implicit in the quest for precision is the notion that there is some "safe" level of greenhouse gases that would "stabilize" the climate.[6] Allen and Frame's point is that we do not know this, we cannot find out whether it is true—and we now have good reasons to suspect that it is *not* true.

Our stakes in history can be high indeed. From family, ethnicity, and nation to holocaust, apartheid, slavery, and war, the facts of the past matter a very great deal. So it is with the history of climate, and the stakes have never been higher. Our climate knowledge is provisional and imperfect. Yet it is real, and it is strong, because it is supported by a global infrastructure. Climate knowledge is built on old, robust observing systems and refined predictive models, on extensive simulation experiments and model intercomparisons. Its large expert community long ago reached a stable consensus on the climate's sensitivity to greenhouse-gas emissions and on the reality of the global warming trend. That consensus has survived many rounds of intensive review from every imaginable quarter.

We have few good reasons to doubt these facts and many reasons to trust their validity. The climate's past and its future shimmer before us, but neither one is a mirage. This is the best knowledge we are going to get. We had better get busy putting it to work.

# Notes

### Introduction

1. N. Dawidoff, "The Civil Heretic," *New York Times Magazine*, March 25, 2009.

2. For periods before instrument records began, one can collect information about climate "proxies": things that change along with the climate, such as tree rings, species ranges (fossils, pollen, etc.), and glaciers. However, this book is concerned almost exclusively with the historical instrument record, which is both the most precise source of climate knowledge and the one most important for detecting anthropogenic (human-caused) global warming.

3. The term "data model" has a technical meaning in computer science, where it refers to an abstract conception of data elements, structure, representation, and relationships within a given application or workflow. My sense of the term has more in common with the "models of data" concept developed by philosophers of science. (See chapter 10.)

4. The Monitoring, Modeling and Memory Project is researching these issues in the context of several large cyberinfrastructure projects now being developed by environmental and Earth system scientists. Project members include numerous colleagues at the University of Michigan, the University of California at Los Angeles, the University of Pittsburgh, and Georgetown University. I gratefully acknowledge their collective contributions to the ideas in this book. For more information, see monmodmem.org.

5. A. Hollingsworth et al., "The Transformation of Earth-System Observations into Information of Socio-Economic Value in GEOSS," *Quarterly Journal of the Royal Meteorological Society* 131, no. 613 (2005); Group on Earth Observations, *The Full Picture* (Tudor Rose, 2007).

### Chapter 1

1. L. H. Tribe, "Technology Assessment and the Fourth Discontinuity: The Limits of Instrumental Rationality," *Southern California Law Review* 46 (1973): 617–60.

2. The similar phrase "think global, act local" first appeared in a 1915 book on city planning: P. Geddes, *The Evolution of Cities* (William and Nordgate, 1915). Brower is usually credited with its first use in the environmental context.

3. S. Jasanoff, "Image and Imagination: The Formation of Global Environmental Consciousness," in *Changing the Atmosphere: Expert Knowledge and Environmental Governance*, ed. C. Miller and P. N. Edwards (MIT Press, 2001).

4. A. C. Hughes and T. P. Hughes, eds., *Systems, Experts, and Computers: The Systems Approach in Management and Engineering, World War II and After* (MIT Press, 2000); P. N. Edwards, *The Closed World: Computers and the Politics of Discourse in Cold War America* (MIT Press, 1996).

5. D. Cosgrove, *Apollo's Eye: A Cartographic Genealogy of the Earth in the Western Imagination* (Johns Hopkins University Press, 2001); D. Held et al., *Global Transformations: Politics, Economics and Culture* (Stanford University Press, 1999).

6. A terabyte is a trillion ($10^{12}$) bytes.

7. Gallup Poll, 2001; Program on International Policy Attitudes, "Americans on the Global Warming Treaty," 2000 (www.pipa.org); IPSOS, "Les Français face au réchauffement de la planète," 2004 (www.ipsos.fr); A. Leiserowitz, "American Opinions on Global Warming," University of Oregon, Survey Research Laboratory, 2003 (osrl.uoregon.edu).

8. S. Jasanoff and M. L. Martello, eds., *Earthly Politics: Local and Global in Environmental Governance* (MIT Press, 2004); S. R. Weart, *The Discovery of Global Warming* (Harvard University Press, 2003); W. C. Clark et al., *Learning to Manage Global Environmental Risks: A Comparative History of Social Responses to Climate Change, Ozone Depletion, and Acid Rain* (MIT Press, 2001); S. Rayner and E. L. Malone, eds., *Human Choice and Climate Change: The Societal Framework* (Battelle Press, 1998); S. Jasanoff, *The Fifth Branch: Science Advisors as Policymakers* (Harvard University Press, 1990); H. N. Pollack, *Uncertain Science, Uncertain World* (Cambridge University Press, 2003).

9. P. N. Edwards, "Infrastructure and Modernity: Scales of Force, Time, and Social Organization in the History of Sociotechnical Systems," in *Modernity and Technology*, ed. T. J. Misa et al. (MIT Press, 2002).

10. Adapted from S. L. Star and K. Ruhleder, "Steps Toward an Ecology of Infrastructure: Design and Access for Large Information Spaces," *Information Systems Research* 7, no. 1 (1996): 111–134.

11. International Telecommunication Union, *Measuring the Information Society: The ICT Development Index* (2009).

12. W. Bijker et al., *The Social Construction of Technological Systems* (MIT Press, 1987); W. E. Bijker and J. Law, eds., *Shaping Technology/Building Society: Studies in Sociotechnical Change* (MIT Press, 1992); P. Blomkvisk and A. Kaijser, eds., *Den*

*Konstruerade Världen: Tekniska System i Historiskt Perspektiv* (Brutus Östlings, 1998); I. Braun and B. Joerges, *Technik ohne Grenzen* (Suhrkamp, 1994); O. Coutard, *The Governance of Large Technical Systems* (Routledge, 1999); O. Coutard et al., *Sustaining Urban Networks: The Social Diffusion of Large Technical Systems* (Routledge, 2004); A. Gras, *Les Macro-Systèmes Techniques* (Brutus Östlings, 1997); T. P. Hughes, *Networks of Power: Electrification in Western Society, 1880–1930* (Johns Hopkins University Press, 1983); T. P. Hughes, *Rescuing Prometheus* (Pantheon Books, 1998); T. P. Hughes, *Human-Built World: How to Think About Technology and Culture* (University of Chicago Press, 2004); A. Kaijser, *I Fädrens Spår: Den Svenske Infrastrukturens Historiska Utveckling och Framtida Utmaningar* (Carlssons, 1994); T. R. La Porte, ed., *Social Responses to Large Technical Systems: Control or Adaptation* (Kluwer, 1991); R. Mayntz and T. P. Hughes, *The Development of Large Technical Systems* (Westview, 1988); J. Summerton, ed., *Changing Large Technical Systems* (Westview, 1994).

13. Star and Ruhleder, "Steps Toward an Ecology of Infrastructure"; P. N. Edwards, "Y2K: Millennial Reflections on Computers as Infrastructure," *History and Technology* 15 (1998): 7–29; G. C. Bowker and S. L. Star, *Sorting Things Out: Classification and Its Consequences* (MIT Press, 1999); Edwards, "Infrastructure and Modernity"; E. van der Vleuten, "Infrastructures and Societal Change: A View From the Large Technical Systems Field," *Technology Analysis & Strategic Management* 16, no. 3 (2004): 395–414.

14. Held et al., *Global Transformations*; S. Graham and S. Marvin, *Splintering Urbanism: Networked Infrastructures, Technological Mobilities and the Urban Condition* (Routledge, 2001); G. C. Bowker, *Memory Practices in the Sciences* (MIT Press, 2005); J. Schot et al., "Tensions of Europe: The Role of Technology in the Making of Europe," special issue, *History and Technology* 21, no. 1 (2005); P. N. Edwards et al., *Understanding Infrastructure: Dynamics, Tensions, and Design* (Deep Blue, 2007); E. van der Vleuten and A. Kaijser, eds., *Networking Europe: Transnational Infrastructures and the Shaping of Europe, 1850–2000* (Science History Publications/USA, 2007).

15. Hughes, *Networks of Power*.

16. Coutard et al., *Sustaining Urban Networks*.

17. P. A. David and J. A. Bunn, "The Economics of Gateway Technologies and Network Evolution: Lessons From Electricity Supply History," *Information Economics and Policy* 3 (1988): 165–202; Edwards et al., *Understanding Infrastructure*.

18. T. Egyedi, "Infrastructure Flexibility Created by Standardized Gateways: The Cases of XML and the ISO Container," *Knowledge, Technology & Policy* 14, no. 3 (2001): 41–54.

19. This phenomenon can readily be re-described using concepts from actor-network theory (ANT). Actors seek to increase their power by extending their alliances, forming networks consisting of both human and non-human entities

(actants). Such networks can include technical as well as social links (what I am calling gateways). ANT emphasizes, correctly, the unpredictable effects of network extension. Despite its name, ANT is relentlessly anti-explanatory, having more in common with ontology than with theory in the usual sense. I do not pursue it further in this book. See J. Law, "Technology and Heterogeneous Engineering: The Case of Portuguese Expansion," in *The Social Construction of Technological Systems*, ed. W. Bijker et al. (MIT Press, 1987); B. Latour, *Reassembling the Social: An Introduction to Actor-Network-Theory* (Oxford University Press, 2005); B. Latour, *Politics of Nature: How to Bring the Sciences into Democracy* (Harvard University Press, 2004); B. Latour, *Science in Action: How to Follow Scientists and Engineers through Society* (Harvard University Press, 1987); M. Callon, "Society in the Making: The Study of Technology as a Tool for Sociological Analysis," in *The Social Construction of Technological Systems*, ed. W. Bijker et al. (MIT Press, 1987); M. Callon, "Some Elements of a Sociology of Translation: Domestication of the Scallops and the Fishermen of St. Brieuc Bay," in *Power, Action, and Belief: A New Sociology of Knowledge?*, ed. J. Law (Routledge and Kegan Paul, 1986); M. Callon and B. Latour, "Unscrewing the Big Leviathan: How Actors Macro-Structure Reality and How Sociologists Help Them to Do So," in *Advances in Social Theory and Methodology: Toward an Integration of Micro- and Macro-Sociologies*, ed. K. D. Knorr Cetina and A. V. Cicourel (Routledge and Kegan Paul, 1981).

20. An assessment of 1990s "national information infrastructure" projects captured the inherent tensions: "Information technology standards have been touted as a means to interoperability and software portability, but they are more easily lauded than built or followed. Users say they want low-cost, easily maintained, plug-and-play, interoperable systems, yet each user community has specific needs and few of them want to discard their existing systems. Every vendor wants to sell its own architecture and turbo-charged features, and each architecture assumes different views of a particular domain (e.g., business forms, images, databases). International standards founder on variations in culture and assumptions—for example, whether telephone companies are monopolies—in North America, Europe, and Asia." M. Libicki, "Standards: The Rough Road to the Common Byte," in *Standards Policy for Information Infrastructure*, ed. B. Kahin and J. Abbate (MIT Press, 1995), 35.

21. Edwards et al., *Understanding Infrastructure*.

22. S. J. Jackson et al., "Understanding Infrastructure: History, Heuristics, and Cyberinfrastructure Policy," *First Monday* 12, no. 6 (2007), www.firstmonday.org.

23. Jean-Louis Fellous, personal communication; World Meteorological Organization and Global Climate Observing System, *Second Report on the Adequacy of the Global Observing Systems for Climate in Support of the UNFCCC* (2003).

24. Science is not the only form of human knowledge, of course. Though I will not explore others here, one could consider (for example) law, medicine, or agriculture.

Also, knowledge infrastructure need not be high-tech. The material basis of traditional knowledge systems is simple, but they still rest on communally accepted, enduring mechanisms for producing, communicating, storing, and maintaining knowledge.

25. T. S. Kuhn, *The Structure of Scientific Revolutions* (University of Chicago Press, 1962).

26. B. Latour and S. Woolgar, *Laboratory Life: the Social Construction of Scientific Facts* (Sage, 1979); Callon and Latour, "Unscrewing the Big Leviathan"; B. Latour, "Give Me a Laboratory and I Will Raise the Earth," in *Science Observed*, ed. K. Knorr Cetina and M. Mulkay (Sage, 1983); Latour, *Science in Action*; B. Latour, *Aramis, or, the Love of Technology* (Harvard University Press, 1996); S. Traweek, *Beamtimes and Lifetimes: The World of High Energy Physicists* (Harvard University Press, 1988); K. Knorr Cetina, *Epistemic Cultures: How the Sciences Make Knowledge* (Harvard University Press, 1999).

27. The phrase "infrastructural inversion" comes from Bowker and Star (*Sorting Things Out*), who call infrastructural inversion a "gestalt switch," reversing figure and ground. The idea is also discussed in A. Clarke and J. H. Fujimura, eds., *The Right Tools for the Job: At Work in Twentieth-Century Life Sciences* (Princeton University Press, 1992).

28. T. C. Peterson et al., "Homogeneity Adjustments of In Situ Atmospheric Climate Data: A Review," *International Journal of Climatology* 18 (1998): 1493–94 (emphasis added).

29. National Research Council, *Future of the National Weather Service Cooperative Observer Network* (National Academy Press, 1998).

30. D. R. Easterling et al., "On the Development and Use of Homogenized Climate Datasets," *Journal of Climate* 9, no. 6 (1996): 1429–40; T. Karl et al., "Long-Term Climate Monitoring by the Global Climate Observing System (GCOS)," *Climatic Change* 31 (1995): 135–47; Peterson et al., "Homogeneity Adjustments"; R. G. Quayle et al., "Effects of Recent Thermometer Changes in the Cooperative Station Network," *Bulletin of the American Meteorological Society* 72, no. 11 (1991): 1718–23.

31. National Research Council, *Adequacy of Climate Observing Systems* (National Academy Press, 1999).

32. A. Persson, "Appendix A: The Early History of NWP," in *Atmospheric Modeling, Data Assimilation, and Predictability*, ed. E. Kalnay (Cambridge University Press, 2003); National Research Council, *Four-Dimensional Model Assimilation of Data: A Strategy for the Earth System Sciences* (National Academy Press, 1991).

33. T. R. Karl et al., eds., *Temperature Trends in the Lower Atmosphere: Steps for Understanding and Reconciling Differences* (US Climate Change Science Program, 2006), 35.

34. M. McLuhan, *The Gutenberg Galaxy: The Making of Typographic Man* (University of Toronto Press, 1962); McLuhan, *Understanding Media: The Extensions of Man* (Routledge & Kegan Paul, 1964).

35. M. Castells, *The Rise of the Network Society* (Blackwell, 2000).

36. Held et al., *Global Transformations*.

37. M. Hewson, "Did Global Governance Create Informational Globalism?," in *Approaches to Global Governance Theory*, ed. M. Hewson and T. J. Sinclair (State University of New York Press, 1999).

38. P. N. Edwards, "Meteorology as Infrastructural Globalism," *Osiris* 21, special issue on Science, Technology, and International Affairs: Historical Perspectives (2006): 229–50.

39. P. N. Edwards et al., "Monitoring, Modeling and Memory: Dynamics of Data and Knowledge in Scientific Cyberinfrastructures (Grant BCS-0827316)," US National Science Foundation, 2008. See also Bowker, *Memory Practices in the Sciences*.

40. P. Haas, *Saving the Mediterranean: The Politics of International Environmental Cooperation* (Columbia University Press, 1990); P. M. Haas, "Obtaining International Environmental Protection through Epistemic Consensus," *Millennium* 19, no. 3 (1990): 347–64.

## Chapter 2

1. M. S. Monmonier, *Air Apparent: How Meteorologists Learned to Map, Predict, and Dramatize Weather* (University of Chicago Press, 1999); E. R. Tufte, *The Visual Display of Quantitative Information* (Graphics Press, 1983); E. R. Tufte, *Visual Explanations: Images and Quantities, Evidence and Narrative* (Graphics Press, 1997); S. Turkle, *The Second Self: Computers and the Human Spirit* (Simon and Schuster, 1984).

2. Cosgrove, *Apollo's Eye*.

3. G. Dohrn-van Rossum, *History of the Hour: Clocks and Modern Temporal Orders* (University of Chicago Press, 1996).

4. A. E. Nordenskiöld et al., *Facsimile-Atlas to the Early History of Cartography With Reproductions of the Most Important Maps Printed in the XV and XVI Centuries* (P. A. Norstedt, 1889).

5. J. von Hann, *Handbook of Climatology* (Macmillan, 1903), 91.

6. E. Halley, "An Historical Account of the Trade Winds, and Monsoons, Observable in the Seas Between and Near the Tropicks, With an Attempt to Assign the Phisical Cause of the Said Winds," *Philosophical Transactions of the Royal Society of London* 16, no. 183 (1686), 153.

Notes to Chapter 2 447

7. B. Varenius, *Geographia Generalis: In Qua Affectiones Generales Telluris Explicantur* (Apud Ludovicum Elzevirium, 1650).

8. G. Hadley, "Concerning the Cause of the General Trade-Winds," *Philosophical Transactions of the Royal Society of London* 39 (1735): 58–62.

9. Example: "clear blew but yellowish in the NE" (R. Hooke, 1667), cited in F. Nebeker, *Calculating the Weather: Meteorology in the 20th Century* (Academic Press, 1995), 15–16.

10. Cosgrove, *Apollo's Eye*, 211.

11. Monmonier, *Air Apparent*, 18–23.

12. M. N. Wise, "What Can Local Circulation Explain? The Case of Helmholtz's Frog-Drawing-Machine in Berlin," *HoST: Journal of History of Science and Technology* 1 (2007), www.johost.eu.

13. Nebeker, *Calculating the Weather*.

14. Latour and Woolgar, *Laboratory Life*; National Research Council, *Bits of Power: Issues in Global Access to Scientific Data* (National Academy Press, 1997); A. Zimmerman, Data Sharing and Secondary Use of Scientific Data: Experiences of Ecologists, Ph.D. dissertation, University of Michigan, 2003; Bowker, *Memory Practices in the Sciences*.

15. Latour, "Give Me a Laboratory."

16. D. Jonsson, "Sustainable Infrasystem Synergies: A Conceptual Framework," *Journal of Urban Technology* 7, no. 3 (2000): 81–104.

17. M. F. Maury and United States Naval Observatory, *Wind and Current Charts of the North and South Atlantic* (National Observatory, 1849); M. F. Maury, *The Physical Geography of the Sea* (Harper, 1855); M. F. Maury, *The Physical Geography of the Sea, and Its Meteorology* (Harper, 1860).

18. See J. R. Fleming, *Historical Perspectives on Climate Change* (Oxford University Press, 1998), 41–43.

19. J. R. Fleming, *Meteorology in America, 1800–1870* (Johns Hopkins University Press, 1990).

20. Khrgian, *Meteorology*, 116–33.

21. F. L. Williams, *Matthew Fontaine Maury, Scientist of the Sea* (Rutgers University Press, 1963).

22. A. Buchan, "The Meteorological Results of the 'Challenger' Expedition in Relation to Physical Geography," *Proceedings of the Royal Geographical Society and Monthly Record of Geography* 13, no. 3 (1891), 138, emphasis added.

23. A. Buchan, *Report on Atmospheric Circulation Based on the Observations Made on Board H. M. S. Challenger During the Years 1873–1876, and Other Meteorological Observations* (HMSO, 1889).

24. Maury, *The Physical Geography of the Sea*.

25. H. W. Dove, *Meteorologische Untersuchungen* (Sanderischen Buchhandlung, 1837).

26. W. Ferrel, "An Essay on the Winds and Currents of the Ocean," *Nashville Journal of Medicine and Surgery* 11, no. 4–5 (1856): 287–301, 375–89.

27. J. Gleick, *Faster: The Acceleration of Just About Everything* (Pantheon Books, 1999).

28. C. Marvin, *When Old Technologies Were New: Thinking about Electric Communication in the Late Nineteenth Century* (Oxford University Press, 1988).

29. T. Standage, *The Victorian Internet: The Remarkable Story of the Telegraph and the Nineteenth Century's On-Line Pioneers* (Walker, 1998).

30. The word 'synoptic' refers to charts, forecasts, or data covering a large area, typically several tens of degrees of latitude and longitude. American Meteorological Society, *Glossary of Meteorology* (Allen, 2000).

31. Monmonier, *Air Apparent*, 41.

32. Khrgian, *Meteorology*.

33. Monmonier, *Air Apparent*, 157–68.

34. International Meteorological Committee, *Codex of Resolutions Adopted at International Meteorological Meetings, 1872–1907* (British Meteorological Office, 1909), 38–39.

35. Navigation at sea was another story. On the open ocean, accurate calculation of longitude depended on precise chronometry (D. Sobel, *Longitude: The True Story of a Lone Genius Who Solved the Greatest Scientific Problem of His Time*, Walker, 1995). Even there, though, you didn't need a universal standard time. You only needed to know the precise time at *some* fixed location, relative to which you could then calculate your present position.

36. J. R. Beniger, *The Control Revolution: Technological and Economic Origins of the Information Society* (Harvard University Press, 1986); A. D. Chandler, *The Visible Hand: The Managerial Revolution in American Business* (Belknap, 1977).

37. "Synchronous" means "same local time" (I. R. Bartky, "The Adoption of Standard Time," *Technology and Culture* 30, no. 1, 1989, 36).

38. The Metrological Society—not to be confused with the unrelated American Meteorological Society—promoted standard units of measure for both scientific and practical uses.

39. Signal Office, United States Army, *International Meteorological Observations Taken Simultaneously* (War Department, Office of the Chief Signal Office of the Army of the United States, 1882).

40. C. Abbe, "The Weather Map on the Polar Projection," *Monthly Weather Review* 42, no. 1 (1914), 37.

41. I. R. Bartky, *One Time Fits All: The Campaigns for Global Uniformity* (Stanford University Press, 2007); C. E. Stephens, *Inventing Standard Time* (National Museum of American History, Smithsonian Institution, 1983).

42. Star and Ruhleder, "Steps Toward an Ecology of Infrastructure."

43. International Meteorological Committee, *Codex of Resolutions*, 30.

44. Typewritten addendum to ibid.

45. N. Shaw, "'Summer Time' and the British Meteorological Office," *Monthly Weather Review* 43, no. 2 (1918), 76.

46. C. F. Marvin, "Diagrams Showing Conditions and Effects of the Daylight-Saving Act," *Monthly Weather Review* 46, no. 2 (1918): 76, XLVI-19–XLVI-21.

47. Shaw, "'Summer Time' and the British Meteorological Office," 77.

## Chapter 3

1. Perhaps coincidentally, timepieces were among the first items to be mass produced using these techniques. In the 1850s the Waltham Watch Company developed interchangeable parts methods that became world famous as the "American system of watch manufacture." The company produced most of the railroad chronometers in use during the time reform movement discussed above, as well as cheap watches for a burgeoning mass market.

2. Beniger, *The Control Revolution*; J. Yates, *Control Through Communication: The Rise of System in American Management* (Johns Hopkins University Press, 1989).

3. Mayntz and Hughes, *The Development of Large Technical Systems*; Hughes, *Networks of Power*.

4. K. Alder, *The Measure of All Things: The Seven-Year Odyssey and Hidden Error That Transformed the World* (Free Press, 2002).

5. T. P. Hughes, "The Evolution of Large Technological Systems," in *The Social Construction of Technological Systems*, ed. W. Bijker et al. (MIT Press, 1987), 51–82.

6. Quoted in Khrgian, *Meteorology*, 133.

7. H. Daniel, "One Hundred Years of International Co-Operation in Meteorology (1873–1973)," *WMO Bulletin* 22 (1973): 156–203.

8. Ibid.

9. Though the name International Meteorological Organization did not come into formal use until after World War I, this body is conventionally called by that name across its entire period of continuous existence, 1879–1951.

10. International Meteorological Committee, *International Meteorological Meetings, 1872–1907*.

11. W. A. McDougall, *The Heavens and the Earth: A Political History of the Space Age* (Basic Books, 1985).

12. *Convention Relating to the Regulation of Aerial Navigation* (Paris, 1919).

13. G. D. Cartwright and C. H. Sprinkle, "A History of Aeronautical Meteorology: Personal Perspectives, 1903–1995," in *Historical Essays on Meteorology, 1919–1995*, ed. J. R. Fleming (American Meteorological Society, 1996), 443–80.

14. Daniel, "One Hundred Years of International Co-Operation," 171–75.

15. For example, the Nobel Prize. See P. Forman, "Scientific Internationalism and the Weimar Physicists: The Ideology and Its Manipulation in Germany After World War I," *Isis* 64, no. 2 (1973), 154–55.

16. Forman (ibid., 154) recognized that "in certain areas of applied or cosmical physics requiring large quantities of data from diverse geographical sites, division of labor and profit sharing become practical necessities, [requiring] . . . communication, personal contact, and coordination at the international level."

17. International Meteorological Committee, *International Meteorological Meetings, 1872–1907*, 30–32.

18. Great Britain Meteorological Office and N. Shaw, *Réseau Mondial, 1910: Monthly and Annual Summaries of Pressure, Temperature, and Precipitation at Land Stations* (HMSO, 1920), iv–v.

19. "Commission for Climatology—Abridged Final Report of the First Session, Washington, 12th–25th March 1953," Commission for Climatology, WMO publication CC-1/WMO-14 (1953), 35.

20. Weather Bureau, US Dept. of Commerce, *World Weather Records*, 1921/1930–1931/1940 (Government Printing Office, 1959); H. H. Clayton, *World Weather Records* (Smithsonian Institution, 1927); H. H. Clayton, *World Weather Records* (Smithsonian Institution, 1934); H. H. Clayton, *World Weather Records* (Smithsonian Institution, 1947); World Meteorological Organization, "World Weather Records" (2003), at www.wmo.int.

21. See Daniel, "One Hundred Years of International Co-Operation."

22. N. Shaw, *Manual of Meteorology* (Cambridge University Press, 1926), 162.

## Chapter 4

1. After Nebeker, *Calculating the Weather*.

2. M. Poovey, *A History of the Modern Fact: Problems of Knowledge in the Sciences of Wealth and Society* (University of Chicago Press, 1998); T. M. Porter, *The Rise of Statistical Thinking, 1820–1900* (Princeton University Press, 1986); T. M. Porter, *Trust in Numbers: The Pursuit of Objectivity in Science and Public Life* (Princeton University Press, 1995).

3. Nebeker, *Calculating the Weather*, 21.

4. J. von Hann, *Handbuch der Klimatologie* (Engelhorn, 1883). The second German edition of this text (1897) was translated into English by Robert D. Ward as J. von Hann, *Handbook of Climatology* (Macmillan, 1903). Citations in the following discussion are of the English translation.

5. Hann, *Handbook of Climatology*, 3–4.

6. H. W. Dove, *The Distribution of Heat Over the Surface of the Globe: Illustrated by Isothermal, Thermic Isabnormal, and Other Curves of Temperature* (Taylor and Francis, 1853).

7. Hann, *Handbook of Climatology*, 4.

8. Ibid., 201–02, 411.

9. J. von Hann, *Die Erde als Ganzes, Ihre Atmosphäre und Hydrosphäre* (Tempsky, 1896).

10. Hann, *Handbook of Climatology*, 401, 403.

11. S. S. Visher, *Climatic Laws: Ninety Generalizations with Numerous Corollaries as to the Geographic Distribution of Temperature, Wind, Moisture, etc.* (Wiley, 1924), 16.

12. Hann, *Handbook of Climatology*, 2.

13. W. G. Kendrew, *The Climates of the Continents* (Oxford University Press, 1922).

14. C. S. Durst, "Climate—the Synthesis of Weather," in *Compendium of Meteorology*, ed. T. F. Malone (American Meteorological Society, 1951), 967.

15. W. G. Kendrew, *Climate: A Treatise on the Principles of Weather and Climate* (Oxford University Press, 1938), quoted in Durst, "Climate—the Synthesis of Weather," 967.

16. Abbe, "The Weather Map on the Polar Projection," 36–37.

17. See, e.g., R. P. Harnack, "An Appraisal of the Circulation and Temperature Pattern for Winter 1978–79 and a Comparison with the Two Previous Winters," *Monthly Weather Review* 108, no. 1 (1980): 37–55.

18. V. Conrad, *Methods in Climatology* (Harvard University Press, 1944); Conrad, *Methods in Climatology* (Harvard University Press, 1950); V. Conrad and L. W. Pollak, *Methods in Climatology* (Harvard University Press, 1962).

19. Conrad, *Methods in Climatology*, 1–2.

20. Ibid., 6.

21. V. Bjerknes et al., *Physikalische Hydrodynamik, mit Anwendung auf die Dynamische Meteorologie* (Springer, 1933).

22. C.-G. Rossby et al., "Relation Between Variations in the Intensity of the Zonal Circulation of the Atmosphere and the Displacement of the Semi-Permanent Centers of Action," *Journal of Marine Research* 2, no. 1 (1939): 38–55.

23. Durst, "Climate—the Synthesis of Weather," 967.

24. Reviewing climatology's received account of its own past, Fleming, Handel, Risbey, and others have corrected a number of common mistakes widely propagated for decades. The dates and attributions given here take account of these corrections. M. D. Handel and J. S. Risbey, "An Annotated Bibliography on the Greenhouse Effect and Climate Change," *Climatic Change* 21 (1992): 97–255; M. D. Handel and J. S. Risbey, "Reflections on More Than a Century of Climate Change Research," *Climatic Change* 21, no. 2 (1992): 91–96; Fleming, *Historical Perspectives on Climate Change*.

25. Weart, *The Discovery of Global Warming*. Other useful treatments from a more popular perspective include W. K. Stevens, *The Change in the Weather: People, Weather, and the Science of Climate* (Delacorte, 1999) and G. E. Christianson, *Greenhouse: The 200-Year Story of Global Warming* (Walker, 1999).

26. J. B. J. Fourier, *Théorie analytique de la chaleur* (Didot, 1822).

27. Cited in Fleming, *Historical Perspectives on Climate Change*, 65–74.

28. S. Arrhenius, "On the Influence of Carbonic Acid in the Air Upon the Temperature of the Ground," *Philosophical Magazine and Journal of Science* 41 (1896): 237–76.

29. Arrhenius's first publication expressing this view did not appear until 1903. See J. Uppenbrink, "Arrhenius and Global Warming," *Science* 272, no. 5265 (1996): 1122.

30. S. Arrhenius, 1896 lecture, cited in ibid.

31. T. C. Chamberlin, "The Influence of Great Epochs of Limestone Formation Upon the Constitution of the Atmosphere," *Journal of Geology* 6 (1898): 609–21; T. C. Chamberlin, "A Group of Hypotheses Bearing on Climatic Changes," *Journal of Geology* 5 (1897): 653–83.

32. For technical explanation, see R. J. Russell, "Climatic Change Through the Ages," in *Climate and Man* (Government Printing Office, 1941), 67–97; C. E. P. Brooks, "Geological and Historical Aspects of Climatic Change," in

*Compendium of Meteorology*, ed. T. F. Malone (American Meteorological Society, 1951), 1004–18.

33. Fleming, *Historical Perspectives on Climate Change*, 89–91, 111–112.

34. See Fleming's discussion of Ellsworth Huntington's climatic determinism in ibid.

35. M. Milanković, *Théorie mathématique des phénomènes thermiques produits par la radiation solaire* (Gauthier-Villars, 1920).

36. R. Muller and G. J. MacDonald, *Ice Ages and Astronomical Causes: Data, Spectral Analysis, and Mechanisms* (Springer, 2000), 7–13.

37. J. R. Fleming, *The Callendar Effect* (American Meteorological Society, 2007).

38. G. S. Callendar, "The Artificial Production of Carbon Dioxide and Its Influence on Temperature," *Quarterly Journal of the Royal Meteorological Society* 64, no. 275 (1938), 229.

39. Fleming, *Historical Perspectives on Climate Change*, chapter 8.

40. Callendar, "The Artificial Production of Carbon Dioxide," 232.

41. Ibid., 236.

42. Ibid., 237–40. Also see Fleming, *The Callendar Effect*, 71–76.

43. D. Brunt, *Physical and Dynamical Meteorology* (Cambridge University Press, 1939), 403, 405, emphasis added.

44. Brooks, "Geological and Historical Aspects of Climatic Change," 1016, emphasis added.

45. G. S. Callendar, "The Composition of the Atmosphere Through the Ages," *Meteorological Magazine* 74 (1939): 33–39; Callendar, "Can Carbon Dioxide Influence Climate?" *Weather* 4, no. 3 (1949): 310–14; Callendar, "On the Amount of Carbon Dioxide in the Atmosphere," *Tellus* 10, no. 2 (1958): 243–48; Callendar, "Temperature Fluctuations and Trends Over the Earth," *Quarterly Journal of the Royal Meteorological Society* 87, no. 371 (1961): 1–12.

46. Fleming, *The Callendar Effect*.

## Chapter 5

1. H. M. Collins, *Artificial Experts: Social Knowledge and Intelligent Machines* (MIT Press, 1990); Collins, "Humans, Machines, and the Structure of Knowledge," *Stanford Humanities Review* 4, no. 2 (1995): 67–83.

2. M. K. Buckland, "Information as Thing," *Journal of the American Society for Information Science* 42, no. 5 (1991): 351–60.

3. V. Bjerknes et al., *Dynamic Meteorology and Hydrography* (Carnegie Institution of Washington, 1910); V. Bjerknes, *Fields of Force: Supplementary Lectures, Applications to Meteorology* (Columbia University Press, 1906).

4. K. C. Harper, *Weather by the Numbers: The Genesis of Modern Meteorology* (MIT Press, 2008).

5. Nebeker, *Calculating the Weather*.

6. J. van den Ende, "Tidal Calculations in the Netherlands, 1920–1960," *IEEE Annals of the History of Computing* 14, no. 3 (1992): 23–33.

7. D. Fultz, "Experimental Models of Rotating Fluids and Possible Avenues for Future Research," in *Dynamics of Climate*, ed. R. L. Pfeffer (Pergamon, 1960); R. Hide, "Some Experiments on Thermal Convection in a Rotating Liquid," *Quarterly Journal of the Royal Meteorological Society* 79 (1953): 161; H.-L. Kuo, "Theoretical Findings Concerning the Effects of Heating and Rotation on the Mechanism of Energy Release in Rotating Fluid Systems," in *Dynamics of Climate*, ed. R. L. Pfeffer (Pergamon, 1960).

8. P. E. Ceruzzi, "When Computers Were Human," *IEEE Annals of the History of Computing* 13, no. 3 (1991): 237–44; D. A. Grier, *When Computers Were Human* (Princeton University Press, 2005).

9. Well into the 1960s, as I argued in chapters 2 and 3 of *The Closed World*, analog computation remained an important alternative to digital computing, for reasons involving cost, speed, reliability, and the lack of numerical methods for complex mathematical problems. For well-developed discussions of analog computing, its scientific applications, and the analog computer industry, see J. S. Small, *The Analogue Alternative: The Electronic Analogue Computer in Britain and the USA, 1930–1975* (Routledge, 2001); Small, "General-Purpose Electronic Analog Computing: 1945–1965," *IEEE Annals of the History of Computing* 15, no. 2 (1993): 8–18; A. S. Jackson, *Analog Computation* (McGraw-Hill, 1960); A. G. Bromley, "Analog Computing Devices," in *Computing Before Computers*, ed. W. Aspray (Iowa State University Press, 1990).

10. R. M. Friedman, *Appropriating the Weather: Vilhelm Bjerknes and the Construction of a Modern Meteorology* (Cornell University Press, 1989).

11. Ibid., 107, 97.

12. Millikan 1919, cited in Nebeker, *Calculating the Weather*, 84.

13. Khrgian, *Meteorology*, 120.

14. W. A. Koelsch, "From Geo- to Physical Science: Meteorology and the American University, 1919–1945," in *Historical Essays on Meteorology, 1919–1995*, ed. J. R. Fleming (American Meteorological Society, 1996).

15. Ibid.

16. N. Shaw, *The Drama of Weather* (Cambridge University Press, 1939), 254.

17. "The more or less continuous conflict between warm, moist currents, usually from the south or west, and cold dry currents from the north or east (in the Northern Hemisphere), so resembled the tide of battle along the western European battle front that the Norwegian school applied the name "front" to the boundary between different air currents—or air masses, as they are now called." F. W. Reichelderfer, "The How and Why of Weather Knowledge," in *Climate and Man* (Government Printing Office, 1941), 131. Friedman similarly remarks that the Bergen meteorologists, on the heels of the Great War, initially employed the explicitly military metaphor of a "battle line."

18. J. R. Fleming, ed. *Weathering the Storm: Sverre Petterssen, the D-Day Forecast, and the Rise of Modern Meteorology* (American Meteorological Society, 2001), 67.

19. Nebeker, *Calculating the Weather*, 57.

20. T. Bergeron, "Methods in Scientific Weather Analysis and Forecasting: An Outline on the History of Ideas and Hints at the Program," in *The Atmosphere and the Sea in Motion: Scientific Contributions to the Rossby Memorial Volume*, ed. B. Bolin (Rockefeller Institute Press, 1959).

21. Richardson's scheme was in fact far more complex than Bjerknes's, incorporating a wide range of physical theory; for a full discussion, see Nebeker, *Calculating the Weather*, chapter 6. Presciently, Richardson contemplated adding an eighth primary variable, for atmospheric dust. Dust contributes significantly to cloud formation and to the optical properties of the atmosphere. Since about 1990, the inclusion of sophisticated models of dust and other particulate aerosols has been critical in bringing the simulations of general circulation models into line with observed temperature trends.

22. L. F. Richardson, *Weather Prediction by Numerical Process* (Cambridge University Press, 1922).

23. Ibid., 219.

24. Nebeker, *Calculating the Weather*, 82.

25. F. W. Reichelderfer, "The Early Years," *Bulletin of the American Meteorological Society* 51 (1970): 206–11.

26. J. B. Kincer, "Climate and Weather Data for the United States," in *Climate and Man* (Government Printing Office, 1941), 687.

27. L. W. Pollak, "Further Remarks on Early Uses of Punched Cards in Meteorology and Climatology," *Bulletin of the American Meteorological Society* 27, no. 5 (1946), 193.

28. Nebeker, *Calculating the Weather*, 92–93; Pollak, "Further Remarks on Early Uses of Punched Cards"; M. C. George, "An Annotated Bibliography of Some Early Uses of Punched Cards in Meteorology and Climatology," *Bulletin of the American Meteorological Society* 26 (1945): 76–85; H. E. Landsberg, "On Czecho-Slovakian Pioneers in the Use of Punch Cards for Climatological Studies," *Bulletin of the American Meteorological Society* 24 (1943): 174ff.

29. Air Force Data Control Unit, *Machine Methods of Weather Statistics* (US Dept. of the Air Force, 1949), 1.

30. R. Seidel, *Los Alamos and the Development of the Atomic Bomb* (Otowi Crossing Press, 1993).

31. G. W. Platzman, "The ENIAC Computations of 1950—Gateway to Numerical Weather Prediction," *Bulletin of the American Meteorological Society* 60 (1979): 302–12.

32. Air Force Data Control Unit, *Machine Methods of Weather Statistics*, 24.

33. Ibid.

34. National Weather Records Center and G. L. Barger, *Climatology at Work: Measurements, Methods, and Machines* (US Weather Bureau, 1960), 4, 31.

35. National Oceanic and Atmospheric Administration, "Managing Legacy Climate Data in the 20th Century," 2007, celebrating200years.noaa.gov.

36. Air Force Data Control Unit, *Machine Methods of Weather Statistics*, 2.

37. Conrad, *Methods in Climatology*, 353.

38. J. C. Bellamy, "Automatic Processing of Geophysical Data," *Advances in Geophysics* 1 (1952), 18.

39. National Weather Records Center and Barger, *Climatology at Work*, 73, emphasis added.

40. The first major effort to combine computing and communication in a real-time transaction processing system was SAGE, the computerized North American continental air defense system that became operational in 1958. See Edwards, *The Closed World*; K. C. Redmond and T. M. Smith, *From Whirlwind to MITRE: The R&D Story of the SAGE Air Defense Computer* (MIT Press, 2000). A SAGE spin-off, the SABRE airline reservation system, began operating in 1964, with teletype equipment used as its principal input device; SABRE was the first commercial real-time transaction processing system. As systems for collecting and processing data from very large geographical areas in real time, SAGE and SABRE resembled what meteorologists began to hope for in that period.

41. R. G. Quayle, interviewed by P. N. Edwards, 1998.

42. G. E. Valley Jr., "How the SAGE Development Began," *Annals of the History of Computing* 7, no. 3 (1985): 196–226.

43. In 1876 James Thomson (brother of Lord Kelvin) invented a differential analyzer. It solved wave harmonic equations, using gears, pulleys, and cables to model formulas such as $y = A\cos(u) + B\cos(v) + C\cos(w)$. A similar machine for predicting tides, built for the US Coast and Geodetic Survey in 1911, solved for 37 terms and made predictions accurate to 1.5 inches. It remained faster than the best digital equivalent until the mid 1960s, when it was replaced by an IBM computer (Williams, *A History of Computing Technology*).

44. C. L. Mitchell and H. Wexler, "How the Daily Forecast Is Made," in *Climate and Man* (Government Printing Office, 1941), 582.

45. National Weather Records Center and Barger, *Climatology at Work*, 18.

46. H. C. Willett, quoted in A. F. Spilhaus, "World Weather Network," in *Compendium of Meteorology*, ed. T. F. Malone (American Meteorological Society, 1951), 709.

47. G. R. Jenkins, "Transmission and Plotting of Meteorological Data," in *Handbook of Meteorology*, ed. F. A. Berry Jr. et al. (McGraw-Hill, 1945), 574.

## Chapter 6

1. In chapter 1 of *The Closed World*, I argued that in a very real sense World War II did not end with Germany's surrender. Instead it simply morphed into the Cold War, an equally stark contest pairing a different set of enemies and involving much of the rest of the world.

2. Edwards, *The Closed World*, chapter 3.

3. J. J. Tribbia and R. A. Anthes, "Scientific Basis of Modern Weather Prediction," *Science* 237, no. 4814 (1987): 493–99.

4. Quoted in J. R. Fleming, "Fixing the Weather and Climate: Military and Civilian Schemes for Cloud Seeding and Climate Engineering," in *The Technological Fix: How People Use Technology to Create and Solve Problems*, ed. L. Rosner (Routledge Taylor & Francis, 2004), 175.

5. "The Weather Weapon: New Race with the Reds," *Newsweek*, January 13, 1958: 54.

6. H. H. Goldstine and J. von Neumann, "On the Principles of Large Scale Computing Machines," in *John Von Neumann: Collected Works*, ed. A. H. Taub (Pergamon, 1961); J. von Neumann et al., "Preliminary Discussion of the Logical Design of an Electronic Computing Instrument: Report Prepared for the Dept. Of Army Ordnance (1946)," in *John Von Neumann: Collected Works*; H. Goldstine, *The Computer From Pascal to Von Neumann* (Princeton University Press, 1972).

7. In addition to his work on shock waves and aerodynamics for Los Alamos, von Neumann was "consultant to BRL [Ballistics Research Laboratory] on hydrodynamical and aerodynamical problems associated with projectile motion." J. G. Brainerd, ENIAC project supervisor, cited in W. Aspray, *John Von Neumann and the Origins of Modern Computing* (MIT Press, 1990), 37.

8. Quoted in ibid., 247. On von Neumann's political views, see S. Heims, *John von Neumann and Norbert Wiener: From Mathematics to the Technologies of Life and Death* (MIT Press, 1980).

9. Priority debates still rage over pioneer computers. The British code-breaking group at Bletchley Park completed Colossus II before the ENIAC was finished; the Colossus II was reputedly more powerful than the American machine ("Colossal Code of Silence Broken," *The Telegraph*, London, July 9, 2000). Derailed by the German defeat, the various computers designed by Konrad Zuse before, during, and after the war failed to take root, and Zuse labored in relative obscurity for decades before being rediscovered by historians (H. Zuse, "The Life and Work of Konrad Zuse," *EPE Online*, www.epemag.com).

10. A. R. Burks and A. W. Burks, "The ENIAC: The First General Purpose Electronic Computer," *Annals of the History of Computing* 3, no. 4 (1981): 310–99.

11. J. von Neumann, "First Draft of a Report on the EDVAC," *IEEE Annals of the History of Computing* 15, no. 4 (1993): 27–75.

12. ENIAC's designers, John Mauchly and J. Presper Eckert Jr., also (correctly) claimed credit for the stored-program idea, which became an object of bitter disputes, including lawsuits. As a result, von Neumann parted ways with Eckert and Mauchly, who left academia to found UNIVAC, one of the first commercial computer companies. N. Stern, *From ENIAC to Univac: An Appraisal of the Eckert-Mauchly Computers* (Digital Press, 1981).

13. See Aspray, *John von Neumann and the Origins of Modern Computing*, 152.

14. Quoted in Nebeker, *Calculating the Weather*, 136, emphasis added.

15. P. L. Galison, "Computer Simulations and the Trading Zone," in *The Disunity of Science: Boundaries, Contexts, and Power*, ed. P. L. Galison and D. J. Stump (Stanford University Press, 1996).

16. J. M. Stagg, *Forecast for Overlord, June 6, 1944* (Allan, 1971); S. Petterssen, ed. J. R. Fleming, *Weathering the Storm: Sverre Petterssen, the D-Day Forecast, and the Rise of Modern Meteorology* (American Meteorological Society, 2001).

17. Nebeker, *Calculating the Weather*, 113–15.

18. Petterssen, *Weathering the Storm*, chapter 18.

19. K. Harper, *Weather by the Numbers: The Genesis of Modern Meteorology* (MIT Press, 2008), 97–99.

20. S. Shalett, "Electronics to Aid Weather Figuring," *New York Times*, January 11, 1946.

21. W. Winter, "We Can Control the Weather: The Electronic Computer Makes It Possible, Says Dr. Zworykin, Scientist," *Mechanix Illustrated*, January 1948: 68–71, 154; J. von Neumann, "Can We Survive Technology?" *Fortune*, June 1955: 106–08, 151–52.

22. C. Kwa, "Modelling Technologies of Control," *Science as Culture* 4, no. 20 (1994): 363–91; Kwa, "The Rise and Fall of Weather Modification: Changes in American Attitudes Towards Technology, Nature, and Society," in *Changing the Atmosphere: Expert Knowledge and Environmental Governance*, ed. C. A. Miller and P. N. Edwards (MIT Press, 2001).

23. K. C. Harper, "Research from the Boundary Layer: Civilian Leadership, Military Funding and the Development of Numerical Weather Prediction (1946–55)," *Social Studies of Science* 33, no. 5 (2003), 673.

24. Source: 1946 IAS grant proposal, reprinted in P. D. Thompson, "A History of Numerical Weather Prediction in the United States," *Bulletin of the American Meteorological Society* 64 (1983), 766.

25. Rossby chaired a 1947 Panel on Meteorology for the Joint Research and Development Board of the Department of Defense, which recommended an "intensive research and development effort" on "artificially induced precipitation" (Fleming, "Fixing the Weather and Climate").

26. J. G. Charney et al., "Numerical Integration of the Barotropic Vorticity Equation," *Tellus* 2, no. 4 (1950), 237.

27. Meteorology Project Report, quoted in Aspray, *John von Neumann*, 302.

28. Platzman, "The ENIAC Computations of 1950."

29. Ibid.

30. A pressure level is not an altitude, but an imaginary isobaric surface of equal pressure, whose height varies.

31. W. C. Palmer, "On Forecasting the Direction of Movement of Winter Cyclones," *Monthly Weather Review* 76, no. 9 (1948): 181–201.

32. Rossby et al., "Relation Between Variations"; Rossby, "Planetary Flow Patterns in the Atmosphere," *Quarterly Journal of the Royal Meteorological Society* 66 (1940): 68–87; Rossby, "The Scientific Basis of Modern Meteorology," in *Climate and Man* (Government Printing Office, 1941).

33. Vorticity is a measure of the vertical velocity component of winds rotating around a vertical axis. The barotropic vorticity equation is a simplified form of the full vorticity equation. For a concise, semi-technical discussion of these ideas, see

A. Persson, "Early Operational Numerical Weather Prediction Outside the USA: An Historical Introduction: Part I: Internationalism and Engineering NWP in Sweden, 1952–69," *Meteorological Applications* 12, no. 3 (2005): 135–59. For a clear nontechnical discussion, see Harper, *Weather by the Numbers*.

34. K. C. Harper, "Bridging the Gap: The US Move From Research to Operational NWP," in *Symposium on the 50th Anniversary of Operational Numerical Weather Prediction* (American Meteorological Society, 2004).

35. A. Wiin-Nielsen, "The Birth of Numerical Weather Prediction," *Tellus* 43AB, no. 4 (1991): 36–52.

36. Charney et al., "Numerical Integration of the Barotropic Vorticity Equation," 245.

37. Platzman, "The ENIAC Computations of 1950," 311.

38. Charney et al., "Numerical Integration of the Barotropic Vorticity Equation," 245.

39. Ibid., 247.

40. Ibid., 245.

41. Source: 1946 IAS grant proposal, reprinted in Thompson, "A History of Numerical Weather Prediction in the United States," 766.

42. P. Thompson, unpublished manuscript, quoted in G. P. Cressman, "The Origin and Rise of Numerical Weather Prediction," in *Historical Essays on Meteorology, 1919–1995*, ed. J. R. Fleming (American Meteorological Society, 1996), 34.

43. F. H. Bushby and M. K. Hinds, "Electronic Computation of the Field of Atmospheric Development," *Meteorological Magazine* 82, no. 977 (1953): 330–31; J. S. Sawyer, "Electronic Computing Machines and Meteorology," *Meteorological Magazine* 81, no. 957 (1952): 74–77; British Meteorological Office, "Meteorological Office Discussion: Dynamical Forecasting By Numerical Methods," *Meteorological Magazine* 83, no. 984 (1954): 175–82.

44. N. Phillips et al., "Numerical Weather Prediction in the Soviet Union," *Bulletin of the American Meteorological Society* 41, no. 11 (1960): 599–617.

45. G. D. Crowe and S. E. Goodman, "S. A. Lebedev and the Birth of Soviet Computing," *IEEE Annals of the History of Computing* 16, no. 1 (1994): 4–24.

46. Phillips et al., "Numerical Weather Prediction in the Soviet Union."

47. On NWP outside the United States, see the following: Persson, "Early Operational Numerical Weather Prediction Outside the USA: Part I"; Persson, "Early Operational Numerical Weather Prediction Outside the USA: An Historical Introduction—Part II: Twenty Countries Around the World," *Meteorological Applications* 12, no. 3 (2005):

269–89; Persson, "Early Operational Numerical Weather Prediction Outside the USA: An Historical Introduction—Part III: Endurance and Mathematics—British NWP, 1948–1965," *Meteorological Applications* 12, no. 4 (2005): 381–413.

48. IAS Meteorology Project report for 1952, quoted in J. Smagorinsky, "The Beginnings of Numerical Weather Prediction and General Circulation Modeling: Early Recollections," *Advances in Geophysics* 25 (1983), 14.

49. F. G. Shuman and J. B. Hovermale, "An Operational Six-Layer Primitive Equation Model," *Journal of Applied Meteorology* 7, no. 4 (1968), 527.

50. K. C. Harper, "The Scandinavian Tag-Team: Providers of Atmospheric Reality to Numerical Weather Prediction Efforts in the United States (1948–1955)," in *Proceedings of the XXI International Congress of History of Science* (Mexico City, 2004); Harper, *Weather by the Numbers*.

51. A. Carlsson, "Computer Technology in Sweden," *Uppsala Newsletter: History of Science* 29 (2000).

52. P. Bergthorsson et al., "Routine Forecasting with the Barotropic Model," *Tellus* 7, no. 2 (1955): 272–76; Staff Members of the University of Stockholm Institute of Meteorology, "Results of Forecasting with the Barotropic Model on an Electronic Computer (BESK)," *Tellus* 6 (1954): 139–49.

53. Smagorinsky, "The Beginnings of Numerical Weather Prediction," 22–25.

54. Cressman, "The Origin and Rise of Numerical Weather Prediction," 30.

55. H. C. Willett, "The Forecast Problem," in *Compendium of Meteorology*, ed. T. F. Malone (American Meteorological Society, 1951), 731.

56. R. J. Reed, "The Development and Status of Modern Weather Prediction," *Bulletin American Meteorological Society* 58, no. 5 (1977), 395.

57. F. G. Shuman, "History of Numerical Weather Prediction at the National Meteorological Center," *Weather and Forecasting* 4 (1989), 291–292, 288.

58. P. J. Roebber and L. F. Bosart, "The Contributions of Education and Experience to Forecast Skill," *Weather and Forecasting* 11 (1996), 39.

59. P. Daipha, Masters of Uncertainty: Weather Forecasters and the Quest for Ground Truth, PhD dissertation, University of Chicago, 2007, 79. See also G. A. Fine, "Ground Truth: Verification Games in Operational Meteorology," *Journal of Contemporary Ethnography* 35, no. 1 (2006): 3; Fine, *Authors of the Storm: Meteorologists and the Culture of Prediction* (University of Chicago Press, 2007).

60. R. R. Turner, "Teaching the Weather Cadet Generation: Aviation, Pedagogy, and Aspirations to a Universal Meteorology in America, 1920–1950," in *Intimate Universality: Local and Global Themes in the History of Weather and Climate*, ed. J. R. Fleming et al. (Science History Publications, 2006), 142.

61. Edwards, *The Closed World*; K. Flamm, *Creating the Computer: Government, Industry, and High Technology* (Brookings Institution, 1988); Small, *The Analogue Alternative*.

62. Quoted in Smagorinsky, "The Beginnings of Numerical Weather Prediction," 19, emphasis added.

63. Turner, "Teaching the Weather Cadet Generation."

64. B. Varney, 1940, quoted in Koelsch, "From Geo- to Physical Science," 528.

65. Ibid., 530.

66. National Research Council Committee on Atmospheric Sciences, "The Status of Research and Manpower in Meteorology," *Bulletin of the American Meteorological Society* 41, no. 10 (1960), 556.

67. National Research Council, "Meteorology on the Move: A Progress Report of the Committee on Atmospheric Sciences," *Bulletin of the American Meteorological Society* 41, no. 12 (1960), 662.

68. Latour, *Science in Action*.

69. Because of price inflation, $1.5 million in 1964 equaled over $10 million in 2008.

70. P. M. Wolff and W. E. Hubert, "Selection of Computer Systems for Economical Meteorological Operations," *Bulletin of the American Meteorological Society* 45 (1964), 643.

71. D. Mackenzie, "The Influence of the Los Alamos and Livermore National Laboratories on the Development of Supercomputing," *IEEE Annals of the History of Computing* 13 (1991): 179–201.

**Chapter 7**

1. R. Revelle and H. E. Suess, "Carbon Dioxide Exchange Between the Atmosphere and Ocean and the Question of an Increase of Atmospheric $CO_2$ During the Past Decades," *Tellus* 9, no. 1 (1957): 18–27.

2. Others, including Callendar, had already used the "experiment" metaphor, but Revelle's statement became the best known.

3. John von Neumann, quoted in Nebeker, *Calculating the Weather*, 136; see also Galison, "Computer Simulations and the Trading Zone."

4. Rossby, "The Scientific Basis of Modern Meteorology."

5. A weaker, subtropical jet stream also appears at around 30° latitude, the boundary between the Ferrel cell and the tropical Hadley cell. The jet streams occur in both hemispheres.

## Notes to Chapter 7

6. Sometimes the abbreviation is also said to stand for "global climate model," but this applies only to long-term integrations.

7. "Proposal for a Project on the Dynamics of the General Circulation," 1955, reproduced in Smagorinsky, "Early Recollections," 31–33.

8. J. Gleick, *Chaos: Making a New Science* (Viking, 1987); E. N. Lorenz, "Deterministic Nonperiodic Flow," *Journal of the Atmospheric Sciences* 20, no. 2 (1963): 130–48; Lorenz, *The Essence of Chaos* (University of Washington Press, 1993).

9. Lorenz, "Deterministic Nonperiodic Flow"; Lorenz, "A Study of the General Circulation and a Possible Theory Suggested By It," in *Scientific Proceedings of the International Association of Meteorology* (Rome, 1954).

10. Lorenz, "Study of the General Circulation," 603–04.

11. V. P. Starr, "What Constitutes Our New Outlook on the General Circulation?," *Journal of the Meteorological Society of Japan* 36, no. 5 (1958), 168–169, sentences re-ordered and emphasis added.

12. Nebeker, *Calculating the Weather*. See also chapter 4.

13. S. Arrhenius, "On the Influence of Carbonic Acid"; Uppenbrink, "Arrhenius and Global Warming."

14. Callendar, "The Artificial Production of Carbon Dioxide."

15. S. H. Schneider and R. E. Dickinson, "Climate Modeling," *Reviews of Geophysics and Space Physics* 12, no. 3 (1974), 456–60. For a more complex version of the model hierarchy, see K. McGuffie and A. Henderson-Sellers, *A Climate Modeling Primer* (Wiley, 2005).

16. S. Manabe and F. Möller, "On the Radiative Equilibrium and Heat Balance of the Atmosphere," *Monthly Weather Review* 89 (1961): 503–32; S. Manabe and R. F. Strickler, "Thermal Equilibrium of the Atmosphere with a Convective Adjustment," *Journal of the Atmospheric Sciences* 21 (1964): 361–85; S. Manabe and R. Wetherald, "Thermal Equilibrium of the Atmosphere with a Given Distribution of Relative Humidity," *Journal of the Atmospheric Sciences* 24 (1967): 241–59.

17. McGuffie and Henderson-Sellers, *A Climate Modeling Primer*, chapter 2.

18. M. Ghil and A. W. Robertson, "Solving Problems with GCMs: General Circulation Models and Their Role in the Climate Modeling Hierarchy," in *General Circulation Model Development: Past, Present, and Future*, ed. D. A. Randall (Academic Press, 2000), 285–326; J. T. Houghton et al., *An Introduction to Simple Climate Models Used in the IPCC Second Assessment Report: IPCC Technical Report* (Intergovernmental Panel on Climate Change, 1997).

19. For a longer discussion, see S. R. Weart, *The Discovery of Global Warming: A Hypertext History of How Scientists Came to (Partly) Understand What People Are Doing to Cause Climate Change* (American Institute of Physics, 2008), www.aip.org.

20. N. A. Phillips, "The General Circulation of the Atmosphere: A Numerical Experiment," *Quarterly Journal of the Royal Meteorological Society* 82, no. 352 (1956): 123–64.

21. J. M. Lewis, "Clarifying the Dynamics of the General Circulation: Phillips's 1956 Experiment," *Bulletin of the American Meteorological Society* 79, no. 1 (1998): 39–60.

22. Phillips, "The General Circulation of the Atmosphere," 123.

23. Ibid., 124, partially paraphrased.

24. These and other questions Phillips raised lead into technical terrain that is beyond the scope of this book, so I will discuss only a few of them, with the goal of imparting the flavor of this work rather than its details. For an excellent semitechnical discussion and assessment, see Lewis, "Clarifying the Dynamics of the General Circulation."

25. Lewis ("Clarifying the Dynamics," 48) notes that, according to Aksel Wiin-Nielsen, Phillips would have had to run his model for a simulated 14 years in order for it to reach equilibrium. Only much later would computer power be sufficient to allow simulations lasting so long.

26. Lewis, "Clarifying the Dynamics," 54.

27. A. Arakawa, "A Personal Perspective on the Early Years of General Circulation Modeling At UCLA," in *General Circulation Model Development*, ed. D. A. Randall (Academic Press, 2000), 10.

28. Smagorinsky, "Early Recollections."

29. J. Smagorinsky, "On the Numerical Integration of the Primitive Equations of Motion for Baroclinic Flow in a Closed Region," *Monthly Weather Review* 86, no. 12 (1958): 457–66.

30. P. N. Edwards, interviews with Syukuro Manabe (American Institute of Physics, 1998) and Richard Wetherald (1995, 1998, unpublished).

31. J. Smagorinsky et al., "Numerical Results from a Nine-Level General Circulation Model of the Atmosphere," *Monthly Weather Review* 93 (1965): 727–68. Manabe and Wetherald, "Thermal Equilibrium of the Atmosphere."

32. Smagorinsky et al., "Numerical Results from a Nine-Level General Circulation Model," 728.

33. S. Shackley, "Epistemic Lifestyles in Climate Change Modeling," in *Changing the Atmosphere: Expert Knowledge and Environmental Governance*, ed. C. A. Miller and P. N. Edwards (MIT Press, 2001); Edwards, interviews.

34. Smagorinsky, "Early Recollections."

35. S. Manabe and K. Bryan, "Climate Calculations with a Combined Ocean-Atmosphere Model," *Journal of the Atmospheric Sciences* 26, no. 4 (1969): 786–89.

36. S. Manabe et al., "A Global Ocean-Atmosphere Climate Model: Part I. The Atmospheric Circulation," *Journal of Physical Oceanography* 5, no. 1 (1975): 3–29.

37. S. Manabe, "The Dependence of Atmospheric Temperature on the Concentration of Carbon Dioxide," in *Global Effects of Environmental Pollution*, ed. S. F. Singer (Reidel, 1970); S. Manabe, "Estimates of Future Change of Climate Due to the Increase of Carbon Dioxide," in *Man's Impact on the Climate*, ed. W. H. Mathews et al. (MIT Press, 1971).

38. Manabe and Bryan, "Climate Calculations with a Combined Ocean-Atmosphere Model."

39. S. Manabe and R. J. Stouffer, "Multiple-Century Response of a Coupled Ocean-Atmosphere Model to an Increase of Atmospheric Carbon Dioxide," *Journal of Climate* 7 (1994): 5–23.

40. "In Memoriam—Yale Mintz, Atmospheric Sciences: Los Angeles," University of California, 1993, www.calisphere.universityofcalifornia.edu.

41. Y. Mintz, "Design of Some Numerical General Circulation Experiments," *Bulletin of the Research Council of Israel* 76 (1958): 67–114.

42. D. R. Johnson and A. Arakawa, "On the Scientific Contributions and Insight of Professor Yale Mintz," *Journal of Climate* 9, no. 12 (1996): 3211–24.

43. A. Arakawa, oral history interview with P. N. Edwards (American Institute of Physics, 1997); Arakawa, "A Personal Perspective on the Early Years of General Circulation Modeling At UCLA," in *General Circulation Model Development: Past, Present, and Future*, ed. D. A. Randall (Academic Press, 2000).

44. C. E. Leith, oral history interview with P. N. Edwards (American Institute of Physics, 1997).

45. C. E. Leith, "Numerical Simulation of the Earth's Atmosphere," in *Methods in Computational Physics*, ed. B. Alder et al. (Academic Press, 1965), 1–28; Leith, oral history interview.

46. Leith, oral history interview.

47. W. M. Washington, oral history interview with P. N. Edwards (American Institute of Physics, 1998.

48. Ibid.

49. This phrase refers to the sigma coordinate system, in which the vertical coordinate is pressure normalized by its value at the surface (as opposed to its value relative to sea level). This system simplifies the model equations compared to using sea-level

pressure or height as the vertical coordinate, making it easier to introduce land surface topography into the models.

50. Kurihara, "Numerical Integration of the Primitive Equations on a Spherical Grid"; R. Sadourny et al., "Integration of the Nondivergent Barotopic Vorticity Equation with an Icosahedral-Hexagonal Grid for the Sphere," *Monthly Weather Review* 96, no. 6 (1968): 351–56.

51. I. Silberman, "Planetary Waves in the Atmosphere," *Journal of the Atmospheric Sciences* 11, no. 1 (1954): 27–34; G. W. Platzman, "The Spectral Form of the Vorticity Equation," *Journal of Meteorology* 17, December (1960): 635–44.

52. Manabe, oral history interview.

53. A. J. Robert, "The Integration of a Spectral Model of the Atmosphere By the Implicit Method," in *Proceedings of the WMO IUGG Symposium on Numerical Weather Prediction in Tokyo, Japan, November 26–December 4, 1968* (Meteorological Society of Japan, 1969), VII-9–VII-24; S. A. Orszag, "Transform Method for the Calculation of Vector-Coupled Sums: Application to the Spectral Form of the Vorticity Equation," *Journal of the Atmospheric Sciences* 27 (1970): 890–95; Orszag, "Fourier Series on Spheres," *Monthly Weather Review* 102, no. 1 (1974): 56–75; E. Eliasen et al., *On a Numerical Method for Integration of the Hydrodynamical Equations with a Spectral Representation of the Horizontal Fields*, Report No. 2, Institut for Teoretisk Meteorologi, Köbenhavns Universitet, 1970; W. Bourke et al., "Global Modeling of Atmospheric Flow By Spectral Methods," in *General Circulation Models of the Atmosphere*, ed. J. Chang (Academic Press, 1977); Bourke, "A Multi-Level Spectral Model. I."

54. E. Eliasen et al., *On a Numerical Method for Integration of the Hydrodynamical Equations with a Spectral Representation of the Horizontal Fields*, Report No. 2, Institut for Teoretisk Meteorologi, Köbenhavns Universitet, Denmark (1970); Orszag, "Transform Method."

55. P. E. Merilees, ed., *Proceedings of the Ninth Stanstead Seminar* (Department of Meteorology, McGill University, 1972).

56. W. Bourke, "History of NWP in Australia—1970 to the Present," in *The Past, Present and Future of Numerical Modeling: The 16th BMRC Modeling Workshop* (Melbourne, 2004), 4, note 1.

57. D. J. Gauntlett and D. R. Hincksman, "A Six-Level Primitive Equation Model Suitable for Extended Operational Prediction in the Southern Hemisphere," *Journal of Applied Meteorology* 10, no. 4 (1971): 613–25.

58. Bourke, "History of NWP in Australia."

59. Bourke et al., "Global Modeling of Atmospheric Flow"; McAvaney et al., "A Global Spectral Model."

60. B. McAvaney, interview with P. N. Edwards, 2001.

61. Ibid.

62. A. Woods, *Medium-Range Weather Prediction: The European Approach* (Springer, 2006), chapter 7.

63. World Climate Research Programme Working Group on Numerical Experimentation, "Atmospheric Model Intercomparison Project" (Lawrence Livermore National Laboratories, Program for Climate Model Diagnosis and Intercomparison: 1999), www-pcmdi.llnl.gov.

64. C. Covey et al., "An Overview of Results from the Coupled Model Intercomparison Project," *Global and Planetary Change* 37, no. 1–2 (2003): 103–33.

65. To make supercomputer power available to academic institutions, the National Science Foundation funded five supercomputer centers in the 1980s. The agency required each center to provide computer time and networking support, free of charge, for any institution that wished to use its facilities. Without this support virtually no academic laboratories could have afforded the most advanced computers. The NSFNET, created for accessing the supercomputer centers, went on to become the backbone of the Internet in the United States. See J. Abbate, *Inventing the Internet* (MIT Press, 1999).

66. W. L. Gates, *AMIP: The Atmospheric Model Intercomparison Project*, Report 7, Program for Climate Model Diagnosis and Intercomparison, Lawrence Livermore National Laboratory, 1997.

67. NCAR Supercomputer Services Group, "Supercomputer Services Group—History" (University Corporation for Atmospheric Research, 2004), www.cisl.ucar.edu.

68. In the 1950s and the early 1960s, most computing facilities developed their own software in house. By the mid 1970s, however, computer users in most fields could assume that computer manufacturers and third-party software developers would provide operating systems and packaged software for many basic operations. Supercomputing was an exception to this rule.

69. B. Elzen and D. MacKenzie, "The Social Limits of Speed: The Development and Use of Supercomputers," *IEEE Annals of the History of Computing* 16, no. 1 (1994), 50–51.

70. See e.g. Digital Atmosphere by Weather Graphics Technologies (www.weathergraphics.com), or nCast and eCast by Atmospheric and Environmental Research, Inc. (www.aer.com). In general such desktop or workstation models use data generated by analysis models at major forecast centers such as ECMWF.

71. B. Henson, "What Next for SCD? Buzbee's Thoughts," 1997, www.ucar.edu.

72. US Federal Circuit Court of Appeals, *98–1020, NEC Corporation and HNSX Supercomputers, Inc. vs. United States and Dept. of Commerce, 98-1020* (1998).

73. R. A. Anthes, "President's Report for October 1999 UCAR Board of Trustees, Member Representatives and University Relations Committee Meetings" (University Corporation for Atmospheric Research, 1999), www.ucar.edu.

74. See e.g. NCAR Scientific Computing Division Digital Information Group, "History of Supercomputing at the National Center for Atmospheric Research" (University Corporation for Atmospheric Research, 2002), www.cisl.ucar.edu.

75. A computer "word" is made up of a number of "bytes." Historically, one byte represented one character (a number, letter, punctuation mark, etc.). Bytes have ranged in length from five to twelve bits, depending on the size of the character set to be represented. Typical personal computers today use a word length of 32 or 64 bits, corresponding respectively to four or eight eight-bit bytes.

76. D. MacKenzie, "Negotiating Arithmetic, Constructing Proof: The Sociology of Mathematics and Information Technology," *Social Studies of Science* 23, no. 1 (1993): 37–65. See also Collins, *Artificial Experts*.

77. D. Price, "Pentium FDIV Flaw: Lessons Learned," *IEEE Micro* 15, no. 2 (1995): 86–88.

78. D. L. Williamson and W. M. Washington, "On the Importance of Precision for Short-Range Forecasting and Climate Simulation," *Journal of Applied Meteorology* 12, no. 8 (1973): 1254–58.

79. The other major focus of modeling efforts was cumulus convection, the process responsible for the well-defined, puffy clouds such as thunderheads. In a tacit division of labor, Manabe's group at GFDL focused on radiative transfer, while Arakawa's group at UCLA specialized in understanding and parameterizing cumulus convection. See A. Arakawa and W. H. Schubert, "Interaction of a Cumulus Cloud Ensemble with the Large-Scale Environment, Part I," *Journal of the Atmospheric Sciences* 31 (1974): 674–701; S. J. Lord and A. Arakawa, "Interaction of a Cumulus Cloud Ensemble with the Large-Scale Environment. Part II," *Journal of the Atmospheric Sciences* 37 (1980): 2677–92; Arakawa, "A Personal History"; Arakawa, "The Cumulus Parameterization Problem: Past, Present, and Future," *Journal of Climate* 17, no. 13 (2004): 2493–525. I will not discuss this further, except to note that the problem of cumulus convection remains a major area of climate modeling research.

80. Callendar, "Can Carbon Dioxide Influence Climate?"; Callendar, "On the Amount of Carbon Dioxide in the Atmosphere"; Callendar, "Temperature Fluctuations and Trends Over the Earth." For a full account, see Fleming, *The Callendar Effect*.

81. G. N. Plass, "Effect of Carbon Dioxide Variations on Climate," *American Journal of Physics* 24 (1956): 376; Plass, "The Carbon Dioxide Theory of Climate Change," *Tellus* 8, no. 2 (1956): 140–54; Plass, "The Influence of the 15μ Carbon-Dioxide Band

on the Atmospheric Infrared Cooling Rate," *Quarterly Journal of the Royal Meteorological Society* 82, no. 353 (1956): 310–24.

82. Plass, "Effect of Carbon Dioxide Variations on Climate," 141–42.

83. Revelle and Suess, "Carbon Dioxide Exchange."

84. Manabe, oral history interview, edited for readability and emphasis added.

85. F. Möller, "On the Influence of Changes in the $CO_2$ Concentration in Air on the Radiation Balance of the Earth's Surface and on the Climate," *Journal of Geophysical Research* 68 (1963): 3877–86.

86. Manabe and Möller, "On the Radiative Equilibrium and Heat Balance of the Atmosphere"; Manabe and Strickler, "Thermal Equilibrium of the Atmosphere"; S. Manabe et al., "Simulated Climatology of General Circulation with a Hydrologic Cycle," *Monthly Weather Review* 93, no. December (1965): 769–98; Manabe and R. Wetherald, "Thermal Equilibrium of the Atmosphere."

87. Manabe, "Estimates of Future Change."

88. S. Manabe and R. T. Wetherald, "The Effects of Doubling the $CO_2$ Concentration on the Climate of a General Circulation Model," *Journal of Atmospheric Sciences* 32, no. 1 (1975): 3–15.

89. S. H. Schneider, "On the Carbon Dioxide–Climate Confusion," *Journal of the Atmospheric Sciences* 32, no. 11 (1975): 2060–66.

90. Kuhn, *The Structure of Scientific Revolutions*.

91. Smagorinsky et al., "Numerical Results from a Nine-Level General Circulation Model."

## Chapter 8

1. A. F. Spilhaus, "Progress in Meteorological Instrumentation, 1920–1950," *Bulletin of the American Meteorological Society* 31 (1950), 363.

2. Edwards, *The Closed World*; G. Hecht and P. N. Edwards, *The Technopolitics of Cold War: Toward a Transregional Perspective* (American Historical Association, 2008).

3. Cited in Kwa, "The Rise and Fall of Weather Modification," 141–42.

4. See ibid.; see also Fleming, "Fixing the Weather and Climate."

5. J. F. Fuller, *Thor's Legions: Weather Support to the US Air Force and Army, 1937–1987* (American Meteorological Society, 1990), 363; National Research Council Committee on Atmospheric Sciences, "The Status of Research and Manpower in Meteorology," *Bulletin of the American Meteorological Society* 41, no. 10 (1960), 555–57.

6. National Weather Records Center, *Climatology at Work*, 3.

7. A. Moyers, "The Backbone of Air Force Forecasting: Air Weather Service Integration of Joint Numerical Weather Prediction Unit Products, 1954–1959," in *Symposium on the 50th Anniversary of Operational Numerical Weather Prediction* (College Park, 2004).

8. Fuller, *Thor's Legions*.

9. R. C. Hall, "A History of the Military Polar Orbiting Meteorological Satellite Program," *Quest* 9, no. 2 (2002): 4–25.

10. R. Jenne, "Observations for China, and US-China Exchanges, Part 2 (RJ0106)," National Center for Atmospheric Research, 2001, dss.ucar.edu.

11. Richard Davis, interview with P. N. Edwards, 1998.

12. C. A. Miller, "Scientific Internationalism in American Foreign Policy: The Case of Meteorology, 1947–1958," in *Changing the Atmosphere: Expert Knowledge and Environmental Governance*, ed. C. A. Miller and P. N. Edwards (MIT Press, 2001), 189, emphasis added.

13. World Meteorological Organization, *Basic Documents (Excluding the Technical Regulations) [excerpts]*, Basic Documents No. 1 (Secretariat of the World Meteorological Organization, 1971), 9.

14. Defined in article 3(c) as a nation's "being fully responsible for the conduct of its international relations."

15. For membership figures, see Daniel, "One Hundred Years of International Co-Operation"; S. A. Davies, *Forty Years of Progress and Achievement* (World Meteorological Organization, 1990).

16. Davies, ibid., 161.

17. For Ramirez, as for other members of the Communist bloc, this category included individual republics of the Soviet Union, which were still arguing (unsuccessfully) for separate representation at the United Nations. It also incorporated the Communist governments of divided nations such as Germany (not admitted to the UN until 1973), Vietnam (1977), and Korea (1991).

18. World Meteorological Organization, "Sixth World Meteorological Congress, Geneva, 5–30 April 1971—Proceedings,"1972, 162–64.

19. K. Suter, *Global Order and Global Disorder: Globalization and the Nation-State* (Praeger, 2003); Held et al., *Global Transformations*.

20. World Meteorological Organization, "Abridged Final Report of the First Session, Washington, 2nd–29th April, 1953," Commission for Synoptic Meteorology, 1953, 49, 41.

21. World Meteorological Organization, "Second World Meteorological Congress," *WMO Bulletin* 4, no. 3 (1955), 95.

22. World Meteorological Organization, *Final Report: First Congress of the World Meteorological Organization, Paris, 19 March-28 April 1951*, 1951, 9–10.

23. In this period, the UNDP was the chief UN agency for technical assistance of all kinds.

24. Miller, "Scientific Internationalism in American Foreign Policy," 215, note 6.

25. Daniel, "One Hundred Years of International Co-Operation," 42.

26. See, for example, World Weather Watch, *Consolidated List of Voluntary Assistance Programme Projects Approved for Circulation in 1970*, 1971.

27. Miller, "Scientific Internationalism in American Foreign Policy"; Miller, "Hybrid Management: Boundary Organizations, Science Policy, and Environmental Governance in the Climate Regime," *Science, Technology & Human Values* 26, no. 4 (2001): 478–500; Miller, "Climate Science and the Making of Global Political Order," in *States of Knowledge*, ed. S. Jasanoff (Routledge, 2004); Miller, "'An Effective Instrument of Peace': Scientific Cooperation as an Instrument of US Foreign Policy, 1938–1950," *Osiris* 21, special issue on Science, Technology, and International Affairs: Historical Perspectives (2006): 133–60.

28. Miller, "Scientific Internationalism in American Foreign Policy."

29. Unlike Forman's earlier, more internalist, more dialectical discussion of scientific internationalism ("Scientific Internationalism and the Weimar Physicists"), Miller's ("'An Effective Instrument of Peace'") emphasizes the new scientific institutions that emerged after 1939, which were often linked more closely to governments than most of their predecessors. Analyzing the 1950s Atoms for Peace Program, John Krige adopts a sense of "scientific internationalism" somewhere between Miller's and Forman's: "When the science concerned is also an affair of state, of immense importance for national strategic interests, international exchange is at once a window and a probe, an ideology of transparency and, by virtue of that, an instrument of control, a viewpoint which looks in and watches over. . . . International scientific exchange deftly reconciled the universalistic appeal to the pursuit of truth with the particularist needs of national security." (J. Krige, "Atoms for Peace, Scientific Internationalism, and Scientific Intelligence," *Osiris* 21, special issue on Science, Technology, and International Affairs: Historical Perspectives (2006), 171, 181) Another useful concept here is Gabrielle Hecht's term "technopolitics," defined as "the strategic practice of designing or using technology to constitute, embody, or enact political goals." (G. Hecht, *The Radiance of France: Nuclear Power and National Identity After World War II*, MIT Press, 1998, 15; Hecht and P. N. Edwards, *The Technopolitics of Cold War: Toward a Transregional Perspective* (American Historical Association, 2007).

30. Miller, "Hybrid Management," 211.

31. J. van Mieghem and World Meteorological Organization, *The Problem of the Professional Training of Meteorological Personnel of All Grades in the Less-Developed Countries* (Secretariat of the World Meteorological Organization, 1963).

32. M. J. Rubin, "Synoptic Meteorology and the IGY," in *Geophysics and the IGY: Proceedings of the Symposium at the Opening of the International Geophysical Year, June 28–29, 1957*, ed. H. Odishaw and S. Ruttenberg (American Geophysical Union, 1958), 157.

33. Greenaway, *Science International*, 150.

34. Miller, "Hybrid Management"; Miller and P. N. Edwards, "Introduction: The Globalization of Climate Science and Climate Politics," in *Changing the Atmosphere: Expert Knowledge and Environmental Governance*, ed. C. A. Miller and P. N. Edwards (MIT Press, 2001).

35. A. A. Needell, *Science, Cold War and the American State: Lloyd V. Berkner and the Balance of Professional Ideals* (Harwood, 2000); R. E. Doel and A. A. Needell, "Science, Scientists, and the CIA: Balancing International Ideals, National Needs, and Professional Opportunities," in *Eternal Vigilance? Fifty Years of the CIA*, ed. R. Jeffreys-Jones and C. Andrew (Frank Cass, 1997).

36. Greenaway, *Science International*, 151.

37. Ibid., 89, 99–101.

38. As Miller observed in "Hybrid Management," this meant that science was the *only* area in which the PRC was permitted to participate fully in the community of nations.

39. Khrgian, *Meteorology*, 348.

40. S. H. S. Jones, "The Inception and Development of the International Geophysical Year," in *Annals of the International Geophysical Year*, volume X (Pergamon, 1959); Comité Spécial de l'Année Géophysique Internationale, "The Fourth Meeting of the CSAGI," in *Annals of the International Geophysical Year*, volume Y (Pergamon, 1958).

41. Davies, *Forty Years of Progress and Achievement*, 75.

42. National Weather Records Center, *Climatology at Work*, 30.

43. World Meteorological Organization, *Catalogue of IGY/IGC Meteorological Data* (World Meteorological Organization, 1962); World Meteorological Organization, *Catalogue of Meteorological Data for Research (Parts I and II)*, 1965); World Meteorological Organization, *Catalogue of Meteorological Data for Research, Part III: Meteorological Data Recorded on Media Usable by Automatic Data-Processing Machines* (1972).

44. Daniel, "One Hundred Years of International Co-Operation," 75.

45. R. D. Launius, "Sputnik and the Origins of the Space Age," www.hq.nasa.gov.

46. W. R. Kennedy Jr., "Fallout Forecasting—1945 Through 1962," LA-10605-MS, UC-11 (Los Alamos National Laboratory, 1986), www.fas.org.

47. M. Eisenbud, "The First Years of the Atomic Energy Commission New York Operations Office Health and Safety Laboratory," *Environment International* 20, no. 5 (1994), 566.

48. H. L. Beck and B. G. Bennett, "Historical Overview of Atmospheric Nuclear Weapons Testing and Estimates of Fallout in the Continental United States," *Health Physics* 82, no. 5 (2002): 591–608.

49. Ibid., 596–98.

50. In general, we now know, nuclear secrecy overrode safety precautions during this period. Decades later the test program's failure to inform and protect "downwinders," many of them Native Americans living near the test sites, led to the Radiation Exposure Compensation Act of 1990, and to a formal government apology issued by President Clinton.

51. L. Machta, "Meteorological Benefits from Atmospheric Nuclear Tests," *Health Physics* 82, no. 5 (2002): 635–43.

52. Ibid.; Beck and Bennett, "Historical Overview."

53. F. Hagemann et al., "Stratospheric Carbon-14, Carbon Dioxide, and Tritium," *Science* 130, no. 3375 (1959): 542–52.

54. Machta, "Meteorological Benefits," 641.

55. W. F. Libby, "Radioactive Fallout," *Proceedings of the National Academy of Sciences* 43, no. 8 (1957): 758–75.

56. E. C. Anderson and W. F. Libby, "World-Wide Distribution of Natural Radiocarbon," *Physical Review* 81, no. 1 (1951): 64–69.

57. H. Waenke and J. R. Arnold, *Hans E. Suess, 1909–1993: A Biographical Memoir* (National Academies Press, 2005), 8.

58. H. E. Suess, "Natural Radiocarbon and the Rate of Exchange of Carbon Dioxide Between the Atmosphere and the Sea," in *Nuclear Processes in Geologic Settings* (National Academy of Sciences, 1953), 52–56.

59. H. E. Suess, "Radiocarbon Content in Modern Wood," *Science* 122, no. 3166 (1955): 415–17.

60. Revelle and Suess, "Carbon Dioxide Exchange." Although the Revelle-Suess paper is often cited as an origin point for global warming studies, Weart notes that "the bulk of the text reflected the pair's original belief that the oceans were

absorbing most of the new $CO_2$. [Revelle's] brief paragraph on the factor of ten stood apart like an isolated thought. *In the archives it is visibly an addition, Scotch-taped onto the original draft.*" (S. R. Weart, 2008, www.aip.org, emphasis added)

61. For a detailed account of Revelle's role in reviving the $CO_2$ theory of climate change, see Weart, *The Discovery of Global Warming*.

62. D. M. Hart and D. G. Victor, "Scientific Elites and the Making of US Policy for Climate Change Research," *Social Studies of Science* 23 (1993): 643–80.

63. As Lester Machta observed, "in 1962 . . . the same Weather Bureau unit involved in fallout studies took over the carbon dioxide program, and as soon as funds became available the [carbon dioxide monitoring] network was expanded. Later, other countries, such as Australia and Germany, added to the network. Of all the aspects of global warming by greenhouse gases, the growth of the gases in the atmosphere is best known. Other greenhouse gases were also measured at the carbon dioxide monitoring stations. The existence of a fallout network of stations during nuclear testing showed that a global network was possible and might be scientifically valuable." ("Meteorological Benefits," 641, emphasis added) Machta himself directed this laboratory, established in 1948 as the US Weather Bureau's Special Projects Section to provide meteorological support to the atomic test program. By 1955 the SPS had prepared a widely read handbook, titled *Meteorology and Atomic Energy*, which underwent ongoing revision and remained in print for more than 20 years. US Weather Bureau Special Projects Section, *Meteorology and Atomic Energy* (Government Printing Office, 1955); D. H. Slade, *Meteorology and Atomic Energy, 1968* (US Atomic Energy Commission, Division of Technical Information, 1969). The SPS later became the Air Resources Laboratory, which was eventually assigned responsibility for studying air pollution of all sorts.

64. Hagemann et al., "Stratospheric Carbon-14, Carbon Dioxide, and Tritium," 550.

65. D. H. Peirson and R. S. Cambray, "Interhemispheric Transfer of Debris from Nuclear Explosions Using a Simple Atmospheric Model," *Nature* 216 (1967): 755–58; R. J. List and K. Telegadas, "Using Radioactive Tracers to Develop a Model of Circulation of Stratosphere," *Journal of the Atmospheric Sciences* 26, no. 5 (1969): 1128–36; K. Telegadas, "The Upper Portion of the Hadley Cell Circulation as Deduced From the 1968 French and Chinese Nuclear Tests," *Journal of Geophysical Research* 76 (1971): 5018–24.

66. Hart and Victor, "Scientific Elites," 647–48; S. R. Weart, *The Discovery of Global Warming: A Hypertext History of How Scientists Came to (Partly) Understand What Humans Are Doing to the Earth's Climate*, Center for the History of Physics, 2008, www.aip.org.

67. Merrill Eisenbud, the first director of HASL, reported that members of his laboratory were able to publish essentially without restriction after 1953 (Eisenbud, "The First Years of the Atomic Energy Commission New York Operations Office Health

and Safety Laboratory," 568). As with any claim regarding the nuclear weapons program, this one should be viewed with a critical eye.

68. US House of Representatives Joint Committee on Atomic Energy, *Effects of Nuclear Explosions on Weather and Health* and *Health and Safety Problems and Weather Effects Associated with Atomic Explosions* (Government Printing Office, 1955).

69. US House of Representatives Joint Committee on Atomic Energy, *Nature of Radioactive Fall-Out and Its Effects on Man (Part 1)* (Government Printing Office, 1957); *Nature of Radioactive Fall-Out and Its Effects on Man (Part 2)* (Government Printing Office, 1957); *Fallout from Nuclear Weapons Tests* (Government Printing Office, 1959).

70. W. Burr and H. L. Montford, "The Making of the Limited Test Ban Treaty, 1958–1963," National Security Archive, Washington, 2003, www.gwu.edu.

71. D. Masters and K. Way, eds., *One World or None: A Report to the Public on the Full Meaning of the Atomic Bomb* (Whittlesey House, 1946).

72. Hecht, *The Radiance of France*, 15.

73. Ibid.; Hecht and Edwards, *The Technopolitics of Cold War*; Callon and Latour, "Unscrewing the Big Leviathan"; Latour, *Science in Action*; Latour, *Reassembling the Social*.

74. Pilot balloons and rawinsondes measure wind speed and direction. Pilot balloons are tracked visually, so their range is quite limited. Rawinsondes are tracked by radar or (today) GPS. Radiosondes measure temperature, pressure, humidity, and sometimes other variables; if suitably equipped, they may also function as rawinsondes.

75. P. D. Thompson, "An Essay on the Technical and Economic Aspects of the Network Problem," *WMO Bulletin* 12, no. 3 (1963), 3.

76. Anonymous, "The Challenge of the World Network Problem," *WMO Bulletin* 12, no. 3 (1963): 144–48.

77. N. A. Phillips, "Numerical Weather Prediction," in *Advances in Computers*, ed. F. L. Alt (Academic Press, 1960), 49.

78. Hall, "A History of the Military Polar Orbiting Meteorological Satellite Program."

79. A. Schnapf, "The TIROS Meteorological Satellites—Twenty-Five Years: 1960–1985," in *Monitoring Earth's Ocean, Land and Atmosphere from Space: Sensors, Systems, and Applications*, ed. A. Schnapf (American Institute of Aeronautics and Astronautics, 1985).

80. M. Sullivan, "Francis Gary Powers: One Man, Two Countries and the Cold War" (Military Advantage, 2004), www.military.com.

81. Fuller, *Thor's Legions*, 248.

82. J. T. Richelson, "National Security Archive Electronic Briefing Book No. 13: US Satellite Imagery, 1960–1999" (National Security Archive, George Washington University, 1999).

83. At one point, the Soviets did stake out the technically correct, yet rather bizarre position that Sputnik did not in fact *fly over* other nations. Instead, they argued, the Earth rotated beneath the satellite in its orbit.

84. McDougall, *The Heavens and the Earth*.

85. Quoted in Greenaway, *Science International*, 157, emphasis added.

86. McDougall, *The Heavens and the Earth*.

87. H. Cleveland, "Keeping Up with Technology: The World Weather Watch and Other Analogies," 2001 National GeoData Forum (Denver, 2001).

88. D. Masters and K. Way, eds., *One World or None: A Report to the Public on the Full Meaning of the Atomic Bomb* (McGraw-Hill, 1946).

89. W. J. Gibbs, *Federation and Meteorology* (Australian Science and Technology Heritage Centre, 2001), 1111.

90. National Research Council Committee on Atmospheric Sciences, *The Atmospheric Sciences 1961–1971: A Report to the Special Assistant to the President for Science and Technology* (1962).

91. M. Bundy, National Security Action Memorandum No. 101, "Follow-Up on the President's Speech to the United Nations General Assembly on September 26, 1961," 3.

92. UN resolution 1721 C (XVI), reproduced in World Meteorological Organization, *First Report on the Advancement of Atmospheric Sciences and Their Application in the Light of Developments in Outer Space* (Secretariat of the World Meteorological Organization, 1962), 2.

93. Letters quoted in K. C. Heidorn, *Weather Cooperation During the Cold War* (Spectrum Educational Enterprises, 2000).

94. World Meteorological Organization, *First Report*.

## Chapter 9

1. This strategy resembles the German sociologist Ingo Braun's concept of "second-order large technical systems," or systems built "on top of" existing, first-order large technical systems. Braun's phrase for this phenomenon, "Vernetzung der Netze," has been translated as "material interlacing," but a better rendering would be "net-

working of networks"—in other words, internetworking. I. Braun, "Zur intersystemische Vernetzung grösser technische Netze," in *Technik ohne Grenzen*, ed. I. Braun and B. Joerges (Suhrkamp, 1994). See also van der Vleuten, "Infrastructures and Societal Change."

2. A. L. Russell, "'Rough Consensus and Running Code' and the Internet-OSI Standards War," *IEEE Annals of the History of Computing* 28, no. 3 (2006): 48–61.

3. National Research Council, *Bits of Power*, 216.

4. D. E. Hinsman, "The Space-Based Global Observing System, Its Present Configuration and Evolution," in *2000 EUMETSAT Meteorological Satellite Data Users' Conference* (Bologna, 2000), 145.

5. N. G. Leonov et al., *Requirements and Specifications for Data-Processing System*, World Weather Watch Planning Report No. 8 (Secretariat of the World Meteorological Organization, 1966), 1.

6. M. E. Courain, Technology Reconciliation in the Remote-Sensing Era of United States Civilian Weather Forecasting: 1957–1987, Ph.D. dissertation, Rutgers University, 1991.

7. Commission for Synoptic Meteorology, *Satellite and Computer Applications to Synoptic Meteorology* (World Meteorological Organization, 1971).

8. M. Halem et al., "Some Experiments on the Effect of Remote Sounding Temperatures Upon Weather Forecasting," in *Remote Sensing of the Atmosphere: Inversion Methods and Applications*, ed. A. L. Fymat and V. E. Zuev (Elsevier, 1978); M. Halem et al., "An Assessment of the FGGE Satellite Observing System During SOP-1," *Bulletin of the American Meteorological Society* 63, no. 4 (1982): 407–26; R. Atlas et al., "Impact of Satellite Temperature Sounding and Wind Data on Numerical Weather Prediction," *Optical Engineering* 24, no. 2 (1985): 341–46.

9. Abbate, *Inventing the Internet*.

10. World Meteorological Organization, "Global Telecommunication System: Methods and Equipment for Automatic Distribution of Information," World Weather Watch Planning Report no. 24, 1968), viii, 56.

11. Ibid., 16.

12. Edwards, *The Closed World*.

13. World Meteorological Organization, "Global Telecommunication System," 2.

14. Ibid., 30.

15. For an excellent overview of these issues in Internet history, see Abbate, *Inventing the Internet*, especially chapter 5.

16. T. Thompson, *Telecommunications Problems in Computer-to-Computer Data Transfer* (World Meteorological Organization, 1966), 16.

17. World Meteorological Organization, *Fifth World Meteorological Congress, Geneva, 3–28 April 1967—Abridged Report with Resolutions* (1967), Annex V.

18. Ibid., 93.

19. Ibid., 189.

20. J. G. Ruggie, "International Responses to Technology: Concepts and Trends," *International Organization* 29, no. 3 (1975), 571.

21. US Department of Commerce et al., *World Weather Program: Plan for Fiscal Year 1972* (Government Printing Office, 1971), 11.

22. Ibid., 29.

23. US Department of Commerce et al., *World Weather Program: Plan for Fiscal Years 1980 and 1981* (Government Printing Office, 1980), 46.

24. Ruggie, "International Responses to Technology," 570.

25. S. Petterssen, *Research Aspects of the World Weather Watch* (Secretariat of the World Meteorological Organization, 1966), 4.

26. Joint Organizing Committee, *The Planning of the First GARP Global Experiment* (Global Atmospheric Research Programme, 1969), Summary.

27. Ibid., 7.

28. J. P. Kuettner, "General Description and Central Program of GATE," *Bulletin of the American Meteorological Society* 55, no. 7 (1974): 712–19.

29. J. L. Rasmussen, "FGGE Operations and Data Management," *Advances in Space Research* 1, no. 4 (1981): 149–64.

30. J. Harrison, "Data Management Overview," in *The Global Weather Experiment: Final Report of US Operations* (National Oceanic and Atmospheric Administration, US Department of Commerce, 1981).

31. L. Bengtsson, *Results of the Global Weather Experiment* (World Meteorological Organization, 1983).

32. Harrison, "Data Management Overview," 204–05.

33. Rasmussen, "FGGE Operations and Data Management"; Harrison, "Data Management Overview."

34. Bengtsson, *Results of the Global Weather Experiment*, 10.

35. WMO-ICSU Joint Organizing Committee, "Modelling for the First GARP Global Experiment," 1974.

36. J. Smagorinsky, "Numerical Simulation of the Global Atmosphere," in *The Global Circulation of the Atmosphere*, ed. G. A. Corby (Royal Meteorological Society, 1970), 28.

**Chapter 10**

1. H. J. de Vries, "Standardization—a New Discipline?" in *Proceedings of the 2nd IEEE Conference on Standardization and Innovation in Information Technology* (Piscataway, 2001).

2. Most meteorological standards under discussion here are *de jure*, so I do not discuss *de facto* standards further.

3. Latour, *Science in Action*.

4. B. Bolin et al., "Numerical Methods of Weather Analysis and Forecasting," WMO Technical Note No. 44 (World Meteorological Organization, 1962), 2–3.

5. Staff Members, Joint Numerical Weather Prediction Unit, "One Year of Operational Numerical Weather Prediction, Part I," *Bulletin of the American Meteorological Society* 38, no. 5 (1957), 266.

6. To make a checksum, some quantity, such as the sum of the bits making up each data record, is derived from the data to be transmitted. The derived quantity (or some part of it, such as the last digit of the sum) is transmitted along with the data. The recipient can then perform the same calculation on the incoming bitstream. If the checksum figures do not correspond, this indicates a transmission error.

7. Staff Members, Joint Numerical Weather Prediction Unit, "One Year of Operational Numerical Weather Prediction, Part I."

8. H. A. Bedient and G. P. Cressman, "An Experiment in Automatic Data Processing," *Monthly Weather Review* 85, no. 10 (1957), 333.

9. Ibid.

10. Thompson, *Telecommunications Problems in Computer-to-Computer Data Transfer*, 1–2.

11. F. G. Shuman, "History of Numerical Weather Prediction at the National Meteorological Center," *Weather and Forecasting* 4 (1989), 290.

12. V. Bjerknes, *Dynamic Meteorology and Hydrography, Part II. Kinematics* (Gibson Bros., Carnegie Institute, 1911); R. Daley, *Atmospheric Data Analysis* (Cambridge University Press, 1991).

13. Mitchell and Wexler, "How the Daily Forecast Is Made."

14. P. Bergthorsson and B. R. Döös, "Numerical Weather Map Analysis," *Tellus* 7, no. 3 (1955), 329.

15. H. A. Panofsky, "Objective Weather Map Analysis," *Journal of Meteorology* 6, no. 6 (1949): 386–92.

16. Persson, "Early Operational Numerical Weather Prediction Outside the USA," 151, second emphasis added.

17. Bergeron, "Methods in Scientific Weather Analysis and Forecasting," 458.

18. Daley, *Atmospheric Data Analysis* , 21, emphasis added.

19. Harper, "Research From the Boundary Layer," 679.

20. C. W. Newton, "Analysis and Data Problems in Relation to Numerical Prediction," *Bulletin of the American Meteorological Society* 35, no. 7 (1954), 289ff.

21. Staff Members, Joint Numerical Weather Prediction Unit, "One Year of Operational Numerical Weather Prediction, Part I."

22. B. Gilchrist and G. P. Cressman, "An Experiment in Objective Analysis," *Tellus* 6, no. 4 (1954): 309–18.

23. As one of the method's designers observed, "straightforward interpolation between observations hundreds or thousands of miles apart is not going to give a usable value." G. P. Cressman, "Dynamic Weather Prediction," in *Meteorological Challenges: A History*, ed. D. P. McIntyre (Information Canada, 1972), 188.

24. Bergthorsson and Döös, "Numerical Weather Map Analysis."

25. Bolin et al., "Numerical Methods of Weather Analysis and Forecasting," 3.

26. L. S. Gandin, *Objective Analysis of Meteorological Fields* (Israel Program for Scientific Translations, 1963).

27. A fuller, more detailed description of these techniques require complex mathematics beyond the scope of this book. Technically competent readers interested in this history might begin with the following: M. A. Alaka and R. C. Elvander, "Optimum Interpolation from Observations of Mixed Quality," *Monthly Weather Review* 100, no. 8 (1972): 612–24; G. J. DiMego et al., "Data Processing and Quality Control for Optimum Interpolation Analyses at the National Meteorological Center," *Workshop on the Use and Quality Control of Meteorological Observations*, Washington, 1985; E. Kalnay, *Atmospheric Modeling, Data Assimilation and Predictability* (Cambridge University Press, 2003); A. C. Lorenc, "Some Historical Background to Data Assimilation for NWP," International Summer School on Atmospheric and Oceanic Sciences, L'Aquila, 2004; A. C. Lorenc, *Atmospheric Data Assimilation* (UK Meteorological Office, 1995).

28. P. Suppes, "Models of Data," in *Logic, Methodology, and the Philosophy of Science: Proceedings of the 1960 Congress*, ed. E. Nagel et al. (Stanford University Press, 1962).

29. F. Suppe, "Understanding Scientific Theories: An Assessment of Developments, 1969–1998," *Philosophy of Science* 67 (2000), 112. See also S. D. Norton and F. Suppe, "Why Atmospheric Modeling Is Good Science," in *Changing the Atmosphere: Expert Knowledge and Environmental Governance*, ed. C. A. Miller and P. N. Edwards (MIT Press, 2001).

30. R. N. Giere, "Using Models to Represent Reality," in *Model-Based Reasoning in Scientific Discovery*, ed. L. Magnani et al. (Springer, 1999), 55.

31. Newton, "Analysis and Data Problems in Relation to Numerical Prediction"; A. Eliassen et al., *Upper Air Network Requirements for Numerical Weather Prediction—Report of a Working Group of the Commission for Aerology*, Technical Note TN-29 (World Meteorological Organization, 1960); P. D. Thompson, "A Dynamical Method of Analyzing Meteorological Data," *Tellus* 13 (1961): 334–49; L. S. Gandin et al., *Design of Optimum Networks for Aerological Observing Stations*, World Weather Watch Planning Report WWW-PR-21 (Secretariat of the World Meteorological Organization, 1967).

32. The $S_1$ score indicates how well a given forecast's horizontal gradient (usually geopotential height or pressure) corresponds to the observed (analyzed) gradient. Lower values mean greater correspondence. Forecasts with scores above about 60 are generally regarded as not useful. According to the World Weather Research Program, "because $S_1$ depends only on gradients, good scores can be achieved even when the forecast values are biased." WWRP-WGNE Joint Working Group on Verification, "Forecast Verification: Issues, Methods and FAQ" (Australian Government Bureau of Meteorology, 2007), www.bom.gov.au.

33. C. S. Ramage, "Prognosis for Weather Forecasting," *Bulletin of the American Meteorological Society* 56, no. 1 (1975), 5, emphasis added.

34. K. G. Bauer and J. E. Kutzbach, "Evaluation Standards for Dynamical Prediction Models," *Journal of Applied Meteorology* 13, no. 4 (1974): 505–06. See also Persson, "Appendix A: The Early History of NWP," 263.

35. Galison, "Computer Simulations and the Trading Zone"; P. L. Galison, *Image and Logic: A Material Culture of Microphysics* (University of Chicago Press, 1997); M. S. Morgan and M. Morrison, *Models as Mediators: Perspectives on Natural and Social Sciences* (Cambridge University Press, 1999); S. Sismondo, "Models, Simulations, and Their Objects," *Science in Context* 12, no. 2 (1999): 247–60; L. Magnani et al., *Model-Based Reasoning in Scientific Discovery* (Kluwer, 1999).

36. Castells, *The Rise of the Network Society*.

37. Roebber and Bosart, "The Contributions of Education and Experience to Forecast Skill"; Daipha, *Masters of Uncertainty*; Fine, *Authors of the Storm*.

38. Bolin et al., "Numerical Methods of Weather Analysis and Forecasting," 3.

39. Smagorinsky, "Early Recollections," 22.

40. Kalnay, *Atmospheric Modeling, Data Assimilation and Predictability*, 13.

41. R. D. McPherson, "Progress, Problems, and Prospects in Meteorological Data Assimilation," *Bulletin of the American Meteorological Society* 56, no. 11 (1975): 1154–66.

42. S. Teweles, "Review of Efforts to Minimize the Data Set Needed for Initialization of a Macroscale Prognosis," *Bulletin of the American Meteorological Society* 53, no. 9 (1972), 856.

43. Alaka and Elvander, "Optimum Interpolation from Observations of Mixed Quality," 617.

44. J. Charney et al., "Use of Incomplete Historical Data to Infer the Present State of the Atmosphere," *Journal of Atmospheric Sciences* 26, no. 5 (1969): 1160–63.

45. Lorenc, *Atmospheric Data Assimilation*, 1.

46. L. Bengtsson and J. Shukla, "Integration of Space and In Situ Observations to Study Global Climate Change," *Bulletin of the American Meteorological Society* 69, no. 10 (1988), 1134–36. For similar examples, see R. J. Reed et al., *An Evaluation of the Performance of the ECMWF Operational Forecasting System in Analyzing and Forecasting Tropical Easterly Disturbances. Part 1: Synoptic Investigation* (European Centre for Medium-Range Weather Forecasts, 1986).

47. Woods, *Medium-Range Weather Prediction*, 91.

48. Weather instruments also fly on geostationary satellites, whose position remains fixed relative to the Earth. These satellites can observe continent- to hemisphere-scale phenomena and are the principal source of weather imagery. Data from several geostationary satellites can be combined to produce a global data image. Because polar orbits lie much closer to the Earth (800–1200 km) than geostationary orbits (36,000 km), polar-orbiting instruments typically obtain higher resolutions.

49. Radar became, of course, extremely important in local and regional weather forecasting. This book does not discuss it because radar has not (yet) become a significant source of climate data.

50. Courain, *Technology Reconciliation in the Remote-Sensing Era of United States Civilian Weather Forecasting*.

51. C. Bristor, quoted in ibid., 184.

52. F. K. Fye, *The AFGWC Automated Cloud Analysis Model* (Air Force Global Weather Central, 1978); A. Henderson-Sellers and N. A. Hughes, "1979 3D-Nephanalysis Global Total Cloud Amount Climatology," *Bulletin of the American Meteorological Society* 66, no. 6 (1985): 626–27.

53. D. C. Norquist, "Alternative Forms of Humidity Information in Global Data Assimilation," *Monthly Weather Review* 116, no. 2 (1988): 452–71.

54. L. Bengtsson, 1989, quoted in Woods, *Medium-Range Weather Prediction*, 161.

55. Committee on NASA-NOAA Transition from Research to Operations, National Research Council, *Satellite Observations of the Earth's Environment: Accelerating the Transition of Research to Operations* (National Academies Press, 2003), 102.

56. UK Met Office, "Satellite Microwave Radiances."

57. Meteorologists distinguish *in situ* measurements—readings taken by direct contact with the measured quantity (thermometers, anemometers, radiosondes, rain gauges, etc.)—from those of remote sensing instruments, such as radar and satellites. Remote sensing instruments generally measure proxy quantities, such as radiances or radio reflectivity. In most cases these proxy measures must then be converted into the desired meteorological variables.

58. Committee on NASA-NOAA Transition from Research to Operations, National Research Council, *Satellite Observations of the Earth's Environment* (2003), 102–06.

59. UK Met Office, "Atmospheric Line-By-Line Models and Fast RT Models," 2007, www.metoffice.gov.uk.

60. R. W. Spencer and J. R. Christy, "Precise Monitoring of Global Temperature Trends From Satellites," *Science* 247, no. 4950 (1990): 1558; R. W. Spencer and J. R. Christy, "Precision and Radiosonde Validation of Satellite Gridpoint Temperature Anomalies. Part I: MSU Channel 2," *Journal of Climate* 5, no. 8 (1992): 847–57.

61. Lorenc, *Atmospheric Data Assimilation*, 1.

62. E. Winsberg, "Sanctioning Models: The Epistemology of Simulation," *Science in Context* 12, no. 2 (1999), 275.

63. N. R. Hanson, *Patterns of Discovery* (Cambridge University Press, 1958); K. R. Popper, *The Logic of Scientific Discovery* (Basic Books, 1959).

64. P. N. Edwards, "Global Climate Science, Uncertainty and Politics: Data-Laden Models, Model-Filtered Data," *Science as Culture* 8, no. 4 (1999): 437–72.

65. S. H. Schneider, "Introduction to Climate Modeling," in *Climate System Modeling*, ed. K. E. Trenberth (Cambridge University Press, 1992).

66. Kuhn, *The Structure of Scientific Revolutions*; I. Lakatos and A. Musgrave, *Criticism and the Growth of Knowledge* (Cambridge University Press, 1970); Latour and Woolgar, *Laboratory Life*; P. Galison, *How Experiments End* (University of Chicago Press, 1987); H. Collins and T. Pinch, *The Golem: What Everyone Should Know About Science* (Cambridge University Press, 1993).

67. D. Stokes, *Pasteur's Quadrant: Basic Science and Technological Innovation* (Brookings Institution, 1997).

68. Winsberg, "Sanctioning Models," 287.

69. Sismondo, "Models, Simulations, and Their Objects," 255–56.

70. Edwards, "Global Climate Science, Uncertainty and Politics."

71. Morgan and Morrison, *Models as Mediators*.

72. M. Morrison and M. S. Morgan, "Models as Mediating Instruments," in *Models as Mediators: Perspectives on Natural and Social Sciences*, ed. M. S. Morgan and M. Morrison (Cambridge University Press, 1999).

73. Norton and Suppe, "Why Atmospheric Modeling Is Good Science," 70, 72, emphasis added.

74. Joint Organizing Committee, *The Planning of the First GARP Global Experiment*, 23.

75. The ethnographer Myanna Lahsen wrote: "During modelers' presentations to fellow atmospheric scientists that I attended during my years at NCAR, I regularly saw confusion arise in the audience because it was unclear whether overhead charts and figures were based on observations or simulations. In one such presentation about the role of clouds in the climate system, observational data were compared against model simulations. I grew confused as to which charts represented empirical data and which represented simulations. I realized that I was not alone in my confusion when scientists in the audience stopped the presenter to ask for clarification as to whether the overhead figures were based on observations or model extrapolations. The presenter specified that the figures were based on models, and then continued his presentation." M. Lahsen, "Seductive Simulations? Uncertainty Distribution Around Climate Models," *Social Studies of Science* 35, no. 6 (2005), 908. My own experience and interviews confirm this impression.

76. Lorenz, "Deterministic Nonperiodic Flow"; E. N. Lorenz, "A Study of the Predictability of a 28-Variable Atmospheric Model (28-Variable Atmosphere Model Constructed by Expanding Equations of Two-Level Geostrophic Model in Truncated Double-Fourier Series)," *Tellus* 17 (1965): 321–33; E. S. Epstein, "Stochastic Dynamic Prediction," *Tellus* 21, no. 6 (1969): 739–59; C. E. Leith, "Theoretical Skill of Monte Carlo Forecasts," *Monthly Weather Review* 102, no. 6 (1974): 409–18; R. N. Hoffman and E. Kalnay, "Lagged Average Forecasting, an Alternative to Monte Carlo Forecasting," *Tellus, Series A—Dynamic Meteorology and Oceanography* 35 (1983): 100–18.

77. Z. Toth and E. Kalnay, "Ensemble Forecasting At NMC: The Generation of Perturbations," *Bulletin of the American Meteorological Society* 74, no. 12 (1993): 2317–30; M. S. Tracton and E. Kalnay, "Operational Ensemble Prediction at the

National Meteorological Center: Practical Aspects," *Weather and Forecasting* 8, no. 3 (1993): 379–98.

78. European Centre for Medium-Range Weather Forecasts, "Description of the ECMWF Forecasting System in February 2006," 2006, www.ecmwf.int.

## Chapter 11

1. T. R. Karl et al., "Critical Issues for Long-Term Climate Monitoring," *Climatic Change* 31, no. 2 (1995): 185–221; Karl et al., "Detecting Climate Variations and Change: New Challenges for Observing and Data Management Systems," *Journal of Climate* 6 (1993): 1481–94; K. E. Trenberth et al., "The Need for a Systems Approach to Climate Observations," *Bulletin of the American Meteorological Society* 83, no. 11 (2002): 1593–602; World Meteorological Organization and Global Climate Observing System, *Second Report on the Adequacy of the Global Observing Systems for Climate in Support of the UNFCCC*; National Research Council, *Adequacy of Climate Observing Systems*; Panel on Climate-Related Data et al., *Atmospheric Climate Data: Problems and Promises* (National Academy Press, 1986).

2. World Meteorological Organization, *Twenty-Second Status Report on Implementation of the World Weather Watch* (2005), 4.

3. Improvements in the Global Telecommunication System made up for some of this loss, however. The number of SYNOP reports actually received on the GTS increased steadily, reaching a historical peak in 2004, as did the total number of stations in the WWW network (including non-RBSN stations). Africa, Central America, South America, and parts of Asia continued to lag the WWW goals for network coverage. Source: ibid.

4. Ibid.

5. European Centre for Medium-Range Weather Forecasts, "General Overview of the Data Assimilation System" (2007), www.ecmwf.int.

6. L. Bengtsson et al., "Global Observations and Forecast Skill," *Tellus* 57, no. 4 (2005): 515–27.

7. Originally called the Global Data-processing System.

8. K. E. Trenberth, "Atmospheric Circulation Climate Changes," *Climatic Change* 31, no. 2 (1995), 306.

9. Doppler radar can detect falling raindrops, hail, and snow, so it is commonly used for short-term precipitation forecasts. However, the amount of precipitation actually reaching the ground can differ from what radar detects in the atmosphere. For climatological purposes, actual ground-level precipitation is usually all that matters.

10. National Research Council, *Future of the National Weather Service Cooperative Observer Network*.

11. K. R. Briffa et al., "Tree-Ring Width and Density Data Around the Northern Hemisphere: Part 1, Local and Regional Climate Signals," *The Holocene* 12, no. 6 (2002): 737; H. Grudd et al., "A 7400-Year Tree-Ring Chronology in Northern Swedish Lapland: Natural Climatic Variability Expressed on Annual to Millennial Timescales," *The Holocene* 12, no. 6 (2002): 657; J. Esper et al., "Low-Frequency Signals in Long Tree-Ring Chronologies for Reconstructing Past Temperature Variability," *Science* 295, no. 5563 (2002): 2250–53; J. R. Petit et al., "Climate and Atmospheric History of the Past 420,000 Years from the Vostok Ice Core, Antarctica," *Nature* 399 (1999): 429–36; T. L. Root et al., "Fingerprints of Global Warming on Wild Animals and Plants," *Nature* 421, no. 6918 (2003): 57–60; T. L. Root and S. H. Schneider, "Ecology and Climate: Research Strategies and Implications," *Science* 269, no. 5222 (1995): 334; I. Chuine et al., "Back to the Middle Ages? Grape Harvest Dates and Temperature Variations in France Since 1370," *Nature* 432 (2004): 289–90.

12. World Meteorological Organization and Global Climate Observing System, *GCOS Annual Report 2007–2008* (2008), 5.

13. J. M. Mitchell, "On the Causes of Instrumentally Observed Secular Temperature Trends," *Journal of the Atmospheric Sciences* 10, no. 4 (1953): 244–61.

14. D. J. Gaffen, "Temporal Inhomogeneities in Radiosonde Temperature Records," *Journal of Geophysical Research* 99, no. D2 (1994), 3669.

15. P. D. Jones et al., *A Gridpoint Surface Air Temperature Data Set for the Northern Hemisphere* (US Department of Energy, Carbon Dioxide Research Division, 1985), 1.

16. I. Auer et al., "Metadata and Their Role in Homogenising," in *Proceedings of the Fourth Seminar for Homogenization and Quality Control in Climatological Databases* (Budapest, 2004), 18.

17. Easterling et al., "On the Development and Use of Homogenized Climate Datasets"; Peterson et al., "Homogeneity Adjustments."

18. R. Boehm et al., "Regional Temperature Variability in the European Alps: 1760–1998 From Homogenized Instrumental Time Series," *International Journal of Climatology* 21, no. 14 (2001): 1779–801; Auer et al., "Metadata and Their Role in Homogenising."

19. D. A. Robinson et al., "Global Snow Cover Monitoring: An Update," *Bulletin of the American Meteorological Society* 74, no. 9 (1993): 1689–96.

20. Karl et al., "Detecting Climate Variations and Change," 1486.

21. National Research Council, *GOALS (Global Ocean-Atmosphere-Land System) for Predicting Seasonal-to-Interannual Climate: A Program of Observation, Modeling, and*

*Analysis* (National Academy Press, 1994); T. R. Karl et al., "Critical Issues for Long-Term Climate Monitoring," *Climatic Change* 31, no. 2 (1995): 185–221; National Research Council, *Adequacy of Climate Observing Systems*; Trenberth et al., "The Need for a Systems Approach to Climate Observations."

22. Trenberth et al., "The Need for a Systems Approach to Climate Observations," 1598. An earlier version of this list appeared in Karl et al., "Critical Issues for Long-Term Climate Monitoring."

23. W. Higgins et al., "NOAA Climate Test Bed (CTB)," National Weather Service Climate Prediction Center, Camp Springs, Maryland, 2007, www.cpc.noaa.gov.

24. R. S. Bradley et al., *A Climatic Data Bank for Northern Hemisphere Land Areas, 1851–1980* (US Department of Energy, Carbon Dioxide Research Division, 1985).

25. Bowker, *Memory Practices in the Sciences*.

26. Clayton, *World Weather Records*.

27. Bradley et al., *A Climatic Data Bank for Northern Hemisphere Land Areas*, 7–8.

28. World Meteorological Organization, *Guide to Climatological Practices* (1960) and *Guide to Climatological Practices* (1983).

29. P. D. Jones et al., "Variations in Surface Air Temperatures: Part 1. Northern Hemisphere, 1881–1980," *Monthly Weather Review* 110, no. 2 (1982), 59.

30. Robock, "The Russian Surface Temperature Data Set."

31. Jones et al., "Variations in Surface Air Temperatures: Part 1," 59–60; Robock, "The Russian Surface Temperature Data Set."

32. Spangler and Jenne, *World Monthly Surface Station Climatology*.

33. Hansen and Lebedeff, "Global Trends of Measured Surface Air Temperature."

34. Bradley et al., *A Climatic Data Bank for Northern Hemisphere Land Areas*, 34.

35. J. P. Palutikof and C. M. Goddess, "The Design and Use of Climatological Data Banks, with Emphasis on the Preparation and Homogenization of Surface Monthly Records," *Climatic Change* 9, no. 1 (1986), 142.

36. Bradley et al., *A Climatic Data Bank for Northern Hemisphere Land Areas*, 29–31.

37. J. M. Mitchell Jr., "Effect of Changing Observation Time on Mean Temperature," *Bulletin of the American Meteorological Bulletin* 39 (1958): 83–89.

38. Karl et al., "Detecting Climate Variations and Change," 1486.

39. Willett, "Temperature Trends of the Past Century"; Mitchell, "On the World-Wide Pattern of Secular Temperature Change."

40. Correlation coefficients of 0.89 (1950–1976, Vinnikov et al.), 0.95 (1881–1960, Budyko et al.), and 0.97 (1881–1960, Vinnikov et al.). K. Y. Vinnikov et al., "Contemporary Variations of the Northern Hemisphere Climate," *Meteorologiya i Gidrologiya* 6 (1980): 5–17.

41. R. J. Slutz et al., *COADS: Comprehensive Ocean-Atmosphere Data Set, Release 1* (University of Colorado and US National Center for Atmospheric Research, 1985); S. D. Woodruff et al., "A Comprehensive Ocean-Atmosphere Data Set," *Bulletin American Meteorological Society* 68, no. 10 (1987): 1239–50.

42. C. K. Folland and D. E. Parker, "Correction of Instrumental Biases in Historical Sea Surface Temperature Data," *Quarterly Journal of the Royal Meteorological Society* 121 (1995): 319–67; N. A. Rayner et al., "Improved Analyses of Changes and Uncertainties in Sea Surface Temperature Measured in situ since the Mid-Nineteenth Century: The HadSST2 Dataset," *Journal of Climate* 19, no. 3 (2006): 446–69.

43. E. C. Kent et al., "WMO Publication No. 47 Metadata and an Assessment of Voluntary Observing Ships Observation Heights in ICOADS," *Journal of Atmospheric and Oceanic Technology* 24, no. 2 (2006): 214–34.

44. D. W. J. Thompson et al., "A Large Discontinuity in the Mid-Twentieth Century in Observed Global-Mean Surface Temperature," *Nature* 453, no. 7195 (2008): 646–49.

45. Brohan et al., "Uncertainty Estimates in Regional and Global Observed Temperature Changes," D12106, emphasis added.

46. D. E. Parker et al., "Interdecadal Changes of Surface Temperature Since the Late Nineteenth Century," *Journal of Geophysical Research-Atmospheres* 99, no. D7 (1994).

47. P. D. Jones et al., "Global Temperature Variations Between 1861 and 1984," *Nature* 322, no. 6078 (1986): 430–34.

48. Rayner et al., "Improved Analyses of Changes and Uncertainties"; Brohan et al., "Uncertainty Estimates in Regional and Global Observed Temperature Changes."

49. Brohan et al., ibid.

## Chapter 12

1. Bengtsson's notes in R. Jenne, "History of Reanalysis (RJ0349)," National Center for Atmospheric Research, 2004, dss.ucar.edu. Bengtsson had taken part in ECMWF's FGGE Level IIIb analysis.

2. J. Shukla, *Executive Summary, NASA GSFC Global Habitability Program Draft, 22 February* (Goddard Space Flight Center, National Aeronautics and Space Administration, 1983), 3.

Notes to Chapter 12

3. K. E. Trenberth and J. G. Olson, "An Evaluation and Intercomparison of Global Analyses From the National Meteorological Center and the European Centre for Medium Range Weather Forecasts," *Bulletin of the American Meteorological Society* 69, no. 9 (1988), 1056.

4. Trenberth, "Atmospheric Circulation Climate Changes."

5. TOGA panel meeting notes, 5–6 February 1986, excerpted in Jenne, "History of Reanalysis."

6. Notes circa 1986, in Jenne, "History of Reanalysis," 77.

7. Bengtsson and Shukla, "Integration of Space and In Situ Observations to Study Global Climate Change," 1139–41.

8. S. D. Schubert et al., "An Assimilated Dataset for Earth Science Applications," *Bulletin of the American Meteorological Society* 74, no. 12 (1993): 2331–42; S. Schubert, "The GEOS-1 Reanalysis: Overview," in *Proceedings of the First WCRP International Conference on Reanalyses* (Silver Spring, 1998), 8–11.

9. J. K. Gibson et al., "ERA-15 Description," *ECMWF Reanalysis Project Report Series* 1 (1997): 72; Woods, *Medium-Range Weather Prediction*.

10. The US National Meteorological Center (NMC) reorganized as the National Centers for Environmental Prediction in 1995. Hence the reanalysis project was known as NMC-NCAR until 1995, and as NCEP-NCAR after that. To avoid confusion, I use the abbreviation NCEP-NCAR in all references to that reanalysis project.

11. For example, D. A. Paolino et al., "A Pilot Reanalysis Project at COLA," *Bulletin of the American Meteorological Society* 76, no. 5 (1995): 697–710.

12. M. Kanamitsu et al., "Overview of NCEP/DOE Reanalysis-2," World Meteorological Organization (2000), dss.ucar.edu.

13. Söderman letter to Jenne, in Jenne, "History of Reanalysis," 91.

14. R. L. Jenne, "Initiative to Prepare Data Inputs for Reanalyses," Data Support Section, Scientific Computing Division, National Center for Atmospheric Research, 1991, dss.ucar.edu, 4.

15. R. L. Jenne, "The Global Data Base for Climatic Research: A Preliminary Report to the WMO-ICSU Joint Organizing Committee for GARP (Original 1977; Revised 1978, 1979, 1983, 1988)," National Center for Atmospheric Research, 1988, dss.ucar.edu; Jenne, "Initiative to Prepare Data Inputs for Reanalyses"; R. Jenne, "Lessons Learned: Datasets Saved from Near Disasters, and Some Not So Lucky," typescript (National Center for Atmospheric Research, 1998): 19; R. L. Jenne, "A Global Reanalysis: Data Inputs and Methods," in *Proceedings of the 13th Annual Climate Diagnostics Workshop* (1988); W. Baum et al., "Report of the Atmospheric Sciences Data Panel," in *Study on the Long-Term Retention of Selected Scientific and*

*Technical Records of the Federal Government: Working Papers* (National Academy Press, 1995).

16. Jenne, "History of Reanalysis."

17. Ibid.

18. E. Kalnay et al., "The NCEP/NCAR 40-Year Reanalysis Project," *Bulletin of the American Meteorological Society* 77, no. 3 (1996): 437–72. At this writing, the ISI Web of Knowledge shows well over 6000 citations of this article.

19. R. Kistler et al., "The NCEP–NCAR 50-Year Reanalysis: Monthly Means CD-ROM and Documentation," *Bulletin of the American Meteorological Society* 82, no. 2 (2001): 247–68.

20. Woods, *Medium-Range Weather Prediction*, 180.

21. K. Onogi et al., "JRA-25: Japanese 25-Year Re-Analysis Project—Progress and Status," *Quarterly Journal of the Royal Meteorological Society* 131, no. 613 (2005).

22. G. P. Compo et al., "Feasibility of a 100-Year Reanalysis Using Only Surface Pressure Data," *Bulletin of the American Meteorological Society* 87, no. 2 (2006): 175–90; J. S. Whitaker et al., "Reanalysis without Radiosondes using Ensemble Data Assimilation," *Monthly Weather Review* 132, no. 5 (2004): 1190–200.

23. R. M. Dole et al., *Reanalysis of Historical Climate Data for Key Atmospheric Features: Implications for Attribution of Causes of Observed Change* (US Climate Change Science Program, 2008), 10.

24. L. R. Lait, "Systematic Differences Between Radiosonde Measurements," *Geophysical Research Letters* 29, no. 10 (2002): 1382.

25. Intergovernmental Panel on Climate Change, *Climate Change 2007: The Physical Science Basis. Contribution of Working Group I to the Fourth Assessment Report of the Intergovernmental Panel on Climate Change* (Cambridge University Press, 2007), 270.

26. D. P. Dee, "Detection and Correction of Model Bias During Data Assimilation," European Centre for Medium-Range Weather Forecasts, 2003, www.ecmwf.int; D. P. Dee, "Bias and Data Assimilation," *Quarterly Journal of the Royal Meteorological Society* 131, no. 613 (2005): 3323–43.

27. R. B. Rood, "Reanalysis," in *Data Assimilation for the Earth System*, ed. R. Swinbank et al. (Kluwer, 2003).

28. L. Bengtsson et al., "The Need for a Dynamical Climate Reanalysis," *Bulletin of the American Meteorological Society* 88, no. 4 (2007): 495–501.

29. A. J. Simmons et al., "Comparison of Trends and Low-Frequency Variability in CRU, ERA-40, and NCEP/NCAR Analyses of Surface Air Temperature," *Journal of Geophysics Research* 109 (2004), 5.

30. Intergovernmental Panel on Climate Change, *Climate Change 2007*, chapter 3 supplementary materials, SM.3–8, www.ipcc.ch.

31. G. A. Meehl et al., "Low-Frequency Variability and $CO_2$ Transient Climate Change," *Climate Dynamics* 8, no. 3 (1993): 117–33; S. H. Schneider and S. L. Thompson, "Atmospheric $CO_2$ and Climate: Importance of the Transient Response," *Journal of Geophysical Research* 86, no. C4 (1981): 3135–47.

32. For example, D. J. Karoly et al., "An Example of Fingerprint Detection of Greenhouse Climate Change," *Climate Dynamics* 10, no. 1 (1994): 97–105; B. D. Santer et al., "Correlation Methods in Fingerprint Detection Studies," *Climate Dynamics* 8 (1993): 265–76.

33. The tropopause marks the boundary between the troposphere, or lower atmosphere, and the stratosphere. Within this shallow horizontal band occurs a relatively abrupt transition from a steady cooling with height (characteristic of the troposphere) to a steady increase of temperature with height (in the stratosphere). Tropopause height varies with latitude as well as with weather conditions, but should maintain a steady climatological average in the absence of forcing. B. D. Santer et al., "Behavior of Tropopause Height and Atmospheric Temperature in Models, Reanalyses, and Observations: Decadal Changes," *Journal of Geophysical Research-Atmospheres* 108, no. D1 (2003): 4002; B. D. Santer et al., "Identification of Anthropogenic Climate Change Using a Second-Generation Reanalysis," *Journal of Geophysical Research* 109 (2004): D21104, 35; Intergovernmental Panel on Climate Change, *Climate Change 2007*.

34. P. Arkin et al., *Ongoing Analysis of the Climate System: A Workshop Report* (University Corporation for Atmospheric Research, 2003).

## Chapter 13

1. M. Sundberg, "Credulous Modellers and Suspicious Experimentalists? Comparison of Model Output and Data in Meteorological Simulation Modelling," *Science Studies* 19, no. 1 (2006): 52–68.

2. J. T. Kiehl, "Atmospheric General Circulation Modeling," in *Climate System Modeling*, ed. K. E. Trenberth (Cambridge University Press, 1992), 338.

3. M. J. Iacono et al., "Radiative Forcing by Long-Lived Greenhouse Gases: Calculations with the AER Radiative Transfer Models," *Journal of Geophysical Research* 113, no. D13103 (2008).

4. J. P. Chaboureau and P. Bechtold, "A Simple Cloud Parameterization Derived from Cloud Resolving Model Data: Diagnostic and Prognostic," *Journal of the Atmospheric Sciences* 59, no. 15 (2002): 2362–72.

5. R. Pincus et al., "A Fast, Flexible, Approximate Technique for Computing Radiative Transfer in Inhomogeneous Cloud Fields," *Journal of Geophysical Research* 108, no. D13 (2003): 4376.

6. Rosinski and Williamson traced this effect to a single line of code in the CCM-2 cloud parameterization. In the model, "the cloud fraction ramps up from 0 at the limiting relative humidity to 1 at 100 percent. The closer the limiting relative humidity ($RH_{lim}$) is to 1, the sharper the ramp. The maximum value of $RH_{lim}$ in the algorithm is 0.999. . . . At one particular point in the test calculation $RH_{lim}$ = 0.999, and RH is 0.999730300596838 and 0.999730335751138 in the two cases . . . , with a difference in the eighth decimal digit. The subtraction of 0.999 in the equation, coupled with the division by $(1 - RH_{lim})$, moves the difference in the cloud fraction between the two cases up from the eighth digit to the fifth. This immediately affects the radiative heating and leads to the jump in the temperature difference at that point. . . . The (resulting) change in the RMS temperature difference is around one order of magnitude, not three, because the upshifting just described occurred at just one point in the domain, and occurred in a variable that affects the temperature, rather than the temperature itself." J. M. Rosinski and D. L. Williamson, "The Accumulation of Rounding Errors and Port Validation for Global Atmospheric Models," *SIAM Journal on Scientific Computing* 18, no. 2 (1997), 558.

7. G. A. Meehl, "Global Coupled Models: Atmosphere, Ocean, Sea Ice," in *Climate System Modeling*, ed. K. E. Trenberth (Cambridge University Press, 1992), 555–81.

8. J. T. Houghton et al., *Climate Change 1995: The Science of Climate Change* (Cambridge University Press, 1996).

9. J. T. Kiehl and P. R. Gent, "The Community Climate System Model, Version 2," *Journal of Climate* 17, no. 19 (2004): 3666–82; R. A. Kerr, "Climate Change: Model Gets It Right—Without Fudge Factors," *Science* 276, no. 5315 (1997): 1041.

10. Intergovernmental Panel on Climate Change, *Climate Change 2007*, table 8.1.

11. R. J. Charlson and T. M. L. Wigley, "Sulfate Aerosol and Climatic Change," *Scientific American* 270, no. 2 (1994): 48–57; S. J. Ghan and S. E. Schwartz, "Aerosol Properties and Processes: A Path from Field and Laboratory Measurements to Global Climate Models," *Bulletin of the American Meteorological Society* 88, no. 7 (2007): 1059–83; J. T. Kiehl and B. P. Briegleb, "The Relative Roles of Sulfate Aerosols and Greenhouse Gases in Climate Forcing," *Science* 260, no. 5106 (1993): 311–14; J. E. Penner et al., "Quantifying and Minimizing Uncertainty of Climate Forcing by Anthropogenic Aerosols," *Bulletin of the American Meteorological Society* 75, no. 3 (1994): 375–400.

12. J. T. Kiehl, "Twentieth Century Climate Model Response and Climate Sensitivity," *Geophysical Research Letters* 34 (2007): L22710; S. E. Schwartz et al., "Quantifying Climate Change—Too Rosy a Picture?," *Nature Reports Climate Change* 707 (2007): 23–24.

13. M. Lahsen, Climate Rhetoric: Constructions of Climate Science in the Age of Environmentalism, Ph.D. dissertation, Rice University, 1998; Lahsen, "Seductive Simulations?"; S. Shackley, "Epistemic Lifestyles in Climate Change Modeling."

14. Lahsen, Climate Rhetoric.

15. Lahsen, "Seductive Simulations?"

16. Shackley, "Epistemic Lifestyles."

17. H. Svensmark and N. Calder, *The Chilling Stars: The New Theory of Climate Change* (Icon Books, 2007).

18. A. Petersen, *Simulating Nature: A Philosophical Study of Computer-Simulation Uncertainties and Their Role in Climate Science and Policy Advice* (Het Spinhuis, 2007), 39.

19. D. A. Randall and B. A. Wielicki, "Measurements, Models, and Hypotheses in the Atmospheric Sciences," *Bulletin of the American Meteorological Society* 78, no. 3 (1997), 403–04.

20. Intergovernmental Panel on Climate Change, *Climate Change 2007*, 596.

21. Gates, *AMIP: The Atmospheric Model Intercomparison Project*; T. M. L. Wigley et al., "Indices and Indicators of Climate Change: Issues of Detection, Validation, and Climate Sensitivity," in *The Science of Climate Change*, ed. I. M. Mintzer (Cambridge University Press, 1992); S. H. Schneider, "Detecting Climatic Change Signals: Are There Any Fingerprints?" *Science* 263, no. 5145 (1994): 341–47; Schneider, "Introduction to Climate Modeling"; Kiehl, "Atmospheric General Circulation Modeling."

22. N. Oreskes et al., "Verification, Validation, and Confirmation of Numerical Models in the Earth Sciences," *Science* 263, no. 5147 (1994): 641–46.

23. Popper, *The Logic of Scientific Discovery*; Popper, *Conjectures and Refutations: The Growth of Scientific Knowledge* (Basic Books, 1962).

24. Houghton et al., *Climate Change 1995*, 235.

25. Lawrence Livermore National Laboratory, *Program for Climate Model Diagnosis and Intercomparison*, 2009, www-pcmdi.llnl.gov.

26. McGuffie and Henderson-Sellers, *A Climate Modeling Primer*, chapter 6.

27. P. N. Edwards, "Global Comprehensive Models in Politics and Policymaking," *Climatic Change* 32 (1996): 149–61.

28. D. A. Stainforth et al., "Uncertainty in Predictions of the Climate Response to Rising Levels of Greenhouse Gases," *Nature* 433 (2005): 403–06.

29. C. Piani et al., "Constraints on Climate Change from a Multi-Thousand Member Ensemble of Simulations," *Geophysical Research Letters* 32 (2005).

## Chapter 14

1. As was discussed in chapter 11, recent re-examination of the surface data indicates that part of the slight cooling trend from 1940 to 1970 may have been apparent rather than real, due to widespread changes in marine measurement techniques that took place between 1939 and 1945 (Thompson et al., "A Large Discontinuity in the Mid-Twentieth Century in Observed Global-Mean Surface Temperature").

2. Plass, "Effect of Carbon Dioxide Variations on Climate"; Revelle and Suess, "Carbon Dioxide Exchange"; Conservation Foundation, *Implications of Rising Carbon Dioxide Content of the Atmosphere*, 1963.

3. See Fleming, "Fixing the Weather and Climate."

4. National Science Foundation, *16th Annual Report* (1966), 52.

5. See Kwa, "The Rise and Fall of Weather Modification"; Fleming, "Fixing the Weather and Climate."

6. Panel on Weather and Climate Modification, *Weather and Climate Modification: Problems and Prospects* (National Academy of Sciences, National Research Council, 1966); Panel on Weather and Climate Modification, *Weather and Climate Modification: Problems and Prospects. Final Report of the Panel on Weather and Climate Modification to the Committee on Atmospheric Sciences* (National Academy of Sciences, National Research Council, 1966).

7. T. F. Malone, "Weather Modification: Implications of the New Horizons in Research," *Science* 156, no. 3777 (1967): 897–901; National Science Foundation, *16th Annual Report*.

8. J. W. Tukey et al., *Restoring the Quality of Our Environment: Report of the Environmental Pollution Panel, President's Science Advisory Committee* (Government Printing Office, 1965).

9. Hart and Victor, "Scientific Elites," 665.

10. Study of Critical Environmental Problems, *Man's Impact on the Global Environment: Assessment and Recommendations for Action* (MIT Press, 1970); Study of Man's Impact on Climate, *Inadvertent Climate Modification: Report of the Study of Man's Impact on Climate* (MIT Press, 1971).

11. B. Webster, "Scientists Ask SST Delay Pending Study of Pollution," *New York Times*, August 2, 1970.

12. J. Mormino et al., "Climatic Impact Assessment Program: Development and Accomplishments, 1971–1975," (1975): xxx, 206.

13. Consortium for International Earth Science Information Network (CIESIN), "Thematic Guide to Integrated Assessment Modeling of Climate Change" (1995), sedac.ciesin.columbia.edu.

14. S. I. Rasool and S. H. Schneider, "Atmospheric Carbon Dioxide and Aerosols: Effects of Large Increases on Global Climate," *Science* 173, no. 3992 (1971), 138, emphasis added. These speculations (which neglected water vapor feedbacks) would return to haunt Schneider, who later became an outspoken public advocate for global warming theories. See e.g. S. H. Schneider, *Global Warming: Are We Entering the Greenhouse Century?* (Vintage Books, 1989); S. H. Schneider, *Laboratory Earth: The Planetary Gamble We Can't Afford to Lose* (Basic Books, 1997).

15. Study of Critical Environmental Problems, *Man's Impact on the Global Environment*, 5.

16. Ibid., 78, 88.

17. Ibid., 6–7, emphasis added.

18. Rasool and Schneider, "Atmospheric Carbon Dioxide and Aerosols."

19. W. W. Kellogg, "Mankind's Impact on Climate: The Evolution of an Awareness," *Climatic Change* 10, no. 2 (1987), 121–22.

20. United Nations, "Report of the United Nations Conference on the Human Environment" (1973), Recommendation 79.

21. F. Elichirigoity, *Planet Management: Limits to Growth, Computer Simulation, and the Emergence of Global Space* (Northwestern University Press, 1999).

22. J. Forrester, *Industrial Dynamics* (MIT Press, 1961); *Urban Dynamics* (MIT Press, 1969); "Industrial Dynamics: A Major Breakthrough for Decision Makers," *Harvard Business Review* 36, no. 4 (1958): 37–66.

23. Forrester, *Urban Dynamics*, 9.

24. E. Pestel, *Beyond the Limits to Growth* (Universe Books, 1989), 24.

25. In 1971 Forrester published a technical report on the world models titled *World Dynamics* (MIT Press). After that, he moved on to other projects and did not stay deeply involved in the System Dynamics Group.

26. D. L. Meadows et al., *Dynamics of Growth in a Finite World* (Wright-Allen, 1974).

27. A volume of canonical critiques is H. S. D. Cole et al., eds., *Models of Doom: A Critique of the Limits to Growth* (Universe Books, 1973). There were many others; a reasonably comprehensive retrospective on the debate is B. P. Bloomfield, *Modeling the World: The Social Constructions of Systems Analysts* (Blackwell, 1986).

28. Meadows interview, in F. Elichirigoity, *Planet Management*, 75–76.

29. V. Smil, "Limits to Growth Revisited: A Review Essay," *Population and Development Review* 31, no. 1 (2005): 157–64.

30. M. Mesarovic and E. Pestel, *Mankind at the Turning Point* (Signet Books, 1974); C. R. Humphrey and F. R. Buttel, *Environment, Energy, and Society* (Wadsworth, 1982); Bloomfield, *Modeling the World*.

31. A. King, "The Club of Rome and Its Policy Impact," in *Knowledge and Power in a Global Society*, ed. W. M. Evan (Sage, 1981).

32. H. Raiffa, "How IIASA Began," International Institute for Applied Systems Analysis, Laxenburg, Austria, 1992, www.iiasa.ac.at, emphasis added.

33. D. Meadows et al., *Dynamics of Growth in a Finite World*; Meadows et al., *Groping in the Dark: The First Decade of Global Modeling* (Wiley, 1982); Meadows et al., *Beyond the Limits: Confronting Global Collapse, Envisioning a Sustainable Future* (Chelsea Green, 1992); Meadows, "Places to Intervene in a System," *Whole Earth* 91 (1997): 78–84; Meadows et al., *Limits to Growth: The 30 Year Global Update* (Chelsea Green, 2004).

34. Council on Environmental Quality, *The Global 2000 Report to the President: Entering the Twenty-First Century* (Government Printing Office, 1980); Pestel, *Beyond the Limits to Growth*.

35. W. W. Kellogg, "Mankind's Impact on Climate: The Evolution of an Awareness," *Climatic Change* 10 (1987): 113–36; W. W. Kellogg, "Effects of Human Activities on Global Climate: I," *WMO Bulletin* 26 (1977): 229–40.

36. J. Ausubel, "Federal Organization for Climate and Energy: A Brief History and Analysis," unpublished typescript (1989), 1; K. Hechler, *Toward the Endless Frontier: History of the Committee on Science and Technology, 1959–79* (Government Printing Office, 1980), 972.

37. The CIA reports "A Study of Climatological Research as It Pertains to Intelligence Problems" and "Potential Implications of Trends in World Population, Food Production, and Climate," released to the White House Domestic Council in 1974, are reproduced in *The Weather Conspiracy: The Coming of the New Ice Age: A Report* (Ballantine Books, 1977).

38. Congressional Record, "National Climate Program Act, Bill Summary & Status for the 95th Congress: H.R. 6669 (Public Law 95-367, 9/17/78)," 1978.

39. Board on Atmospheric Sciences and Climate, "The National Climate Program: Early Achievements and Future Directions," 1986.

40. Hechler, *Toward the Endless Frontier*, 988.

41. Council on Environmental Quality, *The Global 2000 Report to the President*, volume 1, 6.

42. Ibid., volume 3, 184, 176, emphasis added.

43. Geophysics Study Committee, *Energy and Climate* (National Academy of Sciences, 1977).

44. Ausubel, "Federal Organization for Climate and Energy."

45. W. L. Gates and WMO-ICSU Joint Organizing Committee, eds., *Report of the JOC Study Conference on Climate Models: Performance, Intercomparison and Sensitivity Studies* (World Meteorological Organization, 1979).

46. J. Smagorinsky, "Overview of the Climate Modelling Problem," in ibid. 2–3.

47. A. D. Hecht and D. Tirpak, "Framework Agreement on Climate Change: A Scientific and Policy History," *Climatic Change* 29, no. 4 (1995): 371–402.

48. Ad Hoc Study Group on Carbon Dioxide and Climate, "Carbon Dioxide and Climate: A Scientific Assessment," 1979, 2.

49. Global Atmospheric Research Programme and World Climate Research Programme, *Report of the First Session of the Joint Scientific Committee for the World Climate Research Programme and the Global Atmospheric Research Programme* (World Meteorological Organization, 1980), Annex F.

50. Ad Hoc Study Group on Carbon Dioxide and Climate, "Carbon Dioxide and Climate: A Scientific Assessment," 16.

51. J. van der Sluijs et al., "Anchoring Devices in Science for Policy: The Case of Consensus Around Climate Sensitivity," *Social Studies of Science* 28, no. 2 (1998): 291–323; van der Sluijs, *Anchoring Amid Uncertainty* (University of Utrecht, 1997).

52. J. T. Houghton et al., *Climate Change: The IPCC Scientific Assessment* (Cambridge University Press, 1990), xxv.

53. Quoted in van der Sluijs, *Anchoring Amid Uncertainty*, 48.

54. Email to van der Sluijs, quoted in ibid., 44.

55. van der Sluijs, *Anchoring Amid Uncertainty*, chapter 2; van der Sluijs et al., "Anchoring Devices in Science for Policy." See also J. van der Sluijs and J. van Eijndhoven, "Closure of Disputes in Assessments of Climate Change in the Netherlands," *Environmental Management* 22, no. 4 (1998): 597–609.

56. For a wide-ranging comparative history of how climate change became a public issue elsewhere in the world, see the report of the Harvard Social Learning Project, which also covers ozone depletion and acid rain. Country studies include Germany, the UK, the Netherlands, the USSR, Mexico, Canada, Japan, and the United States. The report is W. C. Clark et al., *Learning to Manage Global Environmental Risks: A Comparative History of Social Responses to Climate Change, Ozone Depletion, and Acid Rain* (MIT Press, 2001).

57. Ibid.

58. L. Eden, *Whole World on Fire: Organizations, Knowledge, and Nuclear Weapons Devastation* (Cornell University Press, 2006).

59. S. H. Schneider and C. Mass, "Volcanic Dust, Sunspots, and Temperature Trends," *Science* 190, no. 4216 (1975): 741–46.

60. National Research Council, *Long-Term Worldwide Effects of Multiple Nuclear Weapons Detonations* (National Academy of Sciences, 1975).

61. For a review see M. Dörries, "In the Public Eye: Volcanology and Climate Change Studies in the 20th Century," *Historical Studies in the Physical and Biological Sciences* 37, no. 1 (2006): 87–125.

62. L. W. Alvarez et al., "Extraterrestrial Cause for the Cretaceous-Tertiary Extinction," *Science* 208, no. 4448 (1980): 1095–108.

63. P. J. Crutzen and J. W. Birks, "The Atmosphere After a Nuclear War: Twilight at Noon," *Ambio* 11 (1982): 114–25.

64. R. P. Turco et al., "Nuclear Winter: Global Consequences of Multiple Nuclear Explosions," *Science* 222 (1983), 1283.

65. R. P. Turco et al., "The Climatic Effects of Nuclear War," *Scientific American* 251, no. 2 (1984): 33–43.

66. Turco et al., "Nuclear Winter."

67. R. Scheer, *With Enough Shovels: Reagan, Bush, and Nuclear War* (Random House, 1982).

68. C. Sagan, "Nuclear War and Climatic Catastrophe: Some Policy Implications," *Foreign Affairs* 62, no. 2 (1983): 257–92.

69. P. R. Ehrlich et al., *The Cold and the Dark: The World After Nuclear War* (Norton, 1984).

70. C. Covey et al., "'Nuclear Winter': A Diagnosis of Atmospheric General Circulation Model Simulations," *Journal of Geophysical Research* 90 (1985): 5615–28; Covey et al., "Global Atmospheric Effects of Massive Smoke Injections from a Nuclear War: Results from General Circulation Model Simulations," *Nature* 308, no. 5954 (1984): 21–25; S. L. Thompson et al., "Global Consequences of Nuclear War: Simulations with Three Dimensional Models," *Ambio* 13, no. 4 (1984): 236–43.

71. S. L. Thompson and S. H. Schneider, "Nuclear Winter Reappraised," *Foreign Affairs* 64, no. 5 (1986): 981–1005.

72. Mormino et al., "Climatic Impact Assessment Program," 146.

73. M. J. Molina and F. S. Rowland, "Stratospheric Sink for Chlorofluoromethanes: Chlorine Atomic Catalysed Destruction of Ozone," *Canadian Journal of Chemistry* 52, no. 8 (1974): 1610–15.

74. E. A. Parson, *Protecting the Ozone Layer: Science and Strategy* (Oxford University Press, 2003), 70, 73.

75. J. C. Farman et al., "Large Losses of Total Ozone in Antarctica Reveal Seasonal $ClO_x/NO_x$ Interaction," *Nature* 315, no. 6016 (1985): 207–10.

76. Parson, *Protecting the Ozone Layer*, 84–86 and n. 134.

77. See for example the Wikipedia entry on "Ozone depletion" (version of March 21, 2009), or S. C. Zehr, "Accounting for the Ozone Hole: Scientific Representations of an Anomaly and Prior Incorrect Claims in Public Settings," *Sociological Quarterly* 35, no. 4 (1994), 610.

78. Quoted in B. Sparling, "Ozone Depletion, History and Politics," NASA Advanced Supercomputing Division, 2001, www.nas.nasa.gov.

79. In the run-up to the 1992 Framework Convention on Climate Change, the Montreal Protocol's remarkable success was often cited, but as policy issues the two cases are dramatically different. CFCs were produced by only a handful of manufacturers worldwide, and alternative refrigerants already existed that could be substituted at relatively low financial and technical overhead. These features made the ban readily enforceable and comparatively inexpensive. Anthropogenic carbon dioxide, by contrast, comes mainly from the combustion of fossil fuels. Carbon-free or carbon-neutral substitutes exist, but all of them remain far more expensive than these fuels. Nuclear power, often touted as the least expensive substitute, is far from carbon-free if the fossil fuels required for uranium mining, waste disposal, and plant construction and retirement are included in the accounting, and the unsolved problems of nuclear waste and weapons proliferation would become even more intractable if new plants came online in large numbers. Even if cost-neutral alternatives existed, the gigantic global infrastructure for fossil fuel extraction, refinement, distribution, and use could not be swiftly replaced.

80. R. B. Rood, "Global Ozone Minima in the Historical Record," *Geophysical Research Letters* 13, no. 12 (1986): 1244–47.

81. S. Solomon, "The Mystery of the Antarctic Ozone 'Hole,'" *Reviews of Geophysics* 26, no. 1 (1987), 146.

82. Parson, *Protecting the Ozone Layer*, 171; R. Stolarski, *Report of the International Ozone Trends Panel* (World Meteorological Organization, 1988); R. P. Kane, "Long-Term Variation of Total Ozone," *Pure and Applied Geophysics* 127, no. 1 (1988): 143–54.

83. Haas, *Saving the Mediterranean* and "Obtaining International Environmental Protection through Epistemic Consensus."

84. S. F. Singer, "Ozone, Skin Cancer, and the SST," *Aerospace America* 32, no. 7 (1994): 22; P. J. Michaels et al., "Analyzing Ultraviolet-B Radiation: Is There a Trend?" *Science* 264, no. 5163 (1994): 1341–42.

85. M. Rex et al., "Impact of Recent Laboratory Measurements of the Absorption Cross Section of ClOOCl on Our Understanding of Polar Ozone Chemistry," in *Stratospheric Processes and Their Role in Climate: 4th General Assembly* (Bologna, 2008); Q. Schiermeier, "Chemists Poke Holes in Ozone Theory," *Nature* 449, no. 7161 (2007): 382–83.

86. Schneider, *Global Warming*, 120–32.

87. Ausubel, "Federal Organization for Climate and Energy," 2.

88. National Research Council and Carbon Dioxide Assessment Committee, *Changing Climate: Report of the Carbon Dioxide Assessment Committee* (National Academy Press, 1983).

89. J. S. Hoffman et al., *Projecting Future Sea Level Rise: Methodology, Estimates to the Year 2100, and Research Needs* (Office of Policy and Resource Management, US Environmental Protection Agency, 1983).

90. Ausubel, "Federal Organization for Climate and Energy," 3.

91. World Climate Programme et al., *Report of the International Conference on the Assessment of the Role of Carbon Dioxide and of Other Greenhouse Gases in Climate Variations and Associated Impacts, Villach, Austria, 9–15 October 1985* (Paris, 1986), 1.

92. Quoted in R. A. Pielke, "Policy History of the US Global Change Research Program: Part I, Administrative Development," *Global Environmental Change* 10 (2000), 17.

93. D. Bodansky, "The History of the Global Climate Change Regime," in *International Relations and Global Climate Change*, ed. U. Luterbacher and D. F. Sprinz (MIT Press, 2001), 23.

94. M. C. Nisbet and T. Myers, "Trends: Twenty Years of Public Opinion About Global Warming," *Public Opinion Quarterly* 71, no. 3 (2007), 446.

95. Congressional Record, "Global Climate Protection Act, Bill Summary & Status for the 100th Congress: H.R.1777, Public Law 100-204 (12/22/87)," 1987.

96. Quoted in B. Block, "A Look Back at James Hansen's Seminal Testimony on Climate, Part Two," Worldwatch Institute and Grist Environmental News and Commentary, 2008, gristmill.grist.org.

97. Nisbet and Myers, "Trends," 446.

98. World Commission on Environment and Development, *Our Common Future* (Oxford University Press, 1987).

99. World Meteorological Organization, *The Changing Atmosphere: Implications for Global Security* (1988), 292.

100. M. Molitor, "The United Nations Climate Change Agreements," in *The Global Environment: Institutions, Law and Policy*, ed. N. J. Vig and R. S. Axelrod (Earthscan, 1999).

101. Houghton et al., *Climate Change (1990)*, 89, xxv, xxii.

102. Houghton et al., *Climate Change 1995*, chapter 6.

103. The George H. W. Bush administration was internally divided on the climate issue. Presidential science advisor Allan Bromley thought the problem serious, convening a number of White House conferences to discuss it (D. A. Bromley, *The President's Scientists: Reminiscences of a White House Science Advisor*, Yale University Press, 1994). But Bush's influential chief of staff, John Sununu, remained among the skeptics. Sununu had an engineering background and repeatedly engaged climate scientists in technical discussions. NCAR modeler Warren Washington recalled that Sununu "asked many questions, such as what type of numerical schemes were we using: finite difference, spectral, or finite element? How were we handling radiation in the model? Were we using line-by-line or bands approximations for the radiative emission and absorption? Clearly, these are not the typical White House questions." At one point Sununu requested a climate model he could run on the Compaq 386 personal computer in his office. Washington, with the help of others at NCAR, delivered a simple one-dimensional radiative-convective model; whether Sununu actually used it is not clear. According to Washington, Sununu misunderstood certain basic elements of the climate system but resisted scientists' attempts to clarify them (W. M. Washington, *Odyssey in Climate Modeling, Global Warming, and Advising Five Presidents*, Lulu.com, 2006; Edwards, interview with Washington).

104. The GCRP was known as the Climate Change Science program from 2003 to 2008.

## Chapter 15

1. IPCC Working Group I treats the scientific understanding of climate change. Working Group II covers human and environmental impacts, and Working Group III assesses mitigation and adaptation.

2. See P. N. Edwards and S. H. Schneider, "The IPCC 1995 Report: Broad Consensus or 'Scientific Cleansing'?" *Ecofables/Ecoscience* 1, no. 1 (1997): 3–9; P. N. Edwards and S. H. Schneider, "Governance and Peer Review in Science-for-Policy: The Case of the IPCC Second Assessment Report," in *Changing the Atmosphere: Expert Knowledge and Environmental Governance*, ed. C. A. Miller and P. N. Edwards (MIT Press, 2001).

3. T. Skodvin, "Revised Rules of Procedure for the IPCC Process," *Climatic Change* 46, no. 4 (2000): 409–15; Intergovernmental Panel on Climate Change, "Procedures

for the Preparation, Review, Acceptance, Adoption, Approval and Publication of IPCC Reports, Adopted at the Fifteenth Session (San Jose, 15–18 April 1999) and Amended at the Twentieth Session (Paris, 19–21 February 2003), Twenty-First Session (Vienna, 3 and 6–7 November 2003), and Twenty-Ninth Session (Geneva, 31 August–4 September 2008)," 2008, www.ipcc.ch.

4. Houghton et al., *Climate Change 1995*, 6.

5. Intergovernmental Panel on Climate Change, *Climate Change 2001: Synthesis Report: Summary for Policymakers* (Cambridge University Press, 2001), 5.

6. Intergovernmental Panel on Climate Change, *Climate Change 2007: Synthesis Report: Summary for Policymakers*, 2, 5.

7. O. D. Mjøs, "Presentation of the Nobel Peace Prize 2007," Nobel Foundation, 2007.

8. Intergovernmental Panel on Climate Change, *Climate Change 2007*.

9. N. Oreskes, "The Scientific Consensus on Climate Change: How Do We Know We're Not Wrong?" in J. F. DiMento and P. Doughman, *Climate Change: What It Means for Us, Our Children, and Our Grandchildren* (MIT Press, 2007), 71. Although Oreskes' result has been challenged, none of the criticisms hold up to scrutiny if only the peer-reviewed scientific literature is considered. See J. Cook, "The Skeptic Argument: Naomi Oreskes' Study on Consensus Was Flawed," www.skepticalscience.com.

10. O. Heffernan, "Copenhagen Summit Urges Immediate Action on Climate Change," *Nature News*, March 12, 2009.

11. R. Gelbspan, *The Heat Is On: The High Stakes Battle Over Earth's Threatened Climate* (Addison-Wesley, 1997); R. Gelbspan, *Boiling Point: How Politicians, Big Oil and Coal, Journalists, and Activists Are Fueling the Climate Crisis—and What We Can Do to Avert Disaster* (Basic Books, 2004); C. Mooney, *Republican War on Science* (Basic Books, 2006).

12. H. M. Collins, *Changing Order: Replication and Induction in Scientific Practice* (Sage, 1985); Collins and Pinch, *The Golem*.

13. Jasanoff, *The Fifth Branch*; Jasanoff, "Science, Politics, and the Renegotiation of Expertise at EPA," *Osiris* 7 (1991): 195–217; Jasanoff, *Science at the Bar: Law, Science, and Technology in America* (Harvard University Press, 1995); Jasanoff, "Contested Boundaries in Policy-Relevant Science," *Social Studies of Science* 17 (1987): 195–230.

14. Y. Ezrahi, *The Descent of Icarus: Science and the Transformation of Contemporary Democracy* (Harvard University Press, 1990).

15. S. P. Hays, "Three Decades of Environmental Politics: The Historical Context," in *Government and Environmental Politics*, ed. M. J. Lacey (Woodrow Wilson Center

Press, 1989), 19–79; S. P. Hays, *Beauty, Health, and Permanence: Environmental Politics in the United States, 1955-1985* (Cambridge University Press, 1987).

16. A. Brandt, *The Cigarette Century: The Rise, Fall, and Deadly Persistence of the Product That Defined America* (Basic Books, 2009).

17. D. Michaels, "Doubt Is Their Product: Industry Groups Are Fighting Government Regulation by Fomenting Scientific Uncertainty," *Scientific American* 292, no. 6 (2005): 96–101.

18. N. Oreskes, "You *Can* Argue with the Facts: A Political History of Climate Change," University of Michigan Science, Technology & Public Policy Program, 2008; N. Oreskes and E. Conway, *Fighting Facts* (Bloomsbury Books, forthcoming).

19. Gelbspan, *The Heat Is On*.

20. Western Fuels Association, *Annual Report*, 1993, 5, 13.

21. US House of Representatives Committee on Oversight and Government Reform, "White House Engaged in Systematic Effort to Manipulate Climate Change Science," 2007, oversight.house.gov.

22. A. Revkin, "Industry Ignored Its Scientists on Climate," *New York Times*, April 24, 2009.

23. Luntz Research Companies, "Straight Talk," 2002.

24. M. Oppenheimer and R. H. Boyle, *Dead Heat: The Race Against the Greenhouse Effect* (Basic Books, 1990).

25. Michaels, "Doubt Is Their Product."

26. A. Gore, *The Assault on Reason* (Penguin, 2007); M. Bowen, *Censoring Science: Inside the Political Attack on Dr. James Hansen and the Truth of Global Warming* (Penguin Group, 2008); S. Shulman, *Undermining Science: Suppression and Distortion in the Bush Administration* (University of California Press, 2007); Mooney, *Republican War on Science*.

27. Office of Inspector General, National Aeronautics and Space Administration, *Investigative Summary Regarding Allegations That NASA Suppressed Climate Change Science and Denied Media Access to Dr. James E. Hansen, a NASA Scientist* (2008).

28. Another 42% were either dominated by anthropogenic accounts of warming or gave only those accounts. About 6% represented only the skeptical position. M. T. Boykoff and J. M. Boykoff, "Balance as Bias: Global Warming and the US Prestige Press," *Global Environmental Change* 14 (2004): 125–36.

29. M. T. Boykoff, "Lost in Translation? United States Television News Coverage of Anthropogenic Climate Change, 1995–2004," *Climatic Change* 86, no. 1 (2008): 1–11.

30. M. T. Boykoff and J. M. Boykoff, "Climate Change and Journalistic Norms: A Case Study of US Mass-Media Coverage," *Geoforum* 38, no. 6 (2007): 1190–204. See also L. Antilla, "Climate of Scepticism: US Newspaper Coverage of the Science of Climate Change," *Global Environmental Change* 15, no. 4 (2005): 338–52.

31. Union of Concerned Scientists, "Sound Science Initiative," www.ucsusa.org.

32. G. E. Brown Jr., Rep., "Environmental Science Under Siege: Fringe Science and the 104th Congress," 1996, section IV.D.

33. UAH originally called this channel 2R. The name was later changed to the more suggestive 2LT, a convention I adopt throughout for consistency. R. W. Spencer and J. R. Christy, "Precision and Radiosonde Validation of Satellite Gridpoint Temperature Anomalies. Part II: A Tropospheric Retrieval and Trends During 1979–90," *Journal of Climate* 5, no. 8 (1992): 858-66.

34. J. Hansen et al., "Satellite and Surface Temperature Data At Odds?," *Climatic Change* 30, no. 1 (1995): 104.

35. J. R. Christy, "Temperature Above the Surface Layer," *Climatic Change* 31, no. 2 (1995): 467.

36. Ibid., 472.

37. J. R. Christy, "Testimony of John R. Christy, July 10, 1997," US Senate Committee on Environment and Public Works, 1997, emphasis added.

38. R. Spencer et al., "Global Climate Monitoring: The Accuracy of Satellite Data," NASA Marshall Space Sciences Laboratory, 1997.

39. J. R. Christy et al., "Analysis of the Merging Procedure for the MSU Daily Temperature Time Series," *Journal of Climate* 11 (1998): 2016–2041.

40. The literature related to the MSU controversy is vast. For a detailed, reliable recent overview, see T. R. Karl et al., eds., *Temperature Trends in the Lower Atmosphere: Steps for Understanding and Reconciling Differences* (U.S. Climate Change Science Program, 2006). Major contributions include the following: J. R. Christy, "Error Estimates of Version 5.0 of MSU/AMSU Bulk Atmospheric Temperatures," *Journal of Atmospheric and Oceanic Technology* 20 (2003): 613–29; Christy et al., "Analysis of the Merging Procedure for the MSU Daily Temperature Time Series,"; Christy et al., "Reducing Noise in the MSU Daily Lower-Tropospheric Global Temperature Dataset," *Journal of Climate* 8 (1995): 888–96; Christy et al., "Tropospheric Temperature Change since 1979 from Tropical Radiosonde and Satellite Measurements," *Journal of Geophysical Research* 112, no. D6 (2007); Christy et al., "MSU Tropospheric Temperatures: Dataset Construction and Radiosonde Comparisons," *Journal of Atmospheric and Oceanic Technology* 17, no. 9 (2000): 1153–70; D. H. Douglass et al., "A Comparison of Tropical Temperature Trends with Model Predictions," *International Journal of Climatology* 27 (2007), online, doi: 10.1002/joc.1651; Q. Fu, "Contribution

of Stratospheric Cooling to Satellite-Inferred Tropospheric Temperature Trends," *Nature* 429 (2004): 55–58; D. J. Gaffen, "Falling Satellites, Rising Temperatures?" *Nature* 394, no. 6694 (1998): 615–16; J. W. Hurrell and K. Trenberth, "Satellite versus Surface Estimates of Air Temperature since 1979," *Journal of Climate* 9, no. 9 (1996): 2222–32; J. W. Hurrell and K. E. Trenberth, "Spurious Trends in Satellite MSU Temperatures from Merging Different Satellite Records," *Nature* 386, no. 6621 (1997): 164–66; C. A. Mears and F. J. Wentz, "The Effect of Diurnal Correction on Satellite-Derived Lower Tropospheric Temperature," *Science* 309, no. 5740 (2005): 1548–51; C. A. Mears et al., "A Reanalysis of the MSU Channel 2 Tropospheric Temperature Record," *Journal of Climate* 16, no. 22 (2003): 3650–64; B. D. Santer et al., "Consistency of Modelled and Observed Temperature Trends in the Tropical Troposphere," *International Journal of Climatology* (2008); M. C. Schabel et al., "Stable Long-Term Retrieval of Tropospheric Temperature Time Series from the Microwave Sounding Unit," in proceedings of 2002 IEEE International Geoscience and Remote Sensing Symposium; R. W. Spencer et al., "Global Atmospheric Temperature Monitoring with Satellite Microwave Measurements: Method and Results 1979–84," *Journal of Climate* 3, no. 10 (1990): 1111–28; Spencer and J. R. Christy, "Precision and Radiosonde Validation of Satellite Gridpoint Temperature Anomalies"; Spencer et al., "Global Atmospheric Temperature Monitoring with Satellite Microwave Measurements: Method and Results 1979–84," *Journal of Climate* 3, no. 10 (1990): 1111–28; Spencer et al., "Global Climate Monitoring"; Spencer, "Monitoring of Global Tropospheric and Stratospheric Temperature Trends," in *Atlas of Satellite Observations Related to Global Change*, ed. R. J. Gurney et al. (Cambridge University Press, 1993); Spencer et al., "Cloud and Radiation Budget Changes Associated with Tropical Intraseasonal Oscillations," *Geophysical Research Letters* 34 (2007): L15707; UK Met Office, "Satellite Microwave Radiances," 2007, www.metoffice.gov.uk; K. Y. Vinnikov and N. C. Grody, "Global Warming Trend of Mean Tropospheric Temperature Observed by Satellites," *Science* 302(5643), no. 5643 (2003): 269–72; F. J. Wentz and M. Schabel, "Effects of Orbital Decay on Satellite-Derived Lower-Tropospheric Temperature Trends," *Nature* 394, no. 6694 (1998): 661.

41. J. C. Bergengren and S. L. Thompson, "Modeling the Effects of Global Climate Change on Natural Vegetation" (National Center for Atmospheric Research, 1994); Bergengren et al., "Modeling Global Climate-Vegetation Interactions in a Doubled $CO_2$ World," *Climatic Change* 50, no. 1–2 (2001): 31–75; S. L. Thompson and D. Pollard, "A Global Climate Model (GENESIS) with a Land-Surface Transfer Scheme (LSX). Part I: Present Climate Simulation," *Journal of Climate* 8 (1995): 732–61; Thompson, "Description of the GENESIS Global Climate Model," (1994); Pollard and Thompson, "Use of a Land-Surface-Transfer Scheme (LSX) in a Global Climate Model: The Response to Doubling Stomatal Resistance," *Global and Planetary Change* 10 (1995): 129–61; E. J. Barron et al., "Past Climate and the Role of Ocean Heat Transport: Model Simulations for the Cretaceous," *Paleoceanography* 8, no. 6 (1993): 785–98.

42. J. D. Aber, "Terrestrial Ecosystems," in *Climate System Modeling*, ed. K. E. Trenberth (Cambridge University Press, 1992), 173.

43. Zimmerman, Data Sharing and Secondary Use of Scientific Data; P. Kareiva and M. Andersen, "Spatial Aspects of Species Interactions: The Wedding of Models and Experiments," *Lecture Notes in Biomathematics* 77 (1986): 35–50.

44. J. Rotmans, *IMAGE: An Integrated Model to Assess the Greenhouse Effect* (Kluwer, 1990), 7.

45. Response Strategies Working Group, "Emissions Scenarios of the Response Strategies Working Group of the Intergovernmental Panel on Climate Change," 1989.

46. M. E. Mann et al., "Global-Scale Temperature Patterns and Climate Forcing over the Past Six Centuries," *Nature* 392, no. 6678 (1998): 779–87; Mann et al., "Northern Hemisphere Temperatures During the Past Millennium: Inferences, Uncertainties, and Limitations," *Geophysical Research Letters* 29, no. 6 (1999): 759.

47. P. Huybers, "Comment on 'Hockey Sticks, Principal Components, and Spurious Significance' by S. McIntyre and R. McKitrick," *Geophysical Research Letters* 32, no. 20 (2005): L20705; M. E. Mann et al., "False Claims by Mcintyre and McKitrick Regarding the Mann et al. (1998) Reconstruction," www. realclimate.org; Mann and P. D. Jones, "Global Surface Temperatures over the Past Two Millennia," *Geophysical Research Letters* 30, no. 15 (2003): 1820; S. McIntyre and R. McKitrick, "Corrections to the Mann et al. (1998) Proxy Data Base and Northern Hemispheric Average Temperature Series," *Energy & Environment* 14, no. 6 (2003); McIntyre and McKitrick, "The M&M Critique of the MBH98 Northern Hemisphere Climate Index: Update and Implications," *Energy and Environment* 16, no. 1 (2005): 69–100; H. Von Storch and E. Zorita, "Comment on 'Hockey Sticks, Principal Components, and Spurious Significance,' by S. McIntyre and R. McKitrick," *Geophysical Research Letters* 32 (2005): 20.

48. J. Barton et al., "Letter to Michael Mann," 2005, republicans.energycommerce. house.gov.

49. A. Watts, *Is the U.S. Surface Temperature Record Reliable?* (The Heartland Institute, 2009), 16.

50. At this writing in 2009, Clear Climate Code and a similar open-source coding effort, OpenTemp.org, appeared to be dormant. Emails to the sites went unanswered.

51. Committee on Surface Temperature Reconstructions for the Last 2000 Years et al., *Surface Temperature Reconstructions for the Last 2000 Years* (National Academies Press, 2006), 113.

52. Kuhn, *The Structure of Scientific Revolutions*.

53. Brown, "Environmental Science Under Siege," section II.B.

54. R. F. Bornstein, "The Predictive Validity of Peer Review: A Neglected Issue," *Behavioral and Brain Science* 14, no. 1 (1991): 138–39; D. E. Chubin and E. J. Hackett, *Peerless Science: Peer Review and US Science Policy* (State University of New York Press, 1990); S. Cole, *Making Science: Between Nature and Society* (Harvard University Press, 1992); P. F. Ross, *The Sciences' Self-Management: Manuscript Refereeing, Peer Review, and Goals in Science* (Ross, 1980).

55. P. H. Abelson, "Scientific Communication," *Science* 209, no. 4452 (1980): 60–62; H.-D. Daniel, *Guardians of Science: Fairness and Reliability of Peer Review* (VCH, 1993).

56. Intergovernmental Panel on Climate Change, "Procedures."

## Conclusion

1. Good engineers don't just design devices. They also analyze, order, and manage various social relations involving their workgroups, funders, competitors, and (most importantly) potential users. See Callon, "Society in the Making"; Latour, *Aramis*; Latour, *Science in Action*.

2. Perhaps this is why some environmental historians, including Emmanuel le Roy Ladurie, have moved seamlessly into the study of climate, e.g. by dating the grape harvests in Chateauneuf du Pape back to 1370 as telltales of seasonal conditions. See I. Chuine et al., "Back to the Middle Ages? Grape Harvest Dates and Temperature Variations in France since 1370," *Nature* 432 (2004): 289–90; E. Le Roy Ladurie, *Histoire du Climat depuis L'an Mil* (Flammarion, 1967).

3. Peter J. Taylor inspired much of my thinking about these questions. See especially Taylor and F. Buttel, "How Do We Know We Have Global Environmental Problems?," *Geoforum* 23, no. 3 (1992): 405–16; Taylor, "How Do We Know We Have Global Environmental Problems? Undifferentiated Science-Politics and Its Potential Reconstruction," in *Changing Life: Genomes, Ecologies, Bodies, Commodities*, ed. P. J. Taylor et al. (Minneapolis: University of Minnesota Press, 1997), 149-74; Taylor, *Unruly Complexity: Ecology, Interpretation, Engagement* (University Of Chicago Press, 2005).

4. J. Hartwig, "On Misusing National Accounts Data for Governance Purposes," Working Paper 101, Swiss Institute for Business Cycle Research (Swiss Federal Institute of Technology, Zurich), 2005.

5. H. Collins, "We Cannot Live by Scepticism Alone," *Nature* 458, no. 30 (2009): 30–31.

6. M. R. Allen and D. J. Frame, "Call Off the Quest," *Science* 318, no. 5850 (2007): 582.

# Index

Abbe, Cleveland, 45, 46, 69
Accumulative infrastructure, 34
Aerology, 88, 89
Aerosols, 342–344, 362, 363
Air masses, 90, 91
Airplane soundings, 107
Air travel, 54, 108
Air Weather Service (AWS), 190, 222
Algorithms, 94
Analog devices, 105
Analog-digital conversion, 105–107
Analytical Engine, 64
Arakawa, Akio, 158
Arrhenius, Svante, 73, 74
Atmospheric Model Intercomparison Project (AMIP), 170, 171, 349
Atomic Energy Commission (AEC), 208
Automatic Picture Transmission, 234, 235

Balloons, 107, 217, 279
BARK, 129
Barometers, 40, 41
Bergen School, 90–92, 202, 259
BESK, 129
Biome classification scheme, 419
Bjerknes, Jacob, 155
Bjerknes, Vilhelm, 85–92
Blogs, 422–427
Brückner, Eduard, 67
Brundtland Report, 391, 392

Brunt, David, 79
Buchan, Alexander, 35, 36
Bureau of Meteorology Research Centre (BMRC), 174, 175
Bush, George W., 409, 410

Calibration, of instruments, 297
Callendar, Guy Stewart, 76–81, 179, 183
Cap-and-trade system, 395
Carbon cycle, 74, 75
Carbon dioxide
  and climate change, 73–81, 179, 348, 349, 377, 438, 439
  doubling of, 177–185, 379, 380
  monitoring of, 210, 211
  as pollutant, 360, 361
  rising levels of, 357
Carbon 14, 210, 212
Cartography, 30, 31
*Challenger*, H.M.S., 35, 36
Chamberlin, Thomas, 74, 75
Chaos theory, 141
Charney, Jule, 118, 119, 130, 133, 152
Charney report, 376–380, 396
Chlorofluorocarbons (CFCs), 384–387
Circulation
  air, 29, 108
  atmospheric, 37–39, 71, 72
  general, xx
  global, 207–215

Circulation diagrams, 37–39, 69
Citizen science, 421–427
Clear Climate Code, 425, 426
CLIMAT codes, 310
Climate, xiv
 as global system, 65
 latitude and, 29, 66, 68
Climate Audit, 422–426
Climate change
 carbon dioxide and, 73–81, 179, 348, 349, 359, 360, 377, 438, 439
 error detection and, 254–256
 Hann on, 65, 66
 physical theories of, 72–76
 and policy, 357–359, 391–396
 politics of, 67, 185, 186, 211, 212, 358, 359, 372–377, 388–395
 reanalysis data and, 332–335
 research in, 287, 288, 374–377
 scientific consensus on, 6, 7, 402–404, 427–430
 secular, 65
 as weather modification, 359–361
Climate controversies, 404–411, 427–430, 432
Climate modification, 243
Climate politics, 297–299
 climate change and, 67, 185, 186, 211, 212, 358, 359, 373–377, 388–395
 climate controversies and, 404–411
 nuclear winter and, 380–384
 ozone depletion and, 384–388
Climateprediction.net, 354, 355, 422, 439
Climate record, reconstruction of, 312–317
Climate science, xiv–xix
Climate seers, 345
Climate sensitivity, xx, 348
Climate simulation, 139–141
*Climates of the Continents, The* (Kendrew), 68, 69

Climatological stations, 97
Climatology, xvi, xxiii, 14
 data used in, 97, 98
 forecasting and, 98
 military, 101
 pre-World War II, 61–72
 statistical, 66–71, 98, 99, 116
Closed-form solutions, 86
Cloud physics, 339, 340
Club of Rome, 366–371, 395, 396
Cold War, 3, 112, 113, 297–299
 global data infrastructures and, 224–227
 infrastructural globalism and, 193–202, 206, 207
 International Geophysical Year and, 203, 204
 meteorology and, 189–193
 nuclear weapons tests and, 207–215
Commonwealth Meteorology Research Centre (CMRC), 169
Communication infrastructures, 23, 24
Computer power, 171–177
Computing
 analog, 86–90
 digital, 113–117
 resources for, 137, 139
Conrad, Victor, 69–71
Conservative think tanks, 407, 408
Control revolution, 49
Cooperative Observer Network, 292
Council on Environmental Quality (CEQ), 373, 374, 409, 410
Coupled Model Intercomparison Project (CMIP), 170, 171, 349, 350
Cray computers, 176, 178

Data
 analog-digital conversion of, 105–107
 analysis of, xv, 253, 256–273, 280
 assimilation of, 268–273, 278–280, 283, 291, 292, 324, 325

Index 511

changing meaning of, 282–285
climate, xiv, xv, 14–16, 21, 98–104,
  183–189, 202–207, 251–288,
  292–301–322, 365
 collection of, 6, 21, 205, 206
 consolidation of, 15, 16
 exchange of, 108, 193
 historical, 302, 303, 306
 inhomogeneities in, 295–301
 in situ, 278
 marine, 316–319
 meteorological, 4–6, 32–34, 50, 51, 59
 paleoclimatic, 293
 preservation of, 293
 processing of, 254–256
 production of, 109
 raw, 291, 309
 reconciliation of, 98
 satellite, 299–301
 sensor, 291
 simulated, 264–267, 283
 sounder, 275–278
 surface, 21
 synoptic , 289
 transmission of, 108, 109, 236–242,
  254–256
 upper-air, 107, 108, 215–224
 weather, 21, 22, 27, 28, 32–34, 50, 51,
  89, 287, 288–295
Data Assimilation Office (DAO), 326
Data entry, 254–256
Data guys, 34–40
Data images, 39–40, 283–285, 435
Data resources, xix
Daylight saving time, 46
D-Day, 116
Defense Meteorological Satellite
  Program (DMSP), 191
Deforestation, 67
Department of Defense (DOE), 374,
  375
Digital-analog conversion, 105–107
Digitization, 306

DYNAMO, 371
Dyson, Freeman, xiii

Earth System Modeling Framework
  (ESMF), xxiii, 420
EDVAC (Electronic Discrete Variable
  Automatic Computer), 114–116
Eisenhower administration, 222
Electronic Computer Project, 115
Eliot, T. S., xvii, xviii
Energy crisis, 372
ENIAC (Electronic Numerical Integrator
  and Calculator), 113–116, 119–126,
  132
Environmental impact assessment, 420,
  421
Environmental movement, 361–366
Environmental services, 363, 364
ERA-15, 326, 327, 330, 335
ERA-40, 327–335
Error
 detection of, 254–256
 forecast, 284
 round-off, 176, 177
European Centre for Medium-Range
  Weather Forecasts (ECMWF), 137,
  169, 170, 290, 324, 325, 328–332
Experiments
 First GARP Global (FGGE), 244–250,
  323, 324, 347, 375
 GARP Atlantic Tropical (GATE), 244,
  245, 249
 global, 243–250
 Global Weather, 243–250, 283, 323,
  324, 375
 laboratory, 34
Extreme weather events, 4

Fallout, 208, 209, 212–215
Fax transmissions, 236
Ferrel, William, 38, 39
Fifth World Meteorological Congress,
  239, 240

Fingerprint studies, 334, 335
Finite-difference methods, 93–96
First Assessment Report, 393, 394
First International Meteorological Congress, 51, 52
*First Report* (WWW), 230–235, 242
First World Meteorological Congress, 196
Floating-point arithmetic, 175, 176
Fluid dynamics, 115
Flux adjustments, 341
Forecast-factory, 92–96
Forecasters, human, 132, 257–261, 267, 268
Forecasting and forecasts, xv, xvi, xxiii, 62
  accuracy of, 264–267
  Bergen School of, 90–92
  climatology and, 98
  computerized, 267
  computer models for, 13, 14
  current state of, 431, 432
  data used in, 97, 103, 104
  digital computing and, 113–117
  early, 40, 41
  ensemble, 284, 306, 352–354
  finite-difference methods of, 93–96
  global, 291
  improvements in, 290, 291
  infinite, 141–143
  numerical, 111–137
  predictive ability in early, 130, 134
  steps in, 256, 257
  surface, 266
  synoptic, 264
  telegraphy and, 41–45
Forrester, Jay, 367, 368
Fossil fuels, 74, 77
Four-D data assimilation, 272, 273, 278, 279, 283, 291, 292, 306, 324, 325
Fourier, Joseph, 73

Friction
  computational, xv, 80, 83–85, 109, 110, 121–126, 134, 171–177, 310, 311, 327, 432
  data, xiv, 80, 84, 85, 97, 121–126, 134, 135, 183–186, 193, 206, 274, 306, 311, 327, 432
  metadata, xvii, xix, 317–319, 434, 435
Frontal analysis, 131
Fronts, 90, 91

Gateway technologies, 11
General Circulation Project, 141, 142
GENESIS, 418
Geography, 62, 63
Geology, 72
Geophysical Fluid Dynamics Laboratory (GFDL), 153–157, 167
Giere, Ronald, 263, 264
Global Atmospheric Research Program (GARP), 155, 241–250, 375
Global Change Research Program (GCRP), 394, 395
Global climate, as object of knowledge, 3–6
Global Climate Coalition, 407, 408
Global Climate Observing System (GCOS), 15, 293–295
Global Climate Protection Act (GCPA), 390
Global cooling, 362–364
Global data infrastructures, 224–227
Global data network, 193, 198
Global Data Processing and Forecasting System (GDPFS), 289, 291
Global Earth, 1–3
Global economy, 433–435
Global governance institutions, 24
Global infrastructures, 23–25
Globalism
  informational, 23, 24, 194

Index

infrastructural, xviii, xix, 23–25, 111, 193–202, 206, 207, 231, 249, 250, 434
Globalization, 23
Global Observing System (GOS), 289, 291
Global space, 27–34
Global Telecommunication System (GTS), 289, 291
Global thinking, 1–3, 27, 433
Global warming, xiii, xvi
  evidence of, 6, 7
  extent of, 7
  politics of, xviii, 388–395
  projections of, 377–380
Goddard Earth Observing System (GEOS-1), 326, 327
Gore, Albert, Jr., 7, 389, 402
GOSAT (Greenhouse Gas Observing SATellite), 436
Graphical calculus, 87, 88
Greenhouse effect, 76–81
Greenhouse gases, 73–75, 374, 375, 439
Greenwich Mean Time (GMT), 43, 46
Grids
  Cartesian, 146, 148
  finite-difference, 93, 94
  geospatial, 253
  hemispheric, 127–129
Group on Earth Observations System of Systems (GEOSS), xx
Group velocity, 121

HadCRIT3 data set, 319–322
Halley, Edmond, 29–30
*Handbook of Climatology* (Hann), 63, 65, 68, 75
Hann, Julius von, 63–68
Health and Safety Laboratory (HASL), 208
Heat transfer, 145, 146

Historical Climate Network (HCN), 292, 293
Historical record, 302, 303
Hockey stick controversy, 422–426

IAS computer, 127, 151
Ice ages, 72, 74
Information Council for the Environment, 407, 408
Infrastructural inversion, xvi, xix, 20–23, 252, 432
Infrastructure(s), 8–12
Infrasystem, 34
Institute for Advanced Studies (IAS), 117–126, 129
Integrated Model to Assess the Greenhouse Effect (IMAGE), 420, 421
Intel, 176
Interchangeable parts, 49
Intergovernmental Panel on Climate Change (IPCC), xvi–xviii, 7, 16, 19, 20, 348, 349, 398–404, 428, 438
International Bureau of Weights and Measures, 50
International Commission for Aeronautical Meteorology (CIMAé), 55
International Commission for Air Navigation (ICAN), 54, 55
International Comprehensive Ocean-Atmosphere Data Set (ICOADS), 316, 317
International Council of Scientific Unions (ISCU), 203
International Geophysical Cooperation (IGC), 204
International Geophysical Year (IGY), 52, 185, 202–207
International Institute for Applied Systems Analysis (IIASA), 370
International Meteorological Codex, 53
International Meteorological Committee (IMC), 52, 53

International Meteorological Organization (IMO), 13, 24, 51–59, 194
International Morse Code, 50
International Polar Year, 52, 53, 202, 203
International relations, 196, 197
International standards, 49–51
International Telecommunication Union, 50
International Telegraph Union, 50
Interpolation, 268–273, 279
Isobars, 31, 32
Isolines, 31, 32
Isotherms, 30, 31

Japan Meteorological Agency (JMA), 238, 330, 331
Jet streams, 107, 140, 141
Joint Numerical Weather Prediction (JNWP) Unit, 130
Journalism, 410, 411

Kasahara, Akira, 162
Kasahara-Washington GCM series, 163
Kendrew, W. G., 68, 69
Kennedy, John F., 222–226
Khrushchev, Nikita, 222, 225
*Klimata*, 29, 30
Knowledge
 of climate, xv–xix, 12–16, 301, 302, 418–421, 432, 438
 expert, 397, 398
 infrastructures of, 17–20, 371
 production of, 8, 22, 23, 397, 398, 437, 438
 scientific, 17–20
 social construction of, 437, 438
Kyoto Protocol, 7, 395

Large technical system (LTS), 9–11, 16
Latitude, 29, 66, 68, 253
Leith, Cecil E., 159–162

Libya, 199, 200
Limited test ban (LTB) treaty, 226, 227
*Limits to Growth, The*, 366–374, 395, 396, 421
Livermore Automatic Research Calculator (LARC), 160
Longitude, 253
Luntz, Frank, 409

Manabe, Syukuro, 144, 154
Maps
 analysis, 124, 271, 272
 global, 27
 weather, 30–32, 41–43, 69, 105
MARKFORT, 156
Mathematics, 70, 71, 115, 165, 175, 176
Maury, Matthew, 34–37
Mechanical calculations, 85–88
Media, 410, 411
Meteorological observation networks, 32
Meteorologists, xx
Meter Convention, 50
Meteorology
 aeronautical, 54, 55
 beginnings of, 41
 Cold War politics and, 189–193
 dynamical, 85
 empirical, 62, 63
 as infrastructural globalism, 23–25
 naval, 24
 network structure of, 34–40
 scientific, 90
 study of, 90, 135, 136
 subfields, 61, 62
 theoretical, xxiii, 62, 68, 88
 World War I and, 88–92
*Methods in Climatology* (Conrad), 69–71
Metric system, 34, 35, 50
Microwave radiances, 300, 301
Microwave sounding units (MSUs), 278, 279, 413–418

# Index

Milankovic, Milutin, 75, 76
Military applications, 112–117, 189, 190, 359, 360
Military intelligence satellites, 220, 221
Military weather observations, 190, 191
Mintz, Yale, 155, 158
Model bias, 331, 332
Model-data symbiosis, 281, 282, 433
Model physics, 145, 146, 280, 337, 341, 354
Models
analysis, 21, 22, 272, 273, 280, 283, 290, 291
atmosphere-ocean circulation, 341, 419, 420
atmospheric general circulation (AGCMs), 145, 146, 168
baroclinic, 127, 130
barotropic, 127, 130, 132
climate, xiii–xv, 6, 15, 143–149, 171–177, 337–355, 419–421
Community Climate (CCM), 164–167
computer, xv, xix, xx, xxiii, 13, 14, 111, 119–137, 141, 143, 187–189, 252, 253
data, xix, xx, 263, 264, 272
data-laden, 280, 282
Earth system (ESMs), 146, 418, 421
energy balance (EBMs), 144
general circulation (GCMs), xx, 141–160, 158–162, 167–177, 183–186, 283, 337–341, 364, 395, 396
Global 2000, 374
hemispheric, 134, 135
integrated assessment (IAMs), 420, 421
leaf, 418
line-by-line, 339
Livermore Atmospheric (LAM), 159–162
mathematical, 279, 280
perturbed-physics, 354, 355, 439
physical, 86
radiative-convective (RC), 144, 181
radiative transfer, 278
reanalysis, xv
simulation, xv, xx, 139–143, 280, 281, 358
spectral, 156, 165–167
two-dimensional statistical-dynamical, 144
world dynamics, 366–372
*Monthly Climatic Data for the World*, 309–313

National Aeronautics and Space Administration (NASA), 221–225, 234, 300, 301
National Center for Atmospheric Research (NCAR), 162–167, 173–175
National climate observation systems, 15
National Climate Program, 374, 375
National Climate Program Act (NCPA), 373
National meteorological centers (NMCs), 236
National Science Foundation, 190
National Weather Service, 292
National weather services, xv, xvi, 13, 20, 43
computing at, 136
cooperation among, 56, 193
Natural history, 62, 63, 68
NCEP-NCAR reanalysis, 326–328, 333, 334
Nephanalysis, 274
Networks, 11, 12
Nomographs, 87
Nonlinearity, 93
Nuclear weapons, testing of, 207–215, 226, 227
Nuclear winter, 380–384
Numerical approximation, 87

Numerical weather prediction, 111–137, 257–263

Objective analysis, 252, 253, 256–268
Office of Science and Technology (OSTP), 375, 376
Organization of Small Island States, 405
*Our Common Future*, 391, 392
Ozone depletion, 380, 381, 384–388

Parallel processing, 176
Parameterization, 280, 338–341, 354
Parameters, 338
  tuning, 342–346
  types of, 338–341
Peer review, 428, 429
Phillips, Norman, 150–153
Physics, 85, 86, 115
Plass, Gilbert, 179
Platzman, George, 124
Polar front, 91, 202
Politics, xviii, 67
Pollack, L. W., 99
Pollution, 360, 361
Powers, Francis Gary, 221, 222
Precipitation readings, 299, 300
Predictive ability, in early numerical weather prediction, 130, 134
Primitive equations, 85–88
Program for Climate Model Diagnosis and Intercomparison, 380
Project Feederwatch, 421
Ptolemy, 29
Punch-card technology, 98–104, 109, 205

Radiative equilibrium, 73
Radiatively active gases, 179
Radiative transfer, 178, 179, 339
Radiocarbon, 210, 212
Radiometers, 275
Radiosondes, 105, 107, 261, 279

Radio waves, 202, 203
Railroads, 28, 43, 44
Reanalysis, 16, 22, 253, 254, 288, 291, 292, 323–336
Reanalysis-2, 327
Regional Basic Synoptic Networks (RBSNs), 289
Regional meteorological centers (RMCs), 236
Regulation of Aerial Navigation, 54
Reichelderfer, Francis, 133, 135
Remote Sensing Systems (RSS), 416
Reproductionism, 280, 281
Réseau Mondial, 24, 56–59, 309
Revelle, Roger, 210, 211
Richardson, Lewis Fry, 93–96
Rossby, Carl-Gustav, 117, 118, 121, 129, 133

Satellites, 14, 24, 207, 219–226, 230–234, 273–279, 289, 290, 436
Scanning Multichannel Microwave Radiometer (SMMR), 300, 301
Scientific consensus, 6, 7, 402–404, 427–430
Scientific Integrity hearings, 411, 412, 427, 428
Scientific internationalism, 55, 56
Scientific memory, xix
Scientific nationalism, 55
Second Assessment Report, 401, 402
Second World Meteorological Congress, 52, 53, 196, 198
Signal Service, 42–45, 98
Skyhigh, 156
Smithsonian Institution, 41, 58, 59, 98
Sociotechnical systems, 17
Solar energy, 73–76, 178, 179
Soviet Union, 126, 169, 191–193, 207, 214, 219–226, 311
Space race, 207, 222–226
Sputnik, 207

# Index

Standardization, 49–51, 56, 57, 237–239, 251–253, 322
Standard time, 43–47
Starr, Victor, 141, 142
Study of Critical Environment Problems (SCEP), 361–366, 395, 396
Study of Man's Impact on Climate (SMIC), 361–366, 395, 396
Subjective analysis, 261, 262, 267, 268
Supercomputers, 137, 171–177
Supersonic transport (SST), 358, 362, 363, 381
Supersource, 156
SurfaceStations.org, 424–426
Synchronous observation, 46
Synoptic charts, 69–71
Synoptic stations, 97, 217–219, 289
System dynamics, 368–371
System Dynamics Group, 368, 369
Systems Design and Technical Development (SDTD) program, 241

Technological innovations, 40, 202–207
Technology transfer, 10, 13
Technopolitics
 of altitude, 215–226
 World Weather Watch and, 230–232
Telegraphy, 13, 28, 40–45, 50, 97
Teletype tapes, 103, 104, 255
Television Infrared Observation Satellite (TIROS), 221
Teller, Edward, 160
Temperatures
 average, 4, 303–308, 314, 315, 319–322, 350–352
 changes in, 4, 72–78, 181–186, 295, 296
 global average, 303–308, 314–316, 319–322, 350–352
 sea surface, 316–322
Theory images, 39, 40
Thermoelectric probe, 282

Thermometer, 105
Time, 28, 43–47, 98, 271, 272
Time balls, 44, 45
Time observation bias, 314–316
Toronto Conference on the Changing Environment, 392, 393
Transient forcing, 391
Tropical Ocean/Global Atmosphere (TOGA), 324–326
TTAPS, 382
Tyndall, John, 73

UAH, 413–417
UCLA Department of Meteorology, 155–160, 167
UN Conference on Environment and Development (UNCED), 393, 395
UN Conference on the Human Environment (UNCHE), 365, 366
UN Framework Convention on Climate Change, 15, 16, 392–395
Union of Concerned Scientists, 412
United Nations, 196, 197
United Nations Expanded Program of Technical Assistance for the Economic Development of Under-Developed Countries (UNDP), 199, 200
Universal time, 28, 40–49
US Army Signal Corps, 89
US National Center for Atmospheric Research, 137
US National Meteorological Center, 170
US National Weather Records Center, 101, 102
US Weather Bureau, 89, 90, 98, 130

Vertical structures, 177, 178
Vietnam War, 359, 360
Voluntary Assistance Program, 199–201
Von Neumann, John, 111–120, 130, 133, 135, 141, 152, 189

Washington, Warren, 162
Water vapor, 73–75
Water vapor feedback, 180
Weather averages, 4
Weather control, 112, 113, 117, 189, 190, 243
Weather data network, 108, 109
Weather data systems, 14
Weather information infrastructure, 12–16
Weather modification, 359–361
Weather stations, 6, 20, 21, 295, 297
Weather telegraphy, 41–45, 61
Webs, 11, 12
Wilson, Carroll, 361, 366, 367
Winds
  cloud, 275
  geostrophic, 119
  trade, 29, 30
World Climate Conference, 376
World Climate Programme (WCP), 15, 376, 377
World Climate Research Program (WCRP), 250, 376, 377
World Meteorological Center (WMC), 291
World Meteorological Organization (WMO), xviii, 14, 24, 52, 53, 193–202
World system, 1–3, 40
World War I, 53, 54, 88–92
World War II, 53, 54, 101, 107, 135
*World Weather Records*, 15, 58, 59, 309–314
World Weather Watch (WWW), xvi, 14, 24, 59, 188, 224, 229–248
World Wide Web, 11, 421–427

Zodiac, 156

Printed in the United States
by Baker & Taylor Publisher Services